ROUTLEDGE HANDBOOK OF THE STUDY OF THE COMMONS

Edited by Blake Hudson, Jonathan Rosenbloom and Dan Cole

earthscan
from Routledge

Routledge
Taylor & Francis Group
LONDON AND NEW YORK

First published 2019 by Routledge
2 Park Square, Milton Park, Abingdon, Oxon OX14 4RN
52 Vanderbilt Avenue, New York, NY 10017

First issued in paperback 2020

Routledge is an imprint of the Taylor & Francis Group, an informa business

British Library Cataloguing-in-Publication Data
A catalogue record for this book is available from the British Library

Library of Congress Cataloging-in-Publication Data
Names: Hudson, Blake, editor. | Rosenbloom, Jonathan D., editor. | Cole, Daniel G. (Daniel Gerard), editor.
Title: Routledge handbook of the study of the commons / edited by Blake Hudson, Jonathan Rosenbloom and Dan Cole. Other titles: Handbook of the study of the commons
Description: Abingdon, Oxon ; New York, NY : Routledge, 2019. | Includes bibliographical references and index.
Identifiers: LCCN 2018037489 (print) | LCCN 2018048086 (ebook) | ISBN 9781315162782 (eBook) | ISBN 9781138060906 (hbk) | ISBN 9781315162782 (ebk)
Subjects: LCSH: Common heritage of mankind (International law) | Global commons. | Commons.
Classification: LCC KZ1322 (ebook) | LCC KZ1322 .R68 2019 (print) | DDC 341—dc23
LC record available at https://lccn.loc.gov/2018037489

ISBN 13: 978-0-367-65960-8 (pbk)
ISBN 13: 978-1-138-06090-6 (hbk)

Typeset in Bembo
by Swales & Willis Ltd, Exeter, Devon, UK

CONTENTS

FIGURES

TABLES

CONTRIBUTORS

Eduardo Araral is Associate Professor at Lee Yew School of Public Policy, National University of Singapore and Co-director of the Institute of Water Policy.

Sajida Awan is enrolled in the PhD program "Social and Ecological Sustainability" at University of Waterloo, Canada. Her research area is "role of knowledge co-production in climate smart agriculture". She is also associated with Environmental Change and Governance Group (ECGG) and Graduate Student Association at University of Waterloo. She has worked with the Governance Unit of UNDP Pakistan for almost three years. In her role as Program Associate she has been independently, as well as in support of other team members, working on rule of law, strengthening democracy through parliament, legal empowerment of the poor, and election and other governance-related projects. She has played a key role in conceptualizing and scoping the Rule of Law Project funded by the Netherlands and the Government of Khyber Pakhtunkhwa. Sajida has also successfully worked in the UNICEF for two-and-a-half years on its Water, Environment and Sanitation projects under the Earthquake Emergency Support Program.

Clare Barnes is a lecturer in the Geography and Lived Environment group at the University of Edinburgh, UK. Her research interests are in forest and landscape governance in the Global South, especially in South Asia. She holds a PhD in Environmental Governance from Utrecht University, the Netherlands. Her PhD work was on the roles of civil society organizations in community forestry in India.

Esther Blanco is an Associate Professor of the Department of Public Finance at the University of Innsbruck, Austria, and Affiliated Faculty at the Ostrom Workshop, Bloomington, USA. Her research agenda addresses how individuals make decisions in social dilemmas, focusing on incentives and how these incentives are affected by contextual factors such as institutional change. Her research outcomes have been published in multidisciplinary journals such as *Science Advances* and top field journals in environmental economics and experimental economics. Her research has received international press attention, including from *The Washington Post.*

Stanislava Brnkalakova completed the Master's program Environmental Planning and Management in the Department of Landscape Ecology at the Faculty of Natural Sciences at Comenius University in Bratislava, Slovakia. During her PhD studies, she did research on adaptive management of mountain ecosystem services with the emphasis on climate change mitigation at the Institute of Management at Slovak University of Technology in Bratislava. Currently, she has been working as a researcher in the Department of Strategic Environmental Analysis at the Institute of Forest Ecology, Slovak Academy of Sciences in Bratislava. She also takes part in research at the CETIP Network and Centre of Excellence SPECTRA.

Cheryl Chan holds an MES in Environment and Resource Studies from the University of Waterloo, Canada. Prior to graduate studies, she received a BSc in Biology and a BEd in Secondary Education from Queen's University, Canada. After graduating from her Master's program, she completed a Research Award in the Networked Economies program at the International Development Research Centre, where she has since transitioned into the role of Program Management Officer. Her research activities include governance of both traditional commons (e.g. fisheries) and new commons (e.g. knowledge and digital commons).

Dan Cole is a Professor of Law and of Public and Environmental Affairs at Indiana University, USA. Cole has published 10 books and more than 60 articles on a wide range of topics, including property systems for common-pool resources. A former member of the Ostrom Workshop at Indiana University, he most recently edited (with Michael D. McGinnis) a four volume compendium, *Elinor Ostrom and the Bloomington School of Political Economy*.

Michael Cox is an Environmental Social Scientist and Associate Professor of Environmental Studies at Dartmouth College, USA. He studies community-based natural resource management and sustainability transitions in agricultural systems.

Peter Cronkleton is a Senior Scientist at the Center for International Forestry Research (CIFOR) and holds a PhD in Anthropology from the University of Florida, USA. He is a specialist in community forestry development, forest tenure, institutional change and participatory methods. Based in Peru, he has worked as a researcher and development practitioner in Latin America and Africa for more than 25 years.

Brigham Daniels is a Professor at Brigham Young University's J. Reuben Clark Law School, USA. He holds a PhD from Duke University's Nicholas School of the Environment and his JD from Stanford Law School, both USA.

Daniel A. DeCaro is an Assistant Professor in Urban and Public Affairs and Psychological and Brain Sciences at the University of Louisville, USA. He directs the Social Decision Making and Sustainability Lab. His research examines human motivation and decision-making processes in social dilemmas in both experimental and field contexts. Current research topics include participatory democracy, legal foundations of self-governance, regulatory systems, social learning and public education, and governance of water resources and cities.

Tine De Moor is a Professor in Social and Economics History at Utrecht University, the Netherlands, dealing with, in particular, institutions for collective action in historical perspective. Besides her academic work on commons and cooperatives in the past and present she has been actively involved in the International Association for the Study of the Commons both as council member and president.

Barbara J. Evans, PhD, JD, LLM, is the Mary Ann and Lawrence E. Faust Professor of Law and Director of the Center for Biotechnology & Law at the University of Houston Law Center, USA, and holds a joint appointment as Professor of Electrical Engineering at the University of Houston Cullen College of Engineering.

Sheila R. Foster is a joint Professor of Public Policy and Law at Georgetown University, USA. Professor Foster is the author of numerous publications on land use, property and environmental law. She is one of the country's leading scholars on environmental justice. Her most recent work explores city growth and governance through the lens of the "urban commons". As Co-director with Christian Iaione of the Laboratory for the Governance of the Commons (LabGov), she is currently engaged with a groundbreaking international applied research project, the "Co-Cities Project". Their latest article, "The City as a Commons", is published in the *Yale Law and Policy Review* (2016) and is the basis of a forthcoming book on the Co-City for MIT Press.

Brett M. Frischmann is The Charles Widger Endowed University Professor in Law, Business and Economics at Villanova University, USA. He is also an affiliated scholar of the Center for Internet and Society at Stanford Law School, USA, and a trustee for the Nexa Center for Internet & Society, Politecnico di Torino, Italy. He has published foundational books on the relationships between infrastructural resources, governance, commons, and spillovers, including *Infrastructure: The Social Value of Shared Resources* (Oxford University Press, 2012), *Governing Knowledge Commons* (Oxford University Press, 2014, with Michael Madison and Katherine Strandburg), and *Governing Medical Knowledge Commons* (Cambridge University Press, 2017, with Michael Madison and Katherine Strandburg). His latest interdisciplinary book, *Re-Engineering Humanity* (Cambridge University Press, 2018, with Evan Selinger) explores how we engineer ourselves through the techno-social world we build.

Veronika Gezik obtained her PhD in Ecological Economics from University of Sussex, UK. She currently works for the Department of Strategic Environmental Analyses Institute of Forest Ecology Slovak Academy of Sciences associated with the Faculty of Management Comenius University and the Slovak University of Technology. She is an active member of CETIP Network.

Philipp Grundmann is Senior Scientist at the Leibniz-Institute for Agricultural Engineering and Bioeconomy and Adjunct Associate Professor at the Albrecht Daniel Thaer Institute of Agricultural Sciences at Humboldt University Berlin, Germany. His areas of teaching and research are agricultural and resource economics, institutional analysis of sociotechnical innovations, and transitions in agro-food systems and the bioeconomy.

Konrad Hagedorn is Senior Professor at the Albrecht Daniel Thaer Institute of Agricultural Sciences at Humboldt University Berlin, Director of the Berlin Institute for Co-operative Studies and was head of the Division of Resource Economics. His areas of teaching and research are environmental and resource economics, institutional analysis of social-ecological-technical systems, and post-socialist transition processes.

Nina Hagemann is a postdoctoral researcher and lecturer at the Technische Universität Dresden / International Institute Zittau, Germany, and a postdoctoral researcher at the Helmholtz Centre for Environmental Research – UFZ, Germany. As an institutional economist, her research interests are on the sustainable management of natural resources (especially water and soil), governance of the forest-based bioeconomy, and the assessment of agroecosystem services.

Michael Heller is Lawrence A. Wien Professor of Real Estate Law at Columbia Law School, USA, and has served as the school's Vice Dean for Intellectual Life. He writes and teaches on private law theory. Heller has authored several books, including *The Choice Theory of Contracts* (Cambridge University Press, 2017, with Hanoch Dagan); *The Gridlock Economy: How Too Much Ownership Wrecks Markets, Stops Innovation, and Costs Lives* (New York: Basic Books, 2008); and the two-volume edited collection *Commons and Anticommons* (Elgar, 2009); along with dozens of academic articles on property, contracts, and land use.

Bruce Huber is Professor of Law and the Robert and Marion Short Research Scholar at Notre Dame Law School in South Bend, Indiana, USA. He holds a JD and PhD (Political Science) from the University of California at Berkeley, USA. His research and teaching focus on the law of property and natural resource management in the USA .

Blake Hudson holds the A. L. O'Quinn Chair in Environmental Law at the University of Houston Law Center, USA, where he is a Professor of Law. His research currently spans the analysis of governance structure and private property rights impacts on natural resources management, the intersection of land development and natural capital protection, and the cultural factors influencing natural resources and environmental law and policy. He teaches courses in Natural Resources Law and Policy, Water Law, and Property Law at Houston Law.

Christian Iaione is Associate Professor of Urban Law and Policy, Regulatory Innovation and Land Use at LUISS Guido Carli, Italy, and faculty director of LabGov – LABoratory for the GOVernance of the City as Commons (www.labgov.city). He was the expert on the EU Committee of the Regions who drafted the opinion on the "Local and regional dimension of the sharing economy". He is a member of the Sharing Economy International Advisory Board of the Seoul Metropolitan Government, South Korea, and adviser to several Italian local governments and institutions (Tuscany Region, City of Rome, City of Bologna, City of Reggio Emilia). He is Urban Innovative Actions expert appointed by the European Commission for the Co-City project of the City of Turin, Italy, lead expert of the EU Urbact program, and a member of the Urban Partnership on Innovative and Responsible Procurement within the Urban Agenda for the EU.

Tim C. Jennerjahn is a Senior Scientist at the Leibniz Centre for Tropical Marine Research in Bremen, Germany. His research focuses on the biogeochemical response of coastal aquatic systems to environmental change in tropical regions at present and in the past, in particular on the impact of inputs of nutrients, organic matter, organic pollutants, and suspended sediments on mangroves, seagrasses, and coastal seas. He is Editor-in-Chief of the international journal *Estuarine, Coastal and Shelf Science.*

Fatima Noor Khan holds an MES in Environment and Resource Studies from the University of Waterloo, Canada. Prior to graduate studies, she received a BES from York University, Canada.

Since graduating from her Master's program, she has been working as a sustainability professional in the field of waste management. Her social-ecological systems research has focused primarily on small-scale fisheries where she has explored gender dynamics in relation to environmental change.

Tatiana Kluvankova is a Professor of Management and Head of the Department of Strategic Environmental Analyses Institute of Forest Ecology Slovak Academy of Sciences, associated with the Faculty of Management Comenius University Bratislava and the Slovak University of Technology.

Anne M. Larson is a Principal Scientist at the Center for International Forestry Research (CIFOR), based in Lima, Peru. She obtained her PhD in 2001 from University of California at Berkeley, USA, in Wildland Resource Science, with an emphasis on resource policy and institutions. She has done both more traditional and action research, as well as supporting innovative initiatives such as the design of a diploma course for indigenous communities and leaders. Her current research priorities include opportunities for and challenges to forest tenure reforms; women's rights to land in communal forests; and multilevel governance, REDD+, and low emissions development.

Michal Maco obtained his PhD degree in the study programme Spatial Planning at the Institute of Forest Ecology, Slovak University of Technology in Bratislava, Slovakia. During his studies, he focused on the application of commons theories on public spaces. Thus, his PhD research develops the topic particularly concerning the governance of public and common urban spaces. Now he is an external worker at CETIP Network, Prague.

Michael J. Madison is Professor of Law and Faculty Director of the Innovation Practice Institute at the University of Pittsburgh, USA, and Senior Scholar and Academic Director of the University of Pittsburgh Institute for Cyber Law, Policy, and Security. His research and scholarship address intellectual property law and policy, as well as questions concerning the production and distribution of knowledge and innovation. He is the author of more than 40 journal articles and book chapters, the co-editor of *Governing Knowledge Commons* (Oxford University Press, 2014) and *Governing Medical Knowledge Commons* (Cambridge University Press, 2017), and the co-author of *The Law of Intellectual Property* (5th ed., Wolters Kluwer, 2017). He was elected to membership in the American Law Institute in 2016. He received his law degree from Stanford University and his undergraduate degree from Yale University, both USA.

Mary Anderlik Majumder, JD, PhD, is an Associate Professor of Medicine in the Center for Medical Ethics and Health Policy at Baylor College of Medicine, Houston, Texas, USA. She received an AB magna cum laude from Bryn Mawr College in 1985, a JD from Yale Law School in 1989, and a PhD from Rice University in 1997, all USA. Her current research interests include the ethical, legal, and social implications of new genomic technologies, and ethical and policy questions related to problems of cost, quality, and access in health care.

Frank Matose currently holds the position of Associate Professor and Director of Environmental Humanities South Centre at the University of Cape Town, South Africa, where he has been since 2009. Before that he was Programme Leader for the Community-Based Natural Resource

Management Programme at the Institute for Poverty, Land and Agrarian Studies, University of the Western Cape, USA, from 2004 to 2008. Prior to that he was Adaptative and Collaborative Management Project Leader in Zimbabwe for the Centre for International Forestry Research between 2000 and 2003. In the 1990s Frank worked in Harare, in the Social Forestry Research Unit in the Zimbabwe Forestry Commission. He is currently preparing a monograph titled "Beyond Waiting Politics in Conservation: Forests and the Power of the Marginalised in Southern Africa", for University of Arizona Press.

Bonnie J. McCay is an environmental anthropologist who specializes in the study of marine fisheries based on field research on the fringes of the North Atlantic and the eastern Pacific oceans. Her work is informed by theories concerning common pool resources, and she has held leadership positions in the International Association for the Study of the Commons (IASC). Until her recent retirement she was a distinguished professor at Rutgers the State University of New Jersey, USA, in the Department of Human Ecology. She is a member of the National Academy of Sciences.

Michael D. McGinnis received a PhD in Political Science from the University of Minnesota, USA, and currently serves as Associate Dean for Social and Historical Sciences and Graduate Education in the College of Arts and Sciences at Indiana University, USA. He previously served as Director of The Ostrom Workshop, an inter-disciplinary research and teaching center focused on the study of institutions, resource management, and governance. His current research focuses on institutional analysis as applied to US health policy.

Amy L. McGuire is the Leon Jaworski Professor of Biomedical Ethics and Director of the Center for Medical Ethics and Health Policy at the Baylor College of Medicine, USA. Her research explores the legal and ethical issues in genomics with a particular focus on genomic research ethics, and the ethical and policy issues related to the clinical integration of genomics. Currently, she is on the program committee for the Greenwall Foundation Faculty Scholars Program in Bioethics and is President of the Association of Bioethics Program Directors.

Iliana Monterroso, environmental scientist, recently completed her post-doc fellowship with CIFOR. Since 2014 she has coordinated research activities of the Global Comparative Study on Design and Implementation on Tenure Reforms in Peru and Colombia where she led the analysis of outcomes from land and forest right-recognition processes in the Amazon. She holds a PhD in Environmental Sciences from Autonomous Barcelona University, Spain. Her research focus is on collective tenure, gender, and socio-environmental conflicts in Latin America.

Maitreyee Mukherjee is currently pursuing her PhD in Public Policy at the Lee Kuan Yew School of Public Policy, National University of Singapore. With a background in bio-sciences and environmental management, her broad research interests lie in understanding the dynamics of environmental issues in governmental agenda with special focus on water-related environmental regulation and polices. Her prior work, under the capacity of Research Associate at Institute of Water Policy, LKYSPP, included projects in urban water management, water pricing, and environmental sustainability in water resource development projects. In addition to her research exposure, she is also a trained teacher and loves interacting with young inquisitive minds.

Seema Mundoli is an Associate at Azim Premji University, Bengaluru, India. Her research examines the challenges of ecological sustainability and equity faced by an urbanizing India. Her research interests range from understanding historical use and transformations of nature in cities to advancing governance of ecosystems as urban commons in Indian cities.

James Murombedzi is the Officer in Charge of the Economic Commission for Africa's African Climate Policy Center (ACPC) based in Ethiopia.

Harini Nagendra is a Professor of Sustainability at Azim Premji University, India, where she anchors research examining urban social-ecological sustainability. She is a 2013 Elinor Ostrom Senior Scholar awardee and has published widely in areas of remote sensing, conservation, urbanization, and the commons. Her book *Nature in the City: Bengaluru in the Past, Present, and Future* (Oxford University Press, 2016) examines the impact of urbanization on human-nature relationships from a Global South perspective.

Serik Orazgaliyev is an Assistant Professor at the Graduate School of Public Policy, Nazarbayev University, Kazakhstan, and a Research Fellow at Cambridge Central Asia Forum. He received his doctoral degree in Development Studies from the University of Cambridge, UK. His main areas of research include governments and multinational enterprises (MNEs), institutions and development policies, Caspian energy geopolitics, and Central Asian Studies.

Phil René Oyono is a natural resource senior researcher. He studied sociology, anthropology, and social philosophy. Oyono's research field covers natural resource management decentralization policies, institutional change, resource tenure, resource governance, redistribution, domination, conflict, violence, and community vulnerabilities and adaptation.

Jouni Paavola is Professor of Environmental Social Science and Director of the UK Economic and Social Research Council (ESRC) funded Centre for Climate Change Economics and Policy (CCCEP) in the School of Earth and Environment at the University of Leeds, UK. His research examines the role of institutions and social justice in environmental governance.

Stefan Partelow holds a BA in Environmental Studies from the University of California, USA, an MSc in Environmental Studies and Sustainability Science from Lund University, Sweden, and a PhD in Political Science from Jacobs University, Germany. He currently works as a social scientist at the Leibniz Centre for Tropical Marine Research (ZMT) in Bremen, Germany conducting research on coastal commons, collective action in the marine realm, and social-ecological systems.

Faye Victoria Sit Ying Qi is a PhD Candidate at the Lee Kuan Yee School of Public Policy, National University of Singapore. Her previous work and research interests encompass development planning and collective action regarding common pool resources in urban and peri-urban areas.

Timothy O. Randhir is a Professor and Graduate Program Director with the Department of Environmental Conservation at the University of Massachusetts, USA. He holds a PhD from Purdue University, USA, and has expertise in complex systems, ecological economics, environmental sciences, and ecohydrology. His research covers multiple scales and countries in addressing issues as integrated and dynamic systems.

Katharina Rauchenecker received her PhD at the Division of Resource Economics at Humboldt University of Berlin, Germany, under Dr. Konrad Hagedorn. Her scientific scope was hunting and game management in Germany with special focus on self-organized governance. She publishes as a freelancer. Currently she's working for the Roman Catholic Church in Germany and investigates institutional change within the organization.

Jonathan Rosenbloom is the Dwight D. Opperman Distinguished Professor of Law and Director of Environmental and Sustainability Programming at Drake University Law School, USA. In partnership with the University of Colorado Denver, USA, Jonathan is co-director of the Sustainable Development Code, a model land use code designed to provide local governments with the best sustainability practices in land use. Jonathan received a Bachelor in Architecture from the Rhode Island School of Design, a JD from New York Law School, and an LLM from Harvard Law School, all USA. He was named Drake Law Outstanding Professor of the Year in 2013; University of Oregon, School of Law Environmental & Natural Resource Distinguished Visitor in 2016; Vermont Law School, Distinguished Environment Law Scholar in 2017; and Stevens Faculty Scholar of the Year in 2018.

Christian Schleyer is a Temporary Professor at the Section of International Agricultural Policy and Environmental Governance at the University of Kassel, Germany. He also leads a small research team on forest-related governance innovations at the Institute of Geography at the University of Innsbruck, Austria. As an ecological and institutional economist, he works on sustainable use of natural resources such as water, soil, forests, and biodiversity within social-ecological-technical systems.

Achim Schlüter holds the Chair of Social Systems and Ecological Economics at the Leibniz Institute of Tropical Marine Research in Bremen, and at Jacobs University Bremen, Germany. His interest gravitates around institutions in the sense of rules which allow for an economic, social, and ecological sustainable use of tropical coastal and marine systems. Those aspects have been studied within his lab in various different fields like aquaculture, marine litter, sea level rise, tourism, and fisheries. From a methodological perspective, the main emphasis is, on the one hand, on qualitative case study research and, on the other, on social science/economic experiments.

Falk Schmidt is the Head of the Secretariat of the Science Platform Sustainability 2030, established in 2017 at the Institute for Advanced Sustainability Studies (IASS) Potsdam, Germany. Prior to this position, he was an academic officer at the Executive Office at the IASS Potsdam. He has a broad interest in sustainability issues, the role of the human dimensions of global change and governance in the Anthropocene, and the interface of science–policy–society.

Katherine J. Strandburg is the Albert B. Engelberg Professor of Law at New York University School of Law and Faculty Director of NYU's interdisciplinary Information Law Institute. She teaches and researches in the areas of patent law, innovation policy, information privacy law, and the legal and policy ramifications of data analytics and algorithmic decision making. Prior to her legal career, she was a physicist studying phase transitions and critical phenomena.

Insa Theesfeld PhD is Professor in Agricultural, Environmental and Food Policy at the Martin Luther University Halle-Wittenberg, Germany. She is an agricultural and institutional economist who developed an interest in governance questions of various shared natural resources.

A significant strand of her work explored the specificities in commons management in post-socialist countries. She is a known "Commons" scholar, leading the European branch of the International Association for the Study of the Commons since 2018.

Andreas Thiel holds the Chair of International Agricultural Policy and Environmental Governance at University of Kassel, Germany. His research focuses on the role of institutions and governance and their change in the context of social-ecological systems and natural resource management in the European Union and in East Africa (Ethiopia). He is affiliated with the faculty of the Vincent and Elinor Ostrom Workshop in Political Theory and Policy Analysis. Ongoing research addresses the constitutional determinants of polycentric resource governance and its performance and institutional fit, as well as the determinants of change of polycentric natural resource governance, with particular focus on agriculture, climate change adaptation, water and natural resource management.

Luz Elba Torres-Guevara is an Associate Professor in the International School of Economic and Administrative Sciences at Universidad de La Sabana, Colombia. She is a Business Administrator with a Master's in Rural Development and another Master's in Education, all from the Pontificia Universidad Javeriana in Colombia, and a PhD in Economics from Jacobs University in Bremen, Germany. She focuses her research on the governance of natural resources and the economic, cultural, social, and environmental analysis of farming production systems and agri-food systems, with emphasis on self-sufficiency and reciprocal exchange of agricultural products.

Hita Unnikrishnan is currently a Newton International Fellow at the Urban Institute of The University of Sheffield, UK, and a visiting faculty at Azim Premji University, India. Her research focuses on social-ecological systems in the Global South and she has a number of publications including an upcoming book on the social-ecological transition of historical urban lake systems in Bengaluru. She has also been a 2013 recipient of the Professor Elinor Ostrom Fellowship on Practice and Policy in the Commons.

Frank van Laerhoven obtained a PhD in Public Policy from Indiana University, USA. He currently works at the Copernicus Institute of Sustainable Development of Utrecht University, the Netherlands, where his teaching and research gear around topics related to the governance of shared resource systems. He is a co-editor in chief of *The International Journal of the Commons*. Before joining the ranks of academia he worked for FAO in Senegal and Chile.

Josephine van Zeben is fellow of Worcester College, University of Oxford, UK. She holds law degrees from the University of Edinburgh, UK; Amsterdam, the Netherlands; and Harvard University, USA, as well as a BA in Social Sciences from the University of Utrecht, the Netherlands, and a PhD in Law and Economics from the University of Amsterdam, the Netherlands. Her research centres on the functioning (and malfunctioning) of multilevel and polycentric systems, particularly in the area of environmental regulation and with respect to the European Union.

James M. Walker is Professor of Economics at Indiana University, USA, and a long time faculty affiliate of the Ostrom Workshop. His research focuses on the use of experimental methods in the investigation of individual and group behaviour related to social dilemma environments. He has published in a wide range of academic journals and co-authored with his colleagues Elinor Ostrom and Roy Gardner, the widely cited book, *Rules Games, and Common Pool Resources* (University of Michigan Press, 1994).

Nina Wormbs is an Associate Professor in History of Science and Technology at KTH Royal Institute of Technology, Stockholm. Her main strands of interest are media history, in particular broadcasting, and the knowledge production and communication of climate change in the Arctic, especially the shrinking sea ice.

Oran R. Young is Professor Emeritus at the Bren School of Environmental Science and Management at the University of California, USA. He has spent his career analysing the roles that social institutions play in addressing needs for governance in a variety of settings. His applied work deals with issues of international environmental governance relating to the oceans, the atmosphere, and the polar regions, and with issues of comparative environmental governance as they arise in China and the USA. His most recent book is *Governing Complex Systems: Social Capital for the Anthropocene* (MIT Press, 2017).

Peter D. Zuk is PhD candidate in Philosophy at Rice University and Research Associate at Baylor College of Medicine's Center for Medical Ethics and Health Policy, both USA. He received an MA in Philosophy from Rice University in 2014 and a BA in Philosophy from Pepperdine University in 2012, both USA. His research focuses on ethical theory, social-political philosophy, and neuroethics. He is also co-founder of Ethics and Society at Rice, a summer enrichment program for economically disadvantaged high school students.

1
INTRODUCTION
Commons analytical frameworks and case studies

Blake Hudson, Jonathan Rosenbloom and Dan Cole

The term "commons" has come to mean many things to many people and is used inconsistently within the literature. Understanding what constitutes commons resources and their use has become even more challenging in light of the expanded application of commons study. Although the dilemmas associated with commons resource management had been raised in academic scholarship far earlier, Garrett Hardin's *Tragedy of the Commons* (1968) popularized the management difficulties faced by users of common pool resources (or CPRs). Since 1968 commons scholarship has expanded dramatically, both in depth of theoretical analysis and breadth of subject matter covered. Further, a number of regimes have been utilized to manage CPRs, including common-property regimes, government regulation, and private property. Expansion of commons scholarship coupled with the existence of multiple definitions and perspectives on what constitutes a "commons" make studying CPRs and their management both a theoretical and analytical challenge.

This book endeavors to provide a holistic perspective on the current state of commons study. It provides a broad overview of the arc of commons scholarship to date, paying particular attention to varying definitions and components of particular categories of commons analyses. It identifies cross-cutting themes and deconstructs foundational principles presented in commons literature by addressing the following questions: Who are the actors utilizing resources? What are the activities in which they are engaged? What institutions condition their activities? What are the resources affected by the actors' activities? What assumptions are embedded in the terms "commons"? How is the "commons" label designed to help understand resource appropriation behaviors and how to manage that appropriation? How should those understandings be used to inform management modifications and policy decisions?

Commons scholarship has developed in at least three ways—theoretically, metaphorically, and through empirical case study. Part I presents the increasingly firm foundation of theoretical scholarship on the commons, which has resulted, in part, from the development and application of "frameworks of analysis," or lenses through which to undertake commons study and understand the nature and management of commons resources. Part I explores questions such as: Why develop analytical frameworks of analysis to begin with? What are the core components of analytical frameworks and the fundamental, overarching goals driving their use? What are the most prominent frameworks in the field of commons literature? Which frameworks are best utilized for certain subject matter? Which frameworks are more useful in facilitating the interdisciplinary lens of analysis sought by many commons scholars? Are certain frames of analysis more widely applicable across subject matters?

Chapter 2 explores the implications of scientific communities using analytical frameworks of general applicability to organize research, exploring differences in framing, languages, and heuristics, and assessing the strengths and weaknesses of existing analytical frameworks utilized and how they may be improved to facilitate interdisciplinary study of commons problems. Chapter 3 drills down more deeply into the specific frameworks of Institutional Analysis and Development (IAD) and Social-ecological Systems (SES), assessing the ways in which frameworks steer scholarly attention, shape conclusions, and provide a common, meta-theoretical language to enable scholars to engage with each other and produce comparable findings. Chapter 4 analyzes the theoretical and empirical development of the polycentric system framework, with a view toward facilitating a more nuanced identification of categories of polycentric systems and detecting those polycentric features that need to be strengthened in order for such a system to thrive. Chapter 5 delves into the connection between the conceptual framework of IAD and the empirical subject matter of the commons, and explores important connections across the IAD, SES, and polycentric frameworks.

Chapters 6 and 7 shift gear from the more frequently analyzed analytical frameworks and lay a foundation for understanding emerging commons frameworks—Anticommons and the Knowledge Commons Research Framework, respectively. Chapter 8 introduces the reader to how the metaphorical use of commons stories can itself provide a theoretical lens useful for understanding commons problems and solutions. Chapter 9 uses experimental economics to present an analytical framework for understanding the strategic incentives associated with the use of natural resources based upon the complementary analysis of production externalities and degradation externalities. Chapter 10 concludes the book's theoretical section by developing an analytical framework that departs from narrow self-interest as the primary description of human behavior, and that describes the problem of cooperation in terms of fundamental needs and social cognition. The chapter examines three elements of governance systems—shared decision making, enforcement, and communication—that help to address several persistent questions about cooperation and outline the next generation of behavioral theories of the commons.

The aggregation and analysis of analytical frameworks developed in Part I lays a foundation for Part II's exploration of how various analytical frameworks are applied by scholars from various disciplines. Part II presents case studies across diverse disciplines to demonstrate the modern day breadth of study on the commons. Part II covers traditional natural resource commons, like public lands, water, fisheries, and wildlife (Chapters 11, 12, 13, and 16) as well as emerging areas of natural resource commons study, such as coastal ecosystems, climate change, and ecosystem services generally (Chapters 14, 15, and 17). Part II then moves to the "new commons," with chapters focused on the urban environment and infrastructure as commons (Chapters 18, 19, and 20). It further includes chapters on medical information and genomic data as commons, as well as technological commons (Chapters 21, 22, and 23). Chapter 24 places the "revival" of commons case studies into a historical perspective, while subsequent chapters look at governance structures and commons through the lenses of customary authority, post-socialist societies, civil society organizations and self-governance, and indigenous rights (Chapters 25, 26, 27, and 28). Chapter 29 provides a segue into Part III's analysis of global scale commons by discussing globalization, local commons, and a framework for assessing multiscale ecosystems.

The interdisciplinary approaches presented in Part II within the context of analytical frameworks help facilitate a more complete understanding of both the shared and unique challenges faced by commons resource users and managers across a broad range of disciplines, the usefulness of the commons lens as an analytical tool for studying resource management problems, and the best mechanisms by which to formulate policies to address those problems. Part III provides a discussion of the future of commons study (Chapter 30) and the implications of managing resources on a global scale (Chapter 31).

The structure of this book is meant to both summarize the current state of commons literature and add to the literature by structuring the commons inquiry in a useful manner—that is, to illuminate the usefulness of analytical frameworks for commons study across various disciplines, to reveal which frameworks are most useful for forging interdisciplinarity, to highlight the current breadth of subject matter to which commons study has and can be applied, to indicate the circumstances that make some disciplines more conducive to study through particular frameworks, and to detail how various disciplines utilize specific analytical frameworks, how they speak across disciplines, and how they help us resolve commons resource use and management challenges. We hope this book tells a story that demonstrates a progression through time of the many different ways society conceives of commons resources, the methods used to analyze those resources, management problems identified through theoretical analyses, and solutions to those problems. In doing so, we believe this book highlights the most critical themes in commons scholarship to provide a broad and in-depth understanding of the current state of commons study.

Reference

Hardin, G. (1968) "Tragedy of the Commons" *Science* 162(3859): 1243–1248.

PART I

Theoretical frameworks and alternative lenses for analyzing commons

2

BRIDGING ANALYTICAL FRAMEWORKS AND DISCIPLINES TO WHICH THEY APPLY

Konrad Hagedorn, Philipp Grundmann and Andreas Thiel

Why develop analytical frameworks to begin with?

The behavioral patterns of humans and in particular their reasoning, visioning and economic choices rely to a large extent on the existence of social constructions for mediating interactions between actors and their social, natural and physical environment. Processes of social construction and deconstruction are crucially conditioned and influenced by attributes of the physical and natural environment. Those physical stocks and natural systems show a wide range of properties and are extremely diverse and complex, ever changing and understood by humans only to a limited extent. This equally applies to the tools, technologies and infrastructures developed by humans and set up to use, manage, cope with and also protect those systems. Such diversity and complexity in nature together with associated technical systems which have been developed to use them require corresponding diversity and complexity in institutions and governance modes and structures.

This emphasizes the importance of institutional analysis, where different heuristics and languages have been used in analytical frameworks developed by different scientific communities for organizing research. This gives rise to an important question: What are the implications for the development of analytical frameworks which aim at general applicability?

This chapter is organized as follows: First we will make explicit how we understand the role of analytical frameworks for guiding the arrangement of analysis in research processes. Then we will describe a selection of analytical frameworks and heuristics applied in the field of institutional analysis of linked social-ecological and technical systems based on previous investigations of empirical framework applications (Binder et al. 2012; Grundmann 2017; Hagedorn 2008a). This will reveal similarities and differences in the way they are used in the corresponding scientific communities. Then we will expound on some aspects to show what is behind the differences in framing, languages and heuristics, and we develop an understanding of the emergence and existence of the many analytical frameworks and corresponding communities and disciplines respectively. This allows for some insights into the strengths and weaknesses of existing analytical frameworks utilized, and how they may be improved for facilitating interdisciplinary study of commons problems. Finally, we draw conclusions and recommendations for developing analytical frameworks that stimulate communication and collaboration across scientific communities using different analytical frameworks.

Goals driving the development and use of analytical frameworks

Analytical frameworks are supposed to serve as a research heuristic and a nested system of meaningful variables for different types of analysis, including analysis of institutions and governance structures and commons problems in social-ecological and technical systems. As a point of departure we refer to the inherent objective driving the development of analytical frameworks to provide and use a common scientific language for all disciplines and actors involved in the analysis of institutions and commons problems. A basic premise is that an analytical framework can be "theory-neutral", open to different theoretical approaches and different empirical methodologies, and applicable by different research communities and disciplines. This may be considered as an important objective given the necessity to provide a general, cross-disciplinary and multi-method approach to do justice to the study of institutions and commons problems particularly in social-ecological and technical systems.

In this chapter we are reluctant to share the premise of analytical frameworks being problem-, theory- and empirically or methodologically neutral, because analytical frameworks actually emerge in specific scientific communities which have their own framings, theories and methods. This raises the question of whether or not generalizations across different scientific communities and communities of practice can be reasonable and, if so, what concrete methods of generalization this requires.

We understand an analytical framework as an outcome of a way of framing a research process on a research issue, which is an action situation in itself. "Framing is a process of structuring an action situation, and in that sense it governs 'meaning'" (Lindenberg, 2001, 682). Lindenberg uses "a theory on how the definition of an action situation (a 'frame') affects the selection of knowledge chunks, beliefs, attention to certain situational aspects, recall of situational aspects, as well as choice of action" (2001, 682). In other words, meaning, intention to tackle a problem driven by goals, and relevance to a current or already experienced situation are main elements in this kind of understanding of an analytical framework. An analytical framework in this interpretation cannot be problem-neutral, theory-neutral and empirically neutral.

Any framing of a research problem in terms of an analytical framework has three implicit components—intention, meaning and relevance. When applying an analytical framework scholars need to be explicit in how the framework treats the dimensions of intentions associated with it, meanings it is founded upon and the relevance claims it makes. Relevance implies a normative focus: What problems are addressed and solutions found with the analysis? Meaning is related to the theory which provides an explanation of the elements of the observed reality in which the problem is embedded. Relevance is also related to meaning through the empirical substantiation of theories, both in a positive and a normative meaning—it reveals whether and how the problem is understood and how it may be solved, provided that the research process is able to produce not only substantiated explanations but also feasible solutions.

These components of scientific analysis shape the language used within a research process and the communication of its results. Language itself is an expression of empirically substantiated theory. In other words, the social construction of what is recognized as being part of the real world in a research process requires the construction of language and includes conventions on the terminology a scientific community uses. These insights are important for the interpretation of analytical frameworks and the scientific language used in the frameworks. They are developed in a process of interplay between theoretical approaches, empirical methodologies applied in a process of exploration and investigation that is driven by a specific normative background or framing.

Theories express our always imperfect and fallible reconstructions of reality and influence what part of reality we are able to observe and how we tend to interpret it (Blaikie 2000; Popper 1974; Sayer 1992, 602). Theory development represents the process of interaction between observation, empirical "surprises" and revision of the theory-critical pre-understanding of the problem concerned. These processes are crafted by creating scientific languages which are used in describing the problems, framing the intentions, crafting the theories and communicating the methodologies, and finally, the appraisal of results from which recommendations may emerge with the aim of solving problems. Human language itself is a theory developed by processes of trial and error for identifying elements of the social and physical world and to find an appropriate terminology. Even the very foundation of cognition, that is perception, consists of a process of generating and falsifying hypotheses by means of the system of symbols that constitutes human language (Albert 1977, 114–119). This contradicts the idea of a theory-neutral basis for human experience and the language used for its description.

What therefore can only be collected and systematically arranged in an analytical framework is language that emerged from the processes of theory identification, finding intentions and theorizing insights (being more or less empirically substantiated). Analytical frameworks as research heuristics entail different scientific languages and views on what social-ecological and technical systems are and how they are best analyzed. Research-guiding approaches are developed as the outcome of iterative processes involving interaction between theory and empirical study. Theories are expressing our social constructions of reality and pre-determine what parts of reality we are able to observe and how we tend to interpret them (Blaikie 2000). Accordingly, analytical frameworks grow from the problems, the intentions, and the theoretical and empirical background of their producers. They depend on disciplines and schools scholars belong to, theoretical and empirical process used, the specific object of research, interpretative perspectives internalized by the researchers (mental models), and the scale at which analysis takes place. Therefore, in short, analytical frameworks differ in the intentions, meanings and relevance they carry within themselves.

> A "framework" provides the concepts and terms that may be used to construct the kinds of causal explanations expected of a "theory". A "theory" posits specific causal relationships among core variables, while a "model" constitutes a more detailed manifestation of a general theoretic explanation in terms of the values of particular variables and functional relationships. Just as different models can be used to represent different aspects of a given theory, different theoretical explanations can be built upon a common conceptual framework.
>
> *(McGinnis and Ostrom 2014)*

However, this train of thought could, or even should, be extended. If we want to find a heuristic and a language for guiding research on a certain problem, we have to be aware that such an analytical framework is not only a premise but also an outcome of constructing theories, deriving models and applying empirical methodologies in the respective field of research. Therefore, an analytical framework which has proven useful for framing research in one scientific community may fail when applied to the research issues of another scientific community.

We cannot neglect the question of boundaries as regards theories, models and empirical methodologies that can be reasonably allocated and employed within one analytical framework. This may expose scientific communities using different analytical frameworks for analyzing (maybe their specific domains of) social-ecological and technical systems to obstacles of incommensurability (Feyerabend 1975; Grundmann, 2017; Kuhn 1996). Established analytical frameworks may play a paradigmatic role originating in the above-mentioned categories of intention, meaning and relevance of different communities or schools. As a consequence, it cannot be simply assumed that "frameworks provide a meta-theoretical language that can be used to compare theories" (Binder et al. 2012). Instead, we have to investigate, as an empirical object, why and how research problems, theories, models and methodologies have brought about analytical frameworks and vice versa.

Analytical frameworks with their incorporated heuristics and languages may only have meaning for some research communities and appear to be unreasonable or at least confusing to other research communities also working on social-ecological and technical systems, because they have developed their own analytical frameworks, heuristics and languages in other areas of problem solving, theorizing and empirical work. If this is the case, and this is what this chapter aims to show, "generalizing" an analytical framework will only improve the research capacity of the different scientific communities addressed if they find their problems, intentions, meanings and what is considered relevant by that specific community reflected in it.

In the following section, we will characterize selected analytical frameworks used for the analysis of institutions and commons problems in social-ecological and technical systems based on their core components. As we do not consider analytical frameworks being independent from the theory, models and empirical methodologies, we will pay particular attention to these linkages.

Analytical frameworks for the analysis of institutions and commons problems in social-ecological and technical systems

Two research traditions exist that address, among others, institutions and commons problems in social-ecological and technical systems. The first tradition broadly aims to achieve sustainable use of natural resource systems and effective regulation mainly of environmental pollution by crafting robust institutions and effective governance structures. Focuses of this tradition are, first, withdrawal of resource units from resource systems, and second, emissions into and impact on ecological systems. The second research tradition is characterized by contributions from scholars representing specific branches of agricultural, horticultural, environmental or forestry sciences. These scholars are concerned with the way production and provision in the so-called "Green Sectors" (i.e. nature-related sectors) of the economy are organized and comprise areas such as the farm as an organization, communal and cooperative forest management, land and labor institutions, cooperatives, self-organization and self-help, and the political economy of agriculture. These strands of research share a focus on, first, production seen as transformation based on the use, and sometimes overuse, of ecosystems such as fertile soils by farms (rather than just withdrawal of resource units) associated with other activities institutionalized for channeling the produce to the consumer; and second, the governance and political economy of nature-based production and provision systems, and the resulting biological and environmental implications.

The collection of analytical frameworks presented next is not complete, but some further frameworks are as notable as the ones described here. Paavola and Adger (2005), for example, have introduced "Institutional Ecological Economics". Pahl-Wostl (2009) has developed a very advanced "Management and Transition Framework" which adds to the IAD Framework and extends it by a social learning approach combined with the concept of Adaptive Management Regimes for water (Folke et al. 2005; Pahl-Wostl 2007; Pahl-Wostl et al. 2007a, 2007b, 2010). Kooiman (2008) and Jentoft (2007) use a "Governability Framework". Each of these analytical frameworks, although often similar to, or based on, one of the analytical frameworks described later, would be relevant for this overview. However, in this chapter we decided to focus upon a limited set of frameworks, since our purpose here is to illustrate how different frameworks can enrich the analysis of social-ecological and technical systems, and how research can contribute to their further development in the future. Our findings are based on the investigation of empirical literatures comparing the designing and application of the frameworks described here such as Grundmann (2017) and Binder et al. (2012).

Institutional Analysis and Development (IAD) Framework

The Institutional Analysis and Development (IAD) Framework is centred on action arenas in which actors interact in an action situation and are at the same time affected by exogenous variables—biophysical conditions, community attributes and rules-in-use (see Figure 2.1). The outcomes produced in this way have feedback effects on the actors and the action situation (Ostrom 2005). Action situations can be interlinked and are described by a set of rules and variables.

The IAD Framework emerged from a process of deliberations, among others, on the idea that the "Tragedy of the Commons" cannot be avoided without either privatizing Common Pool Resources (CPR) or managing them by government bureaucracy, as stated by Garret Hardin (1968). The IAD Framework also responded to reflections for theory building on group and organizational behavior (Olson 1965) and, less explicitly, also to theory building on property rights (Demsetz 1967). The authors stress the general applicability of the IAD Framework even though in the past it has frequently been applied with a focus on local CPR scenarios. Empirical evidence from numerous case studies has shown that cooperation works in the case of CPRs, i.e. users of CPRs are able to solve their problems by crafting rules and forming organizations. However, this depends on physical and social conditions. Explaining "the origin of self-governed common-pool resources", Ostrom (2005, 244) points out several "attributes of resources and of appropriators . . . conducive to an increased likelihood that self-governing associations will form". Accordingly, the theoretical and empirical background against which the IAD Framework has been developed (i.e. where the language and perceptions come from) is characterized by CPRs institutionalized as common property and jointly used by groups for collective action (e.g. common grazing land, commonly used irrigation systems).

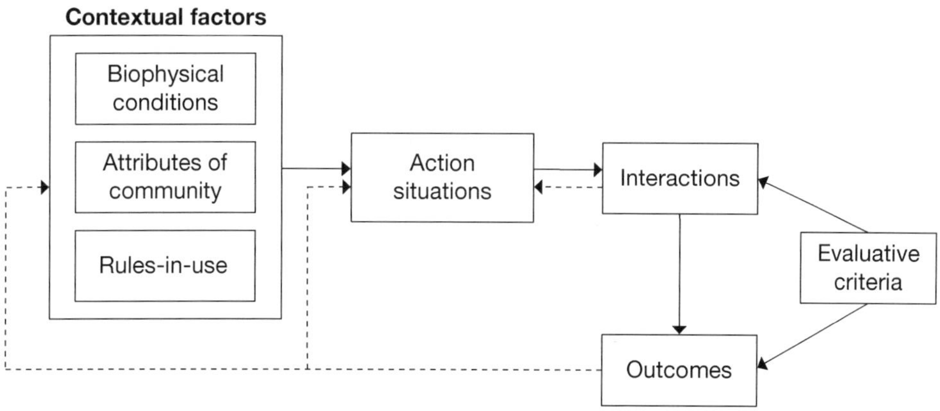

Figure 2.1 Institutional Analysis and Development (IAD) Framework
Source: Adapted from Ostrom (2011, 10).

The Social-Ecological Systems (SES) Framework

The development of what became known as the Social-Ecological Systems (SES) Framework was proposed in an article for the first time by Anderies et al. (2004). The SES Framework was then developed further by Ostrom (2007) and her colleagues from the Ostrom Workshop as well as by discussions in the SES Club (www.cooperationresearch.eu/SES/). The SES Framework was inspired by Ostrom's increasing interactions with ecologists, who objected that the biophysical elements of the IAD framework were impoverished. This resulted in the objective of developing a general framework and common language for analyzing the relationship between social and ecological systems. It implies that scholars from different disciplines and schools, and with different theoretical and empirical approaches, should be able to introduce their terms and perspectives into such an analytical framework. Thus, they are expected to assume language, heuristics and understanding of the SES Framework (Figure 2.2).

The starting point for this work on a cross-disciplinary language was the initial SES Framework published by Ostrom (2007) that builds on the IAD Framework and tries to recognize both social and ecological elements of social and ecological systems more explicitly. It aims to be open-ended and reaches into a great depth of system description by relying on a multi-tiered, nested hierarchy of variables. Due to its origins the SES Framework was intended to subsume the IAD as it grew out of the research experience of the Ostrom Workshop (see McGinnis and Ostrom 2014). Naturally, the SES Framework adopted variables from the IAD Framework which are interlinked by the idea that systems are nested and can be decomposed. The present version is the result of a generalization process, although many elements of the framing use the terminology and main constructs of the IAD Framework; for example, the

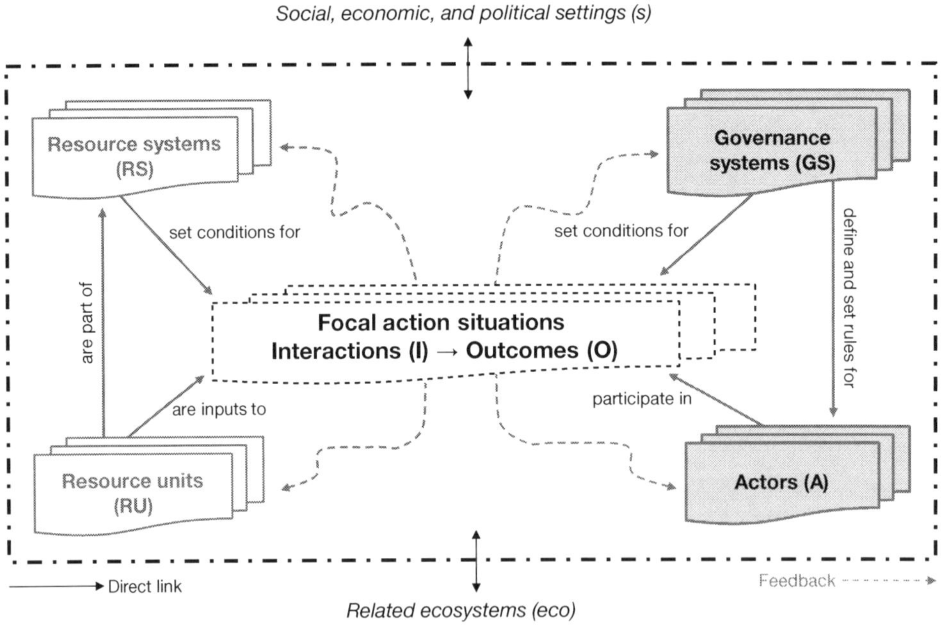

Figure 2.2 Revised SES Framework with multiple first-tier components
Source: McGinnis and Ostrom (2014).

notion of withdrawal of resource units as core physical transaction, well-defined boundaries of resource systems as main spatial property of nature-related transactions or the emphasis of collective governance.

Institutions of Sustainability (IoS) Framework

The Institutions of Sustainability Framework (IoS) (Hagedorn et al. 2002; Hagedorn 2008a) has frequently been applied in institutional research on the "Green Sectors" (i.e. nature-related sectors of the economy such as agriculture, horticulture, fishery and forestry, that interact frequently with natural systems) and focuses on how to regularize human action that leads to transactions affecting the relationship between natural and social systems. Institutions (sets of norms and rules) and governance structures (forms of organization) emerge either spontaneously through self-organization or intentionally by human design. Their social construction has been conceptualized as an outcome of interactions in action situations and depends on the properties of the transactions and the characteristics of the actors involved (see Figure 2.3). The IoS Framework includes elements of the frameworks described previously, but it also adds new concepts.

The under-researched topic of nature-related transactions in interacting social and natural systems (i.e. human-environment interactions) is the core of the analysis within the IoS

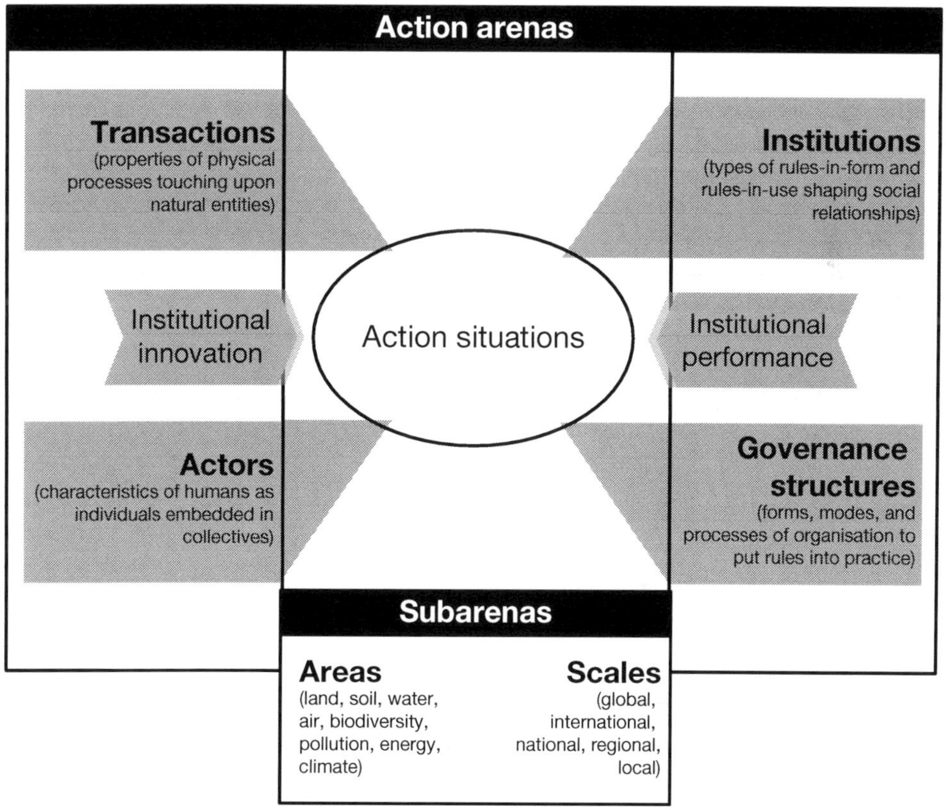

Figure 2.3 Institutions of sustainability

Framework (for details, see Hagedorn 2008a). Transactions such as use, harvest, withdrawal, management and maintaining infrastructure in nature-related sectors have specific properties causing particular forms of interdependence between actors, which are the main cause for the emergence or design of institutions (Paavola and Adger 2005). Institutional analysis in nature-related sectors often focuses on "non-commodities"; that means resources, goods and services whose transactions involve processes of self-organization in ecosystems not completely engineered by humans (Hagedorn 2008a). Institutional analysis in these sectors needs to consider additional properties of transactions and attributes particularly observed in ecosystems, such as "jointness and absence of separability, coherence and complexity, limited standardisability and calculability, dimensions of time and scale, predictability and irreversibility, spatial characteristics and mobility, adaptability and observability, etc." (Hagedorn 2008a, 366). Linkages between activities due to the coherence of the ecological system and the interconnectedness of its parts represent particular reasons why transactions have to be regularized by institutions and organizations. The IoS Framework allows for analyzing the implications of specific relations between characteristics of nature-related transactions and ways of organizing them, because it addresses the micro-foundations of actors' behavior in research. The specificity of nature-related transactions considered within the IoS Framework is also relevant "in many other research areas (although to a different degree) including consumer food economics, supply chain management, agricultural trade and institutional change in transition countries" (Hagedorn 2008a, 358).

Potential incommensurability between analytical frameworks

Incommensurability occurs when scholars are rooted in paradigms which cannot simply be changed by processes of falsifying hypotheses because they originate in different basic views on reality. This section describes characteristics of analytical frameworks which can be considered as potential incommensurability between the outlined analytical frameworks.

The spacial (or physical) dimensions of activities related to a resource system

Language and meaning in some of the analytical frameworks described earlier can be traced back to the context of CPRs institutionalized as common property and governed by collective action, particularly in the case of most elements of the SES Framework. A demand frequently emphasized by scholars applying analytical frameworks is well-defined boundaries of the resource system. However, biophysical boundaries or structures may not be sufficiently identified by their spatial dimension. Ecological systems have many attributes which may be linked with spatial definitions. Accordingly, *transactions between actors mediated through ecological systems* may be equally diverse. Even if we only consider the spatial properties of individual ecosystem attributes, these may already be very diverse so that it is not possible to clearly delineate between separable domains. This can be illustrated by examples like flows of surface and ground water in water systems, the emergence and dispersal of genetic resources, the micro-climate shaped by landscape structures, or the distant and diffuse effects of the many types of nonpoint pollution. The heterogeneous dimensions and attributes of such ecosystem functions can hardly be standardized according to spatial criteria.

CPRs, at least as they are managed in traditional ways without modern inputs, may have a privileged position because the boundaries and properties of the main nature-related transactions (including use, harvest, withdrawal, management and maintaining infrastructure) often coincide with physical and spatial phenomena. However, natural resource systems are often more complex and less transparent, and when they are exposed to human activities other properties of nature-related transactions may become relevant. Therefore it seems necessary to extend the notion of a resource system on which analytical frameworks are built.

Prevalence of non-collective use of natural resources

Although group formation and cooperation play a large role in "Green Sectors" (or nature-related sectors) of the economy, non-collective arrangements dominate. In general and for most of the land area in the world, decisions on farming or grazing are not made by a "group" but by individual farmers operating family farms, partnerships or farm companies, such as limited or joint stock companies. Registered production cooperatives only play a major role in some post-socialist agricultural sectors. However, cooperative arrangements such as marketing and service cooperatives are important elements of the "institutional environment" of family and peasant farms which are as relevant for their historical success. Nevertheless, such cooperatives play a minor role as forms of organization regarding agricultural production itself.

Recognizing the heterogeneity of human activities in resource systems

Actors operating within ecosystems convert, protect, dispose of waste, pollute, cultivate, drain, fertilize, spray, harvest, degrade, increase, etc. natural resources. In contrast, the research heuristic deriving from research on CPRs predominantly refers to "users" who "withdraw" some "resource units" from the resource system (i.e. withdrawal). Accordingly, the main interdependence originating from this is that different users want to acquire the same resource units, representing a cause of conflict which needs to be settled by rules and organization. Generalizing the users towards actors who conduct very different nature-related transactions (intentionally or non-intentionally, voluntarily or non-voluntarily, with or without mutual agreement) seems more adequate.

Nature-related transactions affecting resource systems are often very diverse, e.g. the difficult impact chains in the case of water pollution, or the indirect effects on the diversity of plant genetic resources which may be caused by Genetically Modified Organisms (GMOs). Even so, increased use of a resource may not lead to depletion of a resource but to its preservation and improvement, i.e. have the opposite impact. For example, agro-biodiversity is maintained in this way and decreases by non-use of existing genetic resources. In the case of biodiversity, many actors may never use a resource but insist on having the property right on the option value of wild plant and animal species, for example for future breeding.

Analytical frameworks developed for engineered systems

The way transactions are conceptualized is crucial for the successful application of approaches and frameworks to the institutional analysis of social-ecological and technical systems. In transaction cost economics, "A transaction occurs when a good or service is transferred across a technologically separable interface. One stage of activity terminates and another begins" (Williamson 1985, 1). This focus on the characteristics of a transaction may be too limited, because it explains the need for a transaction to be governed by institutional and organizational arrangements mainly in terms of frictions between activities. By contrast, linkages between activities are equally important reasons why transactions require institutions and governance structures (Hagedorn 2008a). Both frictions and coherence can cause interdependencies between actors that have to be governed by institutional and organizational arrangements that are usually not costless. Not only "friction costs" (Williamson 1985, 18f.) but also "coherence costs" are relevant for institutional choice.

To understand this we must be aware that transaction cost theory originated in research processes that focused on industrial and commercial sectors. Transactions are different in nature-related sectors. Transaction cost theory usually focuses on transactions that can be seen as transfers of "commodities", i.e. goods predominantly produced by engineered processes within designed systems set up by humans. By contrast, institutional analysis in nature-related sectors often focuses on "non-commodities"; that means resources, goods and services whose transactions involve processes of self-organization in ecosystems not completely engineered by humans (Hagedorn 2008a). Conventional Transaction Cost Economics does not sufficiently take into account the properties of transactions that are typical for a natural system. Similarly, the view on governance structures (market, hierarchy and hybrids) seems to be too rigid to cover the heterogeneity of governance structures which can be observed in the area of environmental governance and natural resource institutions.

Reinforcing microfoundations by linking nature-related transactions and actors' interdependence

Nature-related transactions (Hagedorn 2008a) cause particular forms of interdependence between actors, which are the main cause of institutions coming into being (Paavola and Adger 2005). Furthermore, as the properties of transactions related to natural systems differ largely from those related to engineered systems, the institutions and governance structures may differ.

The IoS Framework covers aspects that have not sufficiently been taken into account by other analytical frameworks in the area of institutional analysis in social-ecological and technical systems. Thus it can add elements to a general framework and, above all, develop bridges between existing analytical frameworks. For example, looking at transactions as endogenous driving forces of rulemaking in social-ecological and technical systems helps explain the diversity and dynamics in action arenas. This follows the example of a central element in the IAD Framework which has detailed the micro-foundations regarding the characteristics of actors. We argue that this could be equally done for the *physical* properties of transactions.

There are two main extensions to Transaction Cost Economics which could make it more suitable to contribute to analyzing the interaction between social and ecological systems: (1) the IoS Framework can add the properties of transactions from biophysical and ecological systems including associated infrastructure and technologies and that way make properties of transactions endogenous in a way other frameworks like the SES Framework do not address to date, and (2) it helps in overcoming the rigidity of predefined standard types of organizations and actors usually analyzed in Transaction Cost Economics and opens up interdependence between actors, institutions and polycentric governance.

Integrative institutions in nature-related sectors

Although some analytical frameworks recognize that institutions regularizing human-environment relationships require particular concepts for framing their analysis, micro-foundations in this regard are rare. As transactions are different in nature-related sectors of the economy, the properties of transactions and the actors' interdependence associated with them are likely to be different as well. This suggests that institutions and governance structures for governing nature-related transactions also differ from those existing in other (or less nature-related) sectors. There are some indications that such an alignment can be found following the dichotomy of "integrative and segregative institutions" (Hagedorn 2008b).

Referring to the typical action situation of family farms prevailing in agriculture, Hagedorn (1996, 1997, 2003) argues that the integrating institution "family farm" coordinates decisions which are not related to structural change at lower transaction costs compared to segregative institutions prevailing outside agriculture. However, it coordinates decisions affected by structural change at higher transaction costs than the segregative institutions prevailing outside agriculture. The integrative institution "family farm" influences decision making related to factor reallocation and factor incomes in a particular way because of a series of interwoven elements, including (a) the family of the farmer as a social system, (b) the joint household of the family members being their economic organization, and (c) the farm as their technical production unit. As a consequence, the coordination mechanisms for farming are not separated from those of the family and the household, and involved actors are less able to externalize transactions into their decision-making processes than economic actors in segregating institutional systems. In segregating systems many decisions are taken in a separated way and are correspondingly based on a smaller amount of total transaction costs. This example of framing institutional analysis of nature-related transactions shows the relevance of differences in analytical frameworks and calls for bridging the existing gap between analytical frameworks in general and, in particular, between frameworks for analyzing institutional and commons problems in social-ecological and technical systems.

Facilitating interdisciplinary study using frameworks

Since analytical frameworks are transposing external objects of researchers into language that should have a "proper" internal representation in their minds, then questions arise if certain frameworks are more conducive to collaborative research across subject matters, and how to translate between the different communities and reconcile their cognitive schemata—or in other words, how to bridge the different analytical frameworks including the heuristics and languages they provide. Accordingly, instead of extending the applicability of one overarching framework by bonding, bridging may be at least equally important.

Bonding would mean preserving the original heuristic and language, and extending its influence by popularizing the technical structure but sticking with the terminology and meaning predominantly inherited from the original framework. Bridging would require exploring other frameworks used in other scientific communities, engaging proactively in a process of mutual and interdisciplinary learning and trying to understand languages and heuristics and why they differ. The question arises whether this open approach is more rewarding than developing one overarching framework, although it is probably more demanding. It requires that those research agents in the action situation who intend to construct a general framework first have to know different analytical frameworks, heuristics and languages of other disciplines; and, second, they have to negotiate commonalities and mitigate differences in favor of increased commensurability, provided that this can actually be achieved. The challenge would be to make an overarching analytical framework comprehensible, reasonable and useful for different communities—or to settle with a group of highly compatible analytical frameworks.

From the previous characterization and comparison of analytical frameworks we can conclude, first, that the communities applying these frameworks learned from analyzing the different elements in social-ecological and technical systems. Second, they found different theories and methodologies were adequate. Third, this explains the differences in how they understand and, accordingly, what language they use for understanding the linked elements in social-ecological and technical systems. Fourth, some of the frameworks and languages are similar while others are not. Reasons could be that they are based on different theoretical assumptions and/or an empirical background.

Reasons of differences in framing, languages and heuristics

We argue that the rapidly progressing research stream on institutional analysis of social-ecological and technical systems is specifically prone to such differences between framing, languages and heuristics. Even large communities from disciplines such as agricultural economics, forestry economics and nutrition economics exist which necessarily have to combine knowledge from social, natural and technical sciences; for most disciplines in social sciences this is a novelty. It is unavoidable that social scientists acquire at least some knowledge in fields of natural and technical sciences if they want to engage in the analysis of social-ecological and technical systems.

These reasons may produce a tendency among social scientists working on social-ecological and technical systems to select those technical and natural fields which are relatively easy to understand (e.g. withdrawal of water or fish in contrast to the complex production and provision processes and the environmental implications in intensive cropping or specialized dairy farming). Single research communities may only develop familiarity with a particular area of natural resource use, learn only about selected parts of social-ecological and technical systems and, therefore, develop narrow perspectives resulting in non-commensurable processes of framing which find their expression in contrasting analytical frameworks, languages and heuristics. Against this background any attempt at generalizing may encounter severe obstacles.

The phenomenon described above seems to be the key for developing more commensurable frameworks that link social, ecological and technical systems by properly integrating all three components into the micro-foundations of research. Even this puts significant strain on researchers because it requires lengthy engagement into understanding the underlying intentions, meanings of different communities and their analytical frameworks as well as spelling out what the corresponding research communities consider relevant.

Beyond a single analytical framework: an extended view

Some authors in institutional economics go beyond the conventional framing of institutional change processes (Bromley 2006; Grundmann 2017; Vatn and Bromley 1994). They do not emphasize one analytical framework, but rather show the necessity of several analytical frameworks for organizing research processes. Grundmann (2017, 342) highlights that:

> [t]he combination of analytical frameworks from various fields and communities of research and practice is conducive to the broadening and deepening of individual frameworks used for analysing socio-technical regimes. Individual frameworks are essential, but they need to be deepened and broadened further through combination with other relevant frameworks, as a framework only addressing single regimes or dimensions will most probably overlook existing interdependencies and deliver insufficient outcomes and explanations.

Vatn (2005) argues that the actors' mode of action selection (Ostrom et al. 1994) crucially depends on the institutional context as well as on the framing of decisions as these may either be made for individual or collective gains (Vatn 2005, 2007). For example, increased division of resources into individually owned properties operating in sectors with rapid structural change have facilitated an unprecedented level of economic growth because individuals strived for increases in income. This strategy has produced increasing "externalities" or "cost-shifting practices" because distinct de jure properties cannot avoid de facto interconnections in nature-related transactions. To cope with this ecologically detrimental development Vatn recommends common property structures to internalize shifting of costs between decision units (Vatn 2010). Vatn and Bromley (1994) apply a similar approach regarding the role of property regimes in producing rational choices and efficient outcomes of actors' interactions in natural resource exploitation.

Bromley (2006) reveals the significance of such an understanding of institutional analysis by providing a micro-analysis on how institutional change actually works. Institutions are the outcomes of collective decisions on ways to regularize interdependence between actors, which is associated with nature-related transactions, sets of rules and forms of organization. They are ways actors organize their interdependence in activities of natural resource use—such as extraction, withdrawal, production or provision of goods and services—in a way that follows individually or socially constructed and preferred "visions" of economic organization. Such visions or "created imaginings" can be considered more or less valuable and are sanctioned by epistemic communities (Bromley 2006). Their choice depends on the rules that structure collective decision making on institutions and they change when actors perceive pertinent institutions as nonaligned with the social constructions they prefer. In other words, institutions and their change require explanations exogenous to the realm in which they operate (Bromley 2006).

If we seriously want to take into account such relationships, which are crucial for processes of institutionalization and de-institutionalization, we have to recognize the need for several analytical frameworks for organizing research processes and to "empirically understand both the limits and possible advantages of reworking, supplementing, or even combining such frameworks" (Grundmann 2017, 339). For this "it is necessary to first take a deep look at their heuristics and languages, before indiscriminately favouring and adopting any given framework" (Grundmann 2017, 339). Moreover, any adequate understanding of social-ecological and technical systems needs to integrate the analysis of sectors external to the domain where institutions become effective.

The activating role of the physical link in social-ecological and technical systems as in action situations: the transaction-interdependence-institutions nexus

To illustrate what it means to make the ecological and technical components an integrative part of action situations we look at how the IAD Framework analyzes in depth the influence of actors' characteristics on action situations (Ostrom 2005). The IAD Framework disaggregates an action arena into:

> [t]he set of participants, the positions to be fulfilled by participants, the potential outcomes, set of allowable actions and the function that maps actions into realized outcomes, the control that an individual has in regard to this function, the information available to participants about actions and outcomes and their linkages and the costs and benefits—which serve as incentives and deterrents—assigned to actions and outcomes.

We make a similar attempt to detail the logic of how the properties of natural and technical systems affect an action situation. Even more, as research on social-ecological and technical systems explicitly focus on questions where humans impact natural systems, most frequently via technical systems, and where they are in turn affected (profiting or suffering from) by feedback loops, the consequences of these activities are at stake. In other words, it is the decisions taken by at least one actor which cause nature-related transactions. Within their structured context, nature-related transactions are operationalized through their biophysical material context and the means of technology and infrastructures, which stimulate actors to think about their potential choices and to respond by taking action, i.e. this makes the action situation work.

This is a core relationship, the *transaction-interdependence-institutions nexus*. Actors make decisions which—via technical devices—produce nature-related transactions that cause interdependence between actors which makes them mutually dependent and possibly motivates them to interact and engage in rule-making. If there are no transactions, there is nothing relevant to talk and negotiate about, and there is no reason for designing or crafting institutions.

For example, if the use of natural resources such as land, water, air, atmosphere, biodiversity, etc. in a community does not put any constraints on the claims other citizens have on natural resources, people are rather indifferent. However, if the choices the actors take cause nature-related transactions (e.g. by means of cropping and livestock technologies) and induce, for example, relocation of sediments from soil erosion to cities or rivers, pollute drinking water by pesticides, diffuse odors from livestock husbandry, or reduce biodiversity and landscape amenities, then the actors will enter into interdependence. This may lead to a very lively action situation with fierce debates in the community. If this community has the communicative and coordinative capacity to transform this interdependence into institutional and organizational innovation, this may yield satisfactory results for all. This requires that they are able to actively build institutions and governance structures; in most cases such problems can only be solved by social construction at various levels.

Unpacking the activating role of the transaction-interdependence-institutions nexus in social-ecological and technical action situations

The micro-foundation we propose to introduce is based on detailing the logic of the causal connection between (a) actors who take decisions within specific types of nature-related action contexts, (b) thus initiating transactions that have nature-related attributes, which (c) create intended or unintended nature-related interdependence between actors (d) that may lead to rule making and, in this way, regularize decision making and interdependence. This micro-perspective raises the question of what decisions by actors eventually impact natural systems via infrastructure or technological systems, and what decisions therefore affect other actors in a way that the interdependence created stimulates interaction and communication in the action situation. Action situations can be arranged according to typical properties of physical transactions to enable an understanding of the differences between processes mobilizing action situations. In other words, core activities take place, induced by conscious or unconscious choices of actors, in which humans interact—mediated by resource systems and the use of technological systems—and provoke specific types of action situations. A physical transaction leads to interdependence between actors and may finally cause what could be called a social transaction that means the provision and change of social organization in order to cope with the physical transaction. This concept of sequencing of institutional and organizational analysis builds on the "alignment principle" and "micro-foundations", where actors either individually or from a social welfare perspective try to minimize the perceived transactions and production costs and maximize the benefits involved in the organization of nature-related transactions.

Such a conception of micro-foundations refers to an explicit analysis of physical transformations and transactions, and to the alignment principle which puts forward the idea (inspired by transaction cost economics) that de facto governance structures align with properties of transactions and capacities of institutions and actors. Specifically, the attempt at structuring research that follows this hypothesis addresses the question of how to capture and structure the analysis of relevant transactions (including transformations involved) that are to different degrees "nature-related", and how to describe their characteristics.

Thus, the question arises of how to develop a reasonable ordering of transformations and transactions which stimulate action arenas in linked ecological, technical and social systems. We suggest arranging this according to what we call nature-related core activities of humans. They cause physical transformations and physical transactions in resource systems, eventually with the help of technical systems that contribute to the emergence of interdependencies between actors which in turn may stimulate action situations. The principle of ordering is decreasing deep and active intrusion of human influence in natural systems proceeding from very active behavior of humans to rather passive behavior vis-à-vis natural resources and the natural environment. Nevertheless, different ways of ordering these activities could also be chosen such as categorizations in relation to extractive or non-extractive use of the environment and their relation to environmental flows.

As the most brutal intervention into natural systems we consider the *conversion* of the use of natural resources, for example, from agricultural land use to urban land use for housing or traffic. These activities involve the entire resource system and change its overall properties. Changing the entire resource system is also relevant for *conservation*; however, the direction of influence is the opposite. This is followed by *extraction* of *non-renewable* resources (resource units), for example from units of crude oil or natural gas, and *withdrawal* of units of *renewable* resources, for example, from a water canal or forest. These activities only deal with components of the resource system by taking out resource units which are part of larger resource systems. The following group of activities only affects a part of the resource system, for example the upper layer of the earth's crust, by posing in it units of matter or energy by *waste* disposal and environmental *pollution*. In this case, units "flow" into the resource system and affect it.

The following group of activities preserves and influences the entire resource system or specific stocks of resource units to profit from biological processes which are controlled or even manipulated for that purpose. Activities in this category could be *growing* livestock and *cultivating* crops, which rely on exploitation or even manipulation of ecosystem functions and processes catered for by the resource system. These activities are still in close physical contact with natural systems, i.e. they still rely predominantly on biophysical processes. The next category of activities does not directly rely on ecosystem functions. Thus, activities are largely detached from the resource systems and deal with decomposed commodities. They are gained from the above-mentioned, partly controlled biological processes such as cultivating maize or wheat. *Marketing* and *processing* of food or non-food agricultural commodities, for example, belong to this class of core activities. However, such decomposed commodities will sooner or later return into biological and ecological systems after they are consumed by humans, or even as they produce waste throughout consumption. Therefore, these detached activities also have an important nature-related impact.

However, humans are also *using natural systems directly*, for example by enjoying landscape amenities. The last activity in this attempt to logically derive an ordering of core activities aims at enhancing the *living condition for humans* in resource systems through enhancing natural processes or detaching humans from resource systems or stocks of resource units. This refers to favorable natural conditions in which humans and other living organisms can flourish, such as a micro-climate, and then protecting humans against natural disasters such as earthquake, flooding and heat waves.

These core activities, which are of central importance in linked ecological, technical and social systems, may be grouped according to the logic of decreasingly deep and active intervention of humans in non-human natural systems. Obviously, throughout such an attempt of ordering empirical phenomena in a structured fashion, certain inconsistencies may remain.

Given the complexity and diversity of linked natural, technical and social systems, we consider this unavoidable. Thus we propose to continue by characterizing these activities according to four additional criteria:

1. First, whether these activities are associated with *physical transformations* which can happen either within a *natural system* only being induced by humans but just to a limited extent controlled by them; or whether transformations can occur in *engineered systems* where humans have more or less complete control of transformation processes. In addition, human activities may not cause any major transformation but only undertake physical transfers (transact) within the resource system which lead to the transformation of functionally interlinked parts of it. These physical transformations (or rather transfers) may be incorporated in physical transactions that only implicate one intentional actor.
2. Second, the nature of the dominant structure that determines the *physical transaction* between two or more actors. Such transactions can be *nature-related, production-related* or *provision-related.*
3. This classification of core activities and *nature-, production-* or *provision-related physical transactions*, is complemented by a classification of characteristics of transactions. According to the alignment hypothesis, they provoke specific processes of communication, interaction, conflict resolution or seeking cooperation in an action situation. In addition they are crucially shaped by characteristics of actors.
4. A typology of properties of transactions may fruitfully be based on the criteria of modularity and functional interdependence, as suggested by Hagedorn (2008b). These properties may involve dimensions of jointness and absence of separability, coherence and complexity, limited standardizability and calculability, dimensions of time and scale, predictability and irreversibility, spatial characteristics and mobility, adaptability and observability, etc.

These concepts need further empirical investigation to explore what this means for the actors involved, individually and collectively, and their capacity to use frameworks and to craft institutions. Such research may particularly reveal ways to facilitate interdisciplinary work necessary to cope with the high complexity in social-ecological-technical systems.

Summary

Analytical frameworks used by scientific communities for analyzing social-ecological and technical systems from an institutional and commons problems perspective show major differences. They are neither neutral as regards the theory used by scientific communities nor independent from their empirical field of application and the methodological tools used. They embody differences in intention, meaning and relevance. Characteristics for an analytical framework are the dimensions of intentions associated with the research it guides, the meaning it attributes to the research objects and the relevance claims it makes. Relevance implies a normative focus: What problems should be solved with our analysis? Meaning arises from theories which provide an explanation of the perceived reality in which the problem occurs. Relevance derives from the empirical background of theories as it raises the question whether we actually understand the problem, including context and causalities, and, based on this, whether we can make reliable and feasible recommendations.

This leads to an important strategic finding: We should recognize the reasons for the existing diversity of analytical frameworks. Any attempt at developing a general framework for the most important institutional constructions that shape societies' relationships to its biophysical, non-human environment are deemed to fail, if it is predominantly based on one analytical framework only—including its specific framing, heuristic and language—that just one scientific community or group of scholars has developed and is used to. Furthermore, generalizing an existing framework in this manner necessarily may lead to misunderstandings and resistance by communities that did not participate in its development, as well as it being likely that such a generalized framework will not speak well to the settings these are to analyze. Therefore, we argue in favor of exploring the background of analytical frameworks that are used by a series of communities to analyze institutional settings which in one way or the other are highly relevant to the most important aspects that shape societies' physical relationships to its biophysical, non-human environment.

For illustrative purposes we singled out analytical frameworks used for the analysis of institutional and commons problems by addressing interactions between (human) actors mediated by natural systems, and facilitated by technical ones, either predominantly by nature-related transactions or also by organizing production and provision-related transactions most visible in economic sectors that strongly depend on such linkages, like the "Green Sectors" (or nature-related sectors) of the economy. These include agriculture, forestry, fishery and horticulture. The subsequent treatment of these analytical frameworks elucidated the intentions, meanings and categories of relevance that characterize each of these analytical frameworks, as well as their genesis and some of the main authors and theorists in the field. We subsequently highlighted requirements that current frameworks seem to have specific difficulties in addressing thus far.

First, we highlighted the connectedness of the framing to local social-ecological and technical systems, which largely neglect cross-scale and cross-sectoral interactions between action arenas and governance arrangements, and which have difficulties in capturing dynamics in management of large-scale social-ecological and technical systems. Furthermore, large sample data gathering instead of case-driven data gathering have not been conducted yet, which would allow generalization of the role of categories addressed by currently used frameworks. This is said to undermine the possibility of generalizing its findings.

A second set of weaknesses of our benchmark approach is associated with our observation of spatial heterogeneity and often diffuse boundaries of resource systems and flows of resource units. We emphasized that CPRs at least as they are managed in traditional ways without modern inputs, may have a privileged position because the boundaries and properties of the main nature-related transactions (including use, harvest, withdrawal, management and maintenance) coincide often with physical and spatial phenomena. However, resource systems are often more complex and less transparent, and when they are exposed to human activities, other properties of nature-related transactions requiring different modes of governance may become relevant.

Third, the institutional and organizational forms that affect societies' physical relationships to their biophysical, non-human environment often differ from those inherent in currently used framings. Therefore, the frameworks do not allow for properly capturing these categories, which many scholars consider important. This becomes specifically evident in nature-related sectors of the economy where social-ecological and technical systems are often not institutionalized in a collective way, but where institutions concern interactions between individual actors or between individuals and the state. While cooperation remains important, it often follows a different logic than the type of collaboration presumed in commonly used frameworks, such as the SES Framework.

Fourth, actors affecting natural systems do so in more ways than the mere extraction of resources or the discharge of pollutants into those systems. For example, other activities may be the conservation of ecosystems, the exploitation of productive capacities of natural systems by controlling biological processes for livestock raising and crop cultivation, and the enhancement of or the detachment from social and ecological systems in order to maintain people's living conditions. Thus, we argue that it is difficult to use the current analytical frameworks in order to properly capture such diverse institutional and organizational choices that differ from predominantly collaborative arrangements for overall benefits of voluntary collaborators or described alternative activities. However, we consider it important that analytical frameworks for analyzing institutional and commons problems are able to represent the most important aspects of social-ecological and technical systems, such as types of institutions and forms of organization and activities that shape actors' underlying motivations and purposes of action, and finally the impacts of their actions on the environment. Therefore, we essentially suggest ways to build bridges between analytical frameworks that address different forms of organization and activities.

Fifth, currently used frameworks are strong on detailing the role of actors' characteristics and institutions, but they do not systematically treat properties of human-environment interactions (nature-related transactions) as exogenous. Biophysical conditions, community attributes and rules-in-use are labeled as "exogenous conditions" in frameworks such as the IAD, even though one may argue that feedback loops in the IAD Framework endogenize these conditions to the framework. The SES Framework suggests some categories to address differences in resource systems. However, from the perspective of the focal action situation, they still remain exogenous. We emphasize that properties of transactions have a significant effect on the way coordination problems involving the non-human environment are solved and how they perform. In a last step, this would also require an in-depth analysis of actors and their capacities to shape these interactions and the institutions that regularize them.

This collection of arguments calling for an integrative procedure is surely incomplete, but it may be sufficient to conclude that it seems recommendable to refrain from bonding frameworks by generalizing just one framing of the institutional analysis of social-ecological and technical systems. Instead, a welcome culture towards scientific communities using (different and perhaps very specific) analytical frameworks in their disciplines should be institutionalized by employing a bridging strategy. To facilitate this we recommend mapping analytical frameworks according to the ways activities relate to processes involved in transformation and types of physical transaction. Such a concept of ordering frameworks can help scholars to understand in what fields of activity actors can make choices that cause different types of transactions—which may or may not embody different categories of transformation—where the transactions again have properties which may be arranged according to a reasonable system of classification. Thus, a mapped set of analytical approaches would address the way properties of types of transactions shape governance structures.

Doing research on all fields of the mapped frameworks would exceed not only the cognitive capacity of individual researchers but also the organizational capacity of single scientific communities. Bridging the boundaries between frameworks will only be possible if scientific agents understand the differences and specificities of the described activities, transformations and transactions and the particular properties thereof in terms of underlying intentions, meanings and relevance. Finally, developing a specific "science map of analytical frameworks" could help to explain the locations and networks of knowledge which have shaped particular analytical frameworks. Following these arguments, it seems recommendable to adopt a welcome culture towards all scientific communities using analytical frameworks for the analysis of institutional and commons problems in social-ecological and technical systems and to engage in a bridging strategy. The need for embarking on such a bridging or similar strategy for developing and using analytical frameworks in a context involving several distinct communities has been shown by Grundmann (2017), Binder et al. (2012) and Hagedorn (2008a). The authors emphasize that adopting such a strategy for theory building and methodological innovation requires the involvement of a high-level international academic community in advanced studies and teaching as well as the engagement of communities of practice. The process of transfer, reciprocal understanding, mutual exchange and combining of intensions, languages, meanings and technical structures requires researchers and practitioners with different backgrounds to engage in adapting, mingling and jointly adopting their complementary terms and views in both ways but also understanding their particularities.

References

Albert H. (1977) *Kritische Vernunft und menschliche Praxis*, Reclam, Stuttgart, Germany.

Anderies J. M., Janssen M.A., and Ostrom E. (2004) "A Framework to Analyze the Robustness of Social-Ecological Systems from an Institutional Perspective", *Ecology and Society*, 9(1): 18.

Binder C.R., Hinkel J., Bots P.W.G., and Pahl-Wostl C. (2012) "Comparison of Frameworks for Analysing Social-Ecological Systems", *Ecology and Society*, 18(4): 26.

Blaikie N.W.H. (2000) *Designing Social Research: The Logic of Anticipation*, Polity Press, Malden, MA.

Bromley D. (2006) *Sufficient Reason: Volitional Pragmatism and the Meaning of Economic Institutions*, Princeton University Press, Princeton, NJ.

Demsetz H. (1967) "Toward a Theory of Property Rights", *The American Economic Review*, 57(2): 347–359.

Feyerabend P.K. (1975) *Against Method: Outline of an Anarchistic Theory of Knowledge*, Verso, London.

Folke C., Hahn T., Olsson P. and Norberg, J. (2005) "Adaptive Governance of Social-Ecological Systems", *Annual Review Environmental Resources*, 30: 8.1–8.33.

Grundmann, P. (2017) *Framing Socio-Technical Regimes in Bioeconomy Transitions*, Humboldt Universität zu Berlin, Berlin.

Hagedorn K. (1996) *Das Institutionenproblem in der agrarökonomischen, Politikforschung Schriften zur Angewandten Wirtschaftsforschung*, J.C.B. Mohr (Paul Siebeck), Tübingen, Germany, 72.

Hagedorn K. (1997) "Regelungsdefizite der bäuerlichen Agrarverfassung in der Marktwirtschaft als Begründung für Agrarpolitik? Ein institutionenökonomischer Ansatz", in Behrends S. ed., *Ordnungskonforme Wirtschaftspolitik in der Marktwirtschaft Volkswirtschaftliche Schriften* (474), Duncker und Humblot, Berlin, pp. 123–169.

Hagedorn K. (2003) "Rethinking the Theory of Agricultural Change in an Institution's of Sustainability Perspective", in von Huylenbroeck G., Verbeke W., Lauwers L., Vanslembrouck I. and D'Haese, M. eds, *Importance of Policies and Institutions for Agriculture*, Academia Press, Ghent, Belgium, pp. 33–56.

Hagedorn K. (2008a) "Particular Requirements of Institutional Analysis in Nature-related Sectors", *European Review of Agricultural Economics*, 35(3): 357–384.

Hagedorn K. (2008b) "Integrative and Segregative Institutions: A Dichotomy for Nature-Related Institutional Analysis", in Schäfer C., Rupschus C. and Nagel U.J. eds, *Enhancing the Capacities of Agricultural Systems and Producers*, Marggraf, Weikersheim, Germany, pp. 26–38.

Hagedorn K., Arzt K. and Peters U. (2002) "Institutional Arrangements for Environmental Co-operatives: A Conceptual Framework", in Hagedorn K. ed., *Environmental Co-operation and Institutional Change: Theories and Policies for European Agriculture*, Edward Elgar, Cheltenham, UK, pp. 3–25.

Hardin G. (1968) "The Tragedy of the Commons", *Science*, 162(3859): 1243–1248.

Jentoft S. (2007) "Limits of Governability: Institutional Implications for Fisheries and Coastal Governance", *Marine Policy*, 31: 360–370.

Kooiman J. (2008) "Exploring the Concept of Governability", *Journal of Comparative Policy Analysis*, 10: 171–190.

Kuhn T.S. (1996) *The Structure of Scientific Revolutions*, University of Chicago Press, Chicago, IL.

Lindenberg S. (2001) "Social Rationality versus Rational Egoism", in Turner J.H. ed., *Handbook of Sociological Theory*, Kluwer, New York, pp. 635–668.

McGinnis M.D. and Ostrom E. (2014) "Social–Ecological System Framework: Initial Changes and Continuing Challenges" *Ecology and Society*, 19(2): 30. DOI: 10.5751/ES-06387-190230.

Olson M. (1965) *The Logic of Collective Action. Public Goods and the Theory of Goods*, Harvard University Press, Cambridge, MA.
Ostrom E. (2005) *Understanding Institutional Diversity*, Princeton University Press, Princeton, NJ.
Ostrom E. (2007) "A Diagnostic Approach for Going Beyond Panaceas", *Proceedings of the National Academy of Sciences of the United States of America*, 104(39): 15176–15178.
Ostrom E. (2011) "Background on the Institutional Analysis and Development Framework", *The Policy Studies Journal*, 39(1): 7–27.
Ostrom E., Gardner R. and Walker J. (1994) *Rules, Games, and Common Pool Resources*, University of Michigan Press, Ann Arbor, MI.
Paavola J. and Adger W.N. (2005) "Institutional Ecological Economics", *Ecological Economics*, 53(3): 353–368.
Pahl-Wostl C. (2007) "Transition Towards Adaptive Management of Water Facing Climate and Global Change", *Water Resources Management*, 21(1): 49–62.
Pahl-Wostl C. (2009) "A Conceptual Framework for Analysing Adaptive Capacity and Multi-Level Learning Processes in Resource Governance Regimes", *Global Environmental Change*, 19: 354–365.
Pahl-Wostl C., Craps M., Dewulf A., Mostert E., Tabara D. and Taillieu T. (2007a) "Social Learning and Water Resources Management", *Ecology and Society*, 12(2): 5. www.ecologyandsociety.org/vol12/iss2/art5/.
Pahl-Wostl C., Holtz G., Kastens B. and Knieper C. (2010) "Analyzing Complex Water Governance Regimes: The Management and Transition Framework", *Environmental Science and Policy*, 13: 571–581.
Pahl-Wostl C., Sendzimir J., Jeffrey P., Aerts J., Berkamp G. and Cross K. (2007b) "Managing Change Toward Adaptive Water Management through Social Learning", *Ecology and Society*, 12(2): article 30.
Popper K.R. (1974) *Objektive Erkenntnis*, Hoffmann und Campe, Hamburg, Germany.
Sayer A. (1992) *Method in Social Science: A Realist Approach*, Routledge, London.
Vatn A. (2005) *Institutions and the Environment*, Edward Elgar, Cheltenham, UK.
Vatn A. (2007) "Resource Regimes and Cooperation", *Land Use Policy*, 24(4, SI): 624–632.
Vatn A. (2010) "An Institutional Analysis of Payments for Environmental Services", *Ecological Economics*, 69: 1245–1252.
Vatn A. and Bromley D.W. (1994) "Choices without Prices without Apologies", *Journal of Environmental Economics and Management*, 26(2): 129–148.
Williamson O.E. (1985) *The Economic Institutions of Capitalism*, The Free Press, New York.

3

USING THE OSTROM WORKSHOP FRAMEWORKS TO STUDY THE COMMONS

Michael Cox

The commons and institutional analysis at the Ostrom Workshop

In this chapter, I will discuss two well-known frameworks developed at the Workshop in Political Theory and Policy Analysis at Indiana University, now affectionately referred to as the Ostrom Workshop (https://ostromworkshop.indiana.edu/). These frameworks are known as the Institutional Analysis and Development (IAD) and Social-ecological systems (SES) frameworks. Each reflects one of the primary analytical approaches developed at the Ostrom Workshop: institutional analysis. This involves identifying the formal rules and more informal norms that influence the incentives that participants have in decision-making environments and to what effect for public outcomes.

That institutions are a key part of understanding human behavior in any social context has been reinforced by work that has demonstrated just how norm-driven humans are (Henrich 2015). Human psychology assumes a norm-driven world: in any social setting we look for the rules that guide our behavior. Rules are a universal feature of socially relevant human decision-making, and this is where much of the power of the Ostrom Workshop's institutional approach comes from. This approach has been applied to many areas of human activity (see McGinnis 1999), but its most famous application at the Ostrom Workshop has been to the study of environmental commons governance.

In studies of the commons the primary theoretical motivation is the presence of a social dilemma, or collective-action problem. The ideas of a commons and a collective-action problem are closely connected. A commons is a shared resource. When individuals have an incentive to use a commons such as a forest or a fishery for their own short-term benefit to the detriment of the longer-term group interest, interests at the two social levels diverge and we have a collective-action problem. In many cases individual decisions may lead to the deterioration of the commons, to the point where it is no longer able to support further use. This is commonly known as the "tragedy of the commons" as popularized by Hardin (1968). It is referred to as a tragedy because individual-level decisions lead to collective ruin, rather than having some winners and some losers.

Hardin (1998) subsequently clarified his point by stating that he would have preferred to call this the tragedy of the "unmanaged" commons. In more common terminology we could call this the "open access" commons. An open access regime is one without any property rights that define who can and who cannot access and use a commons, and so does not involve any constraints on commons use. This clarification is made to emphasize that a commons can be used sustainably for the long-term benefit of the group. Much of the institutional work on the commons is based on the argument that collective-action is needed to maintain the commons and the interest of the group that relies on it. If resource users can co-provide public goods in the form of enforced institutional constraints on use, then they can mutually constrain each other's behavior towards their shared resource.

The most well-known work to come out of institutional analysis of the commons is Elinor Ostrom's (1990) *Governing the Commons*, in which she established the importance of a set of institutional design principles for sustainable community-based commons management. This work also helped establish the study of community-based natural resource management, and human communities generally, from an institutional perspective (see Wilson et al. 2013). Interestingly, while he is frequently criticized for advocating for only private or public ownership, Hardin (1968, 1247) includes language that sounds sympathetic to a communal solution as well, writing that "the only kind of coercion I recommend is mutual coercion, mutually agreed upon by the majority of the people affected." Ostrom's work also helped create an academic community now formally known as the International Association for the Study of the Commons, with a well-respected academic journal (www.thecommonsjournal.org/). Thanks in large part to the efforts of Ostrom Workshop scholars, the study of the commons is now a well-established field that engages with concepts and methods from multiple areas of research (Berge and van Laerhoven 2011).

Why develop frameworks to begin with?

The literatures on commons governance and social-ecological systems have a strong interest with frameworks as a scientific object (e.g. see Binder et al. 2013). This interest begs the question, why develop analytical frameworks in the first place? To answer this, we need to understand the role of frameworks in the scientific process. McGinnis and Ostrom (2014, 30) describe the role of frameworks here, along with two other concepts that are popular at the Ostrom Workshop, theories and models:

> A framework provides the basic vocabulary of concepts and terms that may be used to construct the kinds of causal explanations expected of a theory. Frameworks organize diagnostic, descriptive, and prescriptive inquiry. A theory posits specific causal relationships among core variables. In contrast, a model constitutes a more detailed manifestation of a general theoretical explanation in terms of the functional relationships among independent and dependent variables important in a particular setting.

Elsewhere, Ostrom (2011, 8), using the language of institutional analysis, has also related frameworks to the measurement of variables:

> Frameworks identify the elements and general relationships among these elements that one needs to consider for institutional analysis and they organize diagnostic and prescriptive inquiry. They provide a general set of variables that can be used to analyze all types of institutional arrangements.

In this account, frameworks provide the language that is used to express other important scientific objects: namely, variables and theories. A framework should help a scientist know

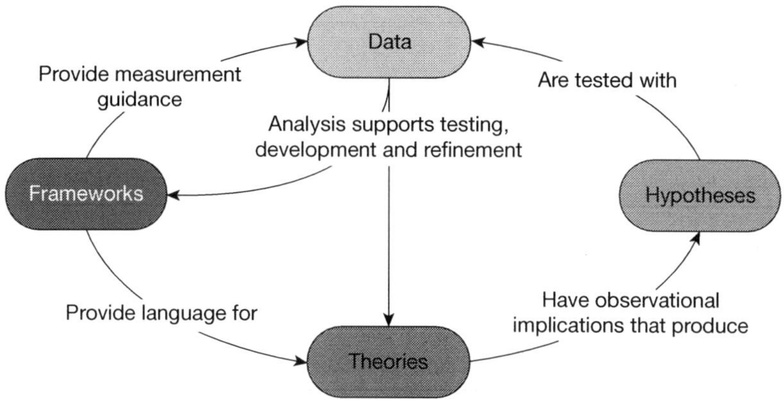

Figure 3.1 Model of the scientific process. Frameworks support the expression of theories and the measurement of variables. Data analysis tests theories and supports the generation of new ones and informs the utility of current frameworks.

(1) what they should be measuring, (2) how they might measure it as a variable, and (3) what theories they would or could be testing or developing by analyzing the relationships among their measured variables.

The distinction between frameworks and theories is fuzzy, particularly given Ostrom's (2011) use of the term "relationships" in the second quotation here. Both frameworks and theories express relationships. McGinnis and Ostrom (2014) are helpful here in clarifying that the primary difference is that theories specify causal relationships among a set of variables, whereas frameworks generally do not, or at least do so in less specific ways. Models then enter the picture primarily at the stage of data analysis to test the hypotheses derived from theories.

Figure 3.1 incorporates this understanding into an otherwise standard formulation of the hypothetico-deductive model of scientific practice. Here we see theories, hypotheses, and data that ideally interact in an iterative cycle to accumulate well-supported scientific theories, the primary goal of the scientific practice from this point of view. Here we see the role of frameworks displayed via its relationships with theories and the data used to test them. With this understanding in hand, we can proceed to discuss the two frameworks.

The IAD and SES frameworks

To discuss and compare the two frameworks, I present a combined depiction of them (Figure 3.2). In each framework, the primary analytical focus is the "action situation," which is a social arena in which interdependent actors make decisions. In the work on the commons the most prominent type of action situation involves a collective-action problem related to the use and management of a shared environmental commons. This underscores an important feature of action situations: while they can represent a unique meeting or discussion that a group of actors has at one point in time, as several examples from Ostrom (2005) illustrate, they also represent an ongoing set of interactions among a consistent set of interdependent actors. In the history of the IAD framework, action situations have played a similar role to that of games in game theory work, and the language used to describe the two also has historically overlapped (see Ostrom et al. 1994). For example, in making the distinction between more transitory and more ongoing action situations, we frequently talk about how many "rounds" will occur in a game.

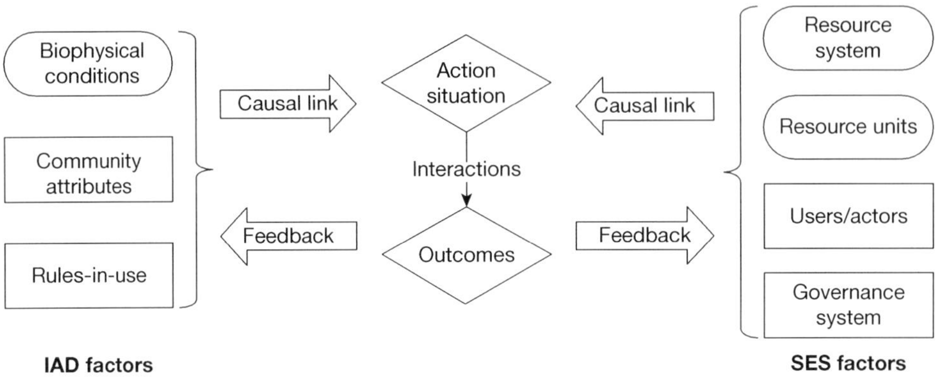

Figure 3.2 A combined representation of the two frameworks. Each contains a set of factors that affect social interactions in an action situation. The decisions in such settings produce outcomes that feed back into the original determining factors. Social factors are shown as rectangles and biophysical factors as rounded rectangles

An additional point to make about both frameworks is that they contain causal content: each contains factors that are assumed to affect human decision-making. While the precise nature of the relationships is not specified (this is left for theories), the frameworks here are not causally neutral. This is an example of the fuzzy distinction between frameworks and theories as I described earlier.

The IAD framework

There are three buckets of factors that are understood to affect decision-making in action situations in the IAD framework. Two of these are social: institutions, or rules-in-use, and the attributes of the communities who create them or are affected by them. The other is biophysical: in most cases this is understood to represent the relevant attributes of a commons. This way of breaking down the types of factors that influence decision-making in a social setting has carried over to much of the work on the commons (Agrawal 2003).

Of the three buckets, the literature associated with the IAD framework has developed the institutional factors most thoroughly. One of the unique aspects of the IAD is an accompanying typology of rules (Crawford and Ostrom 1995). This material reflects a point I made earlier, that human social settings (or now, action situations) are thickly institutional: it is a defining attribute of our species that we are so social and that our sociality is so strongly governed by rules and norms. Rules determine which actors hold which positions (such as a resource user, a religious leader, or a government agent), and they determine what actions are allowed or forbidden for those holding such positions. In the context of commons management, such rules determine who is a legitimate resource user and what types of property rights they have with respect to a shared commons (see Schlager and Ostrom 1992).

The actions that actors then take based on their positions are also affected by the information that they have available to them, which itself is a function of other rules. Moreover, how individual actions and decisions influence outcomes at the group level is influenced by aggregation rules (such as voting rules). Finally, there are rules that require or forbid certain outcomes (e.g. total allowable catches in fisheries governance or total maximum daily loads in water governance), and rules that affect the costs and benefits obtained from achieving certain outcomes. This foundational institutional content has continued to be elaborated in subsequent work (Basurto et al. 2010).

In terms of the relationships represented in Figure 3.1, the IAD framework, and the associated literature[1] that Ostrom and colleagues have developed to articulate the contents of the "rules in use" box, provide important guidance in the measurement of institutional variables that affect outcomes in action situations. With respect to biophysical attributes, the two main developments most closely associated with the IAD framework are the typology of types of goods (private, public, tool, common-pool) and different production functions of such goods (Ostrom 2003). The community attributes that have been identified mostly come from commons work that has developed lists of important attributes of communities of natural resource users, such as resource dependence, group size, and the presence of leadership (Agrawal 2003).

Regarding the IAD framework's relationship to theory, Ostrom (2005, 817–818) makes the following statement: "Microeconomic theory, game theory, transaction cost theory, social choice theory, public choice, constitutional and covenantal theory, and theories of public goods and common-pool resources are all compatible with the IAD framework." Compatibility in this context means that the perspectives that Ostrom mentions make use of variables that can be expressed in terms of the components of the framework. Of these, CPR theory is the most well-known body of theory to have come out of work with the IAD framework, as most famously embodied in Ostrom's (1990) work on the institutional design principles.

This is a helpful statement, although it is difficult to interpret it in the context of the discussion of theories and frameworks earlier. Ostrom is referring to approaches to social science analysis here. Transaction cost theory and public choice theory are arguably not just about theories but also contain what we would call frameworks based on the earlier definition: they point us to concepts to pay attention to in the first place (e.g. transaction costs, bureaucracies, rent-seeking). Theories are then constructed that causally relate such concepts. This points to inconsistencies in the use of the word theory, to refer to causal narratives that generate hypotheses via their observational implications, as well as bodies of literature with at least implicitly stated frameworks, as well as a collection of theories in the first sense.

The SES framework

In Figure 3.2 we see several close similarities between the components of the SES framework and the IAD framework. To begin, we still have the action situation as the primary focus. This is again affected by several buckets of factors, associated with different components of the framework. We have actors, which in the original version were labeled as users, in a move that strongly reflected the historical focus commons settings at the Ostrom Workshop. The actors' bucket from the SES framework roughly corresponds to the community attributes bucket in the IAD framework. We then have governance systems which contain the institutional factors and roughly correspond to the rules-in-rules bucket from the IAD framework.

The most noticeable difference at this level of representation between the two frameworks is that the SES framework divides the biophysical world into two parts: resource systems and resource units. This distinction can be traced back to earlier work at the Ostrom Workshop (Ostrom et al. 1994). What is being imagined in this division is a prototypical situation in which commons users extract an exhaustible pool of resource units, which themselves are part of a larger resource system. In the paradigmatic example of an irrigation system, we have the irrigation infrastructure as the resource system, the provision of which as a public good ensures the availability of the resource unit, in this case water as a common-pool resource. For an in-depth discussion of this distinction and the two associated components of the SES framework, see Hinkel et al. (2015). As these authors point out, the distinction between a resource system and a set of resource units associated with it is not always clear in an empirical setting.

An aspect of the SES framework that is not depicted in Figure 3.2 is Ostrom's (2007) decomposition of what she referred to as the tier 1 components of a SES, or what I am informally calling buckets here, into tier 2 variables. The use of this tiered language and the language of diagnosis that accompanied this hierarchical representation is a second place where the SES framework departs significantly from the IAD framework. Ostrom discussed the idea that the framework would eventually have multiple subsequent tiers, each with a version of objects at the previous tier that were somehow more specific to a smaller subset of cases. By "drilling down" into subsequent tiers, an analyst could diagnose the important factors of a system, or type of system, that they were focusing on.

Several additional points need to be made about the tier 2 objects, which I will also relate to the role of frameworks from Figure 3.1. First and foremost, their scientific identities are unclear. Several objects are readily interpretable as variables. For example, we have attributes of users/actors, such as the number of users (group size), their level of dependence on the resource, and their level of social capital. For the resource units we have their mobility, renewability, and economic value; and for the resource system, we have the clarity of its boundaries and its sector (e.g. water or forests).

Other factors listed by Ostrom (2007) and others are more difficult to interpret in this way, such as what she labels as "interaction among resource units," or the "history of use" of a group of users. These represent broad topics of interest, but additional work is needed to turn them into measurable factors in an empirical analysis. For an excellent example of work that does this for a particular type of system, see Leslie et al. (2015). Finally, we have factors such as human-constructed facilities, and governmental and non-governmental organizations. These are material and social objects in the world and, as such, are most easily thought of as actual subcomponents of an SES, rather than factors that we would measure about a subcomponent of an SES. The literature on the SES framework has mostly avoided this ambiguity, and so it remains unclear how to treat these objects.

Additionally, these objects strongly reflect the collective-action orientation of the Ostrom workshop. The majority of the factors listed are known to play important theoretical roles in the incentives that a set of users sharing a common-pool resource have to cooperate or defect. For example, dependence on a resource and local norms, as well as clarity of system boundaries and property rights systems are all strongly tied to the literature on common-pool resource management and collective-action. As such the main explanatory focus of the SES framework is in how external factors affect human decision-making, which in turn affect outcomes. Put another way, to the extent that the SES framework engages with theory, it is largely social theory that would explain how a set of factors affect human decision-making, rather than ecological theory that explains how ecological systems work to produce important outcomes. Ostrom (2007) herself uses the framework to express the "tragedy of the commons" theory (Hardin 1968). She ties key components of Hardin's argument to objects she labels as variables from her framework to formally express the theory that predicts ecological destruction from large numbers of users using a renewable resource.

To summarize, the SES framework does not provide as much guidance as many have assumed it would with respect to variable measurement, in large part due to the ambiguous identities of the tier 2 objects. Additionally, in spite of its name, I would argue that its theoretical orientation remains primarily social due to its direct inheritance of much of the components of the IAD framework, particularly the action situation as its focal point, and the collective-action perspective implied by the tier 2 objects.

Using the Frameworks

A question that has been often asked of the SES framework is, how exactly should it be used to diagnose a particular SES? Ostrom published two initial papers on the framework (Ostrom 2007, 2009) and subsequently published several articles with colleagues that make minor modifications to it (Ostrom and Cox 2010; McGinnis and Ostrom 2014). But an official manual for the framework has never appeared. I believe this has resulted in a high level of variability in how the framework is implemented. This argument is supported with empirical data presented by Schlager and Cox (2017), who demonstrate that, in terms of measurement protocols, there is little consistency across implementations of the SES framework. I believe this is a problem given the explicit goals of the framework, which includes supporting just this type of consistency, and the fact that a lack of comparability across cases is one of the primary challenges facing scholars of the commons (Poteete et al. 2010).

In order to produce comparable data across the analysis of SESs, several questions must be answered to augment the box-and-arrow representations that are commonly used in describing scientific figures. The first of these is, what is the unit of analysis implied by a framework for analyses that would use it? The unit of analysis is the entity about which a scientist is asking their research questions. It is tempting to say that the unit of analysis implied in the two frameworks is an action situation. But several applications of the frameworks have focused on a case containing multiple interdependent action situations (McGinnis 2011; Ostrom 2011; Cox 2014a), and it is these interactions that are used to understand case-level outcomes. So the role of action situations has often been to unpack a particular case, which is itself the unit of analysis.

This feature is in fact built into the IAD framework and SES framework in two ways. First, we have the distinction between different but interrelated institutional levels and associated action situations (from constitutional to collective-choice to operational) that have consistently been associated with the IAD framework. Second, we have the distinction between resource units and resource systems. Each of these is commonly associated with its own type of collective-action problem and action situation. Moreover, it has been commonly understood, at least in the prototypical irrigation system example, that the provision of the services required to maintain the resource system helps maintain the availability of the resource unit. Ostrom et al. (1994) labeled these provision vs. appropriation situations, respectively.

I believe this interdependence of action situations leads us to an important protocol for implementing either of these frameworks to understand a particular case: starting with a focal action situation, the analyst then considers what other decision-making arenas produce outcomes that affect the incentives faced by the actors in the focal situation. This focal situation is most commonly oriented around the collective use of a shared commons by a set of natural resource users, but it need not be: Fleischman (2016) relies in part on the IAD framework to examine the behavior of foresters in India, while Gibson et al. (2005) apply an institutional analysis to the Swedish development agency, SIDA.

Accompanying this protocol is an answer to the initial question: what is the unit of analysis here? For the SES framework, it is just this: a social-ecological system. For the IAD framework it is commonly a social system with multiple interacting action situations and their participants. I believe that this protocol also implies the need for mixed methods in the analysis of at least the social aspects of the cases examined with these frameworks. In my own work, which does start with a focal action situation oriented around resource use, this has commonly involved conducting formal interviews with a sample of natural resource users (fishers or farmers) and conducting quantitative analyses of the interview data to understand the behavior of these core actors. Then, more informal interviews are conducted with more peripheral actors whose decisions affect the incentives that natural resource users face. Such actors are most often linked to the natural resource users through markets (see Ribot 1998) and formal governance (Andersson and Ostrom 2008). Case-level inference of these interactions is then both qualitative and quantitative.

Having explored the application of these frameworks to a particular case, an additional question we must ask is, how might we use the frameworks to compare across cases? In answering this question, we face the well understood trade-off between complexity and generalizability (Cox 2008): the more specific we are in our descriptions of a case, the harder it is to compare that case to others. In the case of these two frameworks, we can see how their box-and-arrow representations naturally lead us to unpack the complexity of a case. The component-subcomponent structure of the SES framework reinforces this, and I think this point is particularly easy to understand with this framework. For if we accept that the second-tier objects in this framework can be measured, then we see that we are measuring characteristics of components of SESs, not of SESs themselves. But comparison across systems, at least quantitative comparison (and I would argue that many so-called small-n qualitative comparisons are implicitly quantitative in nature), requires that we have SES-level measurements. So we face the challenge of aggregating component measurements to SES-level measurements in this case.

One way to meet this challenge is to provide instructions for how data associated with the use of a scientific framework can be stored, managed, and retrieved. Neither of the original frameworks discussed here originally came with such instructions, and this is the most common scenario with scientific frameworks in the social sciences. I believe that this is a critical problem for the study of the commons. Often the only data model that graduate students and scholars are aware of is a "flat model," or a simple spreadsheet. By data model I mean the specification of the structure of the database that will store data. I would go so far as to argue that a data model should be a required part of the specification of a framework, or at the very least that authors of scientific frameworks should specify how someone might collect empirical data on a set of comparable cases when using the framework. Such a process is far from intuitive based on the box-and-arrow representations that come with most frameworks.

There are two projects to highlight that have done just this for each of the two frameworks. For the IAD framework we have the International Forestry Institutions (IFRI) project, which originated at Indiana University and is now based at the University of Michigan (www.ifriresearch.net/). This project developed a relational database to store information about the interactions between forest user groups and the resources they use and manage (Wollenberg et al. 2007). The IFRI community has developed extensive coding protocols to help researchers reliably measure variables across sites. The unit of analysis in the majority of IFRI studies is the interaction between a user group and a forest. This is arguably the most successful empirical research project on the commons to date, producing many important studies on the forest commons (Chhatre and Agrawal 2009; Coleman and Steed 2009).

The project that has done the most to construct a data model for the SES framework is known as the Social-ecological Systems Meta-analysis Database (SESMAD) project (sesmad.dartmouth.edu) (Cox 2014b; Ban et al. 2017). This project has not had the large-scale empirical success of the IFRI project but nevertheless has produced some important findings and lessons. It is also based on a relational database model that enables a user to construct a model of an SES by adding components and coding information about their interactions. In order to address the two specific roles that frameworks have from Figure 3.1, the SESMAD database includes a table that explicitly describes the variables that can be measured in any system, and what types of components they are relevant for; it also contains a table describing the most important social-ecological theories that could be tested in such cases (Cox et al. 2016).

Conclusions

Frameworks determine what we attend to in our scientific analyses. Given the fuzzy boundary between frameworks and theories, they also enable what is commonly referred to as the "theory-ladenness of observation." As such, they represent the interpretive lens that scholars apply to their empirical cases. Given this importance, we must be self-aware about the ways in which any framework steers our attention and thus our conclusions. This awareness has been expressed in terms of each of these frameworks, particularly with respect to aspects that scholars have argued are under-developed in them. One primary concern has been that they do not capture power relationships and historical dynamics sufficiently (Clement 2010). This has been a concern of institutional analysis more broadly, and, according to a recent review conducted by Kramer et al. (2017), much of the scholarly community studying human-environment interactions does not strongly emphasize these themes. So, it is likely a valid critique, and it reflects a concern among many commons scholars over the temptation to offer technocratic solutions without an awareness of the power dynamics that will largely determine how such solutions are implemented and for whose benefit. Ostrom's own design principles have been criticized in this way (Cox et al. 2010).

I believe one response to this concern is that the language of institutions is an effective way of describing disparities of power and authority, since it is through institutions that such power is codified, most prominently in the language of property rights. At the same time, more work could be done to integrate the IAD framework and institutional analysis with concepts such as path dependence as most prominently developed by Arthur (1994), applied to institutional change by North (1990), and recently applied to commons analyses (Heinmiller 2009; Cody et al. 2015). This language is inherently historical and, because it also addresses issues of vested interests as an inherent obstacle to change, can incorporate power dynamics as well.

An additional concern has been the relatively small amount of ecological content in the frameworks. Arguably much of the motivation for the SES framework is to further develop the ecological content that is not formally captured in the IAD framework. Nevertheless, I believe it is fair to say that the SES framework does not extensively cover many important concepts commonly used by ecologists in their research. Rather, as I indicated earlier, much of the resource-oriented content in the SES framework is present because of its relevance to the collective-action/commons perspective that the SES framework inherited from the IAD framework. There have been efforts to explicitly add ecological content to the framework, or otherwise clarify how the resource-based components should be analyzed (Epstein et al. 2013; Hinkel et al. 2015). But these efforts have not led to formal changes to the framework. Moreover, as far as I know, no protocol exists for making formal updates to the framework in the way implied by the scientific process depicted in Figure 3.1. So, this remains a challenge for the commons community in its continued use of this framework. Frameworks are not tested in the way theories are, but arguably they should be amended and changed based on users' experiences with them.

To conclude, the promise of scientific frameworks is that they can provide a common, meta-theoretical language to enable scholars to engage with each other and produce comparable findings. However, having a framework does not guarantee that these goals are met, particularly if they do not come with protocols that guide their implementation and a literature that further develops them over time. As Poteete et al. (2010) eloquently describe in their final chapter of their excellent book *Working Together*, as a community of commons scholars we face our own collective-action problems in producing more highly comparable data and results with the frameworks that are currently available to us. We should consider applying our social theories to our research community to understand how this has been and should be done. Ostrom herself embodied one of the most important theoretical arguments in work on collective-action when she served as a leader for the community, providing important public goods that the rest of us could benefit from and use to establish a common understanding and basis for mutual engagement. Another key factor is in-person communication that can enable interdisciplinary collaborations to develop organically over time based on a genuine understanding of the separate fields involved. Just like common understandings among resource users, this is a costly, time-consuming process that cannot be easily substituted with novel technologies. Producing more comparable results with highly integrated frameworks will require just as much change in social infrastructure as it will technological infrastructure. In the Ostrom Workshop we have a model for much of this, in the social capital and craft-oriented collective mentality embodied in its name.

Note

1 It is difficult to determine where the framework itself stops and literature loosely associated with it starts.

References

Agrawal, A. (2003). "Sustainable governance of common-pool resources: Context, methods, and politics" *Annual Review of Anthropology*, 32: 243–262.

Andersson, K.P., and Ostrom, E. (2008). "Analyzing decentralized resource regimes from a polycentric perspective" *Policy Sciences*, 41: 71–93.

Arthur, B. (1994). *Increasing Returns and Path Dependence in the Economy*. Ann Arbor, MI: University of Michigan Press.

Ban, N.C., Davies, T.E., Aguilera, S.E., Brooks, C., Cox, M., Epstein, G., . . . Nenadovic, M. (2017). "Social and ecological effectiveness of large marine protected areas" *Global Environmental Change*, 43: 82–91.

Basurto, X., Kingsley, G., McQueen, K., Smith, M., and Weible, C.M. (2010). "A systematic approach to institutional analysis: Applying Crawford and Ostrom's grammar" *Political Research Quarterly*, 63(3): 523–537.

Berge, E., and van Laerhoven, F. (2011). "Editorial: Governing the commons for two decades – A complex story" *International Journal of the Commons*, 5(2): 160–187.

Binder, C.R., Hinkel, J., Bots, P.W.G., and Pahl-Wostl, C. (2013). "Comparison of frameworks for analyzing social-ecological systems" *Ecology and Society*, 18(4): article 26.

Chhatre, A., and Agrawal, A. (2009). "Trade-offs and synergies between carbon storage and livelihood benefits from forest commons" *PNAS*, 106(42): 17667–17670.

Clement, F. (2010). "Analysing decentralised natural resource governance: Proposition for a 'politicised' institutional analysis and development framework" *Policy Sciences*, 43(2): 129–156.

Cody, K.C., Smith, S.M., Cox, M., and Andersson, K. (2015). "Emergence of collective action in a groundwater commons: Irrigators in the San Luis Valley of Colorado" *Society and Natural Resources*, 28(4): 405–422.

Coleman, E.A., and Steed, B.C. (2009). "Monitoring and sanctioning in the commons: An application to forestry" *Ecological Economics*, 68(7): 2106–2113.

Cox, M. (2008). "Balancing accuracy and meaning in common-pool resource theory" *Ecology and Society*, 13(2): n.p.

Cox, M. (2014a). "Applying a social-ecological system framework to the study of the Taos Valley irrigation system" *Human Ecology*, 42(2): 311–324.

Cox, M. (2014b). "Understanding large social-ecological systems: Introducing the SESMAD project" *International Journal of the Commons*, 8: 265–276.

Cox, M., Arnold, G., and Villamayor-Tomas, S. (2010). "A review of design principles for community-based natural resource management" *Ecology and Society*, 15(4): n.p.

Cox, M., Villamayor-Tomas, S., Epstein, G., Evans, L., Ban, N.C., Fleischman, F., . . . Garcia-Lopez, G. (2016). "Synthesizing theories of natural resource management and governance" *Global Environmental Change*, 39: 45–56.

Crawford, S., and Ostrom, E. (1995). "A grammar of institutions" *The American Political Science Review*, 89(3): 582–600.

Epstein, G., Vogt, J.M., Mincey, S.K., Cox, M., and Fischer, B. (2013). "Missing ecology: Integrating ecological perspectives with the social-ecological system framework" *International Journal of the Commons*, 7(2): 432–453.

Fleischman, F. (2016). "Understanding India's forest bureaucracy: A review" *Regional Environmental Change*, 16: 153–165.

Gibson, C., Andersson, K., Ostrom, E., and Shivakumar, S. (2005). *The Samaritan's Dilemma: The Political Economy of Development Aid*. New York: Oxford University Press.

Hardin, G. (1968). "The tragedy of the commons" *Science*, 162: 1243–1248.

Hardin, G. (1998). "Extensions of 'The tragedy of the commons'" *Science*, 280(5364): 682–683.

Heinmiller, B.T. (2009). "Path dependency and collective action in common pool governance" *International Journal of the Commons*, 3(1): 131–147.

Henrich, J. (2015). "The secret of our success: How culture is driving human evolution, domesticating our species, and making us smarter." Princeton, NJ: Princeton University Press.

Hinkel, J., Cox, M., Schlüter, M., Binder, C., and Falk, T. (2015). "A diagnostic procedure for applying the social-ecological systems framework in diverse cases" *Ecology and Society*, 20(1): article 32.

Kramer, D., Hartter, J., Boag, A., Jain, M., Stevens, K., Nicholas, K., . . . Liu, J. (2017). "Top 40 questions in coupled human and natural systems (CHANS) research" *Ecology and Society*, 22(2): n.p.

Leslie, H.M., Basurto, X., Nenadovic, M., Sievanen, L., Cavanaugh, K.C., Cota-Nieto, J.J., . . . Aburto-Oropeza, O. (2015). "Operationalizing the social-ecological systems framework to assess sustainability" *Proceedings of the National Academy of Sciences*, 112(19): 5979–5984.

McGinnis, M. (2011). "Networks of adjacent action situations in polycentric governance" *Policy Studies Journal*, 39(1): 51–78.

McGinnis, M.D. (1999). Polycentricity and Local Public Economies: Readings from the Workshop in Political Theory and Policy Analysis. Ann Arbor, MI: University of Michigan Press.

McGinnis, M.D., and Ostrom, E. (2014). "Social-ecological system framework: Initial changes and continuing challenges" *Ecology and Society*, 19(2): n.p.

North, D. (1990). Institutions, Institutional Change and Economic Performance. Cambridge University Press, Cambridge, MA.

Ostrom, E. (1990). Governing the Commons: The Evolution of Institutions for Collective Action. Cambridge, MA: Cambridge University Press.

Ostrom, E. (2003). "How types of goods and property rights jointly affect collective action" *Journal of Theoretical Politics*, 15(3): 239–270.

Ostrom, E. (2005). *Understanding Institutional Diversity*. Princeton, NJ: Princeton University Press.

Ostrom, E. (2007). "A diagnostic approach for going beyond panaceas" *PNAS*, 104(39): 15181–15187.

Ostrom, E. (2009). "A general framework for analyzing sustainability of social-ecological systems" *Science*, 325: 419–422.

Ostrom, E. (2011). "Background on the institutional analysis and development framework" *Policy Studies Journal*, 39(1): 7–27.

Ostrom, E., and Cox, M. (2010). "Moving beyond panaceas: A multi-tiered diagnostic approach for social-ecological analysis" *Environmental Conservation*, 37(4): 451–463.

Ostrom, E., Gardner, R., and Walker, J. (1994). *Rules, Games, and Common-Pool Resources*. Ann Arbor, MI: University of Michigan Press.

Poteete, A.R., Janssen, M.A., and Ostrom, E. (2010). *Working Together: Collective Action, the Commons, and Multiple Methods in Practice*. Princeton, NJ: Princeton University Press.

Ribot, J.C. (1998). "Theorizing access: Forest profits along Senegal's charcoal commodity chain" *Development and Change*, 29(2): 307–341.

Schlager, E., and Cox, M. (2017). "The IAD Framework and the SES Framework: An introduction and assessment of the Ostrom Workshop frameworks" in Weible C and Sabatier P eds, *Theories of the Policy Process*. Boulder, CO: Westview Press.

Schlager, E., and Ostrom, E. (1992). "Property-rights regimes and natural resources: A conceptual analysis" *Land Economics*, 68(3): 249–262.

Wilson, D.S., Ostrom, E., and Cox, M.E. (2013). "Generalizing the core design principles for the efficacy of groups" *Journal of Economic Behavior and Organization*, 90: S21–S32.

Wollenberg, E., Merino, L., Agrawal, A., and Ostrom, E. (2007). "Fourteen years of monitoring community-managed forests: Learning from IFRI's experience" *International Forestry Review*, 9(2): 670–684.

4
POLYCENTRICITY

Josephine van Zeben

The study of polycentric systems lies at the heart of Vincent and Elinor (Lin) Ostrom's work on governance and democracy, and forms the basis for the Bloomington School of Institutional Analysis and the empirical work undertaken at the Ostrom Workshop (https://ostromworkshop.indiana.edu/). The theoretical and empirical insights provided by the Ostroms and the Ostrom Workshop give scholars important alternative tools for institutional analysis. Polycentric theory provides the theoretical and normative basis on which some of the more empirically oriented tools that originated from the Workshop, such as the Institutional Analysis and Development (IAD) and Social-ecological systems (SES) frameworks, are based.

Polycentric systems are governance systems that encompass many centers of decision-making, which act autonomously and non-hierarchically under a common set of overarching rules (Ostrom, Tiebout and Warren, 1961, 831). Polycentricity seeks to embrace the complexity of governance systems by rejecting the simplification of institutions and their relationships into one-dimensional models. In doing so, it rejects the classic dichotomies of "market" versus "state" (Ostrom, 2010a) and federal versus unitary states (Ostrom, 1996). The recognition of this complexity provides scholars with the opportunity to develop richer institutional theories, but often also requires a more interdisciplinary approach (Poteete, Janssen and Ostrom, 2010) and results in less easily generalizable findings.

This chapter sets out the cyclical theoretical and empirical development of the concept of polycentricity. It starts with a discussion of Vincent Ostrom's work on polycentricity, followed by a discussion of some of the case studies undertaken to test polycentric systems. The results of these empirical studies also informed Lin's work on the governance of environmental commons and citizenship, and the eventual identification of design principles of successful common-pool resource management systems. The third section moves away from existing work in search of an analytical framework that will allow for a more nuanced identification of species of polycentric systems. This in turn will help to detect those polycentric features that need to be strengthened in order for the system to thrive. The final section demonstrates how such an analytical framework may be applied to the European Union (EU).

Polycentricity, self-governance, federalism and democracy

The development of the concept of polycentricity started with a seminal article by Vincent Ostrom, Charles Tiebout and Robert Warren published in the *American Political Science Review* in 1961. In "The organization of government in metropolitan areas: A theoretical inquiry", Ostrom, Tiebout and Warren use the term "polycentric political system" to describe the organization of large metropolitan areas in the United States (Ostrom, Tiebout and Warren, 1961, 831). At the time of writing, the multiplicity of political units – federal, state, county, municipal and semi-private – in metropolitan areas was considered highly problematic; a state of "too many governments and too little government" giving rise to a "duplication of functions" in "overlapping jurisdictions" (ibid., 831). Ostrom, Tiebout and Warren posit that the polycentric nature of these metropolitan areas is not inherently problematic; rather, both polycentric and monocentric systems struggle to accommodate the "multiplicity of interests in various public goods sought by people in a metropolitan region" (ibid., 838). Moreover, they found no theoretical basis for the claim that it would be easier or more efficient for a central regulator to accommodate the interests of the sub-groups of her constituency than for polycentric regulators to accommodate their shared interests.

Although beyond the scope of the 1961 article, their definition of polycentricity already underlines the need to empirically verify the types of relationships between the identified centers of decision-making. Vincent started this research in earlier work on urban governance and natural resource management, which was continued together with Lin and expanded further through the Workshop (Ostrom, 1953a, 1953b, 1959, 1962a, 1962b; Ostrom, Bish and Ostrom, 1988).

Ostrom, Tiebout and Warren state that:

> "Polycentric" connotes many centres of decision-making which are formally independent of each other. Whether they actually function independently, or instead constitute an interdependent system of relations, is an empirical question in particular cases. To the extent that they take each other into account in competitive relationships, enter into various contractual and cooperative undertakings or have recourse to central mechanisms to resolve conflicts, the various political jurisdictions in a metropolitan area may function in a coherent manner with consistent and predictable patterns of interacting behavior. To the extent that this is so, they may be said to function as a "system".
>
> *(Ostrom, Tiebout and Warren, 1961, 831)*

This definition forms the basis of Vincent Ostrom's later work on institutional analysis and already hints at later developments. For example, the "consistent and predictable patterns of interacting behavior" may be seen to refer to Vincent's understanding of democracy as a dynamic process, which requires skill on the part of the participant in order to navigate uncertainty and conflict (Ostrom, 1973).

In 1971, Vincent co-authored an article with Lin in which he summarizes and questions accepted truths in public administration, and specifically public choice scholarship (Ostrom and Ostrom, 1971). This article signaled the start of Vincent's move away from mainstream public choice scholarship, which had become increasingly reliant on rational choice models and other techniques aimed at achieving higher efficiency by minimizing damaging rent-seeking behavior. Polycentric theory developed in part as a response to limitations, as aptly summarized by McGinnis and Ostrom (2012), that Vincent perceived to exist in public choice scholarship.

Ostrom's polycentric systems are complex multi-layered entities, which are constantly evolving in response to the needs of the individuals and communities that shape them. In order to maintain such systems, much is required from their participants and the institutions within them.

One of the key prerequisites of polycentricity therefore is an individual's ability to self-govern. In essence,[1] polycentricity is an expression of the capacity to self-govern, or to engage in collective action, which over time will give rise to a complex system of governance institutions. In obtaining the necessary skills to self-organize, entrepreneurship plays an essential role, making entrepreneurs vital for both public and private enterprise. The cross-fertilization between public and private institutions relates closely to Ostrom's view of the market as a "complex institutional arrangement" (McGinnis and Ostrom, 2012, 20), as compared to the "ideal" type for public choice scholars.

By linking it to theories of democracy and self-governance, polycentricity also normatively diverged from existing public choice theory. This divergence reflects a fundamentally different understanding of human motivation in the creation and maintenance of institutions. Vincent considered individuals to be intrinsically motivated to self-organize, provided that they were taught the skills needed to craft institutional arrangements. This motivation is multi-faceted, whereas the rationality assumption that underlies public choice models only captures one dimension of motivation (Ostrom, 1989; 1997, 96–99). Similarly, Vincent found the benchmark of efficiency as externally imposed to have limited utility since individuals or communities may decide to prioritize other goals (Ostrom, 1959, 1993).

The foundations of Ostrom's conception of self-governance are laid in De Tocqueville's treatise *De la Démocratie en Amérique* (*Democracy in America*) (1835 and 1840). Tocqueville emphasized the importance of smaller societal units and the dangers of excessive centralization, in light of their undermining effect on the people's ability to govern themselves. To Tocqueville, one of the great virtues of US society was the rich variety in townships, churches and other non-state centric forms of organization that preceded the federal United States. He traced the success of US democracy back to generations of social learning on the practice of self-governance and considered complacency one of the key dangers of an "overinvolved" government (De Tocqueville as quoted in Ostrom, 1994, 203). Ostrom's view on the role of the individual in democracy similarly relies heavily on responsibility, resilience, ownership and learning (Ostrom, 1994):

> When the cares of thinking and the troubles of living are left to others, self-government is abandoned, democracy withers away, autocracy emerges, and people begin preying upon one another in the name of liberty and equality. The future belongs to those whose covenants are bonds of mutual trust grounded in principles of self-governance and who learn to use processes of conflict and conflict resolution to elucidate information, clarify alternatives, stimulate innovation, and extend the frontiers of inquiry to open new potentials for human development.
>
> *(Ostrom, 1994, 272)*

Though weary of the potential excesses of consolidated power, Ostrom's view of polycentricity through self-governance should not be understood as hostile towards, or incompatible with, governmental action. Much like entrepreneurs, public officials are potential allies and teachers in the process of self-governance by supporting communities to identify and solve common problems (McGinnis and Ostrom, 2012, 21). This speaks to the importance of practicing polycentricity in all arenas of life; the delicate balancing act that polycentric governance requires is more likely to succeed when reinforced by polycentricity in political processes, judicial affairs and economic affairs (Ostrom, 1972 in McGinnis, 1999a, 56). The conditions needed to maintain such a system – for example learning, freedom of entry and exit, enforcement of general rules (Ostrom, 1972 in McGinnis, 1999a), peaceful method of revising the general rules, and the constraint of government officials (Ostrom, 1972) – are typically associated with democratic rule. However, Ostrom considered democracy to be contingent on polycentric governance, rather than as a prerequisite for polycentricity (Wagner, 2005, 184).

Relatedly, Vincent's study of US federalism in *The Political Theory of a Compound Republic* emphasizes the polycentric nature of the US federal system (Ostrom, 2008b). Ostrom finds that federalism – understood as an expression of polycentric governance that limits the potential negatives of democracy – constitutes a breeding ground for democratic governance, allowing people to participate in several types of association. This may be contrasted with the assumption that democratic government naturally decentralizes and becomes federal (Wagner, 2005, 184). To Ostrom, some of the polycentric features of US democracy were being worryingly eroded, such as the ability for individuals to contest authority at all levels of government (Wagner, 2005, 185).

In essence, polycentrism implies the diffusion of sovereignty over several levels of governance and numerous institutions. This is also a feature of many federal systems, which increases the analytical overlap between the two kinds of systems. However, ultimately, polycentricity and federalism are not identical, nor interchangeable. Rather, federalism is one form or type of polycentric governance, the latter being a broader concept. As compared with federalism, polycentricity allows for an additional dimension of crosscutting "issue-specific" jurisdictions and envisages an explicit role for non-governmental bodies, such as private and community-based organizations (McGinnis and Ostrom, 2012, 15).

When successfully maintained, polycentric systems could result in "increased liberty, a wider array of choices available to citizens seeking assistance from public authorities, and easier access to quasi-market arrangements that more closely approximate the efficiency of competitive markets" (McGinnis and Ostrom, 2012, 22). However, the information costs in maintaining such a system may be prohibitive, and political stalemate, as well as conflict, is likely. Thus, as quickly realized by the Ostroms in the context of metropolitan governance, the "better" approach to governance could not be identified through theoretical debate alone; empirical testing of the hypothesis resulting from these different approaches was needed (Ostrom and Ostrom, 1965, 135–136).

Assessing polycentric systems

In the 1960s and 1970s, the Ostroms set up an empirical research agenda to test claims of the "pathology" of US metropolitan areas. They found that, for several large metropolitan areas in the American Midwest, "polycentric arrangements with small, medium, and large departmental systems generally outperformed cities that had only one or two large departments" (Ostrom, 1972, in McGinnis, 1999a, 148). Thus, on the basis of economic efficiency, no *prima facie* preference should be expressed in favor of monocentric over polycentric systems.[2] The functioning of polycentric systems has been subsequently tested through a multitude of other studies, including water management in the United States (V. Ostrom, 1953b, 1953b, 1962a, 1962b; E. Ostrom, 1964, 2010a), and the commons systems studied by Lin Ostrom that crystalized into the governance rules set out in *Governing the Commons* (1990). The findings of these studies are well-documented elsewhere (e.g. Ostrom, 2008) and will not be repeated in similar detail here. However, it is worth highlighting some of the hypotheses tested and results found.

Within the Workshop, the Ostroms designed urban area studies in order to verify a number of basic assumptions of polycentric theory. As summarized by Lin, these include:

(1) Public goods and services differ substantially regarding their production functions and scale of effects.
(2) Individuals with relatively similar preferences for public goods and services tend to cluster in neighbourhoods. Preferences will tend to be more homogeneous within neighbourhoods than across an entire metropolitan area.
(3) Citizens who live in multiple jurisdictions learn more about the performance of any one jurisdiction by seeing or hearing about how other jurisdictions handle problems.
(4) Multiple jurisdictions with different scopes and scales of organization allow citizens more effective choice in selecting packages or services most important to them, in articulating their preferences and concerns, and if necessary, in moving to other jurisdictions.
(5) The presence of large numbers of potential producers of urban goods and services in a metropolitan area allow elected officials more effective producer choices.
(6) Producers who must compete for contracts are more likely to search for innovative technologies, to encourage effective team production, as well as citizen coproduction, to enhance their own performance (Ostrom, 2008a, 2008b; Ostrom, Parks and Whitaker, 1978, as cited in Ostrom, 2008, 4).

When applied to the polycentric system of metropolitan policing in the United States, studies clearly showed that small- to medium-sized police agencies performed more effectively, and often at lower costs, than large police departments (McGinnis, 1999b). Moreover, areas with a larger number of police agencies received more police presence for their tax expenditure (Parks, 1985). Similar studies have been conducted for school districts (Hanushek, 1986; Teske et al., 1993), further challenging the validity of the assumption that centralized government in municipal areas is preferable to polycentric governance. Contrary to the pathology assumed by political scientists in the 1960s, the shared finding of these studies of polycentric systems is that collective action problems can be "effectively, equitably and efficiently" resolved by citizens in their local jurisdictions (Ostrom, 2008, 7).

Lin's later work on locally designed regimes for the regulation of common-pool resources flows directly from this earlier empirical work. She identified eight design principles that were present in all successful systems:[3]

(i) Well defined boundaries, so as to allow for a defined set of relationships;
(ii) Benefits are allocated proportionally to inputs;
(iii) Wide participation in rule creation;
(iv) Rule monitoring and enforcement;
(v) Use of graduated sanctions by robust governance arrangements;
(vi) Access to rapid, low cost, local arenas for conflict resolution;
(vii) External recognition of rule systems;
(viii) For larger systems: allow for several layers of nested enterprise (Ostrom, 2008, 15–16).

These design principles were based on her own empirical work undertaken through the Workshop (Ostrom, 2005) and later confirmed by many others through additional field work (see, for example, Abernathy and Sally, 2000; Crook and Jones, 1999; Merrey, 1996; Weinstein, 2000). Ostrom shows that collective action problems with respect to common-pool resources can also be successfully managed through polycentric systems (Ostrom, 2008). However, it is never suggested that the mere presence of polycentric governance can resolve all underlying problems of resource management or public good provision (Ostrom, 2007); it can only nurture and sustain the self-governing abilities of local communities (McGinnis and Walker, 2010).

Moreover, as Ostrom's principles are derived from research on locally managed resources, scaling up these principles to polycentric systems designed to manage global resources remains an outstanding task (Ostrom et al., 1999; Ostrom, 2010b).

Developing the analysis of polycentric systems

The last two sections have set out the normative theoretical framework shaped by the Ostroms' work on polycentricity, democracy and self-governance, and the empirical research done on existing polycentric systems. This empirical research had led to the development of the design principles listed earlier, as well as the Institutional Analysis and Development (IAD) (McGinnis, 2011)[4] and the Social-ecological systems (SES) frameworks (Ostrom, 2009; McGinnis and Ostrom, 2014). The aim of this research has been to identify in what ways polycentric systems can continue and strengthen self-governance, and thereby ensure more effective management of common-pool resources. Ultimately, this should amount to "more effective, equitable, and sustainable outcomes at multiple scales" (Toonen, 2010, as cited in Ostrom, 2010a).

As highlighted by Aligica and Tarko (2012), the concept of polycentricity could be used more powerfully to draw comparisons between complex institutional structures. At present, most research has focused on the success – by any given benchmark – of polycentric systems. However, as demonstrated by the empirical work, there are many different types of polycentric systems, all of which achieve success in their own ways. By paying closer attention to the features of the polycentric systems under inspection, we may add another layer of analysis to polycentric inquiry. This would focus more on the polycentric features of specific systems – and its relative success at "being polycentric" – than on its success in terms of efficiency or sustainability. This approach echoes Vincent's insistence on the centrality of self-governance and the role polycentricity plays in maintaining it, rather than the other way around. It would also allow a clearer analytical distinction between polycentricity and other multi-centered systems, such as multi-level governance, federalism and network theory (McGinnis and Ostrom, 2012).

Working off the core features from the Bloomington School definition of a polycentric system – many centers of decision-making, ordered relationships that persist in time; many legitimate rules enforcers; single system of rules; centers of power at different organizational levels; spontaneous order resulting from free entry and exit; the alignment between rules and incentives (rules are considered useful); and the public involvement in rule design (rules about changing rules) – Aligica and Tarko propose a logical structure of polycentricity that results in 288 different possible types of polycentric systems (2012, 257–258). They create this structure by identifying three basic features of polycentric systems, which are maintained through a number of necessary conditions (see Table 4.1). These features and necessary conditions are supplemented by non-necessary conditions, such as different forms of rule design, jurisdictions, and entry and exit conditions (2012, 256–257). Through different combinations of the features and conditions in Table 4.1, different types of polycentric systems are created.

The greatest contribution of Aligica and Tarko's proposed framework is their identification of essential and non-essential features, and the recognition of the diversity with which these conditions are met by different polycentric systems. The exercise of identifying these elements is a vital step in the further operationalization of polycentric theory, as it allows for the identification of pathologies within polycentric structures, i.e. malfunctioning or absent features that undermine the polycentric nature of the system. As noted already by Vincent, polycentric systems must remain balanced in order to function successfully and to continue to be polycentric; otherwise the systems will revert to monocentric rule or a type of federalism (Aligica and Tarko, 2012, 258).

Table 4.1 Basic features and necessary conditions of polycentric systems (based on Aligica and Tarko, 2012)

Basic features	*Necessary conditions*	*Non-necessary conditions*
Multiplicity of decision-making centres	Active exercise of diverse opinions and preferences	• Common goals/individual goals • Territorial/non-territorial jurisdictions
Overarching system of rules	Alignment between rules and incentives	• Internal/external rule design • Collective choice through consensus/individual decisions/majority rule
Spontaneous order through evolutionary competition	Autonomous decision-making layers	• Free/constraint entry/exit • Public/private information

In their framework, Aligica and Tarko define "necessary conditions" as those found in all polycentric systems, and "non-necessary conditions" as variations in polycentric governance (2012, 255, 256). Building on Aligica and Tarko's important contribution, recent work by van Zeben has tried to further nuance the identification of essential and non-essential features of polycentric systems by centralizing the potential for self-governance within polycentric systems (2013, forthcoming). This alternative analytical system distinguishes between "attributes" – necessary but non-exclusive features of polycentric systems – and "institutional essentials" – the conditions that ensure the continued existence of "self-generating" polycentricity by ensuring that "individuals acting at all levels will have the incentives to create or institute appropriate patterns of ordered relationships" (Ostrom, 1972, 246).

These conditions are in turn aided by three "prerequisites": access to information, capacity to learn and access to justice (van Zeben, forthcoming). Different from Aligica and Tarko's framework, these prerequisites are not organized by their relationship to specific necessary conditions; rather, these prerequisites are vital to each institutional essential in different ways, making them "necessary" to the system as a whole and specifically the maintenance of institutional essentials.

A detailed discussion of each of these aspects and their operation, both in general and in terms of their specific expression in different polycentric systems, is beyond the scope of this chapter (see van Zeben, forthcoming). By means of example, the interrelationship between attributes, institutional essentials and prerequisites can be illustrated as follows: Vincent Ostrom emphasized that in a polycentric order, competition and conflict resolution is part of a broader process of continuous readjustment under a generally accepted body of over-arching rules:

> [t]he many elements are capable of making mutual adjustments for ordering their relationships with one another within a general system of rules where each element acts with independence of other elements.
>
> *(Ostrom, 1972, 57)*

The *attribute* of continuous competition and conflict resolution needs peaceful contestation and enforcement of a shared system of rules (both *institutional essentials*) in order to be maintained. In turn, access to information, the capacity to learn and access to justice are all *prerequisites* for these two institutional essentials to function and to prevent the reversal of a polycentric system to another type of pluri-centric system (see also Table 4.2).

Table 4.2 Attributes and institutional essentials of polycentric systems

Attributes	(i) Multiple independent centres of decision-making
	(ii) Overarching shared system of rules
	(iii) Continuous competition and conflict resolution
Institutional essentials	(i) Freedom and ability to enter and exit
	(ii) Enforcement of shared system of rules
	(iii) Peaceful contestation among different (interest) groups
Prerequisites	(i) Access to information
	(ii) Capacity to learn
	(iii) Access to justice

Polycentric potential

The Ostroms' contributions to the development of polycentric inquiry through the Workshop cannot be overstated. Nevertheless, the impact of polycentric theory on other pluri-centric theories, regardless of similarities and complementarity,[5] is nowhere near to reaching its potential. Similarly, the empirical applications of polycentric theory have been largely restricted to the United States and small, local communities and their management of common-pool resources. This may partially be explained by the continued terminological confusion that surrounds "polycentricity" as a concept. When it was introduced by Michael Polanyi, polycentricity focused on the importance of trial and error in certain systems, such as science and law. Polanyi emphasized the impossibility, and undesirability, of imposing morality, or achieving justice by a central authority, and thus the desirability of having many actors involved in these processes (Polanyi, 1951). In legal scholarship, this idea was applied to contrast the relative strengths of judicial versus governmental decision-making (Fuller, 1978; King, 2006). However, as will have become clear from the above, the Ostroms' concept of polycentricity goes much beyond the presence of multiple actors, both descriptively and normatively.[6]

The further development of analytical frameworks for polycentric systems (as those suggested earlier) could go some way to resolving this confusion. It would position polycentricity more clearly vis-à-vis other pluri-centric theories, work towards a clearer language of polycentric analysis, and open up the possibility for "non ad-hoc analogies between different forms of self-organizing complex social systems" (Aligica and Tarko, 2012, 260). A polycentric study of the EU provides a useful starting point. In terms of descriptive and normative theories of European governance, polycentricity has much to offer (for a full discussion, see van Zeben and Bobic, forthcoming). However, at the moment, even theories that arguably share many common features with polycentricity underuse the existing insights.

The "functional, overlapping, and competing jurisdictions" (FOCJ) theory as proposed by Frey and Eichenberger (1996) is an example of this. FOCJ are defined as a form of federal governance that is created as a bottom-up response to citizen preferences and foresees direct access to federal level courts for all citizens, and independence of local communities in order to constitute local governance systems, which would also have the right to levy taxes to finance themselves (Frey and Eichenberger, 1996, 315). FCOJ theory differs from multi-level governance theory (arguably the most influential theory of EU governance)[7] insofar as it would require the establishment of a "fifth freedom" in the EU, which would allow every community and citizen to appeal directly to the Court of Justice of the EU and to freely establish governance units independent from the nation-state in direct contact with the EU institutions (Frey and Eichenberger, 1996, 316).

This prescriptive recommendation is not a feature of multi-level governance theory, which is generally more descriptive. While promising, FOCJ theory remains underdeveloped with respect to more practical concerns regarding exit and voice and (the presence or absence of) constitutional rule-design, and the interaction between public and private actors.

A key difference between FOCJ and polycentricity is FOCJ's emphasis on efficiency, whereas polycentricity allows for a broader set of motivations. This is also an important contrast with the normative claims of multi-level governance theory, which centers around the concept of "good governance". The concept of good governance emphasizes elements such as transparency, accountability and citizens' participation but also has an important efficiency component (see EU Commission, 2001; Riedel, 2012, 65). Both FCOJ and multi-level governance theory sidestep self-governance as a governance goal in favor of efficiency and good governance, even though a close reading of Vincent's work suggests that by ensuring the polycentric functioning of a system, self-governance could help achieve these goals.

Shifting our focus to the polycentric features of the governance system instead of, or at least before, considering its functioning, provides the opportunity to analyze governance systems from different perspectives. For example, holistic study of EU governance is complicated by the differences in institutional structure in individual areas of competence. Rather than embracing this complexity, studies tend to look for general themes or approaches across policy areas, or to focus exclusively on small parts of the system. Notwithstanding the value of this work, by considering European governance through the lens of polycentric governance, this complexity may be embraced more easily, thereby providing a fuller picture of the institutional framework (see van Zeben and Bobic, forthcoming).

Expanding the existing theoretical foundations provided by the Ostroms also allows for broader empirical application of polycentric theory. While the number of studies is vast, particularly in the wake of Lin's Nobel prize, variation in the *types* of polycentric systems that have been studied has been limited. For example, common-pool resources tend to focus only on the local institutions, to the exclusion of other scales of governance involved (Andersson and Ostrom, 2008). Similarly, outside of resource regimes, the study of polycentric systems has largely focused on the United States.

The application of polycentric theory to the EU has also been very limited. Vincent commented on the EU several times but always in an ad hoc fashion (see e.g. Ostrom, 1996, 303). The most complete attempt by other members of the Bloomington School is the unpublished work by McGinnis and Hanisch (2005). Other, such as Wind (2003) and DeHousse (2003), have used the term "polycentricity" in reference only to the multiplicity of actors and processes within the EU, to the exclusion of the other attributes identified in Ostrom's work and the normative implications of polycentric theory. Similarly, the call for, and observance of, increased polycentricity in international relations and international law, particularly with respect to climate change (Cole, 2015; Ostrom, 2010b), has not yet translated to in-depth study of these systems from a polycentric perspective. Or rather, in-depth studies of these systems' polycentricity. While it is undoubtedly true that these systems display polycentric features and/or have polycentric potential, this does not mean that they already function in a polycentric manner.

Through the first attempt of a complete study of the EU through the lens of polycentric theory (van Zeben and Bobic, forthcoming), it has become clear that the study of large multi-jurisdictional, multi-issue systems requires specification of the existing theory to these systems. For example, the study of "peaceful contestation" within such a system is only possible if the concept of peaceful contestation is defined in a manner that reflects the particularities of a given system. This is not to say that the Ostroms' conceptualization should have been more specific. Rather, as Lin herself was quick to point out, polycentric systems are not a panacea for complex problems of governance (Ostrom, 2010b, 555). It is to say that in profiting from the Ostroms' insights on the presence and potential of polycentric systems, much theoretical and empirical work remains to be done in harnessing that potential towards a more complete understanding of governance and application of their framework to new (especially larger) polycentric systems.

Notes

1 With thanks to Mike McGinnis for capturing our discussions on this topic in this one phrase.
2 It bears repeating that while efficiency is a valid external benchmark for system performance, the individuals and communities involved in polycentric systems are presumed to set their own goals, which may lead to the sacrifice of efficiency for the pursuit of other goals, such as sustainability or equity (see also earlier McGinnis and Ostrom, 2012, 20–21).
3 These rules can be found throughout Lin's work and commentary on her work in different forms (e.g. Ostrom, 1990; McGinnis and Walker, 2010). This list is used as it was a recent restatement by Lin herself.
4 See also Chapter 3 (M. Cox) in this Handbook.
5 With respect to networking theory, see also McGinnis and Ostrom (2012).
6 In fact, even though Polanyi's coining of the phrase preceded Ostrom, Tiebout and Warren's 1961 article, they were not aware of his contribution at the time of writing.
7 *Inter alia* Marks (1993) and Hooghe and Marks (2010).

References

Abernathy, C.L. and Sally, H. (2000) "Experiments of some government-sponsored organizations of irrigators in Niger and Burkina Faso, West Africa" *Journal of Applied Irrigation Studies* 35(2): 177–205.

Aligica, P. and Tarko, V. (2012) "Polycentricity: From Polanyi to Ostrom, and beyond" *Governance* 25(2): 237–262.

Andersson, K. and Ostrom, E. (2008) "Analyzing decentralized resource regimes from a polycentric perspective" *Policy Sci* 41: 71–93.

Cole, D. (2015) "Advantages of a polycentric approach to climate change policy" *Nature Climate Change* 5: 114–118.

Crook, D.S. and Jones, A.M. (1999) "Design principles from traditional mountain irrigation systems (Bisses) in the Valais, Switzerland" *Mountain Research and Development* 19(2): 79–122.

DeHousse, R. (2003) "Beyond Representative Democracy: Constitutionalism in a Polycentric Polity" in Weiler J. and Wind M. (eds) *European Constitutionalism beyond the State* Cambridge, UK: Cambridge University Press, 135.

De Tocqueville, A. (1835) *Democracy in America*. New York: Anchor.

EU Commission (2001) *European Governance: A White Paper*, available at http://europa.eu/rapid/press-release_DOC-01-10_en.htm.

Frey, B. and Eichenberger, R. (1996) "FOCJ: Competitive governments for Europe" *International Review of Law and Economics* 16: 315–327.

Fuller, L. (1978) "The forms and limits of adjudication" *Harvard Law Review* 92: 353–409.

Hanushek, E.A. (1986) "The economics of schooling: Production and efficiency in public schools" *Journal of Economic Literature* 24: 1141–1177.

Hooghe, L. and Marks, G. (2010) "Types of multi-level governance" in Enderlein H., Wälti S. and Zürn M. (eds) *Handbook on Multi-Level Governance*. Cheltenham, UK: Edward Elgar, 17.

King, J. (2006) *Polycentricity and Resource Allocation: A Critique and Refinement*. Oxford Jurisprudence Discussion Group, Oxford.

Marks, G. (1993) "Structural policy and multilevel governance in the EC: The state of the European Community" in Cafruny A. and Rosenthal G. (eds) *The State of the European Community: The Maastricht Debate and Beyond*. Boulder, CO: L. Rienner, 391–411.

McGinnis, M. (ed.) (1999a) *Polycentric Governance and Development: Readings from the Workshop in Political Theory and Policy Analysis*. Ann Arbor, MI: University of Michigan Press.

McGinnis, M. (ed.) (1999b) *Polycentricity and Local Public Economies: Readings from the Workshop in Political Theory and Policy Analysis*. Ann Arbor, MI: University of Michigan Press .

McGinnis, M. (2011) "An introduction to IAD and the language of the Ostrom Workshop: A simple guide to a complex framework" *Political Studies Journal* 39: 169–183.
McGinnis, M. and Hanisch, M. (2005) "Analysing problems of polycentric governance in the growing EU" Paper presented at the Transcoop Workshop on Problems of Polycentric Governance in the Growing EU Humboldt University Berlin (on file with the author).
McGinnis, M. and Ostrom, E. (2012) "Reflections on Vincent Ostrom, public administration, and polycentricity" *Public Administration Review* 72(1): 15–25.
McGinnis, M. and Ostrom, E. (2014) "Social-ecological system framework: initial changes and continuing challenges" *Ecology and Society* 19(2): article 30. www.ecologyandsociety.org/vol19/iss2/art30/ES-2013-6387.pdf.
McGinnis, M. and Walker, J. (2010) "Foundations of the Ostrom workshop: institutional analysis, polycentricity, and self-governance of the commons" *Public Choice*, 143: 293–301.
Merrey, D.J. (1996) "Institutional design principles for accountability on large irrigation systems" Research Report no. 8, Colombo, Sri Lanka: International Irrigation Management Institute.
Ostrom, E. (1964) *Public Entrepreneurship: A Case Study in Ground Water Basin Management*. Unpublished PhD manuscript, available at http://citeseerx.ist.psu.edu/viewdoc/download?doi=10.1.1.196.3254&rep=rep1&type=pdf.
Ostrom, E. (1990) *Governing the Commons: The Evolution of Institutions for Collective Action*. New York: Cambridge University Press.
Ostrom, E. (2005) *Understanding Institutional Diversity*. Princeton, NJ: Princeton University Press.
Ostrom, E. (2007) "A diagnostic approach for going beyond panaceas" *Proceedings of the National Academy of Sciences* 104(39): 15181–15187.
Ostrom, E. (2008) "Polycentric systems as one approach for solving collective action problems" *W08–6 9/2/08*.
Ostrom, E. (2009) "A general framework for analyzing sustainability of social-ecological systems" *Science* 235: 419–422.
Ostrom, E. (2010a) "Beyond markets and states: Polycentric governance of complex economic systems" *American Economic Review* 100: 641–672.
Ostrom, E. (2010b) "Polycentric systems for coping with collective action and global environmental change" *Global Environmental Change* 20: 550–557.
Ostrom, E., Burger, J., Field, C., Norgaard, R. and Policansky, D. (1999) "Revisiting the commons: Local lessons, global challenges" *Science* 284(5412): 278–282.
Ostrom, E., Parks, R. and Whitaker, G. (1978) *Patterns of Metropolitan Policing*. Cambridge, UK: Ballinger.
Ostrom, V. (1953a) "State administration of natural resources in the West" *American Political Science Review* 47(2): 478–493.
Ostrom, V. (1953b) *Water and Politics: A Study of Water Policies and Administration in the Development of Los Angeles*. Los Angeles, CA: Haynes Foundation.
Ostrom, V. (1959) "Tools for decision-making in resource planning" *Public Administration Review* 19(2): 114–121.
Ostrom, V. (1962a) "Political economy of water development" *American Economic Review* 52(2): 450–458.
Ostrom, V. (1962b) "Water economy and its organization" *Natural Resources Journal* 2(1): 55–73.
Ostrom, V. (1971) "Public choice: A different approach to the study of public administration" *Public Administration Review* 31(2): 203–216.
Ostrom, V. (1972) "Polycentricity" in McGinnis M (ed.) (1999b), *Polycentricity and Local Public Economies: Readings from the Workshop in Political Theory and Policy Analysis*. Ann Arbor, MI: University of Michigan Press, 52–74.
Ostrom, V. (1973) "Can federalism make a difference?" *Publius* 3(2): 197–237.
Ostrom, V. (1989) "Some developments in the study of market choice, public choice, and institutional choice" in Rabin, J., Bartley Hildreth, W. and Miller, G. (eds) *Handbook of Public Administration*. New York: Marcel Dekker, 861–882.
Ostrom, V. (1993) "Epistemic choice and public choice" *Public Choice* 77(1): 163–176.
Ostrom, V. (1994) *The Meaning of American Federalism: Constituting a Self-Governing Society*, San Francisco, CA: ICS Press.
Ostrom, V. (1996) "Faustian bargains" *Constitutional Political Economy* 7: 303–308.
Ostrom, V. (1997) *The Meaning of Democracy and the Vulnerability of Democracies: A Response to Tocqueville's Challenge*. Ann Arbor, MI: University of Michigan Press.

Ostrom, V. (2008a) *The Intellectual Crisis in American Public Administration*, 3rd ed. Tuscaloosa, AL: University of Alabama Press.
Ostrom, V. (2008b) *The Political Theory of a Compound Republic: Designing the American Experiment*, 3rd ed. Lanham, MD: Lexington Books.
Ostrom, V., Bish, R. and Ostrom, E. (1988) *Local Government in the United States*. San Francisco, CA: ICS Press.
Ostrom, V. and Ostrom, E. (1965) "A behavioral approach to the study of intergovernmental relations" *Annals of the American Academy of Political and Social Science* 359: 137–146.
Ostrom, V. and Ostrom, E. (1971) "Public choice: A different approach to the study of public administration" *Public Administration Review* 31(2): 203–216.
Ostrom, V., Tiebout, C.M. and Warren, R. (1961) "Organization of government in metropolitan areas: A theoretical inquiry" *American Political Science Review* 55(4): 831–842.
Parks, R. (1985) "Metropolitan structure and systemic performance: The case of police service delivery" in Hanf K. and Toonen A.J. (eds) *Policy Implementation in Federal and Unitary Systems*. Dordrecht, the Netherlands: Martinus Nijhoff, 161–191.
Polanyi, M. (1951) *Logic of Liberty: Reflections and Rejoinders*. Chicago, IL: University of Chicago Press.
Poteete, A., Janssen, M. and Ostrom, E. (2010) *Working Together: Collective Action, the Commons, and Multiple Methods in Practice*. Princeton, NJ: Princeton University Press.
Riedel, R. (2012) "Silesian representations in Brussels: How the sub-national authorities utilise opportunity structures in the multi-level governance of the EU" in Grzeszczak R. and Karolewski I. (eds) *The Multi-level and Polycentric European Union: Legal and Political Studies*. Münster, Germany: Lit Verlag, 57.
Teske, P., Schneider, M., Mintrom, M. and Best, S. (1993) "Establishing the micro foundations of a macro theory: Information, movers, and the competitive local market for public goods" *American Political Science Review* 87(3): 702–713.
Toonen, A.J. (2010) "Resilience in public administration: The work of Elinor and Vincent Ostrom from a public administration perspective" *Public Administration Review* 70(2): 193–202.
van Zeben, J. (2013) "Research agenda for a polycentric European Union" Ostrom Workshop Working Paper Series, W13-13.
van Zeben, J. (forthcoming) "Polycentric features of the European Union" in van Zeben J. and Bobic, A. (eds) *Polycentricity in the European Union*. Cambridge, UK: Cambridge University Press.
van Zeben, J. and Bobic, A. (forthcoming) *Polycentricity in the European Union*. Cambridge, UK: Cambridge University Press.
Wagner, R. (2005) "Self-governance, polycentrism, and federalism: Recurring themes in Vincent Ostrom's scholarly oeuvre" *Journal of Economic Behavior & Organization* 57: 173–188.
Weinstein, M. (2000) "Pieces of the puzzle: Solutions for community-based fisheries management from native Canadians, Japanese cooperatives, and common property researchers" *Georgetown International Environmental Law Review* 12(2): 375–412.
Wind, M. (2003) "The European Union as a polycentric polity: Returning to a neo-medieval Europe?" in Weiler J. and Wind M. (eds) *European Constitutionalism beyond the State*. Cambridge, UK: Cambridge University Press, 103.

5

CONNECTING COMMONS AND THE IAD FRAMEWORK

Michael D. McGinnis

The Institutional Analysis and Development (IAD) framework provided the analytical foundation upon which Elinor Ostrom built a research program on community-based management of natural resources. For that work she was named a co-recipient of the 2009 Nobel Memorial Prize in Economic Sciences. She was also the primary driving force behind the initial development and subsequent revisions of the IAD framework, which remains one of the most prominent theoretical perspectives in scholarly research on public policy (Weible and Sabatier 2018; Cole and McGinnis 2017). Although she makes no explicit use of IAD in her highly influential 1990 book, *Governing the Commons*, she does use an informal discussion of its basic components to frame her mode of analysis (Ostrom 1990, 45–57). In later work Ostrom (2005, 2010) more fully articulates the broad scope of this analytical framework, which she envisioned as a means whereby scholars from multiple disciplines could more effectively communicate with each other, as they used diverse perspectives to better understand complex policy settings.

In this chapter I investigate this deep connection between the conceptual framework of IAD and the empirical subject of the commons. I suggest ways in which a deeper realization of this connection might inspire promising directions for future research. The first section introduces the concept of a commons as a configuration of resources, rules, and the people whose access to those resources is determined by those rules. The following section summarizes the IAD framework and highlights a remarkable congruence between its key components and the nature of a commons, which helps explain why IAD proved to be such a solid foundation for commons research. Then I use the design principles from Ostrom (1990) to illustrate how the IAD framework helps contextualize the conditions necessary to insure the long-term sustainability of a commons.

Thereafter I locate both IAD and the commons within broader perspectives on social-ecological systems and systems of polycentric governance. Late in her career Ostrom (2007, 2009) introduced a Social-Ecological Systems (SES) framework that fine-tuned the IAD framework to be more directly applicable to the study of resource systems. Polycentric governance requires a multitude of decision centers, each with a limited range of authority which, to some extent, overlaps with the jurisdiction of other decision centers, thus requiring these authorities to interact with each other to achieve their own goals. Ostrom highlighted the centrality of polycentricity to her Nobel-winning research on the "study of governance, especially the commons" by selecting

"The Polycentric Governance of Complex Economics Systems" as the title for her Nobel Memorial Lecture (Ostrom 2010). This chapter concludes that a fuller appreciation of close connections among the IAD framework, commons, social-ecological systems, and polycentric governance can help lay a firmer foundation for future research in all of these areas.

Commons as a configuration of resources, rules, and people

Informally, the word *commons* refers to resources to which members of some group share access. Resource commons occur in many different forms and are familiar to people from all cultures. Typical examples include common grazing land, fisheries, lakes, or forests. Other commons are humanly constructed, ranging from irrigation systems to complex infrastructures for transportation, communication, and the generation and distribution of electric power.

More technically, a commons may include one or both of two kinds of goods: (1) *public goods*, for which one person's enjoyment does not preclude others from also enjoying it, and (2) *common-pool resources*, for which consumption by one makes that particular resource unit unavailable to others, even while the broader resource pool remains available to all eligible users (Ostrom and Ostrom 1977). Difficulties grounded in the ubiquitous temptation to free-ride on the contributions of others towards the production of public goods are universally recognized, and common-pool resource is a technical term for something that happens every day whenever fish are caught, firewood is collected, or farmers draw upon an irrigation system to water their crops. Beneath this mundane façade, however, lie subtle connections between our understanding of private and public. When a particular resource unit (a fish caught in a lake) is extracted from a pool of commonly shared resources (the population of fish swimming in that lake), that fish becomes the private property of some individual or corporate entity, whereas the rest of those fish remain available for use by others.

Individuals extract resources for their own use, but if too many people extract too much in too short a period of time, the common pool may be degraded or destroyed. This *tragedy of the commons* (Hardin 1968) is especially likely in an "open access" commons, from which anyone can draw. This tragic outcome can be avoided only if someone takes responsibility for insuring the replenishment or maintenance of that resource. Hardin concluded that there were only two possible answers: either the commons should be (1) managed by some central authority assigned the task of acting as its steward, or (2) divided into parcels of private property, since owners can reasonably be expected to look after their own property.

In 2009 Elinor Ostrom was awarded a Nobel Prize for demonstrating the relevance of a third answer to this question (Ostrom 2010). A commons may be effectively managed as *common property*, an institutional arrangement though which a specific group of individuals shares responsibility for jointly producing, consuming, and/or managing a common resource. In *Governing the Commons* (Ostrom 1990) and many other publications, Ostrom drew upon examples from countries throughout the world to demonstrate how local communities dependent on continued access to natural resources can, in some circumstances, work together to craft, monitor, enforce, and revise rules limiting their own behavior and thereby manage to keep those resources available for long periods of time. These rules typically specify how many resource units can be extracted, and when, as well as requiring contributions towards maintaining any needed infrastructure. By transforming common resources into common property, a group of resource users act as their own stewards.

Ostrom's research inspired others to carefully examine the concept of a commons. An especially useful insight is provided by Frischmann et al. (2014: 2):

> Commons refers to a form of community management or governance. It applies to resources, and involves a group or community of people, but commons does not denote the resources, the community, a place, or a thing. Commons is the institutionalized arrangement of these elements.

In their effort to extend Ostrom's mode of analysis to knowledge commons, such as intellectual property rights systems, they identify the core analytical task as understanding, "How people interact with rules, resources, and each other" (Frischmann et al. 2014: 19).

This conceptualization of a commons as a mutually sustaining configuration of resources, rules, and people provides an especially rich foundation for analysis. A commons consists not just of resources or the people with access to those resources, or even to the rules that restrict how these people may use those resources; rather, a commons should be understood as a holistic configuration of resources, rules, and people.

A framework for understanding institutional processes

Resources, rules, and people are all located within broader biophysical, political-legal, and social-cultural contexts. This brings into play the IAD framework, which was designed to be applicable to all kinds of policy settings (Ostrom 1986, 2005, 2010, 2011). In her words:

> If a wide diversity of institutional forms exist not only side by side but nested within one another, behavior cannot be explained, nor guided, controlled and evaluated, through reliance on any limited sets of pure theories. We need to ask whether similar conceptual characteristics underlie all hierarchies, markets, courts, electoral contests, collegial fora, and solidaristically organized communities. Is there a common set of variables that can be used to analyze all types of institutional arrangements?
>
> *(Ostrom 1986: 459–460)*

The IAD framework posits a few key categories of potentially explanatory variables from which common sets of variables could be constructed and used to explain all kinds of policy situations. As shown in Figure 5.1, the IAD framework represents institutional processes by a series of boxes within which different causal determinants or processes are located. At the heart of the IAD framework is an *action situation*, in which individuals (acting on their own or as agents of formal organizations) observe information, select actions, engage in patterns of interaction, and evaluate outcomes. Choices and outcomes are influenced by the beliefs and incentives of individual actors, as shaped by the responsibilities and social expectations attached to any official position they may hold and by the information available to them.

Each action situation is shaped by preexisting *contextual conditions*, grouped for analytical purposes into three categories: (1) "nature of the good" under consideration, including all relevant biophysical conditions; (2) "rules-in-use," which includes the entire body of laws, regulations, rules, norms, and shared understandings held by the participants to be relevant to deliberations on that policy area; and (3) "attributes of the community," that is, the social ties and cultural attributes that characterize the individuals directly affected by that policy problem.

Each action situation denotes a nexus where a group of decision makers jointly confront important decisions related to some particular policy concern. As is typical in strategic interactions, potential outcomes are differentially valued by actors with only partial control over the final determination of results. Ostrom (1986) frames an action situation as a generalization of standard game models. To define a game, modelers must specify the actors involved, the actions

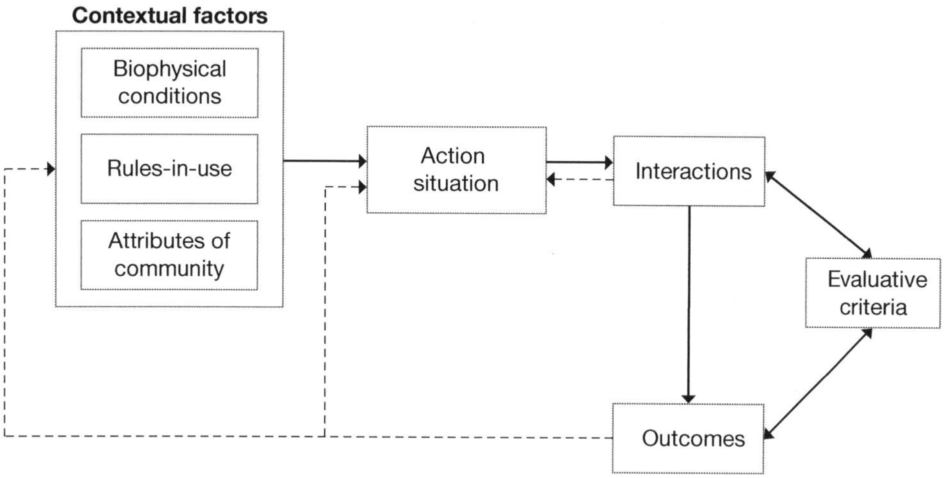

Figure 5.1 Institutional Analysis and Development (IAD) framework
Source: Adapted from Ostrom (2011: 10).

available to them, and how these actions jointly generate alternative outcomes that are differentially valued by the actors, who may have access to different sources of information as well as different types or levels of resources they can use to influence the actions of other players.

In the IAD framework, each action situation is configured by interlocking "working components," which are related in the following manner:

> *Participants*, who can either be individuals or any of a wide diversity of organized entities, are assigned to *positions*. In these positions, they choose among *actions* in light of their *information*, the *control* they have over *action-outcome linkages*, and the *benefits and costs* assigned to actions and outcomes.
>
> *(Ostrom 2005, 188; my emphasis)*

Implicitly, the values taken by each of these italicized components is determined through processes occurring in other settings for strategic interaction, that is, in other action situations.

The IAD framework differentiates among three arenas of choice, or conceptual levels of analysis: (1) *operational-choice* settings in which the choices of the relevant actors directly impact tangible outcomes, (2) policymaking or *collective-choice* settings in which the actors shape the rules that constrain actors in operational-choice arenas, and (3) settings for *constitutional-choice* in which decisions are made concerning which actors have standing in different choice situations as well as which kinds of alternative institutional mechanisms are available to them as they make their collective deliberations and operational-level choices (Ostrom 2005: 58–62).

Individuals and collective actors generate patterns of *interaction* from which specific *outcomes* emerge, which they (or other actors) compare against *evaluative criteria* that seem pertinent to the actors doing the evaluations. All these evaluations (and the outcomes that triggered them) feed back into the entire set of preexisting conditions, thereby setting the stage for the next round of action situations. Although evaluative criteria are placed in a box off to the side of Figure 5.1, a more complete representation would show how these criteria were determined by processes of collective or constitutional choice occurring in other action situations.

Concurrent action situations operating at the same or different levels of analysis interact in subtle ways. In particular, the contextual factors which define any given action situation will themselves have been determined by outcomes generated by other action situations. Ostrom and Ostrom (2004: 134) illustrate how processes of constitutional choice can shape the general context under which collective policy decisions are made, and the outcomes generated by action situations at the collective choice level can determine the specific conditions under which operational choices are implemented. McGinnis (2011b: 54) provides an alternative graphical representation in which different kinds of collective or constitutional choice processes can influence the working components that constitute an action situation at the operational level.

Strictly speaking, this determination of the working components of a focal action situation may be no different from a complete specification of a game model, but in practice this concern with simultaneous consideration of multiple choice arenas that endogenously generate conditions for choice in other arenas inspires IAD scholars towards a more inductive mode of analysis (Mitchell 1988). Formal mathematical analyses are rendered problematic because of the complexity of the full network of interconnected action situations. However, in some settings the same actors may participate in the most critical action situations (McGinnis 2011b). In a self-organized community of resource users who live in a remote area and only rarely experience interference from outside actors, a complex network of adjacent action situations could distill down to a small number of tightly interwoven action situations. This is one reason why this mode of analysis has proven so useful for analysts studying community management of common resources and other forms of self-governance.

The details of IAD have undergone changes over the three decades of its use (Kiser and Ostrom 1982; Ostrom 1986, 1989, 2010, 2011; Oakerson 1992; Ostrom et al. 1994; Ostrom and Ostrom 2004; Clement 2010; Schlager and Cox 2018). For example, the action situation was initially divided into actors and the action arena, which included all the rest of the working components. These changes are discussed in detail in Ostrom (2011), which was her last solo-authored evaluation of its impact on policy studies in general.

Despite minor changes in terminology, the same basic structure has remained in place from the beginning: diverse kinds of choice processes that are operating concurrently or sequentially at different conceptual levels of analysis and that interact to shape the contexts under which all action situations are conducted. This enduring vision permeates *Understanding Institutional Diversity* (Ostrom 2005), the most comprehensive and authoritative explication of the full analytical repertoire of the IAD framework.

IAD and the design principles for sustainable commons

Although Elinor Ostrom intended IAD to be generally applicable, in practice it tends to be most frequently used for certain kinds of policy settings (Heikkila and Cainey 2018). IAD explicitly presumes that human choice matters and that the contexts within which those choices occur can be shaped by individual choice and collective action. It is not difficult to imagine policy settings in which the choices of ordinary citizens have very little direct impact on ultimate outcomes.

For example, the IAD framework may not be well suited for understanding decision situations within complex hierarchical organizations in which ultimate authority is concentrated at the top levels of that hierarchy. Still, the IAD framework can be modified to improve its ability to answer a broader array of questions. For example, analysts working within this tradition tend to focus on the problem-solving aspect of policy problems and downplay the extent to which policy outcomes are driven by whoever has the most power to shape policy. Clement (2010) expanded the IAD framework in a natural manner by adding two additional categories of contextual variables to highlight the potential consequences of the social-economic foundations for political power and the ways in which public discourse can be shaped to favor the interests of influential actors.

The IAD framework is widely recognized as one of the most influential perspectives in the research literature on public policy, and Heikkila and Cainey (2018) systematically compare IAD to others on that list. Briefly, the multiple streams approach highlights the tendency of Presidents and other national authorities to react sequentially to newly emerging crisis situations, with their decisions being driven to a great extent by the chance concurrence of specific problems and policy solutions that are the most politically expedient at that time, given the (US) partisan configuration of the three branches of government or the electoral calendar. The advocacy coalition and narrative policy frameworks emphasize political disputes between coalitions of actors with similar beliefs towards a given issue or in which the terms of discourse can be instrumentally shaped by clever slogans or framing contexts. Models of policy diffusion can help explain how policy ideas or experiments are transmitted from one state or nation to another, based on the communication networks among different jurisdictions.

Although the IAD framework cannot provide a universal language to study all of the political landscape, there are some circumstances for which it is ideally suited. Individuals who find themselves in situations in which they have the capacity to work together to meaningfully shape the conditions under which they interact, are likely to find it easier to move from operational decisions to reconsideration of collective choices or to renegotiation of fundamental questions concerning the constitution of their own societies, and back again. The primary value to researchers of the IAD framework is the guidance it provides concerning the kinds of factors that need to be examined in situations of such pervasive endogeneity, by structuring the types of questions researchers should ask as they work their way through the complexities of those policy settings (Cole and McGinnis 2017).

The IAD framework requires analysts to realize that biophysical, legal-institutional, and socio-cultural factors interact in complex ways to shape patterns of interactions among relevant actors as well as policy outcomes. By highlighting the fundamentally configural nature of causal relationships in social settings, this framework implies that the effects of any policy intervention will necessarily ramify throughout complex institutional systems in manners that are unlikely to be immediately apparent. In that way, IAD forces policy analysts to dig deeper into the underlying nature of the problem and sensitizes them to follow through the likely consequences of any policy intervention on the subsequent incentives facing relevant actors, as well as the institutional processes which may shape the ways in which they act upon those new incentives.

It was this persistent inclination to dig deeper that helped Elinor Ostrom see more clearly into the fundamental nature of a commons and to help the rest of us see these issues in a new light. Where other analysts saw only general questions of policy or legal context, Ostrom kept pushing to understand the ultimate linkages among the resources, rules, and people that constitute the configurations we have come to call a commons.

The IAD framework directs analysts to focus their attention on a small number of key action situations in which resources, rules, and peoples mutually influence each other (McGinnis 2017). Each of these focal processes can then be subjected to the forms of analysis prompted by the IAD framework. Despite Ostrom's early intentions as quoted earlier, it is more appropriate to say that the IAD is directed towards identifying a common set of *processes*, rather than variables. Ostrom (2005) frames all kinds of institutional arrangements as processes, and the IAD framework detailed there has helped many analysts to focus their attention on identifying certain key processes. Only secondarily does attention shift to identifying particular sets of variables for operationalization.

The way in which IAD nudges analysts to focus their attention on a small number of core processes can be illustrated by considering Ostrom's analysis of common property as a means of sustainable management of a resource commons (Ostrom 1990; McGinnis 2017). She begins by distinguishing two fundamental processes found in any common-pool resource setting: *appropriation* (extraction of a resource unit from a common pool) and *provision* (the process of arranging for the production of public goods, the replenishment of a common resource, or the construction and maintenance of infrastructure used in either setting). In many kinds of resource commons, different groups of individuals or corporate entities are primarily responsible for appropriation and provision. But in most cases studied by Ostrom, the same individuals were directly engaged in both appropriation and provision.

Ostrom also directed attention to other related processes, especially rule-making, monitoring, and sanctioning. Her appreciation of their importance was reflected in her experimental research on common-pool resources. Ostrom et al. (1994) use the term provision to encompass all replenishment and maintenance activities because this part of the CPR situation corresponds to a public good, at least for members of the user group. When they set up experimental versions of this generic model, they determined the rules that the participants had to follow, as well as any rules concerning how those participants may select different types of monitoring and sanctioning regimes. These experimental studies explore how the same subjects behave differently when given different options for monitoring and sanctioning. Even in this highly abstracted version of a CPR, processes of appropriation, provision, rule-making, and monitoring and sanctioning are deeply interconnected.

It is my contention that a similar intertwining of focal action situations provides a key to understanding how Elinor Ostrom developed the "design principles," the single most important take-away from *Governing the Commons*. Ostrom posited a list of eight conditions that were satisfied, in one way or another, in every one of the cases she examined in which local communities successfully managed, in a sustainable fashion, resources needed for their own survival. Although few if any of these communities explicitly set out to satisfy these conditions, in combination they provide a solid foundation for long-term sustainability of a commons.

My claim is that these design principles emerged from Ostrom's intense engagement with her case studies, because her thought processes were shaped, in a fundamental sense, by the categories and relationships expressed in the form of the IAD framework on which she had been working for years (McGinnis 2017). I am not trying to reproduce the exact thought processes through which she reasoned her way to these conclusions but am only interested in using the design principles to show how the IAD framework and the concept of a commons line up in a more fundamental way than is generally recognized.

I find it useful to rearrange and rephrase these design principles in the form of a single, but multiply compounded sentence, as follows:

> A community of resource users dependent on an identifiable body of resources located in well-defined region (*clear social and resource boundaries*), which has sufficient formal authority or effective de facto opportunity to manage these resources (*local autonomy*), can do so in a sustainable manner by collectively crafting rules and procedures regarding levels and modes of resource extraction and the construction and maintenance of necessary infrastructure (*wide participation in rule-making*), sharing information generated through routine monitoring of user actions and resource outcomes, either by themselves or by monitors whom they hold responsible for honest oversight (*responsible monitoring*), imposing sanctions on rule-breakers in a way that imposes significant costs for repeated offenses while still enabling them to return to the good graces of the community

> (*graduated sanctions*), resolving the disputes that will inevitably arise either directly or with the help of trusted intermediaries (*conflict resolution procedures*), and by soliciting assistance from outsiders or delegating important tasks to sub-groups when that is needed to find solutions to specific problems (*nested enterprises*), as long as the rules and procedures that emerge from ongoing interactions among the relevant parties turn out to be appropriate for local circumstances and have consequences that distribute the costs and benefits of their collective action in a manner considered equitable by most members of that community (*rule congruence with changing social and environmental conditions*).

This sentence highlights the many critical interactions among all items in that list (as modified by Cox et al. 2010) and illustrates how these principles operate in conjunction to generate and sustain a mutually reinforcing configuration of specific properties of the resources themselves, the actions of the people within that community, and the rules, norms, and procedures that shape those actions. In sum, the design principles define conditions for a mutually reinforcing configuration of resources, rules, and communities, that is, for a sustainable commons.

Elsewhere (McGinnis 2017) I argue that Ostrom concentrated her analysis on four critical (or focal) action situations in her case studies: appropriation, provision, rule-making, and monitoring and sanctioning. A few related action situations that also contributed to her formulation of the design principles include constitutive processes (the construction of collective actors capable of making a common decision), dispute resolution, evaluations, and the slow accumulation of indigenous knowledge. The framing established by her IAD mind-set worked so well in this analysis because in most of the cases she examined, essentially the same set of actors was directly involved in all four focal action situations as well as these supplementary (or adjacent) action situations.

Each of the design principles can be interpreted as an outcome generated by one or more of these core and supplemental action situations. As an overall configuration, these outcomes will be mutually reinforcing in some empirical settings and not in others. More specifically,

(1) *Clear social and resource boundaries* may emerge from constitutive processes and the accumulation of local knowledge in some circumstances, but in other cases deep fissures between competing groups may have prevented them from arriving at a consensus, or regular patterns of resource growth and decline may have been too frequently disrupted by unpredictable shocks.
(2) A minimal level of *local autonomy* may have been effectively guaranteed in remote areas of little interest to central authorities, but any area which has attracted the attention of national leaders or global corporations may lose its ability to maintain traditional practices.
(3) Groups systematically excluded from *participation in rule-making* would have been much less likely to comply with those rules or contribute towards their enforcement, compared to those groups which helped craft rules that served their own interests.
(4) Many rule violations would have been easily observed within the dense social networks in these communities, but in those circumstances when explicit *monitoring* was required, the people most directly affected by the continued availability of these critical resources would have been better able to resist bribes to overlook certain violations.
(5) *Sanctions* that failed to be applied in a consistent manner would have undermined the legitimacy of the rules, and a lack of any opportunities for second or third chances would, eventually, have generated poisonous levels of resentment.
(6) If processes for the *resolution of disputes* were not available, or if existing processes did not take into account the long-term effects of settlements on the community as a whole, then fewer members of the community would have been willing to continue to comply with or monitor the rules.

(7) If the formation or operation of *nested teams* (each assigned a limited task) has long been hampered by deep social fissures, or the lack of trusted intermediaries, some important functions could not have been accomplished in any consistent manner.
(8) Appropriation and maintenance rules that failed to be *congruent* with local biophysical conditions would, eventually, lead to resource degradation, and it is unlikely that rules which generated unacceptable levels of inequities would have continued to be widely observed or enforced by members of disadvantaged groups.

Each design principle explicitly connects to factors from one or more of the three categories of contextual variables: biophysical conditions of the resources, rules-in-use, and attributes of the relevant community. As a whole, these design principles emerged as the outcomes of multiple arenas of choice at the operational, collective, and constitutional levels. Continued operation of all of these design principles would have the effect of generating patterns of dynamic changes that exhibit a strong tendency towards co-evolution of the resources, rules, and communities that constitute a commons. In this way, factors within each category influence factors in all of the other categories to maintain a dynamic sustainability of the entire configuration. The IAD framework helps draw the attention of analysts to the full range of factors and dynamic processes that are necessarily involved in the operation and sustainable management of a resource commons.

Extensions to social-ecological systems and polycentric governance

As argued earlier, no one commons exists in isolation, since its constituent resources, rules, and people each connect outwards to broader contexts. The IAD framework directs analysts to focus attention on a small number of key action situations which serve as the nexus where resources, rules, and communities mutually influence each other. No action situation occurs in isolation; instead each action situation points outward to other action situations in which the defining components of that focal action situation have been determined.

Since there is no such thing as an institution-free context (Cole et al. 2014), no policy reform can be applied to a completely blank slate. Policy advocates necessarily introduce purposeful interventions into an already complex ecosystem of institutional arrangements. Unfortunately, the dynamism embedded in this line of argument is not clearly represented in the canonical depiction of Figure 5.1, which might convey the mistaken impression that any one policy situation can be understood in isolation from the many other policy situations with which it is, in the real world, complexly interrelated.

Later in her career, Elinor Ostrom (2007, 2009) developed a more complex framework intended to provide a more comprehensive approach to the study of closely coupled systems of complex human-environment interactions, or social-ecological systems. This *SES Framework* was designed to give equal weight to both the social and ecological sides, whereas the IAD framework focused most of its attention on the social-institutional side of policy problems.

The SES framework was optimized for application to a relatively well-defined domain of common-pool resource management situations in which *resource users* extract *resource units* from a *resource system*. The resource users also provide for the maintenance of the resource system according to rules and procedures determined by an overarching *governance system* and in the context of *related ecological systems* and *broader social-political-economic settings*. Processes of resource extraction and maintenance were identified as the most important forms of the *interactions and outcomes* located in the very center of this framework (Ostrom 2007, 2009), but consideration was also given to rule-making, monitoring, sanctioning, dispute resolution, and the formation of organizations.

In practice, subsequent researchers have tended to focus on identifying lists of variables included in the first-tier categories of resource users, resource units, resource systems, and governance systems. The action-situation-based category of interactions and outcomes has received considerably less attention, despite its location at the center of all SES-based diagrams. Unfortunately, the IAD's characteristic emphasis on identifying a few key action situations for particular attention was somehow lost in the transition from IAD to SES.

In hopes of re-integrating these related approaches to the study of resource commons, Cole et al. (2014) draw upon both the IAD and SES frameworks to demonstrate that there was a lot more going on in Hardin's story of the tragedy of the commons than the overgrazing of an open-access commons.

> If the grass on the pasture was not subject to appropriation, the cattle were not privately owned, or property- and contract-enforcement institutions supporting market exchange were absent, then the "tragedy of the commons" would not have arisen regardless of the open-access pasture.
>
> *(Cole et al. 2014: 353)*

In other words, the tragedy that Hardin considered inevitable would only be logically possible for settings characterized by a particular arrangement of ownership rights over different but interrelated types of resources. Neither was Hardin's prediction of an inevitably tragic outcome the result of a lack of consideration of the possibly remedial effects of institutions, as argued by Ostrom (2007). Cole et al. (2014: 365–366) argue that Hardin's conclusion required an unfortunate combination of:

> [m]aladapted institutions that support incentives for overexploitation . . . The open-access pasture does not exist in splendid isolation but operates within a larger universe of interacting resources and institutions. Hardin's "tragedy of the commons" is not just about the pasture; it is equally about the grass, the cows, the herders, and the human society.

The governance and resource utilization parts of the SES framework are generated by dynamic iterations of the complex network of interrelated action situations that should be examined in any analysis based on the IAD framework. No action situation stands alone because its contextual conditions have to have been determined by the outcomes of other, related action situations. Similarly, no commons exists in splendid isolation; each is embedded within broader contexts of social, political, economic, legal, and other forms of institutions which, in combination, assign different configurations or responsibility for governance functions to actors with different kinds of property rights to different resources. When seen as an interrelated network of action situations, attention shifts to the range of authority that is being enacted by the actors-in-position within each action situation.

To analyze the ways in which action situations, as partially autonomous decision centers, are interconnected, it is necessary to step up to a higher level of aggregation and to consider the nature of *governance*. Governance is concerned primarily with the processes of provision, production, financing, and evaluation of public goods, as well as the management of common resources and public infrastructures. Governance necessarily requires tough decisions involving tradeoffs among alternative goods, many of which are high priority items for different parts of society. Since the benefits of enjoying public goods or shared resources cannot be easily appropriated exclusively by the individuals who invest their time and resources in producing or procuring those goods, some kind of authority is critical for solving the problems of free riding. Typically, authorities need to be able to enforce at least a minimal level of legitimate coercion to gather the resources needed to support public purposes, by requiring individuals to pay taxes or charging fees to individuals who seek access to restricted commons.

Elsewhere (McGinnis 2011a: 171), I defined governance as a "process by which the repertoire of rules, norms, and strategies that guide behavior within a given realm of policy interactions are formed, applied, interpreted, and reformed." The basic point is that governance must be understood as a *process* and that both government officials and non-governmental actors play critical roles in that process.

The IAD framework is naturally congruent with a particular understanding of governance as being driven by a multitude of decision centers, each with a limited range of authority which typically overlaps with the jurisdictions of other decision centers, thus requiring these authorities to interact with each other to realize shared aspirations, or even to achieve their own goals. Each decision center typically has jurisdiction over only a few action situations, and the conditions under which one decision center operates have been shaped by the operation of interrelated decision centers.

In an ideal typical system of *polycentric governance*, a diverse array of public and private authorities with overlapping domains of responsibility interact in complex and ever changing ways, and out of these seemingly uncoordinated processes of mutual adjustment emerges a resilient system of social ordering that can support and sustain capacities for self-governance (Cole and McGinnis 2015). Originally introduced by Ostrom et al. (1961) as a vision of governance that embraced the potentially positive consequences of governmental fragmentation in U.S. metropolitan areas, this conceptualization inspired empirical analyses of police services and other aspects of metropolitan governance (McGinnis 1999; Oakerson 1999; Oakerson and Parks 2011). The most widely known application of polycentricity to real-world settings remains the work of Elinor Ostrom (1990, 2010), who concluded that community-based commons management was likely to be sustainable only if its rules were nested within a broader system of polycentric governance that allowed for alternative mechanisms of collective decision making and conflict resolution at all levels of aggregation.

Ostrom's analysis of southern California groundwater management in chapter 4 of *Governing the Commons*, based in part on her PhD dissertation (1965), demonstrates that conditions similar to the design principles summarized earlier in this chapter can be observed in less isolated and more intensely politicized settings. In this case critical contributions were made by a diverse array of organizational actors operating in constitutive, judicial, and scientific arenas – experts from the U.S. Geological Society; state laws granting special water districts and other inter-local entities the authority to impose taxes on their members; equity court proceedings directed towards fact-finding rather than blame-setting; and the establishment of the role of watermaster as a reliable source of information and dispute resolution, but not regulation or enforcement. New institutions were built incrementally in a sequential and self-transforming process of learning and joint discussions, within a supportive regime of state law. Finding the right balance of small- and large-scale operations was critical to success.

A key puzzle in the literature on polycentric governance remains how well polycentric systems can generate effective matches between the functional scope of the jurisdictions of specific governance entities and the geographical and temporal extent of the ecological processes that need to be monitored and governed (Cole and McGinnis 2015; DeCaro et al. 2017). Following the line of argument developed here, policy reformers should strive to identify the kinds of decision centers that can realistically be established and operated in a manner conducive to the effective governance of the interconnected webs of resource systems that constitute the context within which specific resources deemed critical to human survival are generated, utilized, replenished, and governed.

This chapter has revealed deep connections between the IAD framework and the commons. Ostrom's design principles define conditions under which co-evolutionary processes operating at all levels of analysis generate sustainable matches among the IAD's three categories of contextual variables corresponding to the resources, rules, and communities that jointly define a commons. A complex network of action situations forms the ever changing context within which exogenous and endogenous drivers push towards the emergence of polycentric governance in social-ecological systems. The rich vein of concepts and analytical tools associated with the IAD framework and the commons remains ripe for further exploration and exploitation by future researchers.

References

Clement, F. (2010) "Analysing Decentralised Natural Resource Governance: Proposition for a 'Politicised' Institutional Analysis and Development Framework" *Policy Sciences*, 43, 129–156.

Cole, D.H., G. Epstein, and M.D. McGinnis (2014) "Digging Deeper into Hardin's Pasture: The Complex Institutional Structure of 'The Tragedy of the Commons'" *Journal of Institutional Economics*, 10, 353–369.

Cole, D.H., and M.D. McGinnis, eds. (2015) *Elinor Ostrom and the Bloomington School of Political Economy, Volume 1: Polycentricity in Public Administration and Political Science*. Lanham, MD: Lexington.

Cole, D.H., and M.D. McGinnis (2017). *Elinor Ostrom and the Bloomington School of Political Economy, Volume 3: A Framework for Policy Analysis*. Lanham, MD: Lexington.

Cox, M., G. Arnold, and S.V. Tomás (2010) "A Review of Design Principles for Community-Based Natural Resource Management" *Ecology and Society*, 15, article 38. www.ecologyandsociety.org/vol15/iss4/art38/.

DeCaro, D.A., B.C. Chaffin, E. Schlager, A.S. Garmestani, and J.D. Ruhl (2017) "Legal and Institutional Foundations of Adaptive Environmental Governance" *Ecology and Society*, 22, 32. https://doi.org/10.5751/ES-09036-220132.

Frischmann, B., M. Madison, K. Strandburg, eds. (2014) *Governing Knowledge Commons*. New York: Oxford University Press.

Hardin, G. (1968) "The Tragedy of the Commons" *Science*, 162, 1243–1248.

Heikkila, T., and P. Cainey (2018) "Comparisons of Theories of the Policy Process" in C.M. Weible and P.A. Sabatier, eds., *Theories of the Policy Process*, 4th ed. Boulder, CO: Westview, 301–327.

Kiser, L.L., and E. Ostrom (1982) "The Three Worlds of Action: A Metatheoretical Synthesis of Institutional Approaches" in E. Ostrom, ed. *Strategies of Political Inquiry*. Thousand Oaks, CA: Sage, 179–222.

McGinnis, M.D., ed. (1999) *Polycentricity and Local Public Economies: Readings from the Workshop in Political Theory and Policy Analysis*. Ann Arbor, MI: University of Michigan Press.

McGinnis, M.D. (2011a) "An Introduction to IAD and the Language of the Ostrom Workshop: A Simple Guide to a Complex Framework" *Policy Studies Journal*, 39, 163–177. Updated version http://php.indiana.edu/~mcginnis/iad_guide.pdf.

McGinnis, M.D. (2011b) "Networks of Adjacent Action Situations in Polycentric Governance" *Policy Studies Journal*, 39, 45–72.

McGinnis, M.D. (2017) "The IAD Framework in Action: Understanding the Source of the Design Principles in Elinor Ostrom's Governing the Commons" in D.H. Cole and M.D. McGinnis, eds., *Elinor Ostrom and the Bloomington School of Political Economy, Volume 3: A Framework for Policy Analysis*. Lanham, MD: Lexington, 87–108.

McGinnis, M.D., and E. Ostrom (2014) "Social-Ecological System Framework: Initial Changes and Continuing Challenges" *Ecology and Society*, 19, 30. http://dx.doi.org/10.5751/ES-06387-190230.

Mitchell, W.C. (1988) "Virginia, Rochester, and Bloomington: Twenty-Five Years of Public Choice and Political Science" *Public Choice*, 56, 101–119.

Oakerson, R.J. (1992) "Analyzing the Commons: A Framework" in D.W. Bromley et al., eds., *Making the Commons Work: Theory, Practice, and Policy*. San Francisco, CA: ICS Press, 41–59.

Oakerson, R.J. (1999) *Governing Local Public Economies: Creating the Civic Metropolis*. San Francisco, CA: ICS Press.

Oakerson, R.J., and R.B. Parks (2011) "The Study of Local Public Economies: Multi-Organizational, Multi-Level Institutional Analysis and Development" *Policy Studies Journal*, 39, 147–167.
Ostrom, E. (1965) *Public entrepreneurship: a case study in ground water basin management*. Unpublished PhD thesis, University of California, Los Angeles.
Ostrom, E. (1986) "An Agenda for the Study of Institutions" *Public Choice* 48, 3–25.
Ostrom, E. (1989) "Microconstitutional Change in Multiconstitutional Political Systems" *Rationality and Society*, 1, 11–50.
Ostrom, E. (1990) *Governing the Commons: The Evolution of Institutions for Collective Action*. New York: Cambridge University Press.
Ostrom, E. (2005) *Understanding Institutional Diversity* Princeton, NJ: Princeton University Press.
Ostrom, E. (2007) "A Diagnostic Approach for Going Beyond Panaceas" *Proceedings of the National Academy of Sciences USA*, 104, 15181–15187.
Ostrom, E. (2009) "A General Framework for Analyzing Sustainability of Social-Ecological Systems" *Science*, 325, 419–422.
Ostrom, E. (2010) "Beyond Markets and States: Polycentric Governance of Complex Economic Systems" *American Economic Review*, 100, 641–672.
Ostrom, E. (2011) "Background on the Institutional Analysis and Development Framework" *Policy Studies Journal*, 39, 7–27.
Ostrom, E., R. Gardner, and J. Walker, with A. Agrawal, W. Blomquist, E. Schlager, and S.Y. Tang (1994) *Rules, Games, and Common-Pool Resources*. Ann Arbor, MI: University of Michigan Press.
Ostrom, E., and V. Ostrom (2004) "The Quest for Meaning in Public Choice" *American Journal of Economics and Sociology*, 63, 105–147.
Ostrom, V., and E. Ostrom (1977) "Public Goods and Public Choices" in E.S. Savas, ed., *Alternatives for Delivering Public Services. Toward Improved Performance*. Boulder, CO: Westview, 7–49.
Ostrom, V., C.M. Tiebout, and R. Warren (1961) "The Organization of Government in Metropolitan Areas: A Theoretical Inquiry" *American Political Science Review* 55, 831–842.
Schlager, E., and M. Cox (2018) "The IAD Framework and the SES Framework: An Introduction and Assessment of the Ostrom Workshop Frameworks," in C.M. Weible and P.A. Sabatier, eds., *Theories of the Policy Process*, 4th ed. Boulder, CO: Westview, 215–252.
Weible, C.M. and P.A. Sabatier, eds. (2018) *Theories of the Policy Process*, 4th ed. Boulder, CO: Westview.

6
ANTICOMMONS THEORY

Michael Heller

Introduction

This chapter offers a concise overview of anticommons theory (see Heller 1998, 2008, 2011, collecting articles). The anticommons thesis states that when too many people own pieces of one thing, nobody can use it. Usually private ownership creates wealth. But too much ownership can have the opposite effect—it leads to wasteful underuse. This is a free-market paradox that shows up all across the global economy. If too many owners control a single resource, cooperation breaks down, wealth disappears, and everybody loses.

The field of anticommons studies is now well established, with thousands of scholars detailing examples across the innovation frontier, including drug patenting, telecom licensing, climate change, eminent domain, oil field unitization, music and art copyright, and post-socialist transition. Addressing anticommons tragedy is a key challenge for any legal system committed to innovation and economic growth.

Commons as a mirror

To understand the dilemma of resource underuse in an anticommons, it is helpful to start with overuse in a commons. Aristotle was among the first to note how shared ownership can lead to overuse: "That which is common to the greatest number has the least care bestowed upon it . . . each thinks chiefly of his own, hardly at all of the common interest; and only when he is himself concerned as an individual" (1996, 33).

Ecologist Garrett Hardin captured this dynamic well when he coined the phrase *tragedy of the commons* (1968, 1243–1244). In 1968 he wrote, "Ruin is the destination toward which all men rush, each pursuing his own best interest in a society that believes in the freedom of the commons. Freedom in a commons brings ruin to all" (p. 1244). Since Hardin wrote these lines, thousands have identified additional areas susceptible to overuse and commons tragedy.

In addition, Hardin's metaphor inspired a search for solutions. Most solutions revolve around two main approaches: regulation or privatization. Suppose a common lake is being overfished. Regulators can step in and decide who can fish, when, how much, and with what methods. By contract, privatizing the lake may create powerful personal incentives to conserve.

Harvest too many fish in your own lake today, starve tomorrow; invest wisely in the lake, profit forever. Extrapolating from such experience, legislators and voters reason—wrongly—that if some private property is a good thing, more must be better. In this view, privatization can never go too far.

The moral justifications for private ownership are controversial for philosophers, but as a practical matter, moving to private property often does prevent overuse in a commons. Harold Demsetz, author of the leading economic theory of ownership, argues that this "conservation effect" is the main reason private property emerges in, and provides a benefit to, society (1967, 354–359).

Until now, ownership, competition, and markets—the guts of modern capitalism—have been understood through the opposition suggested by Figure 6.1. Private property solves the tragedy of the commons. Privatization beats regulation. Market competition outperforms state control. Capitalism trounces socialism. But these simple oppositions mistake the visible forms of ownership for the whole spectrum. The assumption is fatally incomplete.

Privatizing a commons may cure the tragedy of wasteful overuse, but it may inadvertently spark the opposite. English lacks a term to denote wasteful underuse. To describe this type of fragmentation, I coined the phrase *tragedy of the anticommons* (Heller 1998, 667–669; see also Michelman 1982). The term covers any setting in which too many people can block each other from creating or using a scarce resource. Rightly understood, the opposite of overuse in a commons is underuse in an anticommons.

This concept makes visible the hidden half of our ownership spectrum, a world of social relations as complex and extensive as any we have previously known (see Figure 6.2). Beyond normal private property lies anticommons ownership. As one commentator notes, "To simplify a little, the tragedy of the commons tells us why things are likely to fall apart, and the tragedy of the anticommons helps explain why it is often so hard to get them back together" (Fennell 2004).

Though the anticommons concept refers at its core to fragmented ownership, the idea extends to fragmented decision-making more generally. Resource use often depends on the outcome of some regulatory process. If the regulatory drama involves too many uncoordinated actors—neighbors and advocacy groups; local, state, and federal legislators; agencies and courts—the sheer multiplicity of players may block use of the underlying resource.

Often, we think that governments need only to create clear property rights and then get out of the way. So long as rights are clear, owners can trade in markets, move resources to higher valued uses, and generate wealth. But clear rights and ordinary markets are not enough.

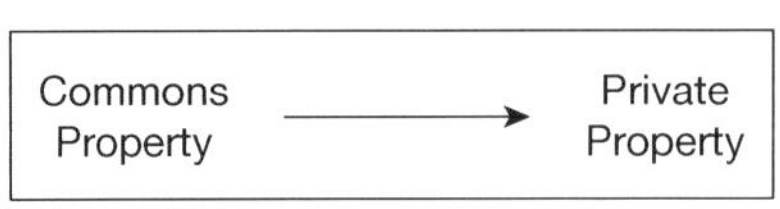

Figure 6.1 The standard solution to commons tragedy

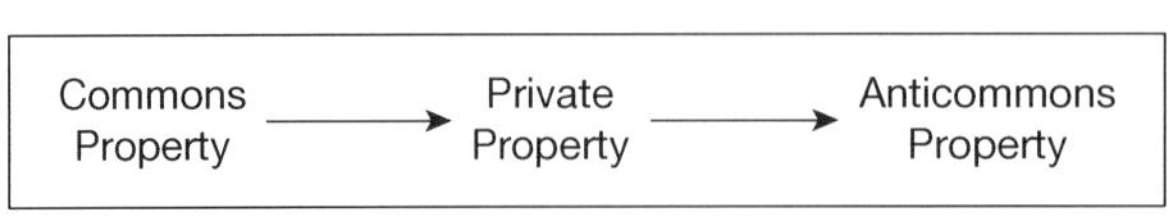

Figure 6.2 Revealing the hidden half of the ownership spectrum

The anticommons perspective shows that the *content* of property rights matters as much as the *clarity*. Wasteful underuse can arise when ownership rights and regulatory controls are too fragmented.

Making the tragedy of the anticommons visible upends our intuitions about private property. Private property no longer can be seen as the end point of ownership. Privatization can go too far, to the point where it destroys rather than creates wealth. Too many owners paralyze markets because everyone blocks everyone else. Well-functioning private property is a fragile balance poised between the extremes of overuse and underuse.

The new world of use

Adding the concepts of underuse and anticommons makes visible a new frontier for private bargaining, political debate, and wealth creation. But underuse and anticommons are still "squiggly"—until recently, my Word spellchecker rejected them by underlining each with red squiggles, and instead suggested *undersea* and *anticommunist* as replacements. These squiggles are a signal: the nonexistence of a word can be as telling as its presence. When we lack a term to describe some social condition, it is because the condition does not exist in most people's minds. So, it should be no surprise that we have overlooked the hidden costs of anticommons ownership. We cannot easily fix the problem until we have created a shared lexicon to spot tragedies of the anticommons.

Besides highlighting the language problem, the squiggles prompted me to look around the Internet at overuse and underuse. Googling *overuse* yielded about 10 million hits recently, while *underuse* generated under 1 million. (*Commons* had 800 million hits and *anticommons* had 25,000.) To me, the data immediately suggest two possibilities: either overuse is about ten times more important a social problem than underuse is, or we are only about 10% as aware of underuse as we should be. You will not be surprised that I believe the latter to be correct.

To understand the Google results, start with overuse. According to the *Oxford English Dictionary (OED), overuse* entered the language as a verb in the early 1600s. One of the earliest usages is as apt today as it was centuries ago: "When ever we overuse any lower good we abuse it." By 1862, the noun form was well recognized: "The oyster beds are becoming impoverished, partly by over-use."

Overuse continues to mean "to use too much" and "to injure by excessive force," definitions that have been stable for hundreds of years. Many of Google's top links for *overuse* come from medicine. Doctors diagnose "overuse syndrome" and dozens of "overuse injuries"—injuries from too much tennis, running, violin playing, book reading, whatever. So what is the opposite of overuse?

Ordinary use. The opposite of injuring yourself through too much use and excessive force is staying injury free by using an ordinary amount of force. Instead of abandoning an activity, do it in a reasonable, sustainable way. In medicine as in everyday language, the opposite of overuse is ordinary use (Figure 6.3). Since the 1600s, overuse and ordinary use have been an either-or proposition. Either you will feel pain in your elbow or you will be able to play happily, if not well. When you overuse a resource, bad things happen. It is much better to engage in ordinary use.

How do we achieve ordinary use? The usual solutions to tragedies of the commons are, as I have mentioned, privatization, cooperation, and regulation. These three solutions map onto

Figure 6.3 Ordinary use as the end point

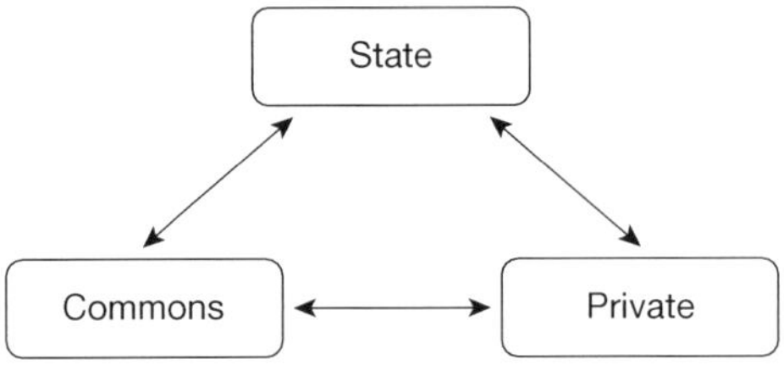

Figure 6.4 The trilogy of ownership

the traditional view that ownership can be organized into three basic types of property: private, commons, and state (Figure 6.4) (Heller 2001, 82–92).

We all have strong intuitions about private property, but the term is surprisingly hard to pin down. A good starting point is William Blackstone, the foundational eighteenth-century British legal theorist. His oft-quoted definition of *private property* is "that sole and despotic dominion which one man claims and exercises over the external things of the world, in total exclusion of the right of any other individual in the universe" (Blackstone 1765–1769). In this view, private property is about an individual decision maker who directs resource use.

Commons property refers to shared resources, resources for which there is no single decision maker. In turn, the commons can be divided into two distinct categories. The first is *open access*, a regime in which no one at all can be excluded, like anarchy in the parking lot or on the high seas. Mistakenly, the legal and the economics literature have long conflated the commons with open access, hence reinforcing the link between commons and tragedy. The second type of commons has many names, but for now let's call it *group access*, a regime in which a limited number of commoners can exclude outsiders but not each other. If the ocean is open access, then a small pond surrounded by a handful of landowners may be group access. Group access is often overlooked even though it is the predominant form of commons ownership and often not tragic at all (Eggertsson 2003, 74–85; see also Dagan and Heller 2001, advocating use of "liberal commons" to describe group access).

State property resembles private property in that there is a single decision maker, but it differs in that resource use is directed through some process that is, in principle, responsive to the needs of the public as a whole. In recent years, state property has become less central as a theoretical category: the Cold War is over; most socialist states have disappeared; intense state regulation of resources has dropped from favor; and privatization has accelerated. Today, for many observers, the property trilogy can be reduced to an opposition of private and commons property, what one scholar calls simply "all and none" (Figure 6.5) (Barzel 1989, 71).

I believe a substantial cause of our cultural blindness to the costs of fragmented ownership arises from the dominance of this too simple image of property. Note how the commons–private opposition tracks the overuse–ordinary use opposition. The former implies that there is nothing beyond private property; the latter suggests that we cannot overshoot ordinary use. Together, these oppositions reinforce the political and economic logic of the global push toward privatization. We assume, without reflection, that the solution to overuse in a commons is ordinary use in private ownership. This logic makes it difficult to imagine underuse dilemmas and impossible to see the uncharted world beyond private property.

Figure 6.5 The familiar split in ownership

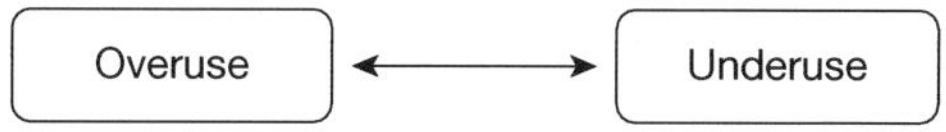

Figure 6.6 The new spectrum of use

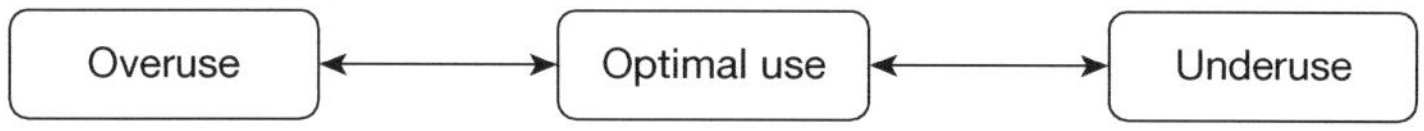

Figure 6.7 Goldilocks' quest for the optimum

According to the *OED, underuse* is a recent coinage. In its first recorded appearance, in 1960, the word was hedged about with an anxious hyphen and scare quotes: "There might, in some places, be a considerable 'under-use' of [parking] meters." By 1970, copyeditors felt sufficiently comfortable to cast aside the quotes: "A country can never recover by persistently under-using its resources, as Britain has done for too long." The hyphen began to disappear around 1975.

In the *OED*, this new word means "to use something below the optimum" and "insufficient use." The reference to an "optimum" suggests to me how *underuse* entered English. It was, I think, an unintended consequence of the increasing role of cost-benefit analysis in public policy debates. What happens when we slot underuse into the opposition in Figure 6.3? Although the result seems simple, it leads to conceptual turmoil (Figure 6.6).

In the old world of overuse versus ordinary use, our choices were binary and clear-cut: injury or health, waste or efficiency, bad or good. In the new world, we are looking for something subtler—an "optimum" along a continuum. Looking for an optimum level of use has a surprising twist: it requires a concept of underuse and surreptitiously changes the long-standing meaning of overuse. Like Goldilocks, we are looking for something not too hot, not too cold, not too much or too little—just right. Figure 6.7 suggests how underuse changes our quest.

How can we know whether we are underusing, overusing, or optimally using resources? It is not easy and not just a matter for economic analysis. Searching for an optimum between overuse and underuse sets us on the contested path of modern regulation of risk, an inquiry that starts with economic analysis but quickly implicates our core beliefs. We have to put dollar values on human lives and on the costs of overuse and underuse behaviors—a process filled with moral and political dilemmas. I note this difficult topic to show that finding the optimum requires the idea of *underuse* and that this new word in turn transforms the meaning of *overuse*. Overuse no longer just means using a resource more than an ordinary amount. The possibility of underuse reorients policy making from relatively simple either-or choices to the more contentious trade-offs that make up the modern regulation of risk.

The tragedy of the anticommons

Adding the idea of "underuse" sets the stage for the anticommons. Looking back at Figures 6.3–6.7, there is a gap in our labeling scheme. We have seen the complete spectrum of *use* but not the analogous spectrum of *ownership*. What form of ownership typically coincides with squiggly underuse? The force of symmetry helped reveal a hidden property form. Figure 6.8 shows my path to the anticommons.

I coined the term *tragedy of the anticommons* to help make visible the dilemma of too fragmented ownership beyond private property. Just as the idea of "underuse" transforms the

Figure 6.8 An ownership puzzle

Figure 6.9 The full spectrum of ownership

continuum of resource use, "anticommons" transforms the continuum of ownership. It shows that the move from commons to private can overshoot the mark (Figure 6.9). When privatization goes too far, resources can end up wasted in an unfamiliar way.

Seeing the full spectrum of ownership has another benefit. Because of our private-property focus, we tend to overlook cooperative solutions to overuse dilemmas. Cooperative solutions are often small-scale, context specific, local, and not reliant on legal structures—thus invisible. In *Governing the Commons*, Ostrom delineated the factors that communities around the world have used to manage group property without tragic outcomes (1990, 1–28). There are thousands of stories of successful cooperation that preserve contested resources and promote overall social welfare.

An understanding of commons ownership informed by Ostrom's contributions helps also to inform solutions to anticommons tragedy. To start, consider the distinction between open access (anarchy open to all) and group access (property that is commons to insiders and private to outsiders). This distinction can do some work on the anticommons side of the spectrum as well. The conventional wisdom has often overlooked group access, but we do not have to. Under the right conditions, groups of people succeed at conserving a commons resource without regulation or privatization (Ellickson 1991, 167–183). Cooperation can get us to optimal use. Under what conditions does cooperation work, and what does that teach us about fixing underuse dilemmas?

At the extreme of open access, group norms do not stick. For example, anyone can fish for tuna on the high seas. Tuna fleets work in relative isolation, and their catches can be sold anonymously to diverse buyers. Conservation norms, such as voluntary limits on fishing seasons, may gain little traction. Gossip and other low-cost forms of policing do not work for wide-ranging international fleets. Unless states intervene, overuse is hard to avoid. Whales were saved from extinction more through naval powers enforcing international treaties than through gossip at the harbor bar.

The state can sponsor hybrid solutions. What if the state asserted ownership over lobsters and fish, and then created private rights (such as licenses and tradable quotas) that complement cooperative solutions? Often, such hybrid regimes lead to fairer and higher-yielding results than informal group access can achieve. For example, in Australia, the government issues licenses for a sustainable number of lobster traps and enforces strict harvesting limits. Lobster catchers can wait to harvest until the lobsters mature, or they can sell their government-created rights, secure in their markets and property. With far less fishing effort, this system yields more and bigger lobsters than U.S. lobster catchers harvest either in coastal harbor gangs or on the open ocean (Acheson 1988; Tierney 2000).

Hybrid systems are the cutting edge of natural resource management: examples include not only tradable fishing quotas but also carbon-emission markets and transferable air-pollution permits (Rose 1999). These solutions can work far beyond lobsters and tuna, even beyond natural resources generally. They may reach the edge of high-tech innovation.

Solutions to commons property dilemmas give clues to solving anticommons tragedy. For *open access*, like the high seas, states must command resource use directly or create hybrid rights, such as fishing quotas. The anticommons parallel to open access is *full exclusion* in which an unlimited number of people may block each other. With full exclusion, states must expropriate fragmented rights or create hybrid property regimes so people can bundle their ownership. Otherwise, the resource will be wasted through underuse. There is, however, one important respect in which full exclusion differs from open access: an anticommons is often invisible. You have to spot the underused resource before you can respond to the dilemma.

Group access in a commons also has an anticommons parallel: *group exclusion* in which a limited number of owners can block each other. Recall the multiple owners of our magical parking lot. For both group access and group exclusion, the full array of market-based, cooperative, and regulatory solutions is available. Although self-regulation may be more complex for anticommons resources (Depoorter and Vanneste 2007), close-knit fragment owners can sometimes organize to overcome anticommons tragedy. For group exclusion resources, the regulatory focus should be support for markets to assemble ownership and removal of roadblocks to cooperation.

Group property on the commons or anticommons side of private ownership is exponentially more important than the rare extremes of open access or full exclusion. Much of the modern economy—corporations, partnerships, trusts, condominiums, and even marriages—can be understood as legally structured group property forms for resolving access and exclusion dilemmas (Dagan and Heller 2001, 552–554; Dagan and Frantz 2004). We live or die depending on how we manage group ownership. Now, we can see the full spectrum of property, as shown in Figure 6.10.

The economics of the anticommons

After I proposed the possibility of anticommons tragedy, economist and Nobel laureate James Buchanan and his colleague Yong Yoon undertook to create a formal economic model. They wrote that the anticommons concept helps explain "how and why potential economic value may disappear into the 'black hole' of resource underutilization" (2000, 2). According to their model, society gets the highest total value from a resource—say, the magical parking lot—when a single decision maker controls its use. As more people can *use* the lot independently, the value goes down—a tragedy of the commons. And as more people can *block* each other from the lot, the value also goes down symmetrically—a tragedy of the anticommons. Figure 6.11 shows their graphic summarizing this finding.

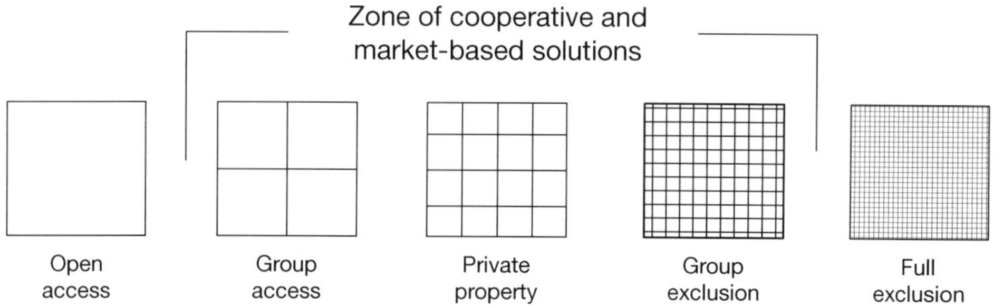

Figure 6.10 The full spectrum of property, revealed

Source: Adapted from Heller (1999, 1194–1198).

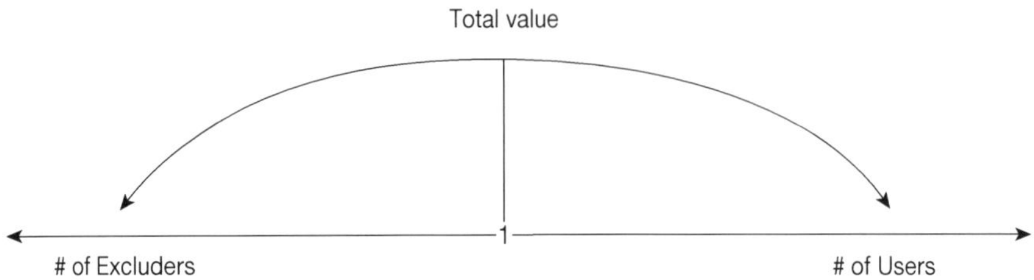

Figure 6.11 Value symmetry in an anticommons and a commons
Source: Buchanan and Yoon (2000, 8).

After developing their proof and showing how the anticommons construct may apply to a wide range of problems, Buchanan and Yoon concluded, "the anticommons construction offers an analytical means of isolating a central feature of sometimes disparate institutional structures . . . [People] have perhaps concentrated too much attention on the commons side of the ledger to the relative neglect of the anticommons side" (2000, 12).

In recent years, economic models of the anticommons, including game theory approaches, have become quite sophisticated (Parisi 2002; Parisi et al. 2003; see also Fennell 2004, modeling anticommons as a chicken game). At the simplest level, anticommons theory can be understood as a legal twist on the economics of "complements" first described by Antoine-Augustin Cournot in his 1838 *Researches into the Mathematical Principles of the Theory of Wealth* (and discovered independently by Charles Ellet in 1839 in his work on railway tariffs). Anticommons theory is a partial corrective for modern economic models that focus on "substitutes" and often neglect the role of "complements" (Varian et al. 2005, 43–45; see also Dari-Mattiacci and Parisi 2006, modeling interaction of substitutes and complements in the anticommons). What's the difference?

In Figure 6.12, Railways A, B, and C are substitute ways to get from here to there. Say the fare is 9. If railway A finds a way to provide the service for 8, it will win riders. B and C must become more efficient to keep up. In markets with robust substitutes, competitors have incentives to innovate, lower prices, and thereby indirectly benefit society as a whole. By contrast, Railways D, E, and F are complements. When inputs are complementary, generally you want all or none of them.

Again, assume the fare from here to there is 9. D, E, and F each charge 3. Railway D knows that if you want to ride, you must buy its ticket. So why innovate? Instead, D may *raise* its fare

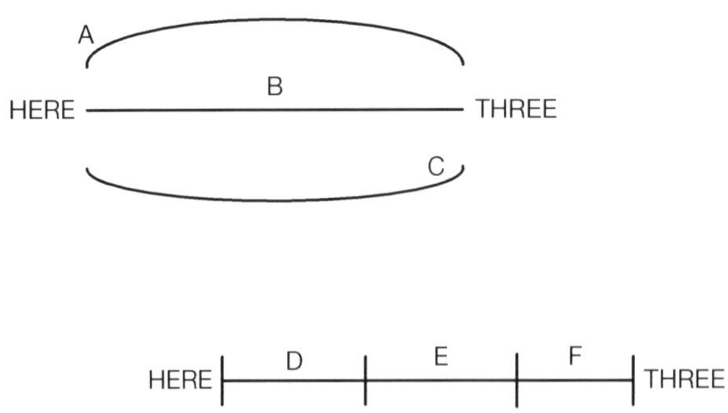

Figure 6.12 Substitutes versus complements

to 5, hoping that E and F lower theirs to 2 each. But why would E and F go along? More likely, they too would raise fares, so the total exceeds 9, and ridership falls below the optimal level. With complementary competition, incentives to innovate are blunted: if D did lower fares, then E and F just might raise theirs.

It is the same problem if D, E, and F are complementary patents instead of railways. Then innovators face what economist Carl Shapiro calls a "patent thicket," many phantom tollbooths on the route to commercializing new technology (Shapiro 2001, 119; cf. Burk and Lemley 2009, 75–78, 86–92, distinguishing anticommons and patent thickets in the patent context). Cournot proved that in markets dominated by complements—whether railways or patents—we can get higher overall social welfare if D, E, and F merge. Here, monopoly trumps competition. Anticommons theory, in turn, moves from railways and patents to ownership and regulation generally. All these concepts describe facets of the same dilemma: too many uncoordinated owners or regulators blocking optimal use of a single resource.

Anticommons puzzles

There are a near infinity of everyday puzzles that share this common structure—one whose solution could jump-start innovation, release trillions in productivity, and help revive the global economy.

To date, the most debated application of anticommons theory has built on the drug patent example. Numerous potential drugs may be lost to anticommons ownership. Since the late 1980s, drug R&D has been going steadily up, but discoveries of major new classes of drugs have been declining. Instead, drug companies focus on minor spinoffs of existing drugs for which they have already assembled the necessary property rights. How did this new drug discovery gap happen? Patent anticommons. Paradoxically, more biotech patent owners can mean fewer life-saving innovations. Drugs that should exist, could exist, are not being created.

The field was sparked in part by my article with Rebecca Eisenberg (1998) on the anticommons in biomedical research, an article that has since been cited several thousand times (Heller 2008, 49–78). There has been a flurry of follow-on papers, studies, and reports, many of which conclude that patents should be harder to obtain, in part to avoid potential anticommons tragedy effects. The 2011 reforms to the Patent Act, a string of Supreme Court patent decisions, and a number of pending bills before Congress all aim to ameliorate potential anticommons tragedies (Barnett 2015, 3)

In their recent book, *The Patent Crisis*, Dan Burk and Mark Lemley review recent literature on patents and biotech innovation and conclude that, "the structure of the biotechnology industry seems likely to run high anticommons risks," particularly when companies are attempting to bring products to market (2009, 89). However, the empirical basis for finding that anticommons effects are stifling innovation remains inconclusive. Scholars such as Jonathan Barnett are skeptical (2015; Burk and Lemley 2009, 5). This remains an important area for empirical work.

In addition, the anticommons framework has been used to analyze ownership across the high-tech frontier, ranging from Thomas Hazlett's work on broadcast spectrum ownership, what he calls the "tragedy of the telecommons," to Rosemarie Ham Ziedonis's study of anticommons effects in technology patenting (Hazlett 2005; Heller 2008, 79–106; Ziedonis 2004). In my *Gridlock Economy* book, I show that it is not just biotech that is susceptible to anticommons tragedy. Cutting-edge art and music are about mashing up and remixing many separately owned bits of culture (Heller 2008, 9–16, 27–30; Parisi and Depoorter 2003). And, as I said earlier, even with land the most socially important projects often require assembling multiple parcels (Heller 2008, 107–142).

Anticommons theory is now well established, but empirical studies have yet to catch up. How hard is it to negotiate around ownership fragmentation? How much does ownership fragmentation slow down technological innovation? Does the effect vary by industry? (Barnett 2015, collecting empirical studies). It is difficult to measure discoveries that should have been made but were not, industries that could exist but do not. We are just starting to examine these conundrums.

In 2006, a team of law, economics, and psychology researchers reported experimental findings that rejected the presumed symmetry of commons and anticommons. They found that anticommons dilemmas "seem to elicit more individualistic behavior than commons dilemmas" and are "more prone to underuse than commons dilemmas are to overuse." The researchers conclude that "if commons leads to 'tragedy,' anticommons may well lead to 'disaster'" (Vanneste et al. 2006; Depoorter and Vanneste 2007, 21–23).

These preliminary findings of bargaining failure around anticommons ownership may help provide some insight into otherwise puzzling economic phenomena. For example, some of the world's biggest energy companies have for years failed to agree on joint management of oil and gas fields they own together. If one company pumps the oil too fast, it can wreck the pressure in the gas field; if the other extracts gas too fast, it traps the oil. U.S. law has offered them an effective legal tool, called "unitization," to overcome anticommons tragedy and smooth joint management of divided oil and gas interests. Yet firms block each other year after year (Libecap and Smith 2002; see also Thompson 2000, 241–250, discussing groundwater pumping).

How can this be? Oil units are not a case of two spiteful neighbors arguing over a broken backyard fence. They involve arm's-length business negotiations between savvy corporations. Everyone has good information about the underlying geologic and technical issues. The gains from cooperation are in the billions of dollars. Why doesn't one firm sell its interest to the other? Why don't the firms merge? What's going on? The experimental studies are beginning to give us explanations rooted in the psychology of the anticommons. Even the most sophisticated businesspeople can fail to reach agreement when a negotiation is framed in anticommons terms.

Rounding out the lexicon

In rounding out the anticommons lexicon, there is a caveat and a caution. First, the caveat. My term focuses on one form of underuse, the tragedy that arises when ownership is too fragmented. Here, *multiple* owners block each other from using a scarce resource. Underuse can also arise in the monopoly context when a *single* owner blocks access to a resource. This situation may be tragic, but it is not an anticommons tragedy in my sense of the term.

In the old economy, many companies held monopolies: Ma Bell (the U.S. telephone monopoly), railways, local water utilities, and others. Society gained the economic benefits of scale and scope from allowing these sectors to be monopolized. The state policed against abuse of monopoly power through complex rate regulation and oversight. Phone lines were cheaper and more available than in many other countries. The costs of these monopolies were often invisible, costs such as deferred and dampened innovation.

In an information economy, any piece of intangible property, such as a patent, is also a monopoly. We award patents because monopoly profits create incentives to invent and because patents give inventors incentives to disclose their discoveries (without patents people might prefer to invent things they could keep secret). On the other hand, drugs would be cheaper and lives could be saved if competitors could make generic copies at will. To balance the values of innovation, disclosure, and competition, the U.S. Congress keeps shifting the bundle of rights that a patent confers.

The dilemmas of any individual monopoly in the old or new economy are a great topic but are not relevant here. For better or worse, these quandaries are familiar. However, we do not have much experience dealing with the interaction of ownership fragments or an array of blocking patents. The anticommons lexicon addresses not monopoly per se but ownership multiplicity.

Next, a caution: when talking about an anticommons, stay away from absolutes. First, you should not assume that anticommons ownership is inevitably tragic (Heller 1998, 676). If we lived in a world where people had perfect information and could bargain with each other at no cost, they could avoid anticommons tragedy every time (just as, in a perfect world, there would be no commons tragedy, or for that matter, tragedy of any sort) (Coase 1960). In practice, however, bargaining is never free: people shirk duties and hold out for payoffs, and there are cognitive limits that shape owners' decisions. In the real world, anticommons ownership is not necessarily tragic, but it does tend that way.

Second, it is theoretically possible that an anticommons may face overuse instead of underuse (Fennell 2004, 934–937). For example, consider real estate development along the California coast. It is a mess. Multiple community groups, environmentalists, neighbors, and government agencies may each prefer different versions of a project. Even in that regulatory morass, though, *overbuilding* may occur if it is sufficiently costly to exercise each right to veto development. Every opponent of development may prefer to go surfing and hope the others sit through the boring public hearings. If enough people opt for a free ride, a project might face *too little* opposition, not too much. It is an empirical question whether the California coast tips toward over- or underbuilding. That said, in most cases I have seen, anticommons regulation tends to be associated with too little economic development, not too much (Buzbee 2005).

Finally, there is a caveat that comes from the legal theorist Carol Rose. Certain resources, such as roads and waterways, are sometimes owned most efficiently in common. As Rose points out, creating and enforcing private-property rights is itself costly; sometimes these costs exceed the gains, not just economically but also socially. Village greens and town halls may strengthen communities in ways that are socially valuable but hard to quantify in monetary terms. Rose (1986) coined the phrase "the comedy of the commons" to describe such social and economic benefits that can flow from group access (see also Ellickson 1993, 1336–1338).

Rose's insight is equally true on the anticommons side: there are both economic and social reasons why we may prefer group exclusion to sole ownership. For example, it's possible that creating multiple vetoes may help preserve a treasured resource against transient political pressures for development: for instance, Central Park in New York City or Indian burial grounds in Arizona (Bell and Parchomovsky 2003, 3–4, 31–36, 60–61). Similarly, "conservation easements" intentionally use anticommons ownership to foster environmental goals (Mahoney 2002, 739–785). (With a conservation easement, the owner sells or gives away the right to develop land, gets a tax break, and retains the right to continue a current use such as farming.) The underuse created by split ownership may be justifiable if the environmental gains exceed fragmentation costs. On balance, though, I am skeptical. What happens a generation from now when communities want to reduce sprawl but face a patchwork of easements that make "in-fill" development prohibitively difficult? Many conservation easements look to me like potential anticommons tragedies (see also Bell and Parchomovsky 2003: 785–86).

The "comedy of the anticommons" insight suggests that sometimes, for some resources, we should promote little or no use. Most of the time, for most resources, however, some positive level of use will be socially most valuable. Underuse is rarely the optimum.

Toward a non-squiggly language

We have millennia of practice in spotting tragedies of the commons. When too many people fish, fisheries are depleted. When too many people pollute, we choke on dirty air. Then, we spring into action with market-based, cooperative, and legislative solutions. Similarly, we have a lot of experience spotting underuse caused by a particular monopoly owner. We have created regulatory bodies that know (more or less) what to do with such dilemmas.

But underuse caused by multiple owners is unfamiliar. There has been an unnoticed revolution in how we create wealth. In the old economy, during the 1980s and 1990s, you invented a product and got a patent; you wrote a song and got a copyright; you subdivided land and built houses. Today, the leading edge of wealth creation requires assembly. From drugs to telecom, software to semiconductors, anything high-tech demands the assembly of innumerable patents. And it is not just high-tech that has changed—today, cutting-edge art and music are about mashing up and remixing many separately owned bits of culture. Even with land, the most socially important projects, like new runways, require assembling multiple parcels. Innovation has moved on, but we are stuck with old-style ownership that is easy to fragment and hard to put together.

A tragedy of the anticommons may be as costly to society as the more familiar forms of resource misuse, but we have never noticed, named, debated, or learned how to fix underuse. How do we stumble into the problem of too many owners? How do we get out? As a first step, underuse in a tragedy of the anticommons should be squiggly no more.

Fixing anticommons tragedies is a key challenge for our time. Some solutions are entrepreneurial; for example, people can profit from finding creative ways to bundle ownership. Philanthropists can assemble patents for disease cures. Political advocacy and legal reform will be needed to secure solutions. But, I want to stress that the first and most important step to solving an anticommons is to name it and make it visible. With the right language, we can all spot links among ownership puzzles, and we can all come together to fix them. Nothing is inevitable about an anticommons. Every ownership puzzle results from choices we make (and thus can change) about how to control the resources we value most.

References

Acheson J. (1988) *Lobster Gangs of Maine*. Lebanon, NH: University Press New England.

Aristotle (1996) *The Politics and the Constitution of Athens*. Everson S ed., Jowett B trans. Cambridge, UK: Cambridge University Press.

Barnett J. (2015) "The Anti-Commons Revisited" *Harvard Journal of Law & Technology* 1: 16–57.

Barzel Y. (1989) *Economic Analysis of Property Rights*. Cambridge, UK: Cambridge University Press.

Bell A. and Parchomovsky G. (2003) "Of Property and Antiproperty" *Michigan Law Review* 102: 1–61.

Blackstone W. (1765–1769) *Commentaries on the Laws of England: In Four Books*. Bk. 2. Oxford, UK: Clarendon Press.

Buchanan S. and Yoon Y. J. (2000) "Symmetric Tragedies: Commons and Anticommons" *Journal of Law and Economics* 43: 1–12.

Burk D. L. and Lemley M. A. (2009) *The Patent Crisis and How the Courts Can Solve It*. Chicago, IL: University of Chicago Press.

Buzbee W. W. (2005) "The Regulatory Fragmentation Continuum, Westway, and the Challenges of Regional Growth" *Journal of Law and Politics* 21: 323–364.

Coase R. H. (1960) "The Problem of Social Cost" *Journal of Law and Economics* 1: 1–44.

Cournot A. (1897 [1838]) *Researches into the Mathematical Principles of the Theory of Wealth*. Bacon N. T. trans. London: Macmillan Co.

Dagan H. and Frantz C. J. (2004) "Properties of Marriage" *Columbia Law Review* 104: 75–133.

Dagan H. and Heller M. A. (2001) "The Liberal Commons" *Yale Law Journal* 110: 549–623.
Dari-Mattiacci G. and Parisi F. (2006) "Substituting Complements" *Journal of Comparative Law and Economics* 2: 333–347.
Demsetz H. (1967) "Toward a Theory of Property Rights" *American Economic Review, Papers and Proceedings* 57: 347–359
Depoorter B. and Vanneste S. (2007) "Putting Humpty Dumpty Back Together: Pricing in Anticommons Property Arrangements" *Journal of Law, Economics, and Policy* 3: 1–28.
Eggertsson T. (2003) "Open Access versus Common Property" in Anderson T. and McChesney F. eds, *Property Rights: Cooperation, Conflict, and Law*. Princeton, NJ: Princeton University Press, 74–75.
Ellet C. Jr. (1839) *An Essay on the Laws of Trade in Reference to the Works of Internal Improvement in the United States*. Richmond, UK: P. D. Bernard.
Ellickson R. C. (1991) *Order without Law: How Neighbors Settle Disputes*. Cambridge, MA: Harvard University Press.
Ellickson R. C. (1993) "Property in Land" *Yale Law Journal* 102: 1315–1338.
Fennell L. A. (2004) "Common Interest Tragedies" *Northwestern University Law Review* 98: 907–937.
Hardin G. (1968) "The Tragedy of the Commons" *Science* 162: 1243–1244.
Hazlett T. W. (2005) "Spectrum Tragedies" *Yale Journal of Regulation* 22: 242–272.
Heller M. A. (1998) "The Tragedy of the Anticommons: Property in the Transition from Marx to Markets" *Harvard Law Review* 111: 621–688.
Heller M. A. (1999) "The Boundaries of Private Property" *Yale Law Journal* 108: 1163–1223.
Heller M. A. (2001) "The Dynamic Analytics of Property Law" *Theoretical Inquiries in Law* 2: 79–92.
Heller M. A. (2008) *The Gridlock Economy: How Too Much Ownership Wrecks Markets, Stops Innovation, and Costs Lives*. New York: Basic Books.
Heller M. A. ed. (2011) *Commons and Anticommons*. Cheltenham, UK: Elgar Publishing.
Heller M. A. and Eisenberg R. S. (1998) "Can Patents Deter Innovation? The Anticommons in Biomedical Research" *Science* 280: 698–701.
Libecap G. D. and Smith J. L. (2002) "The Economic Evolution of Petroleum Property Rights in the United States" *Journal of Legal Studies* 31: S589–S608.
Mahoney J. D. (2002) "Perpetual Restrictions on Land and the Problem of the Future" *Virginia Law Review* 88: 739–786.
Michelman F. J. (1982) "Ethics, Economics and the Law of Property" *Nomos* 24(3): 667–669.
Ostrom E. (1990) *Governing the Commons: The Evolution of Institutions for Collective Action*. Cambridge, UK: Cambridge University Press.
Parisi F. (2002) "Entropy in Property" *American Journal of Comparative Law* 50: 595–632.
Parisi F. and Depoorter B. (2003) "Fair Use and Copyright Protection: A Price Theory Explanation" *International Review of Law and Economics* 21: 453–473.
Parisi F., Schulz N., and Depoorter B. (2003) "Fragmentation in Property: Towards a General Model" *Journal of Institutional and Theoretical Economics* 159: 594–613.
Rose C. M. (1986) "The Comedy of the Commons: Custom, Commerce, and Inherently Public Property" *University of Chicago Law Review* 53: 711–781.
Rose C. M. (1999) "Expanding the Choices for the Global Commons: Comparing Newfangled Tradable Allowance Schemes to Old-Fashioned Common Property Regimes" *Duke Environmental Law and Policy Forum* 10: 45–72.
Shapiro C. (2001) "Navigating the Patent Thicket: Cross Licenses, Patent Pools, and Standard Setting" in Jaffe A. et al. eds, *Innovation Policy and the Economy*. Cambridge, MA: MIT Press.
Thompson B. H. Jr. (2000) "Tragically Difficult: The Obstacles to Governing the Commons" *Environmental Lawyer* 30: 241–250.
Tierney J. (2000) "A Tale of Two Fisheries" *New York Times Magazine*, August 27.
Vanneste S. et al. (2006) "From 'Tragedy' to 'Disaster': Welfare Effects of Commons and Anticommons Dilemmas" *International Review of Law and Economics* 26: 104–122.
Varian H. R., Farrell J. V. and Shapiro C. (2005) *The Economics of Information Technology: An Introduction*. Cambridge, UK: Cambridge University Press.
Ziedonis R. H. (2004) "Don't Fence Me In: Fragmented Markets for Technology and the Patent Acquisition Strategies of Firms" *Management Science* 50: 804–820.

7
KNOWLEDGE COMMONS

Michael J. Madison, Brett M. Frischmann and Katherine J. Strandburg

Introduction

This chapter provides an introduction to and overview of the knowledge commons research framework. Knowledge commons refers to an institutional approach (commons) to governing the production, use, management, and/or preservation of a particular type of resource (knowledge). The research framework supplies a template for interrogating the details of knowledge commons institutions on a case study basis, generating qualitative data that may be used to support comparative analysis.

The framework was introduced to the literature by Madison et al. (2010) as a framework for researching 'constructed cultural commons,' a shorthand for shared resources composed primarily of products of the human mind, namely knowledge and information in scientific domains, domains related to arts and culture, and resource domains defined largely by their human-generated character and their intangibility. That shorthand has been refined in later work into the phrase 'knowledge commons' (Frischmann et al. 2014; Strandburg et al. 2017), aligning the focus of the research with an earlier introduction by Ostrom and Hess (2007) to themes raised by 'knowledge commons.' Other research on related topics has characterized such resources as 'information commons' (Bollier and Helfrich 2012), 'intellectual commons' (Boyle 2003), 'cultural commons' (Bertacchini et al. 2012), 'socio-technical' commons (de Moor 2015; Hinkel et al. 2015), and 'new commons' (Hess 2008). Subdomains of research in these domains include scholarship on 'data commons' (Borgman 2015), 'research commons' (Reichman and Uhlir 2003; Reichman et al. 2016), 'spectrum commons' (Weiser and Hatfield 2005), and 'scientific commons' (Nelson 2004). The word 'commons' has been applied for rhetorical effect to a variety of ideological and institutional forms, such as the 'Creative Commons' licensing enterprise.[1] Those rhetorical applications of 'commons' terminology are not addressed here.

For the purposes of this chapter, and to render the research framework inclusive and meaningful in a broad, comparative sense, the term knowledge refers to all of the domains identified in the previous paragraph and therefore to a broad set of intellectual, scientific, technical, and cultural resources. Differences among the domains and among the resources within them may be significant. Knowledge, information, and data may be different from each other in meaningful ways. The research framework described here is sufficiently flexible to permit researchers to capture both their commonalities and their differences in their respective ecological contexts.

For similar reasons related to inclusiveness, potentially narrower definitions of relevant goods are avoided. Neither this chapter nor the framework limit its approach to precise distinctions among private goods, public goods, club goods, and/or toll goods.

Commons refers to a form of community management or governance of a shared resource. Governance involves a group or community of people who share access to and/or use of the resource and who manage their behavior via an established set of formal and information rules and norms. 'The basic characteristic that distinguishes commons from noncommons is institutionalized sharing of resources among members of a community' (Madison et al. 2010, 841). Commons does not denote the resource, the community, a place, or a thing. Commons is the institutional arrangement of these elements and their coordination via combinations of law and other formal rules and social norms, customs, and informal discipline. Community or collective self-governance of the resource by individuals who collaborate or coordinate among themselves effectively is a key feature of commons as an institution. Self-governance may be and often is linked to other formal and informal governance mechanisms. Technological and other material constraints may play important roles in the constitution and governance of knowledge commons.

Knowledge commons thus refers to the institutionalized community governance of the creation, sharing, and preservation of a wide range of intellectual and cultural resources. This chapter briefly describes the motivations for creation of the knowledge commons research framework. It reviews the content of the framework, indicating where appropriate how the framework builds on, but is distinguishable from, its most important antecedent, Ostrom's Institutional Analysis and Development (IAD) framework. This chapter briefly summarizes tentative conclusions from case study research completed to date and indicates important directions for future research.

Motivations

The knowledge commons research framework owes its origins in part to the insight that knowledge, in its broadest and most general form, suggests social dilemmas that can be addressed by commons governance (Ostrom and Hess 2007). In an important early work, Hess (2008) suggested the extension of Ostrom's commons research to specific institutional settings beyond natural resource domains that she denominated so-called 'new' commons. These included knowledge-related domains. However, these early efforts did not advance beyond the suggestion that perspectives on commons, based on Ostrom's work, might be useful in knowledge and information contexts.

From a different beginning but at approximately the same time, legal scholars studying the Internet questioned systems of exclusive intellectual property (IP) rights (patent and copyright in particular) as appropriate comprehensive legal and governance systems for cultural and scientific resources, particularly digital resources circulating in networks. Information and knowledge sharing as a modality of production appeared to be especially salient, exemplified by open source software production and Wikipedia. Because of the apparent absence of firm-style hierarchy in these contexts, this perspective on governance—drawing a contrast between intellectual production by firms in markets, relying on exclusive rights, and intellectual production by actors organized non-hierarchically, and not relying on exclusivity—was crystallized by Benkler (2004, 2006) in the phrase 'commons-based peer production.' However, attention to commons-based peer production did not advance beyond the conceptual modeling of a handful of salient institutions.

Madison, Frischmann, and Strandburg (2010) initiated the knowledge commons research framework in part by building on the instinct to critique IP systems that motivated Benkler.

The problem of providing empirical justifications for IP law regimes has vexed modern researchers for at least 50 years (Machlup 1958). Intellectual property products such as patentable inventions and copyrighted works are conventionally modeled as or assumed to be public goods, which are neither excludable nor depletable, in contrast to private goods, which are both. That assumption leads to framing production of knowledge and information resources as a classic tragedy of the commons. It is typically assumed by researchers that likely over-consumption of intellectual property goods by uncoordinated, self-interested consumers and users in some 'common' domain, i.e., free riders, will lead to diminishment of the incentive to invest in their production, because the inventors and creators of those goods will be discouraged by their inability to recover necessary investments in and returns on innovation (Hardin 1968; Landes and Posner 2003). That diagnosis leads to the conclusion that such a tragedy may be solved, if at all, by the creation and assignment of marketable property rights to individual owners of specific resources within what was formerly held 'in common,' or by production and governance of the resource by a state-based authority.

The theoretical assumptions behind the conventional tragedy of the commons model have long been belied by the fact that knowledge-generating institutions based on successful coordination and collaboration among knowledge producers and users have existed for centuries, often despite the absence of IP rights owned by individual creators or inventors. In colloquial terms, knowledge and information resources are naturally shareable because of their intangibility as a conceptual matter. Institutions founded on precepts of knowledge sharing are as well known as those founded on precepts of exclusivity of right. Universities have long served as knowledge-generating and knowledge-sustaining institutions despite faculty researchers often exercising few conventional market-based IP interests (Madison et al. 2009). In short, openly accessible knowledge resources do not necessarily pose dilemmas leading to so-called 'tragic' outcomes. Social dilemmas involving knowledge need not be solved by the delineation of private rights exchanged in markets, nor left to strong coordination by the state.

The knowledge commons research framework marries that insight to one of the key lessons of Ostrom's work: the fact that commons research is and should be empirical, rather than simply conceptual (Frischmann 2013). It advances via the premise that details of governance of knowledge-sharing institutions matter as much as standard theoretical templates do. The knowledge commons research framework fleshes out that premise by borrowing the empirical orientation of Ostrom's research and by borrowing specifically from the IAD framework, as a well-tested approach to studying commons governance other than in knowledge and information settings.

The knowledge commons research framework

As research subjects, cases of knowledge and information sharing are typically characterized by three features: (i) the production of knowledge resources (especially innovative and creative things) via one or more modes of collective action, (ii) institutions and other formal and informal structures for sharing information and knowledge resources, and (iii) governance processes that depend significantly on openness (open access to resources and/or open participation by creators and innovators) or that are characterized by a relative absence of the conventional exercise of formal IP rights. The knowledge commons research framework described here unifies these cases under the heading 'commons' in order to enable research to proceed in a systematic way across different specific cases. Research in related domains has also developed under other headings, at times using other conceptual frameworks, at times advancing more via theory than via empirical investigation, and at times not organized in a systematic way. As noted earlier, 'commons-based peer production' is one such category. 'The public domain' in intellectual property law is a second (Boyle 2008). 'Open innovation' (Chesbrough 2003), 'User Innovation' (von Hippel 2005), and 'Intellectual Production without Intellectual Property' ('IP without IP') (Perzanowski and Darling 2017) is a third.

To date, the fragmentation of these fields, both across disciplines (law, management, and economics, for example) and within them (copyright and patent, for example) has placed certain limits on their effectiveness from an empirical standpoint. There has been little systematic effort to identify the virtues and drawbacks of collective action in the knowledge production context from an empirical standpoint, and there has been no comprehensive intellectual framework for doing so. The knowledge commons research framework guides that effort toward systemization.

Ostrom and researchers working in her tradition collected extensive evidence of commons governance used by a wide variety of communities to manage many different types of resources, particularly natural resources, referred to collectively in this body of work as 'common-pool resources' (CPRs). CPRs are distinguished by the twin propositions that the resource is depletable or rival, and that it is difficult if not impossible to exclude users from consuming the resource. In CPR contexts and elsewhere, establishing and sustaining community-based governance confronts various obstacles to sharing and cooperation. Some of those obstacles derive from the nature of the resources and others derive from other factors, such as the nature of the community or external influences. Ostrom was awarded the Nobel Prize in Economic Sciences in 2009 for her pioneering research demonstrating that self-governed communities can and often do overcome obstacles through purpose-built or constructed commons as well as emergent commons. Her co-Nobelist, Oliver Williamson, is also celebrated for his work on economic institutions, confirming that the recognition of Ostrom is noteworthy not only for her advocacy of commons governance as such but also for her contributions to the field of comparative institutional analysis.

The knowledge commons research framework draws on Ostrom's comparative institutionalism as well as on her research on natural resource commons. The empirical challenge addressed by the framework is constructing and applying a technique that accomplishes three goals. One goal is to capture appropriately the range of potential knowledge commons cases. A second goal is to respect the tradition of open and inclusive inquiry that Ostrom established. A third goal is to distinguish Ostrom's work when appropriate. Knowledge and information resources typically do not exhibit the characteristics of CPRs, for example. The fact that they are believed to exist primarily in intangible form(s) means that knowledge resources are not necessarily depletable.

Some examples illustrate the variety of institutional arrangements and resources that are governed via knowledge commons. Most obvious may be scientific and research commons, given the importance of sharing and collaboration norms within scientific communities (Merton 1973). Reichman and Uhlir (2003) examined scientific data commons, pressures on the 'sharing ethos' within various scientific communities, and institutional means for reconstructing commons. Cook-Deegan and Dedeurwaerdere (2006) mapped out relationships between the structure and function of resource commons in the life sciences. Sharing the products of genomics research has been a fruitful area of knowledge commons research (Contreras 2010, 2011; Contreras and Reichman 2015). Reichman et al. (2016) dive deeply into the laws, regulations, and practices governing the scientific communities that they collect under the heading 'microbial research commons.' Strandburg et al. (2017) collect qualitative case studies of knowledge commons governance in medical and public health research. Borgman (2015) synthesizes a large number of small qualitative studies of research projects relying on data sets under a commons rubric. Open source software development communities have been examined as knowledge commons (Schweik and English 2012).

Madison et al. (2010) discuss the following less obvious examples, some of which are discussed in the research literature mentioned earlier: intellectual property pools, in which owners of patents in a technological domain license their patents to a common 'pool' from which producers of complex products can obtain all of the permissions needed to make and sell goods that use the patents (Merges 1996; Shapiro 2000); Wikipedia, which offers users of this Internet encyclopedia the power to add to and edit its contents (Hoffman and Mehra 2009); wire services for journalism, which allow individual member media outlets the opportunity to publish work produced by other members; and 'jamband' fan communities, which record, share, and comment on musical performances of their favorite groups, with the permission of the artists themselves (Schultz 2006).

The primary challenge of qualitative case studies such as these has been how to approach the process of thick description of each case in a systematic way. Some of the research just described has been conducted using the knowledge commons research framework, but much of it, and in particular older research, has not. Prospectively, such a framework offers obvious virtues. A standardized yet flexible research framework permits research and data collection to proceed under a common set of assumptions and questions, even if specific research methods and disciplinary foundations may vary from researcher to researcher or field to field. In cases of knowledge and information commons, researchers in social sciences, the humanities, and multiple professional fields (law, management) are interested in knowledge commons problems. That diversity of background means that the framework should be usable by researchers in each of those fields. It also means that the framework is neither theory nor model, as an initial matter; it does not, in itself, provide a taxonomy of governance institutions or suggest specific hypotheses to be tested. (For related work categorizing such institutions that build on collective or communal self-governance, see Dusollier 2010; Van Overwalle 2009.) It is sufficiently detailed to enable data collection that supports comparative analysis and permits hypotheses to be generated, but it is not so detailed or prescriptive that researchers are guided to overlook important attributes of cases or to exclude possibly relevant fields of further research.

The knowledge commons research framework builds on the IAD framework pioneered by Ostrom and her colleagues (Ostrom 2005). The IAD framework has been used principally to structure analysis of solutions to collective action problems in natural resource contexts (so-called action arenas, or action situations, in which commons participants resolve social dilemmas by applying formal and informal 'rules-in-use'), such as forests, fisheries, and irrigation systems. Empirical research has largely confirmed Ostrom's hypothesis that successful commons governance involves attention to a handful of 'design principles' (Ostrom 1990, 90) that focus on clarity of the attributes of the resource, identity and membership of the relevant community, and the community's adoption and application of proportionate disciplinary rules on resource consumption.

IAD analysis is premised largely on choice-processing and goal-oriented behavior by self-interested individuals. It looks to explain sustainable collective action relative to CPRs, which produces measurable, productive results. The insight from applying the IAD framework to a large number of governance institutions and resources is that commons solutions can be as stable and robust as market-oriented solutions to classic 'tragedy of the commons' overconsumption dilemmas involving depletable natural resources. Shared governance of these CPRs by community members can lead to sustainable fisheries and forests and to regular supplies of usable water.

The knowledge commons framework differs from the IAD framework in certain key respects. First, it does not assume the agency of choice-selecting, self-interested individuals, as the IAD framework tends to do. Second, it accepts the role of historical contingency in the creation and maintenance of commons governance and of both inward-directed (selfish) and outward-directed (selfless, pro-social, or other-oriented) agents in the evolution of collective or commons institutions. Third, at the front end of the analysis, it requires understanding the contingency and human-created character of the underlying resources themselves. The resources subject to natural resource commons are largely provided by biophysical processes: fish, trees, water, and the like. Natural or physical forces are variable, and human agency is often part of producing these resources, but natural resources have an unavoidable biophysical foundation. By contrast, knowledge commons research identifies resource design and creation as variables to be described and analyzed; there is no necessary biophysical substrate. As intellectual resources (that is, as forms of knowledge and information), patents, copyrights, and underlying inventions, creations, and data are shaped by a variety of institutional forces, including the state (via the design of IP systems) rather than by nature. Fourth, critically, the knowledge commons framework does not assume that the relevant resources are rival and depletable; these are not, typically, common-pool resources. The knowledge commons framework generally begins by assuming precisely the contrary: that intangible information and knowledge resources are nonrival, nondepletable public goods.

Given these distinctions, for knowledge and information resources the basic social dilemma to be solved by a governance institution turns out not to be primarily a classic 'tragic commons' overconsumption problem. Instead, it is more likely (in part) a free rider dilemma leading to an underproduction problem, in the sense that the prospect of consumption by free riders may discourage creators and inventors from producing intellectual resources, and (in part) a coordination problem, in the sense that productive use of an intellectual resource may require the participation of more than one user or consumer concurrently or sequentially. But these social dilemmas are only starting points. In applying the framework to any particular case, as the later description makes clear, care must be given to describing the authentic character of the social dilemmas present.

Against that background, the knowledge commons framework proposes to undertake comparative institutional analysis by evaluating cases of commons resources via a series of questions, or clusters of questions, to be applied in each instance. The inquiry is conducted in the spirit of Ostrom's IAD framework, with attention to contextual and ecological detail. The ecology is cultural rather than natural (Benkler 2013). Different modes of empirical research may be appropriate, depending on the availability of contemporary and historical documentary sources and living interview subjects.

The case study investigation begins with a general overview description of the history and character of the problem or problems addressed by governance in the specific case or context. One begins, in short, with a summary of the social dilemma posed by the knowledge or information resource. This may be an explanation that is internal to the governed institution(s), in the sense that problems and explanations may emerge from stories told by participants, either today or historically, or both. Or, it may be an explanation that is external to the governed institution, such as the public goods story of the rise of IP law in general and/or with respect to specific markets (copyright for books) or technologies (patents for mechanical devices).

The researcher should ask whether the relevant resource or case is characterized from the outset by patent rights or other proprietary rights, as in the case of a patent pool, or by a legal regime of formal or informal openness, as in the case of public domain data or information collected in a government archive. In other words, one should consider the extent to which a social dilemma might be addressed in part by IP rights. Producing socially valuable inventions in the university setting may entail permitting university researchers to obtain patents. Alternatively, the presence of IP rights might create or exacerbate a social dilemma involving a knowledge resource. In that sense, collaborative open source production of computer programs is complicated by the fact that each programmer receives a copyright in their separate contribution. A particular regime might involve sharing data and information, or sharing rights in information, or sharing both. The character of the commons solution might involve coordinating holders of different IP interests or holders of different public domain knowledge resources, for example. In many respects, this cluster of queries parallels the investigation of the biophysical attributes of the natural resource that is the first part of examining a common-pool resource in the natural environment.

Answering that question sets a baseline against which a commons governance regime has been constructed. Within that regime, one next asks more specific definitional questions. First, what are the relevant resource and subsidiary resource units, taking into account both intangible and tangible resources and their individual or social character? Knowledge and information resources are presumptively intangible, but it is possible that their production and use depends significantly on related material resources. What are the relationships among these resources? What is the baseline with respect to what is characterized as a resource in a given setting, and what modifications are made by any relevant legal regime? For example, what a scientist considers to be an invention, what patent law considers to be an invention, and the boundaries of the patent itself are three related but distinct things, managed by different communities. Second, what are the boundaries and constitution (membership) of the community or communities that manage access to and use of those resources? How is membership acquired? This may be informal, formal, or a blend of the two. How is membership and participation governed? What is good behavior within the group, what is bad behavior, who polices that boundary, and how?

Next are questions concerning explicit and implicit goals and objectives of commons governance, if any such goals and objectives exist. It is possible that commons governance regimes emerge from historical contingency rather than via planning, and it is possible that commons governance lacks an explicit account of the goals of governance via modes of knowledge sharing. It is also possible that goals and objectives may change over time. Is there a particular resource development or management dilemma or collection of dilemmas that commons governance is intended to address, and what commons strategies have been used to address those dilemmas?

How 'open' are the knowledge and information resources and the community of participants that create, use, and manage them? The details of the relevant aspects of 'openness' should be specified, along with their contributions to the effectiveness of commons. Some commons and commons resources have precise and fixed definitions of both resources and community membership. Either resources or membership or both may be more fluid, with boundaries defined by flexible standards rather than by rules.

A large and critical cluster of questions concerns the dynamics of commons governance. Following Ostrom, the knowledge commons research framework refers to these as the 'rules-in-use' of commons: the interactions of commons participants and resources, implementing and adapting both formal and informal rules. Included in this cluster of questions are: (1) details of stories of the origins, histories, and operations of commons; (2) formal and informal (norm-based) rules and practices regarding distribution and coordination of commons resources among participants, including rules for appropriation and replenishment of commons resources; (3) the institutional setting(s), including the character of the regime's possibly being 'nested' in larger-scale institutions and being dependent on other, adjacent institutions; (4) relevant legal regimes, including but not limited to IP law; (5) the structure of interactions between commons resources and participants and institutions adjacent to and outside the regime; and (6) dispute resolution and other disciplinary mechanisms by which commons rules, norms, and participants are policed.

Ostrom's IAD framework directs attention to the action arena in which commons participants interact via the rules-in-use. The framework also directs attention to the significance of polycentricity, in which resource governance is exercised concurrently by multiple communities with overlapping and/or interlocking spheres of responsibility. Researchers using the knowledge commons research framework should be alert to the possible existence of multiple action arenas in Ostrom's sense in addition to the presence of polycentric governance arrangements, including the fact that smaller commons governance institutions may be 'nested' within larger governance institutions. The knowledge commons research framework does not emphasize the dynamics of single or concentrated action arenas to the same extent that the IAD framework may be understood to do.

In principle, at this point, it becomes possible to identify and assess outcomes. In the IAD framework, outcomes are typically assessed in terms of the resources themselves. Has a fishery been managed in a way that sustains fish stocks over time? Do commons participants, such as the members of a fishing community, earn returns in the commons context that match or exceed returns from participation in an alternative governance context? In knowledge commons, resource-based outcome measures may be difficult to identify and assess. Sustaining the resources and their uses, individually or in combination, may be the point. In a patent pool, pooled resources may constitute components of larger, complex products that could not be produced but for the pooling arrangement that reduces transactions costs among participants. Outcomes take different forms. It may be the case that patterns of participant interaction constitute relevant outcomes as well as relevant inputs. Sustaining the community itself, via its relationship to particular resources, may be the point of knowledge commons governance. Agency, in a manner of speaking, may be less important than identity; the group and its participants, in a particular institutional setting, may be ends as well as means (Kelty 2008). In addition, different levels and types of interaction and combination matter. Participant interaction in the context of a shared resource pool or group may give rise to (or preserve or modify) an industrial field or a technical discipline. In that specific case, such spillovers may be treated as relevant outcomes.

Having identified relevant outcomes, it is important to look back at the social dilemma(s) that defined knowledge commons governance in the first place. Has the regime solved those problems, and if not, then what gaps remain? How do the outcomes produced by commons governance differ from outcomes that might have been available if alternative governance had been employed? Has commons governance created costs or risks that should give policy makers and/or institution designers pause? Costs of administration might be needlessly high; costs of participation might be high. A collection of industrial firms that pool related patents in order to produce complex products may engage in anticompetitive, collusive behavior. Commons governance may facilitate innovation; it may also facilitate stagnation. If commons governance appears to lead to negative outcomes or problems, researchers should consider re-examining their diagnosis of the social dilemmas associated with the resource. It is possible that negative outcomes produced by a commons solution to one social dilemma are, instead, productive outcomes produced by a different social dilemma associated with the same resource. Or it is possible that commons governance is less productive or socially beneficial than alternative resource management schemes. Systems of exclusive IP rights and schemes that depend on state participation in the supply or management of knowledge resources may be superior to commons governance or may be coordinated effectively with commons. Governance is not necessarily an either/or proposition.

The foregoing summary of the knowledge commons research framework may be presented in bullet-point as shown in the box.

Background environment

- What is the background context (legal, cultural, etc.) of this particular commons?
- What is the 'default' status, in that background context, of the sorts of resources involved in the commons (patented, copyrighted, open, or other)?

Attributes

Resources

- What resources are pooled and how are they created or obtained?
- What are the characteristics of the resources? Are they rival or non-rival, tangible or intangible? Is there shared infrastructure?
- What technologies and skills are needed to create, obtain, maintain, and use the resources?

Community members

- Who are the community members and what are their roles?
- What are the degree and nature of openness with respect to each type of community member and the general public?

Goals and objectives

- What are the goals and objectives of the commons and its members, including obstacles or dilemmas to be overcome?
- What are the history and narrative of the commons?

Governance

- What are the relevant action arenas and how do they relate to the goals and objective of the commons and the relationships among various types of participants and with the general public?
- What are the governance mechanisms (e.g., membership rules, resource contribution or extraction standards and requirements, conflict resolution mechanisms, sanctions for rule violation)?
- Who are the decision makers and how are they selected?
- What are the institutions and technological infrastructures that structure and govern decision making?
- What informal norms govern the commons?
- How do nonmembers interact with the commons? What institutions govern those interactions?
- What legal structures (e.g., intellectual property, subsidies, contract, licensing, tax, antitrust) apply?

Patterns and outcomes

- What benefits are delivered to members and to others (e.g., innovations and creative output, production, sharing, and dissemination to a broader audience, and social interactions that emerge from the commons)?
- What costs and risks are associated with the commons, including any negative externalities?

(Adapted from Frischmann et al. 2014)

Results and implications

Knowledge commons case studies conducted so far consist primarily though not exclusively of those collected in Frischmann et al. (2014) and Strandburg et al. (2017). The subject matter of those case studies is both broad and diverse, extending from the production of modern free and open source software (Schweik 2014) to the early technological development of powered flight during the 19th century (Meyer 2014), and from citizen science groups in contemporary astrophysics (Madison 2014) to peer production of online visual content (Morell 2014). The entirety of Strandburg et al. (2017) is directed to knowledge commons case studies regarding research in medicine and public health, building on the example documented in Strandburg et al. (2014). Taken as a group, the studies collected in those two volumes suggest a number of promising but provisional research results with respect to the eventual goal of producing a more or less comprehensive understanding of the mechanisms for effective or successful and less effective or unsuccessful knowledge commons. Those results include the following points of analysis:

Research counsels an expansive view of the problems to be solved by governance institutions. Knowledge commons may confront diverse obstacles or social dilemmas, many of which are not well described or reducible to a tragedy of the commons, to a free rider dilemma, or to some other generic collective action problem. Close analysis of relevant obstacles tends to suggest multiple social dilemmas that create demand for governance institutions. Strandburg et al. (2014), in a case study of a medical research consortium housed in the Rare Diseases Clinical Research Network, describe multiple dilemmas addressed by governance of a shared knowledge resource. In that case, the resource consisted of the research results of medical research on diseases that affect small patient populations. The results of that case suggest the importance of casting a wide net in general. Relevant social dilemmas may include:

(i) In the case of scientific research, dilemmas attributable to the nature of the research and/or the research problem. Researchers and their subjects may be few in number and/or widely distributed, other research inputs such as funding and time may be scarce or, in the case of large datasets, unmanageable via traditional analysis conducted by human beings.

(ii) Dilemmas attributable to the need to coordinate knowledge sharing among multiple constituencies and stakeholders that collaborate with respect to creation and management of the resource. With respect to scientific research, interests to be accounted for include researchers, their subjects, funders, commercial partners, and the public.

(iii) Dilemmas arising from the need to manage rivalrous or depletable resources that are necessary inputs into production and use of the shared knowledge resources. These may include funding, time, and labor.

(iv) Dilemmas arising from (or mitigated by) the broader systems within which a knowledge commons institution is situated or nested. For example, knowledge production in the modern research university is situated in the broader context of knowledge production to serve the interests of society at large. Commons approaches and reliance on formal IP systems offer different modes of governance that may complement each other or that may conflict.

Relationships among governance systems are critical. Researchers should be alert to potentially complex relationships between knowledge commons and the systems within which they operate and/or are nested. The knowledge commons framework suggests a focus on the background legal rights associated with commons resources. Those legal rights may influence the shape of commons governance and/or interact with other framework inquiries in diverse ways. In some cases, the background contexts (that is, the presence or absence of formal IP rights) seem to act as external constraints on commons governance much as biophysical characteristics of the resource do in the natural resource context. In others, background contexts shape participants' goals and objectives, participants' roles, and action arenas in diverse and dynamic ways. For example, knowledge commons and market-based institutions (or institutions for knowledge production based on different hierarchies or histories, such as government, or the military) may operate as complements, as stages in the evolution of a product or technology, or in opposition to each other.

Shared infrastructure may play a key role. Shared infrastructure appears to be often central to the success of knowledge commons institutions. In some cases, technical or technological infrastructures (such as technology platforms, databases, and bibliographies) may substitute for formal rule-based governance and discipline, easing, though perhaps also obfuscating, decision-making processes. Organizational infrastructures, such as shared coordination mechanisms (task forces, steering committees, technical standards and the like), may lower the costs of participation, collaboration, and research among commons participants. Ownership and/or control of infrastructure or platforms that support knowledge commons may have significant impact on knowledge commons governance. The state itself may supply important infrastructural resources for knowledge commons, consisting of funding processes, regulatory processes, or coordination institutions.

Informal governance institutions, and especially trusted leadership and shared normative framing, contribute a great deal. Informal or 'social' governance, especially involving trusted leaders or decision makers, complements and at times may substitute for formal or public disciplinary institutions in many knowledge commons cases. Reliance on informal governance may grow out of relationships or norms predating the emergence of commons governance, such as norms among scientific researchers, or norms of university-based researchers and teachers, and it may evolve further, toward greater formality. A close relationship may exist between informal governance mechanisms and the need to tailor governance to the needs of small and/or local commons communities. Notwithstanding the frequent characterization of commons governance as a mode of community self-governance or collective action, knowledge commons researchers should be alert for the roles and impacts of social hierarchies on commons.

Commons governance often evolves over time, and commons may play an especially important role in the early stages of some industries and technologies. Commons governance may evolve as the number of participants grows or as innovation affects the nature of the shared knowledge or the balance between competition and cooperation within the community. The pattern of evolution may not necessarily follow a path from more informal to more formal governance mechanisms. Smaller or larger-scale feedback loops between smaller-scale governance institutions and larger-scale systems may be implicated.

Knowledge commons governance does not necessarily depend on one strong type or source of individual motivations for cooperation. Knowledge commons entail cooperation in the building, sharing, and preservation of knowledge resources, but the reasons individuals cooperate in particular knowledge commons may vary. Different individuals cooperate for different reasons, and sometimes a single individual has multiple motivations for cooperating. Motivations may be partly intrinsic and partly extrinsic, or social. Participants may have both competitive and cooperative motives, and the balance between the two may vary between individuals or change over time, depending in part on participants' overlapping roles as creators, maintainers, and/or users of shared knowledge resources. This variety of motives seems to be partially responsible for the variety of social dilemmas that arise in governing knowledge commons.

Conclusion

The knowledge commons research framework is intended to be an evolving research tool and may be improved with use over time. The framework may be refined, or its application improved, or both, via considering the following perspectives. These are necessarily preliminary directions for future research, given the relative youth of this field of research.

First, the knowledge commons research framework may be applied usefully to some institutions that are not core or conventional examples of 'knowledge commons.' Researchers may cast a wide net in defining proper subjects for study. Because the framework is primarily methodological rather than normative, it has proved useful to date in guiding study of a broad range of cases, some that were closer to what many researchers would identify as 'typical' institutionalized knowledge sharing regimes (such as scientific research consortia) and others that may seem, at first glance, to be unusual subjects for a study of knowledge commons. One notable example considers law-making processes as knowledge commons (Daniels and Hudson 2015).

Second, researchers using the framework should take a broad approach to identifying relevant resources and participants. The framework helps researchers to avoid tunnel vision in identifying relevant resources and participants merely by prompting researchers to ask explicitly 'What are the resources?' and 'Who are the participants?' Research to date includes case studies that report on a broader range of resources and participants than one might associate with a typical (or stereotypical) knowledge commons.

Third, researchers should account explicitly for evolution of knowledge commons governance over time. It is to be expected that knowledge commons should change over time as resources, communities of participants, and relevant formal and informal rules evolve through the decisions and actions of actors in various action arenas. Research to date suggests a broader question about how the character and stability of some knowledge commons may be affected by changing interactions with the background environment or changes in the knowledge resources themselves.

Fourth, researchers may vary the sequence in which clusters of research questions are asked and may focus from the beginning on goals and objectives and identifying action arenas. The basic summary of the knowledge commons research framework does not fully anticipate the potential complexity in identifying resource attributes and defining action arenas for knowledge commons. In the natural resource context, the primary operational action arena for a commons regime generally is the use of a specified natural resource by a community defined by geographic proximity. (Other action arenas operate at a rule-making or governance level.) Because knowledge resources are intangible and often are created by a self-selected group of commons participants, knowledge commons often form around particular goals and objectives rather than around preexisting resources tied to particular communities or particular geographies. When that is the case, several primary action arenas may exist at the operational level, and the most important action arenas may not be immediately evident at the outset of research. To analyze a knowledge commons regime it may be most sound analytically to begin with goals and objectives, rather than resources, then to identify action arenas related to those goals and objectives, and then to identify resources, participants, rules, and so forth associated with each action arena. In practice, use of the framework is likely to be an iterative process, in which collecting data about particular knowledge resources may lead to the identification of additional goals and objectives, which may lead to the identification of additional participants or additional shared resources and so on.

Fifth, the framework anticipates identifying multiple social dilemmas. Knowledge commons governance responds to a wide variety of social dilemmas in addition to traditional tragic commons or free rider problems. To analyze an action arena, it is helpful to identify the social dilemmas faced by participants. To understand the social dilemmas faced by a group of commons participants, it is also useful to study their motivations, especially since a theme that frequently appears in the case studies documented so far is diversity of participant motivation.

Sixth, research results so far suggest the importance of highlighting shared infrastructure as an attribute of knowledge commons that is not otherwise prioritized in the framework in its basic form. Future case studies may focus specifically on identifying infrastructural resources created or used by the commons institution. In some cases, such as open source software, it will be important to include infrastructural constraints in the analysis of an action arena's 'rules-in-use' in order to get a complete picture of commons governance.

Seventh, both nonrivalrous and rivalrous resources should be included and described. Although the study of knowledge commons focuses on the sharing of intangible, nonrivalrous resources, it is important to identify any rivalrous resources that are important to a particular action arena. Social dilemmas for knowledge commons governance can and do arise from competition or conflict over the allocation of scarce or rivalrous resources.

Finally, because of the potential multiplicity of communities, social dilemmas, and action arenas, boundary management concerns are important topics of study. Knowledge commons may have different types and degrees of 'openness.' In particular, because knowledge resources are nonrivalrous, knowledge commons likely must deal with multiple constituencies, including users, creators, managers, curators, and subjects of the knowledge resources. These different constituencies may make different and sometimes conflicting demands on commons resources. It is important when identifying goals and objectives and action arenas to be aware of the possibility that important action arenas may be devoted to managing overlaps among these interests and boundary conflicts among different participants.

Note

1 Creative Commons, https://creativecommons.org.

References

Benkler, Yochai (2004) 'Sharing Nicely: Shareable Goods and the Emergence of Sharing as a Modality of Economic Production' *Yale Law Journal* 114: 273–358.

Benkler, Yochai (2006) *The Wealth of Networks: How Social Production Transforms Markets and Freedom*. New Haven, CT: Yale University Press.

Benkler, Yochai (2013) 'Commons and Growth: The Essential Role of Open Commons in Market Economies' *University of Chicago Law Review* 80: 1499–1555.

Bertacchini, Enrico, Bravo, Giangiacomo, Marrelli, Massimo and Santagata, Walter eds. (2012) *Cultural Commons: A New Perspective on the Production and Evolution of Cultures*. Cheltenham, UK: Edward Elgar Publishing.

Bollier, David and Helfrich, Silke eds. (2012) *The Wealth of the Commons: A World Beyond Market & State.* Florence, MA: Levellers Press.
Borgman, Christine L. (2015) *Big Data, Little Data, No Data: Scholarship in the Networked World.* Cambridge, MA: MIT Press.
Boyle, James (2003) 'The Second Enclosure Movement and the Construction of the Public Domain' *Law and Contemporary Problems* 66: 33–74.
Boyle, James (2008) *The Public Domain: Enclosing the Commons of the Mind.* New Haven, CT: Yale University Press.
Chesbrough, Henry (2003) *Open Innovation: The New Imperative for Creating and Profiting From Technology.* Brighton, MA: Harvard Business Review Press.
Contreras, Jorge L. (2010) 'Data Sharing, Latency Variables, and Science Commons' *Berkeley Technology Law Journal* 25: 1601–1672.
Contreras, Jorge L. (2011) 'Bermuda's Legacy: Policy, Patents, and the Design of the Genome Commons' *Minnesota Journal of Law, Science & and Technology* 12: 61–125.
Contreras, Jorge L. and Reichman, Jerome H. (2015) 'Sharing by Design: Data and Decentralized Commons' *Science* 350: 1312–1314.
Cook-Deegan, Robert and Dedeurwaerdere, Tom (2006) 'The Science Commons in Life Science Research: Structure, Function and Value of Access to Genetic Diversity' *International Social Science Journal* 58: 299–317.
Daniels, Brigham and Hudson, Blake (2015) 'Our Constitutional Commons' *Georgia Law Review* 49: 995–1065.
De Moor, A. (2015) 'Communities in Context: Taking Control of Their Tools in Common(s)' *The Journal of Community Informatics* 11(2). http://ci-journal.org/index.php/ciej/article/view/1152.
Dusollier, Séverine (2010) 'Scoping Study on Copyright and Related Rights and the Public Domain' WIPO Study No. CDIP/4/3/REV./STUDY/INF/1.
Frischmann, Brett M. (2013) 'Two Enduring Lessons from Elinor Ostrom' *Journal of Institutional Economics* 9: 387–406.
Frischmann, Brett M., Madison, Michael J., and Strandburg, Katherine J. eds. (2014) *Governing Knowledge Commons.* Oxford, UK: Oxford University Press.
Hardin, Garrett (1968) 'The Tragedy of the Commons' *Science* 162: 1243–1248.
Hess, Charlotte (2008) 'Mapping the New Commons' Presented at 'Governing Shared Resources: Connecting Local Experience to Global Challenges,' the 12th Biennial Conference of the International Association for the Study of the Commons, University of Gloucestershire, Cheltenham, England (July 14–18), http://hdl.handle.net/10535/304.
Hinkel, Jochen, Cox, Michael E., Schlüter, Maja, Binder, Claudia R., and Falk, Thomas (2015) 'A Diagnostic Procedure For Applying the Social-Ecological Systems Framework in Diverse Cases' *Ecology and Society* 20(1): 32. http://dx.doi.org/10.5751/ES-07023-200132.
Hoffman, David A. and Mehra, Salil K. (2009) 'Wikitruth Through Wikiorder' *Emory Law Journal* 59: 151–209.
Kelty, Christopher M. (2008) *Two Bits: The Cultural Significance of Free Software.* Durham, NC: Duke University Press.
Landes, William M. and Posner, Richard A. (2003) *The Economic Structure of Intellectual Property Law.* Cambridge, MA: Harvard University Press.
Machlup, Fritz (1958) 'An Economic Review of the Patent System' Study No. 15 of Comm. on Judiciary, Subcommittee on Patents, Trademarks, and Copyrights, 85th Cong., 2nd Sess. (Committee Print).
Madison, Michael J., Frischmann, Brett M., and Strandburg, Katherine J. (2009) 'The University as Constructed Cultural Commons' *Washington University Journal of Law & and Policy* 30: 365–403.
Madison, Michael J., Frischmann, Brett M., and Strandburg, Katherine J. (2010) 'Constructing Commons in the Cultural Environment' *Cornell Law Review* 95: 657–709.
Merges, Robert P. (1996) 'Contracting into Liability Rules: Intellectual Property Rights and Collective Rights Organizations' *California Law Review* 84: 1293–1393.
Merton, Robert K. (1973) 'The Normative Structure of Science' in Robert K. Merton ed., *The Sociology of Science.* Chicago, IL: University of Chicago Press.
Meyer, Peter B. (2014) 'An Inventive Commons: Shared Sources of the Airplane and its Industry' in Brett M. Frischmann, Michael J. Madison, and Katherine J. Strandburg eds. *Governing Knowledge Commons.* Oxford, UK: Oxford University Press.

Morell, Mayo Fuster (2014) 'Governance of Online Creation Communities for the Building of Digital Commons: Viewed Through the Framework of the Institutional Analysis and Development' in Brett M. Frischmann, Michael J. Madison, and Katherine J. Strandburg eds. *Governing Knowledge Commons*. Oxford, UK: Oxford University Press.
Nelson, Richard R. (2004) 'The Market Economy, and the Scientific Commons' *Research Policy* 33(3): 455–471.
Ostrom, Elinor (1990) *Governing the Commons: The Evolution of Institutions for Collective Action*. Cambridge, UK: Cambridge University Press.
Ostrom, Elinor (2005) *Understanding Institutional Diversity*. Princeton, NJ: Princeton University Press.
Ostrom, Elinor and Hess, Charlotte (2007) 'A Framework for Analyzing the Knowledge Commons', in Charlotte Hess and Elinor Ostrom, eds. *Understanding Knowledge as a Commons: From Theory to Practice*. Cambridge, MA: MIT Press.
Perzanowski, Aaron and Darling, Kate eds. (2017) *Creativity Without Law: Challenging the Assumptions of Intellectual Property*. New York: NYU Press.
Reichman, Jerome H. and Uhlir, Paul F. (2003) 'A Contractually Reconstructed Research Commons for Scientific Data in a Highly Protectionist Intellectual Property Environment' *Law and Contemporary Problems* 66: 315–462.
Reichman, Jerome H., Uhlir, Paul F., and Dedeurwaerdere, Tom (2016) *Governing Digitally Integrated Genetic Resources, Data, and Literature: Global Intellectual Property Strategies for a Redesigned Microbial Research Commons*. Cambridge, UK: Cambridge University Press.
Schultz, Mark F. (2006) 'Fear and Norms and Rock & Roll: What Jambands Can Teach Us About Persuading People to Obey Copyright Law' *Berkeley Technology Law Journal* 21: 651–728.
Schweik, Charles M. (2014) 'Toward the Comparison of Open Source Commons Institutions' in Brett M. Frischmann, Michael J. Madison, and Katherine J. Strandburg eds. *Governing Knowledge Commons*. Oxford, UK: Oxford University Press.
Schweik, Charles M. and English, Robert C. (2012) *Internet Success: A Study of Open-Source Software Commons*. Cambridge, MA: MIT Press.
Shapiro, Carl (2000) 'Navigating the Patent Thicket: Cross Licenses, Patent Pools, and Standard Setting' in Adam B. Jaffe, Josh Lerner, and Scott Stern eds. *Innovation Policy and the Economy*. Cambridge, MA: National Bureau of Economic Research.
Strandburg, Katherine J., Frischmann, Brett M., and Cui, Can (2014) 'The Rare Diseases Clinical Research Network and the Urea Cycle Disorders Consortium as Nested Knowledge Commons' in Brett M. Frischmann, Michael J. Madison and Katherine J. Strandburg eds. *Governing Knowledge Commons*. Oxford, UK: Oxford University Press.
Strandburg, Katherine J., Frischmann, Brett M., and Madison, Michael J. eds. (2017) *Governing Medical Knowledge Commons*. Cambridge, UK: Cambridge University Press.
Van Overwalle, Geertrui, ed. (2009) *Gene Patents and Collaborative Licensing Models: Patent Pools, Clearinghouses, Open Source Models and Liability Regimes*. Cambridge, UK: Cambridge University Press.
Von Hippel, Eric (2005) *Democratizing Innovation*. Cambridge, MA: MIT Press.
Weiser, Philip J. and Hatfield, Dale N. (2005) 'Policing the Spectrum Commons' *Fordham Law Review* 74: 663–694.

8

COMMONS STORYTELLING

Tragedies, comedies, and tragicomedies[1]

Brigham Daniels

Introduction

While scholarship on the commons grows out of many disciplines and employs many methods, much of the literature relies heavily on storytelling. The stories told include those that explain the plight facing a particular commons resource, those that describe how a particular set of resource users avoided commons dilemmas, and those that explain how a particular set of strategic interactions have resulted in problems typical for the commons.

To be sure, increasingly, these stories are often paired with advanced social science inquiry, including economic analysis, in-depth case study development and other qualitative work, game theory, experimental games in the lab and in the field, application of Ostrom's (2005) Institutional Analysis and Development (IAD) framework, and even quantitative methods like statistical analysis and complex predictive modeling. There are frontiers yet to be explored on all these fronts and on many others. This exploration will not only help us to better understand particular cases and problems but also to more fully see how various cases relate with each other. All of this will give us insights into how the commons in the broadest sense works. There is every reason to believe that more expertise, not less, is the future of the cutting-edge work on commons.

Yet, no matter what is added to the literature on the commons, it is difficult to imagine a world in which stories do not play an important, perhaps even a central role, in its study. Looking back, it is also important to recognize the importance of storytelling in pushing the literature forward. Much of the most important work on the commons has relied on storytelling, at least in part. These stories come in many forms. As one might suspect, storytelling often frames presentation of qualitative and quantitative empirical research. However, the role of stories is much broader than this. It includes what might be considered the folk wisdom and speculation of scholars. Storytelling allows us to flesh out suspected connections between the problems faced in different commons or the approaches taken in different commons to stave off problems. Storytelling is usually the first step in making generalizations about the commons. Storytelling is what carries the literature's rhetoric flares and its passionate pleas. And, while storytelling might allow those concerned about the commons to speak from the heart, it is also the frame for its best thoughts and theories.

Given the importance of storytelling to the commons literature, it is not surprising that stories make up some its most recognized and celebrated aspects. I begin my chapter by discussing a handful of the most renowned stories of the commons and point out that in a number of ways these stories run in different directions, and even counter to each other—the seminal stories of the tragedy of the commons (Hardin 1968), the comedy of the commons (Rose 1986), and well-governed commons (Ostrom 1990). We see in these stories, for example, the commons characterized as a place where we should expect crowds rushing to ruin along with tragedies of the commons as well as comedies of the commons; we are made to understand that without outside intervention, problems are unavoidable and also that problems are best worked out without outside intervention. In the next section, I provide some brief thoughts about the connections between the stories of the commons that Hardin, Rose, and Ostrom tell and discuss how these connections relate to other problems found in the commons. Then, I add few thoughts about the importance of storytelling in the commons more generally before concluding in the final section.

From tragedy to comedy

While scholarship on the commons finds a toehold on a diverse set of problems, the promise of understanding these diverse problems as a field of study is that, though commons resources are themselves diverse, the sorts of problems faced (for example, curbing consumption and coordinating collective action) present similar sets of challenges. It is the similarities of the problems faced within the commons that provides the possibility that what we learn from studying one type of commons might help us gain leverage in another.

Despite the promise of the commons literature, when we take a survey of the broader literature we certainly find themes, but we also find that these themes run in different directions. This at times makes translating the lessons of the commons like comparing apples to orangutans. For example, should I worry about tragedies of the commons, look for comedies, or invest in governing the commons? Of course, these divides in the commons literature—appropriately—are perhaps best illustrated by some of its most celebrated stories. In this section, I discuss the themes that arise out of the seminal stories of the tragedy of the commons (Hardin 1968), the comedy of the commons (Rose 1986), and well-governed commons (Ostrom 1990).

Let's begin with *The Tragedy of the Commons* (Hardin 1968), a landmark contribution that ignited an explosion of scholarship focused on the commons. While others had previously identified the incentives that fuel the tragedy of the commons (Gordon 1954, 135; Demsetz 1967, 354; Ostrom 1990, 2–3), Garrett Hardin's work proved foundational by identifying the commons as a class of problems (Thompson 2000, 242). Specifically, Hardin looked at resources where, first, benefits were distributed to users of the resource narrowly while costs were distributed broadly and, second, where access to resource was free from restrictions. When resources had these traits, Hardin considered them commons and saw the potential for a resource free-for-all—something that he called *the tragedy of the commons*. To make his point, Hardin told a story. His story about forlorn cattle grazers has become a mainstay to the introduction to the commons. In his story, each grazer had the option of adding more cows to an open pasture, a commons. Each time a grazer added a cow to the field, that grazer later brought the cow home for slaughter, whereas the pressure placed on the pasture by adding the cow was a cost shared across all potential users of the commons (Hardin 1968, 1244).

By identifying a number of situations where the same incentives exist and then classifying these situations as a particular type of problem, Hardin brought an unprecedented amount of attention to the commons and energized scholars by challenging them to find ways to solve the resulting tragedies. Through this work, commentators have identified a host of problems that lead to such tragedies. This scholarship first focused on the archetypal commons resources, like crashing fisheries and overproduction of oil fields. Scholars continue to find new ways to study the tragedy of the commons. Recent examples include overuse of asbestos and the race among states to hold the first presidential primary. These stories follow a familiar arc to a dreary end that Hardin explained this way:

> Each man is locked into a system that compels him to increase his herd without limit—in a world that is limited. Ruin is the destination toward which all men rush, each pursuing his own best interest in a society that believes in the freedom of the commons. Freedom in a commons brings ruin to all.
>
> *(Hardin 1968, 1244)*

Dismal as it seems, the virtue of the arc is its predictable end. The problem with it is that people tell stories with arcs that contradict it.

The greatest bulk of the literature on the commons, particularly in the social sciences, pushes back on Hardin's version of the commons story. The real jewel of this literature was written roughly two decades after Hardin recounted the fictionalized tragedy facing the herdsmen. Elinor Ostrom combed through the empirical evidence and identified unifying principles that characterized those cases where commons users overcame the tragedy of the commons. In her preeminent work, *Governing the Commons* (Ostrom 1990, 88–102), she referred to these as "design principles of long-enduring institutions." These principles are much admired: the Nobel Committee cited these principles before any other aspect of her work in justifying Ostrom's receipt of the Nobel Prize (Royal Swedish Academy of Sciences 2009, 13); fellow Nobel Laureate Douglass North puts so much faith in this aspect of her work that he considers Ostrom's design principles commandments rather than principles (North 1999, 10). Despite the design principles' many virtues, it is likely that from Ostrom's view, calling these commandments would likely be problematic—unless of course one has a fairly relaxed relationship with commandments.

Ostrom's design principles are not prescriptive and are not a recipe for designing a successful commons-management system. Rather, they are the unifying traits she observed, more or less in force, in the successful common-property regimes she studied. That said, most of the principles she illustrates with stories have been found to be robust in subsequent analyses. And, while management systems that best embody Ostrom's principles could take different forms, an irrigation system in Spain is a well-known example she provided (Ostrom 1990, 61–69). Ostrom explains that in parts of Spain, we find irrigation systems in defined areas, known as *huertas*. She reports that these huertas have been around for at least 550 years, and perhaps as long as 1,000 years. Up until the past century, this water allocation system has worked in a number of water basins without the benefit of dams and despite times of extreme drought. To give a bit of perspective, consider that this water allocation system was around before Columbus sailed under the Spanish flag and before the Spanish Inquisition, and may have been around at the time the Moors ruled Spain. The huertas survived the Industrial Revolution, the world wars, and still live on to this day. The take-home message from Ostrom's examination of the history of the huertas, she explains this way:

> Instead of presuming that the individuals sharing a commons are inevitably caught in a trap from which they cannot escape, I argue that the capacity of individuals to extricate themselves from various types of dilemma situations varies from situation to situation. The cases . . . illustrate both successful and unsuccessful efforts to escape tragic outcomes.
>
> *(Ostrom 1990, 14)*

In looking at the commons, Carol Rose saw something quite different: not only do commons not need to be tragic, at times the specter of tragedy is not even the right problem to worry about. Rather than finding that each additional user of the commons gained at the expense of the crowd, she identified a subset of cases where each additional user contributed to the net benefit of the crowd. Given that what she found was quite different from—even the opposite of—what Hardin told us to expect, she called this situation *the comedy of the commons*:

> In a sense, this is the reverse of the "tragedy of the commons": it is a "comedy of the commons," as is so felicitously expressed in the phrase, "the more the merrier." Indeed, the real danger is that individuals may "underinvest" in such activities, particularly at the outset. No one, after all, wants to be the first on the dance floor, and in general, individuals engaging in such activities cannot capture for themselves the full value that their participation brings to the entire group. Here indefinite numbers and expandability take on a special flavor, relating not to negotiation costs, but to what I call "interactive" activities, where increasing participation enhances the value of the activity rather than diminishing it. This quality is closely related to scale economies in industrial production: the larger the investment, the higher the rate of return per unit of investment.
>
> *(Rose 1986, 768)*

If there is a story that speaks to the comedy of the commons, we can find it at the beach: swimming is more fun with friends, and beach volleyball and Frisbee are not entertaining games without the company of others. As Rose noted, "Recreation is often carried on in a social setting, and therefore it clearly improves with scale to some degree: one must have a partner for chess, two teams for baseball, etc." (Rose 1986, 779). But perhaps the phenomenon can best be reduced to a simple insight: for those who go to the beach to strut their stuff, others must be there to see said stuff strutted.

While the comedy of the commons may owe some of its intellectual genesis to the beach, the concept has increasingly become associated with intellectual and cultural resources (Benkler 2013, 1499; Rose 2014). Certainly, libraries are at their best when they are used; Wikipedia and open source software do not even exist without users coming together to create them.

These three different characterizations of the commons, rooted in arguably the three most important works focused on the commons, provide alternative realities through which to view the commons.

Tragicomedies: relationships between characterizations of the commons

In this section I sketch out some connections between the three characterizations of the commons discussed earlier, along with other problems often associated with commons resource management. These interrelationships illustrate the difficulties in governing the commons in a way that resolves problems in the long run. The narratives used in this section may not apply to every commons. However, these stories weave together in a way that allows us to see the commons literature not as divergent but that, more often than not, in a way we can trace the threads of an interrelated narrative. While many of these stories highlight an inertia that slump towards problems, their arc might better be characterized as tragicomedies. Tragedy is not certain, even if the stories left unchecked tend in that direction. In the end, in the commons we are presented many instances where positive and negative commons futures can coexist.

Stories of unacknowledged limits

The first type of situation that weaves the three stories discussed earlier arises when there is nothing stopping the free-for-all of the tragedy of the commons, but the negative consequences of crowding have not manifested themselves. This sort of situation is particularly prone to occur where the robustness of a resource seems to far exceed demand. In such a case, it may seem that we have a well-governed commons or even a comedy of the commons, but a tragedy of the commons is lurking in the shadows.

In his classic article "The Economic Theory of a Common-Property Resource: The Fishery," which served as an important precursor to *The Tragedy of the Commons*, H. Scott Gordon explains:

> During the latter part of the last century, the Scottish fisheries biologist, W. C. MacIntosh, and the great Darwinian, T. H. Huxley, argued strongly against all restrictive measures on the basis of the inexhaustible nature of the fishery resources of the sea. As Huxley put it in 1883: "The cod fishery, the herring fishery, the pilchard fishery, the mackerel fishery, and probably all the great sea fisheries, are inexhaustible: that is to say that nothing we do seriously affects the number of fish. And any attempt to regulate these fisheries seems consequently, from the nature of the case, to be useless."
> *(Gordon 1954, 126)*

It does not take much to see that MacIntosh and even the great Darwinian Huxley missed the mark by some distance. Because fisheries are crashing throughout the world, it may be tempting to dismiss this passage as nothing more than an example of antiquated ignorance. That would be a mistake, however, because we continue to exceed the capacity of resources that seemed beyond our ability to exhaust.

Take space, for example, "the final frontier." While it is easy to think that space is a place where no one has gone before, we find conflict in space (and not just with Klingons and Sith lords). While space law might seem odd outside of the context of the Imperial Senate, it is actually an emerging field of law and for good reason. Consider, for instance, satellites in geostationary orbit traveling directly above the equator. Because these satellites must be spaced, there are only a limited number of orbital "slots" available. We find countries jockeying to secure a piece of orbit; countries at the equator have even attempted to assert their control over the space above them. Similarly, the Convention on International Liability for Damage Caused by Space Objects, also known as the Space Liability Convention, along with the Outer Space Treaty of 1967, set out liability rules meant to resolve liability for accidents and harm. The Convention was relied upon to arbitrate fault related to the crash of the nuclear-powered Soviet satellite Cosmos 954 (Burke 1985, 270–274). There is no way around it; we are starting to see the closing of the final frontier.

"Sure," you might say. That is a particular problem for near space, but not deep space—that is truly unlimited. That argument may make sense today, but a century from now it might seem like the antiquated ignorance of MacIntosh and Huxley a century before. It is just this sort of blindness that makes commons resources prone. We do not see the need for management, and in fact the idea of management may seem foolish, until we see that we have already caused a problem.

In short, crowding of the commons is what matters and *crowded* is not an on-and-off switch, but a spectrum. The crowdedness of a commons is an interaction between the number of users and the consumptiveness of their use (Daniels 2007, 536). The extent to which crowding matters depends on the perceived robustness of the commons resource.

Returning to Rose's depiction of the beach as a place where we are apt to find comedies of the commons, part of what is going on there is that beaches, generally speaking, are not so crowded that we worry about crowding. That is not to say that beaches are never crowded; they can be, and when they are, we start to see the tragedy of the commons—no space for Frisbee, no place for laying out towels, just wall-to-wall people getting in each other's way. However, this is not the general experience we have on beaches, and when the costs of adding another person on the beach are negligible (which is often the case), we might find that the added benefit of another person on the beach will tend to outweigh those costs.

It is also important to note that when we see rapid shifts in the number of users that access the commons or their ability to consume, we might find ourselves quickly pushing up against a tragedy of the commons in areas that we did not assume had problems. As I have stated elsewhere:

> Several factors influence the number of users of a commons, including the size of a population, technological change, levels of wealth, and market demands. While these factors are self-explanatory, importantly, increases in all of these factors tend to increase strain on commons, yet there are some notable exceptions. For example, technology can make it less costly to access the commons (e.g., transportation improvements) or less costly to exclude others (e.g., the barbed-wire fence). Likewise, wealth might mean more consumption or the willingness to invest to protect a commons.
>
> *(Daniels 2007, 536)*

Furthermore, "technology changes may increase or decrease the consumptiveness of a use of a commons" (Daniels 2007, 536). Not far from my hometown, Provo, Utah, is a huge open-pit mine, the Bingham Copper Mine. The story of the rate of extraction of the mine is largely a story of technology. The story began with a few miners, but once the railroad came to the mine, many more miners came as well. Over time, technology made mining efforts more and more productive. I remember visiting the mine when I was younger and standing beside mining truck tires that were so large that my head barely reached the rim of the tire. Some of the trucks in the mining fleet today are almost 30 feet tall and more than 50 feet long, with a hauling capacity of more than 300 tons in a single load. This progression of technology is responsible for transforming a mountain into a pit so large that it can be seen from space.

Stories of interconnected commons

A second thread of our interrelated narrative includes situations where complex connections in the commons inevitably result in externalities—even in very well-managed commons. For an illustration of this, consider a personal story.

A number of years ago, I took my family to Disneyland. My favorite ride of the day was the ride leaving the parking lot. Don't get me wrong, we had a great time. But we left the park along with tens of thousands of other people just as the park closed. As the crowd made its way to the parking lot, I saw all the elements of misery on the horizon: a long wait, exhausted kids, and a compact car. In fact, I dreaded this scenario so much that in the past I had stayed in one of the hotels across the street from Disneyland just to avoid this very situation. Bracing for misery, we all raced to the car and crammed in.

To my surprise, we were out of the parking lot within minutes. The exuberance and thrills the rides had given my children all day were all mine in that moment. It was not dumb luck either. I had Disney engineers and planners to thank for my quick getaway. While it might seem like there could never be such a thing, I would argue that Disney had built a masterpiece parking lot. It allowed traffic to easily flow out of the lot—all in a way that was completely painless for me. Even though traffic picked up dramatically once we had left the lot, I determined I would never again patronize the hotels across from Disneyland. How could I resist the temptation to stay at cheaper hotels further away, and then hop in the car to travel to the paved parking paradise that beckoned me?

Often, when I am leaving a concert, a ballgame, or a crowded shopping center's lot, I think longingly about Disneyland's parking lot. Far too often I find myself stuck in traffic thinking that Disney's parking creation, if implemented widely, could revolutionize the parking experience. I have told my students that Disney manages potential congestion so well that—if properly understood—its ability to manage the parking lot very well could be the envy of resource managers the world over. After all, too much demand for resources is often a central problem facing those managing not only parking lots but also fisheries, rivers, public roads, radio bandwidth, and a myriad of other high-demand commons resources. And the free-for-all that we often face in crowded parking lots is a textbook example of the tragedy of the commons (Hardin 1968, 1245).

Disney seems to have avoided this all too familiar tragedy of the commons. But still, Disney's management of its parking lot might be seen in a dimmer light. Its creation of the greatest parking place on Earth seemingly adversely affected the public and commercial infrastructure surrounding the park. It seems likely that the reason we hit traffic congestion once outside of the parking lot is that the public infrastructure was designed to facilitate Disney's parking as if traffic flowed like a spigot and not a fire hose. Additionally, if patrons calculate the relative values of staying near the park or further away, the restaurants, hotels, and shops surrounding the park no longer have the allure they had when people could be certain of the misery that awaited them as they exited Disneyland. With this in mind, we might wonder what sort of future this commercial district faces, particularly if this area does not become far more pedestrian-friendly and walkable in order to convince patrons that cars are unnecessary once they arrive.

Additionally, despite my periodic longing for the proliferation of Disneyland-like parking lots, this observation might give some pause: if Disney's scheme were widely adopted, it might give people yet another incentive to drive. It could reduce incentives to take the train or bus and perhaps work to kill walkable shopping districts in favor of megastores and malls. It might spell the end of downtowns across the country—other than "downtowns" found in places like lovely amusement parks that feature Main Street as a quirky blast from the past. Perhaps, it is best to leave Disney's parking magic to Disney's Magic Kingdom.

The stories that have made Disney into the children's propaganda behemoth that it is often strain for a moral ending. Given this, we might ask, what sort of heavy-handed message might we glean from this story about the commons? If there is a lesson to be learned about the commons here, perhaps it as simple as this: solving any particular tragedy of the commons does not mean our work is done. After all, if an army of Disney's engineers doing their best work at managing traffic leaves us both in awe of their successes and worried about the implications of applying these same successes more widely, we should proceed with our work on the commons with more than a modicum of caution and humility.

Beyond this personal narrative, consider a couple of examples of commons problems rooted in policy. Another textbook example of a commons is pollution of the atmosphere. When Congress passed the Clean Air Act in 1970, it saw the problem of air pollution as mainly a localized problem (Roelofs 1993, 435). In other words, Los Angeles' air pollution was considered a problem for airsheds around Los Angeles, but Los Angeles' pollution was not seen as a problem for Phoenix. Many of the provisions eventually implemented in the Act provided emitters of air pollution with an incentive to reduce their impact on local air quality. Some larger emitters of air pollution—particularly coal-fired power plants—complied with Clean Air Act regulations in a way probably not foreseen by Congress when it passed the Act. Rather than reducing emissions, these larger emitters converted localized air pollution into regional air pollution (McCormick 1997, 73). They did this by building super-sized smoke stacks that pumped pollution higher into the atmosphere. While this practice relieved local air pollution, it also substantially increased regional air pollution and its associated problems, including acid rain. In fact, one of the principal reasons regional air pollution became such a problem in the 1980s was that the commons solution for individual airsheds provided in the 1970 Clean Air Act overlooked the threats of regionalized pollution (Bosse 1995, 197–212). In addition to aggravating regional air pollution, the political dynamic associated with the Clean Air Act that enticed businesses to make major outlays to comply with the act also created entrenched interests with a stake in continuing to emit regional air pollution, making the problem all the more difficult to address.

Consider a second example. The period since the late 1990s has seen an explosion of wireless technologies that rely on radio bandwidth to function (Hazlett 2001). These wireless devices include wireless laptops, cell phones, GPS technologies, and even garage door openers. One hurdle in getting wireless technologies on the market is the scarce availability of radio spectrum due to generous licensure of the spectrum to technological innovations of the past, particularly broadcast radio (Hazlett 2005). It is important to recognize that the reason broadcast radio holds so much of the spectrum is that the government wanted order on the dial. To do that, it divided up the scarce bandwidth. Before this was instituted, stations bled into each other and listeners would jump around the dial to get a better signal. As Justice Frankfurter said, "With everybody on the air, nobody could be heard" (National Broad. Co. v. U.S. 1943, 212). Licenses issued in the past to meet yesterday's problems make dealing with the challenges of the present more difficult.

Stories of problems of resources with multiple dimensions

The way that we usually conceive the commons is often overly simplistic (Daniels 2007, 521). Let me justify this assertion with an illustrative story.

A few years ago, I attended a conference on the commons in Cheltenham, England. Cheltenham sits at the edge of the Cotswolds, a beautiful landscape of rolling hills. At the close of the conference, my spouse and I went on a field trip to see the landscape.

The highlight of the field trip was when the bus driver announced we were approaching a commons. As the bus slowed down, we saw cows on the commons. There were actual cows on the commons! I could not believe my luck. Beyond the cows, we saw people on the commons in the distance. I wondered as I stood out of my seat, "Are those the herders?!" As the bus moved on and got closer to the people on the commons, these people where no ordinary herders. They had no shepherds crooks, no sheep dogs; instead I saw that they had plaid hats and golf clubs. In a moment of mental confusion, all I could think of was, "Why would herders golf?! Wait . . . These are not herders, these are golfers." It was hard to see it any other way. Golfers were right there . . . with the cows . . . on the commons. I was surprised to say the least, but the tour guide—witnessing my excitement followed by confusion—explained it was really nothing out of the usual for golfers and cows to share a commons.

When Hardin introduced us to the commons, he set me up for confusion. Hardin did not describe a field, he asked us to "picture a *pasture* open to all" (Hardin 1968, 1244). There is a difference between fields and pastures, even though a field has geographic dimensions similar to those of a pasture. The main difference between the two is that implicit in the idea of a pasture is the assumption of a particular use—grazing animals. Grazing happens on the pasture, golfing on the green, and both—apparently—on a commons in the Cotswolds.

While Hardin's environmental storytelling is much celebrated and helpful in getting his point across, he tells the story exclusively through the eyes of the herders, ignoring—as much of the commons literature is apt to do—that there might be competing uses of the field that will have to go unfilled because the herdsmen laid claim to the field. Despite some exceptions, and even a call from the IAD framework to analyze various interests, the commons literature is replete with suggestions about how the herdsmen and other commons users may avoid the tragedy of the commons. But we ignore other potential dimensions of the field that others might find appealing: a place for golfers, a space for picnickers, watershed managers, wildlife hunters, or policymakers in search of a place to grow a forest as a greenhouse gas sink.

Even when a resource does not have multiple uses associated with it, it often has multiple dimensions. Even just focusing on preserving air quality, we find that there are local, regional, and even global dimensions to air resources. Furthermore, people value them for different reasons and to different degrees. For some, their concerns might be protecting forests from acid rain whereas others are concerned about the impact of pollution on asthmatics. These different dimensions of the commons frequently lead to connections between tragedies of the commons, comedies of the commons, and well-governed commons.

Stories that mistakenly assume something is a pure public good resource

When we find resources that have elements of public good resources, we are likely to also find comedy. The main difference between a commons and a public good is that using a public good does not consume or diminish the resource (Keohane & Ostrom 1995, 13). Very few goods (perhaps none) are truly public goods, but frequent examples are national defense (Keller 2002, 321) and information (Carrier 2004, 32). While there are certainly resources that have public good dimensions to them, there are no resources where rivalry does not play at least some role. Because of this, pure public good resources are so rare they are nearly nonexistent.

Often, when we see mention of a public good, what really is at play is a commons good that is not currently facing a lot of pressure from congestion. For example, one of the most common examples of a public good is a highway without much traffic (Cornes & Sandler 1996, 273–77). At least at some level this makes sense: adding more traffic to the highway might have little or no effect on other potential users. Still, highways, and many other resources that serve as an adequate proxy for public goods in many instances, are of course prone to crowding. Regardless, the major point is that we overlook the possibility of crowding often at our own peril.

Where we find a resource with elements of a public good, it captures our attention. The problem is that we might be tempted to overlook aspects of the resource that, while perhaps with less allure, are likely to be a source of problems. At the end of the day, the assertion that a resource cannot be consumed or be diminished is simply not true.

Resources often have multiple dimensions, some of which have public good aspects to them. Take the example of the gases in the Earth's atmosphere that create the greenhouse effect. Emissions of greenhouse gases have often been characterized as a classic commons problem (Stewart 1993, 2099; Thompson 2000, 253). This makes sense because as an individual adds greenhouse gases to the global atmosphere, she experiences all the benefits associated with emitting the gases but only bears a minuscule fraction of the costs. On the other hand, gases in the global atmosphere are also a textbook example of a public good. Greenhouse gases form a fairly uniform blanket around the planet that warms the entire Earth. One person's experience of this additional warmth does nothing to diminish the greenhouse effect. As an illustrative example about how tricky this can be, consider an essay by Professor Daniel Cole arguing that the Industrial Revolution changed the nature of the climate good from a public good to a commons (Cole 2015).

Now, consider what might be the most classic example of a public good: a recipe. It could be easily argued that this is a textbook public good because no matter how many times it is copied, it still remains undiminished for the owner. No doubt, this is true in many respects. Some aspects of a recipe, however, can easily be diminished—such as the ability of a person to make money from a recipe—if it is broadly shared. In other words, we see that aspects of the tragedy of the commons apply to recipes as well. In other work, I made the point this way:

> While it is a bit ridiculous, consider the example of the potential value of the fried chicken recipe at the heart of a memorable speech delivered by one of Mike Myers' characters in *So I Married an Axe Murderer*. The character could hardly contain his scorn as he explained at Kentucky Fried Chicken that the Colonel "puts addictive chemicals in his chicken, making you crave it fortnightly." Had the Colonel blabbed the recipe far and wide, when the craving kicked in, getting out the fryer would be an alternative to paying a visit to the Colonel. In reality of course, addictive chemical or not, KFC goes to great lengths to keep the "Colonel's Secret Recipe" a secret, hiding away the recipe in a vault protected by various motion detectors and surrounded by concrete. If the recipe got out, every potential customer who opted not to buy his chicken would illustrate for the Colonel the sting of rivalry.
>
> *(Daniels 2014, 451–452)*

For the Colonel, the resource of his recipe would be quite diminished mainly because of the way he values its commons dimensions. There would certainly be those who would appreciate the public good aspect of the recipe being exploited more.

Stories of tragic institutions

As mentioned earlier, one of the biggest problems facing the commons is not really something focused on by scholars on the commons—conflicting uses of resources.

Compounding that one person's use might conflict that of another is all the more troubling because institutions often work to lock in gains and losses of those with competing values. While what resource managers are often looking for when solving the tragedy of the commons is stability, when values change, stability becomes rigidity. As an example of this, consider a recently penned op-ed:

> California's worst-ever drought, now entering its fourth year, demonstrates with dramatic force the many deficiencies of the legal system by which the state allocates shrinking water supplies among 39 million Californians . . .
>
> The fundamental problem is that California's water rights system was created over a century ago, when the population of the entire state was less than 3 million residents, and relatively stable water supplies were more than adequate to support the agrarian economy of the state at that time. California law created—and still relies on—a "first in time, first in right" system, under which those who secured their water rights first obtained priority over later, "junior," water claimants.
>
> In an era of relatively few people and limited statewide water demand, California's water rights system worked reasonably well. Now, 100 years later, with the same legal rules intact, California's population growing steadily, and state water supplies shrinking due to the effects of climate change, that system has become antiquated, dysfunctional and unresponsive to 21st century conditions.
>
> *(Frank 2015)*

The reason that California faces problems with its system that favors incumbent users over newer users is that it was designed that way. The system is doing its job; its job just is not valued any more. There are reasons to suspect that where we find stability in staving off the tragedy of the commons, we might find the potential for rigidity lurking.

Why? Any scholar, resource manager, or commentator who concerns herself with the future of a commons or the well-being of a group of commons users would love to have a toolbox that helps ferret out potential exposure to the tragedy of the commons along with tools to address these weaknesses. This is exactly what the scholars who have concerned themselves with synthesizing the volumes of case studies on various commons have provided us—the most well-known of these, of course, is Ostrom's principles of long-enduring institutions (Ostrom 1990, 88–102); but similar compilations by others are also important (Wade 1988; McKean 1992; Baland & Platteau 1996; Agrawal 2001).

While much effort by other scholars has gone into uncovering these principles, these various compilations differ somewhat from scholar to scholar. What seems quite useful, but is not currently available in the literature, is to think about what—if anything—unifies the principles that have been identified. It seems three important threads bind these principles together.

First, many of the factors identified earlier can be seen as an attempt to provide commons users with a credible commitment that they will reap future benefits for any sacrifices they make today (Ostrom 1990, 43–45). For example, we might think about rules that clearly determine at the outset who has rights (Wade 1988, 215; Ostrom 1990, 91; McKean 1992, 258; Agrawal 2001, 1654) to what (Ostrom 1990, 91–92; McKean 1992, 264–266) and where (Wade 1988, 215; Ostrom 1990, 91–92; Baland & Platteau 1996, 344; Agrawal 2001, 1643). We also may not be surprised if the institutions created give incumbent users the ability to participate meaningfully in the governance in the commons and lock out others (Ostrom 1990, 93–94; Baland & Platteau 1996, 344; Agrawal 2001, 1654). It makes sense that this would help because we are asking commons users to trust that the promises made at the outset ultimately come to fruition.

Providing credible commitments in the commons is often difficult because when it comes to cutting back in the commons, "temptations to free-ride and shirk are ever present" (Ostrom 1990, 15) Without credible commitments, it makes sense that users of a commons are squeamish to cut back. As a perceptive federal judge observed more than 100 years ago, "no person would engage in [labor in the commons] if the fruits of his labor could be appropriated by any chance finder" (Ghen v. Rich 1881, 162). Scholars of the commons have recognized that without credible commitments, any conservation only leaves more for others, making them a "sucker" (Ostrom 1990, 17) or a "patsy" (Thompson 2000, 242). In contrast, this risk is part of what has caused property rights scholars to argue for privatization of the commons—due to inadequate private ownership rights individuals are unable to enforce conservation decisions against others on the commons.

A second theme that binds many of the factors that make a difference in warding off the tragedy of the commons is that they attempt to provide credible threats aimed at those tempted to circumvent the rules of the game. Credible threats are just the flip side of the coin used to make credible commitments (Schelling 1960).

The need for compliance and the importance of excluding those without rights are themes that run through the principles scholars have identified. It is not surprising that some of the principles include making monitoring as easy as possible and devoting resources to ensure monitoring is effective (Ostrom 1990, 94; McKean 1992, 272–274; Baland & Platteau 1996, 345), escalating sanctions (Ostrom 1990, 94–100; McKean 1992, 275; Baland & Platteau 1996, 345), and setting out the rules of dispute resolution (Ostrom 1990, 100–101; Baland & Platteau 1996, 375;

Agrawal 2001, 1654). If consequences do not follow the violation of any commitment made in the commons, it is hard to say how we could call such a promise "credible;" credible threats are part and parcel of making commitments (Schelling 1960, 43).

The third thread that runs through these factors is the value of low transaction costs in securing cooperation in the commons. The importance of transaction costs in determining the success of collective action is well documented (Olson 1965). Note that some of the factors identified by commons scholars are matters of institutional design and within the grasp of those attempting to govern the commons to change. Examples include the use of nested enterprises within large and complex commons resources (Ostrom 1990, 101–102; Agrawal 2001, 1654) and giving resource users management responsibilities over the commons (McKean 1992, 258–261; Agrawal 2001, 1654).

In sum, credible commitments and threats help commons users meet the challenges posed by the tragedy of the commons; they work together to assure that sacrifices made by users today will benefit them specifically in the future and not just another user of the commons. Coordination of collective action helps promote cooperation in a circumstance where many would be tempted to free ride. They also reduce the costs of users governing a commons in the first place.

So, we are apt to find conflicts in the commons when values change precisely because solving the tragedy of the commons in the first instance often calls for providing the sort of stability that in many commons resources future users come to regret. More and more, fights in the commons concern not only the problem of overconsumption by a particular kind of user but also the push and pull of entirely different competing uses. Whenever we commit to manage a commons for the benefit of a particular set of users, we often lock out potential rivals and make change more costly. In this way, stability can become rigidity. Because commons resources are complex and can be used for many purposes, it is typical that when we attempt to solve a problem in the commons, we often find that laws and policies put in place to solve past problems complicate our efforts and may in fact be the very source of our frustrations (Daniels 2007).

Managers of the commons are well acquainted with attempts to push the commons in one direction or pull it in another. Should we protect owl habitat or allow loggers to cut down trees? Should our sidewalks be open to vendors, reserved for walkers, or a place for the homeless to sleep? Should coastal wetlands be drained for development, protected for wildlife, or valued for flood control? The number of competing pressures vying for the commons within our forests, rivers, cityscapes, the radio spectrum, and even the global atmosphere highlight such tensions. The problem of competing uses of the commons is not as simple as it appears. There is more going on than an us-against-them problem. The nature of the commons actually lends itself not only to competing values but also entrenched interests. What we find often is that the methods that allowed for governance over a commons resource in one era now serve to complicate governing the commons today. In the commons, the seeds of conflict are often sown in the accomplishments of the past.

Importance of storytelling more generally in the commons

Storytelling is a vital part of the commons literature. As illustrated earlier, some of the most foundational works on the commons rely on storytelling in important ways. It is hard to say what contributed to the commitment to storytelling in the commons, but it is there. Perhaps because the boom in the literature was ignited by a fictional account of herders, perhaps because its earliest standard bearers in large part relied on qualitative research and relayed stories, perhaps it is because the central problem of the commons was labeled a tragedy. Still, historians, anthropologists, and political economists had relied on stories in discussing the commons long before Hardin penned his first essay on the topic. Maybe it is something about commons themselves that lend themselves to stories. Regardless of the source of inspiration, in this section, I present five reasons that the commons literature is better off because of storytelling. This does not just apply to the role of stories in many of the greatest contributions to the literature but also in the more modest contributions as well.

My central claim in this short section is that commitment to storytelling in the commons literature has ended up being one of the literature's strengths. The question is, why? One reason for this is that storytelling is accessible and invites participation not only from the expert scholars but also expert practitioners. So even while the idea of the commons is just that—an idea—it is accessible enough and useful enough that it has attracted the attention of a wide variety of resource managers and advocates. A gathering of people who study the commons often presents a diversity not found in many scholarly circles.

Second, storytelling makes drawing parallels intuitive and invites application in the field by resource users and other decision makers. There are very few fields that would invite comparisons of fisheries on the one hand and allocation of the radio spectrum on the other, but in the commons it not only seems natural to draw the analogy but also to spur thoughts about whether solutions in one area of the commons might help solve problems in the other. There is not much that is easier to relate to in this world than a story. Stories suck us in, keep our attention, invite reflection, and inspire retellings.

Third, stories also invite considered opinions and early impressions in explaining causal interferences of stories. It is not that the commons literature is just shooting the breeze, but storytelling does invite a kind of speculation that might not be welcome when using other tools. This form of hypothesis generation provides fertile ground for the sort of social science research that has come to increasingly make up the scholarship on the commons.

Fourth, storytelling provides a low cost of entry for a broad range of interdisciplinary scholars, who will view stories differently and bring their expertise to bear on important questions raised by storytelling. Without the emphasis of stories, it seems unlikely that it would have captured the attention of so many scholars. But as it is, articles about and references to the commons can be found in a very broad set of fields. The idea of the commons, introduced to budding scholars at the outset, provides fertile ground for future growth of the literature.

Lastly, it is important that storytelling remains part of the conversation on the commons. Even though they are simple to write and understand, they are also valuable because they provide ample room for context and complexities that other methods cannot handle. They provide the foundation for what will grow into the next chapters of the commons literature.

Without storytelling the literature on the commons would be much different and seemingly, for the worse. In a scholarly endeavor that relies on storytelling so much, the importance of storytelling to the endeavor is something that is not discussed nearly enough.

Conclusions

Storytelling plays a large role in the scholarship of the commons. This role, though rarely recognized explicitly, can be found across the literature—from the foundational pieces to those only making a modest contribution. I have argued that this is a good thing. I have also made the argument that though some of these stories present very different views of the commons—from tragedy to comedy—understanding the connections of such stories is vital in improving management of the commons.

Note

1 Significant parts of this chapter are a revision of my article, "The Tragicomedy of the Commons," *2014 BYU L. Rev.* 1347. I thank *BYU Law Review* for allowing me to publish this revised draft in this volume.

References

Agrawal, A. (2001) "Common Property Institutions and Sustainable Governance of Resources" *World Development* 29: 1654–1672.
Baland, J. M. and Platteau, J. P. (1996) *Halting Degradation of Natural Resources: Is There a Role for Rural Communities?* Food and Agriculture Organization of the United Nations, Rome.
Benkler, Y. (2013) "Commons and Growth: The Essential Role of Open Commons in Market Economies" *The University of Chicago Law Review* 80: 1499–1555.
Bosse, M. R. (1995) "George J. Mitchell: Maine's Environmental Senator" *Maine Law Review* 47: 179–223.
Burke, J. A. (1985) "Convention on International Liability for Damage Caused by Space Objects: Definition and Determination of Damages After the Cosmos 954 Incident" *Fordham International Law Journal* 8: 255–285.
Carrier, M. A. (2004) "Cabining Intellectual Property through a Property Paradigm" *Duke Law Journal* 54: 1–135.
Cole, D. (2015). "The Problem of Shared Irresponsibility in International Climate Law" in J. Schechinger (Author) & A. Nollkaemper & D. Jacobs (eds.) *Distribution of Responsibilities in International Law (Shared Responsibility in International Law).* Cambridge University Press, Cambridge, UK, 290–320.
Cornes, R. and Sandler, T. (1996) *The Theory of Externalities, Public Goods and Club Goods.* Cambridge University Press, Cambridge, UK.
Daniels, B. (2007) "Emerging Commons and Tragic Institutions" *Environmental Law* 37: 515–571.
Daniels, B. (2014) "Legispedia" in Frischmann B. M., Madison M. J. and Standburg K. J. (eds.) *Governing Knowledge Commons.* Oxford University Press, Oxford, UK, 445–468.
Demsetz, H. (1967) "Toward a Theory of Property Rights" *American Economic Review* 57: 347–359.
Frank, R. M. (2015) "Another Inconvenient Truth: California Water Law Must Change" *San Francisco Chronicle*, April 10.
Ghen v. Rich, 8 F. 159 (D. Mass. 1881).
Gordon, H. S. (1954) "The Economic Theory of a Common-Property Resource: The Fishery" *Journal of Political Economy* 62: 124–142.
Hardin, G. (1968) "The Tragedy of the Commons" *Science* 168: 1243–1248.
Hazlett, T. W. (2001) "The Wireless Craze, the Unlimited Bandwidth Myth, the Spectrum Auction Faux Pas, and the Punchline to Ronald Coase's 'Big Joke'" *Harvard Journal of Law and Technology* 14: 335–567.
Hazlett, T. W. (2005) "Spectrum Tragedies" *Yale Journal on Regulation* 22: 242–274.
Keller, D. (2002) "A Gaudier Future that Almost Blinds the Eye" *Duke Law Journal* 52: 273–321.
Keohane, R. O. and Ostrom, E. (1995) "Introduction", in Keohane R. O. and Ostrom E. (eds.) *Local Commons and Global Interdependence.* Sage, London.
McCormick, J. (1997) *Acid Earth: The Politics of Acid Pollution.* Earthscan, London.
McKean, M. A. (1992) "Success on the Commons: A Comparative Examination of Institutions for Common Property Resource Management" *Journal of Theoretical Politics* 4(3): 247–281.
National Broad. Co. v. U.S., 319 U.S. 190 ff. (1943).
North, D. C. (1999) "Dealing with a Non-Ergodic World: Institutional Economics, Property Rights, and the Global Environment" *Duke Environmental Law and Policy Forum* 10: 1–12.
Olson, M. (1965) *The Logic of Collective Action.* Harvard University Press, Cambridge, MA.
Ostrom, E. (1990) *Governing the Commons.* Cambridge University Press, Cambridge, UK.
Ostrom, E. (2005) *Understanding Institutional Diversity.* Princeton University Press, Princeton, NJ.
Roelofs, J. L. (1993) "United States-Canada Air Quality Agreement: A Framework for Addressing Transboundary Air Pollution Problems" *Cornell International Law Journal* 26: 435–454.
Rose, C. M. (1986) "The Comedy of the Commons: Custom, Commerce, and Inherently Public Property" *University of Chicago Law Review* 53: 711–786.

Rose, C. M. (2014) "Surprising Commons" *BYU Law Review* 2014: 1247–1282.
Royal Swedish Academy of Sciences (2009) "Scientific Background on the Sveriges Riksbank Prize in Economic Sciences in Memory of Alfred Nobel 2009: Economic Governance," www.nobelprize.org/nobel_prizes/economic-sciences/laureates/2009/advanced-economicsciences2009.pdf.
Schelling, T. C. (1960) *The Strategy of Conflict*. Harvard University Press, Cambridge, MA.
Stewart, R. B. (1993) "Environmental Regulation and International Competitiveness" *Yale Law Journal* 102: 2039–2106.
Thompson, Jr. B. H. (2000) "Tragically Difficult: The Obstacles to Governing the Commons" *Environmental Law* 30: 241–278.
Wade, R. (1988) *Village Republics: The Economic Conditions of Collective Action in India*. Cambridge University Press, Cambridge, UK.

9

COMMON-POOL RESOURCE APPROPRIATION AND CONSERVATION

Lessons from experimental economics

Esther Blanco and James M. Walker

Introduction

The literature on common-pool resources (CPRs) can be viewed initially from its place in the larger literature of collective action problems often referred to as social dilemmas. As discussed in the foundational work of Mancur Olson (1965), "unless the number of individuals is quite small, or unless there is coercion or other special devices to make individuals act in their common interest, *rational, self-interested individuals will not act to achieve their common or group interest.*" Social dilemmas occur when individuals, as part of a group, select strategies that generate suboptimal outcomes from the perspective of the group. The problem for collective action is finding institutional and normative alternatives to overcome these undesired outcomes.

In this chapter, we focus primarily on experimental studies designed to investigate collective action in CPR settings, in particular as they apply to natural resources. We argue that an analytical framework for understanding the strategic incentives associated with the use of natural resources entails the complementary analysis of *production* externalities (the primary focus of the CPR literature on the pecuniary loss of economic rents to appropriators) and *degradation* externalities (the loss of public good benefits of conservation of natural resources in the form of ecosystem services). In this sense, the CPR literature and the public goods literature are complementary, and both necessary, in the understanding of the management of natural resources. In both literatures, the underlying research questions relate to the degree to which behavior leads to suboptimal outcomes and how that behavior is shaped by norms, parametric conditions, and institutional arrangements.

More formally, one can view the link between production externalities and degradation externalities using the notation from the appropriation setting of Ostrom et al. (1994). Let x_i denote appropriator *i*'s appropriation from the CPR, where $0 \leq x_i \leq e$. The group return to appropriation from the CPR is given by the production function $F(\Sigma x_i)$, where F is a concave function, with $F(0) = 0$, $F'(0) > w$, and $F'(ne) < 0$. The production function $F(\Sigma x_i)$ defines the level of appropriation resulting from effort. Initially, appropriating from the CPR yields a higher pecuniary return than the opportunity cost $[F'(0) > w]$, but at some level of investment (x_h) the outcome is counterproductive $[F'(x_h) < 0]$. Appropriation beyond the optimal level captures the *production* externality in this stylized model, the loss of economic rents to those who have appropriation access rights to the resource.

In the expanded version of the model, $g_i\left(T - \sum x_i\right)$ represents the total loss in conservation value of the resource to individual *i* based on total appropriation from the CPR, capturing the *degradation* externality. Consequently, g_iT represents the maximum value individual *i* can receive from the conservation component of the CPR, in the form of a public good. The maximum value of conservation occurs in the case of no appropriation, and the conservation value decreases in aggregate group appropriation from the CPR, $\sum x_i$. Viewing conservation as a public good, degradation externalities affect individuals who actively appropriate, as well as those who could appropriate but choose not to do so. Further, ecosystem degradation affects individuals beyond those that may be involved in appropriation (see the section 'Recent directions in natural commons research' later for a detailed discussion).

Now, let $x = (x_1, \ldots, x_n)$ be the vector of individuals' appropriation from the CPR. The payoff to an individual *i*, π_i, is given by:

$$we + g_i\left(T - \sum x_i\right) \qquad \text{if } x_i = 0$$

$$w(e-x_i) + (x_i/\sum x_i)F(\sum x_i) + g_i\left(T - \sum x_i\right) \qquad \text{if } x_i > 0$$

Based on this more general version of the CPR appropriation game, and using the logic of non-cooperative game theory, optimal group appropriation and the Nash Equilibrium level of appropriation for self-interested payoff-maximizing agents will depend on the payoff parameters initialized across treatment conditions. Of course, for simplicity, one can create decision settings in the laboratory that separate appropriation externalities from conservation externalities. This in fact is the primary approach taken in experiments to date.

Although complementary, there is one aspect of the two social dilemma settings that should be emphasized. First, despite the fact that CPRs and public goods are characterized by non-excludability (individuals can access/consume these goods without playing a role in their provision), they differ in regard to "subtractability or rivalry" in consumption. This difference has important implications for collective action. By definition, pure public goods are non-rival, and therefore once provided, one person's consumption does not subtract from another's. This is not true for CPRs. One person's appropriation has an adverse effect on other appropriators (the production externality). This aspect of CPRs is critical for understanding the implications for cooperative behavior. Specifically, if a subset of appropriators cooperate in appropriation, the effect is to enhance the returns from the CPR, referred to in the literature as resource "rents." Importantly, such rents can be "captured" by non-cooperators through their appropriation. In this sense, collective action in CPR settings implies an additional degree of complexity as compared to collective action in providing public goods. See Schmitt et al. (2000) for an experimental study closely related to this issue.

The chapter is organized as follows. We first examine the CPR research that focuses primarily on production externalities that lead to rent dissipation. Second, we address lessons learned from the public goods literature that relate most directly to the conservation of natural resources. Our discussion is based on an overview of studies that reveal the diversity in social dilemma issues in the experimental literature. Of course, this overview cannot be comprehensive. Instead we hope the reader gains valuable insights into how experimental economics has played an important role in providing a foundation for understanding how individuals respond to alternative contexts in dealing with issues of the commons. In some cases, when feasible, this overview provides a brief discussion of key behavioral findings. In all cases, the reader is urged to go to the primary sources for details of the decision setting faced by subjects and for behavioral results.[1]

Appropriation and production externalities

Since the 1940s, an extensive program of research on the efficiency in use of CPRs has focused primarily on production externalities. The concern for rent dissipation in CPRs was formulated in the early work of Gordon (1954) and Hardin (1968), who argue that when individuals appropriate from CPRs, each confronts incentives to withdraw more of the resource than is optimal from a group perspective (ignoring the negative externality on other resource users), resulting in inefficient economic outcomes.

In the late 1980s and early 1990s field researchers began to document the geographical and historical prevalence of successful systems of community management in limited access CPRs (e.g., Berkes et al., 1989; Bromley, 1992; Feeny et al., 1990; McCay and Acheson, 1987; Netting, 1981; Ostrom, 1990; Sengupta, 1991). They found that communities as diverse as small-scale fishing and horticultural societies, as well as those embedded in large-scale industrial societies, had developed complex institutions and rules-in-use for managing local resources that had survived for long periods of time, under changing economic environments. The seminal work of Ostrom (1990), *Governing the Commons*, is viewed as the cornerstone for developing an analytical approach to examining natural and institutional arrangements that foster cooperation in the commons.[2]

This early literature fostered the now modern approach to examining CPR situations that encompasses new field studies, experimental methods, mathematical modeling, and agent-based simulations (see Poteete et al., 2010). Herein, we focus on the experimental evidence. Some of the earliest experimental economics literature examining appropriation from CPRs, began with the development of a CPR game by Elinor Ostrom, with her colleagues James Walker (one of the authors of this chapter), and Roy Gardner (e.g. Herr et al., 1997; Ostrom et al., 1992, 1994; Walker et al., 1990).[3] This literature, combined with extensive field evidence on the use and management of CPRs, was instrumental in understanding under what conditions users of CPRs can increase cooperation levels through institutional change. Primary results from this early laboratory research can be summarized as follows. First, as predicted by the non-cooperative self-interested payoff-maximizing game theoretic model, subject groups placed in institutional settings where there are no opportunities to coordinate their actions, make appropriation decisions well above the level that would generate maximum resource rents. Face-to-face and reoccurring opportunities to communicate lead to significant improvements in efficiency. Sanctioning also leads to more optimal resource use. However, the gains are diminished significantly when one accounts for the cost of sanctioning. Extending the game setting, Walker and Gardner (1992) investigated how subjects would respond to a repeated game setting where there was an endogenous probability of ending play of the game (and future earnings). In this setting, there existed a "safe" maximum yield, beyond which the probability of ending the game increased in total group appropriation. In every case, appropriation decisions led to a premature ending of the game, often quite quickly. This finding highlights the need for institutional arrangements to promote cooperation.

This early research was followed by a broad set of studies that examined CPR issues in contexts motivated by field settings. The examples below are suggestive of the range of topics.

Institutions revisited: A broad literature has flourished focusing on arrangements for facilitating more efficient use of the commons, with communication, sanctioning, and voting the most studied (e.g. Bischoff, 2007; Buckley et al., 2008; Fischer et al., 2004; Margreiter et al., 2005; Ostrom, 2006; Schott et al., 2007; Vyrastekova and van Soest, 2003; Walker et al., 2000).

Non-standard populations: More recently, this literature examines the robustness of early results to alternative subject populations, focusing on users of natural CPRs in developing countries (e.g. Cardenas, 2011, 2009; Cardenas et al., 2000; Fehr and Leibbrandt, 2011; Hayo and Vollan, 2012; Rodriguez-Sickert et al., 2008; Velez et al., 2009; Vollan, 2008; Vollan and Ostrom, 2010). Cardenas (2011) reviews the results for the relevance of social norms in understanding the management of the local-commons by its users.

Asymmetry: The early experimental CPR research began with decision settings with symmetric incentives among individual appropriators. The question of how asymmetries in individuals' abilities, costs, capacities, etc. was a necessary extension to examine the robustness of the early findings. Margreiter et al. (2005) found appropriation in CPR settings with asymmetric appropriation costs are (on average) not statistically different from appropriation in symmetric groups with intermediate appropriation costs. Studies such as Hackett et al. (1994) examined CPR settings with communication or voting opportunities to facilitate cooperation, where subjects receive differential returns from agreements that are linked to the differences in capacity to appropriate. Motivated by many field CPR settings, Cardenas (2011) and Janssen and Rollins (2012) examined settings where CPR users face sequential decisions, introducing asymmetries through the position (order) each subject has in making appropriation decisions.

Ecological variations: Building on the early studies, researchers began to study how the degree of scarcity (resource capacity relative to the capacity to appropriate) would impact appropriation decisions (e.g. Botelho et al., 2013, 2015; Osés-Eraso et al., 2008; Rutte et al., 1987). In an across-group laboratory setting using university students, where different groups face different levels of scarcity, Osés-Eraso and Viladrich-Grau (2007) report a monotonic *decrease* in appropriation levels under greater scarcity of a shared resource. More recently, however, Blanco et al. (2015) complement these studies by addressing the responses of peasants in rural Colombia to lab-induced changes in the availability of resources in a within-subject design. Interestingly, results in this study show that subjects react to strong reductions in resource availability, by *increasing* appropriation from the resource. In addition, subjects that experience a reduction in the resource availability followed by a rebound to the initial resource size appropriate *more* than those subjects who did not experience any change in the resource availability. This is in line with the results of Pfaff et al. (2015) showing durable selfishness resulting from high early scarcity. Although the Blanco et al. (2015) and Pfaff et al. (2015) studies were not designed to replicate the laboratory decision setting of Osés-Eraso and Viladrich-Grau (2007), the differences in behavior between the two studies leads naturally to the question of whether users of natural resources respond to scarcity of resources in a different way than individuals whose livelihood is not tightly related to the availability of natural resources.

Appropriation and conservation of public goods

Focusing primarily on production externalities, the early experimental literature on the commons largely neglected the relevance of ecosystem services provided by the units of the resource that are not appropriated. As discussed earlier, appropriation from CPRs may generate negative degradation externalities, by reducing the value of the public goods provided by ecosystem services. Research efforts are being undertaken to quantify the economic relevance of such ecosystem services, such as climate change mitigation and adaptation or poverty alleviation (see, for example, Costanza et al., 1997; Naidoo et al., 2008; The Economics of Ecosystems and Biodiversity[4] global initiative; the UK National Ecosystem Assessment;[5] or the Millennium Ecosystem Assessment[6]). In monetary units, ecosystem services contributed more than twice as much to human well-being as global GDP in 2011 (Costanza et al., 2014). In situations where the degradation of natural resources is a result of insufficient human incentives to maintain environmental pubic goods, these degradation externalities represent a social dilemma.

As discussed in Blanco et al. (2016), degradation externalities can be deterministic or probabilistic. Adding conservation externalities as a component to natural resource management is particularly salient for contexts of probabilistic degradation externalities where the appropriation from the natural resources creates the possibility of catastrophic outcomes. In such contexts, degradation of the remaining resources would imply the loss of economic rents that would accrue from future appropriation, and also the benefits of the ecological system dependent upon sustaining a sufficiently high level of quality of the commons. The role of the fragility of resources shared in common is increasingly relevant as changing climatic conditions are expected to affect the resilience of resources held in common (see, for example, Schröter et al., 2005) and, consequently, the complex endogenous relationship between resource use and resource degradation.

A growing number of experimental studies focus on issues related to probabilistic losses associated with the provision or maintenance of public goods. Motivated by issues of climate change, several previous studies have examined the "collective-risk social dilemma" (introduced by Milinski et al., 2008). In this setting, subjects face a known probability of group losses. Subjects can avoid this loss through collective action associated with a provision-point public good. Other similar studies have addressed the relevance of asymmetries in wealth (Milinski et al., 2011; Tavoni et al., 2011) and asymmetries in damages (Burton-Chellew et al., 2013), as well as alternative technologies for abatement, communication, and the type of uncertainty (Barrett and Dannenberg, 2012; Milinski et al., 2008, 2011; Tavoni et al., 2011). For example, in Barrett and Dannenberg (2012), each subject's payoff was reduced if contributions did not exceed a threshold. Depending on the decision setting, the payoff "impact" of contributions and the threshold were known with certainty or were uncertain. The primary finding was that treatments where the threshold was known led to greater contributions to avoid losses than the treatments with threshold uncertainty. Whether or not payoff impact was certain or uncertain did not have an important effect on contributions to avoid losses.

In related work, with a more direct focus on public goods provision, Dickinson (1998) and Gangadharan and Nemes (2009) examine provision-point public goods settings where the probability of provision of a public good increases in contributions. Extending this research, Blanco et al. (2016) used a "menu-game setting" to examine behavior in settings with probabilistic degradation of a public good from appropriation. In this type of setting, subjects make multiple one-shot decisions without feedback. They observe that introducing probabilistic degradation increases average cooperation. This increased cooperation, however, is not large enough to offset the expected losses. Blanco et al. (2017) build on the findings in Blanco et al. (2016) to address the implications from variations in the size of the potential loss. Additionally, they explore the role of expectations in shaping decisions. They find that subgroups of optimistic and pessimistic subjects, regarding the capacity of the group to maintain the shared resource, respond *qualitatively* different to the variations in the potential losses. In particular, pessimistic subjects respond systematically and significantly to changes in their expectations of others' appropriation. Surprisingly, optimistic subjects decrease appropriation relative to a benchmark game without probabilistic losses but do not systematically modify their behavior based on the expectations of group conservation.

Recent directions in natural commons research

The diversity in the experimental literature investigating the sustainable use of natural CPRs is growing. Cardenas (2009) and Brent et al. (2016) present overviews of lab, lab-in-the-field, field experiments, and random trials as applied to environmental economics. These surveys are good starting points for the interested reader. Due to space limitations, herein we focus on two topics we feel are particularly relevant to policies that address the design of collective action institutions.

Insiders-outsiders

As the policy world turns to finding solutions to environmental social dilemmas, it is apparent that such policies will often have to deal with the reality that the provision of ecosystem services often occurs in settings where efforts that only some can undertake (insiders) result in benefits to a larger group (outsiders). Examples include efforts to conserve biodiversity or forest management to retain soil and water. These settings suggest the need for institutions that allow for reciprocity between insiders and outsiders.

Compared to the large literature on social dilemma interactions, there are relatively few studies that involve actions by a group of insiders that create externalities on a group of passive outsiders. Among the existing studies, Engel and Rockenbach (2011) examine public good settings where contributions impose negative, positive, or zero externalities on outsiders, maintaining the condition that provision of the public good is pro-social at the aggregate level. The sign of the externality varies in combination with the initial endowment of outsiders such that insiders might be initially richer, poorer, or equally endowed as outsiders. The results of this study suggest that the presence of an outside group significantly reduces insiders' contributions to the public good if they face a risk of falling behind the earnings of outsiders. The authors attribute this finding to an interaction of conditional cooperation and inequity aversion. Additional studies address the relevance of social distance and communication in settings where insiders impose negative externalities on outsiders (Delaney and Jacobson, 2014; Schwartz-Shea and Simmons, 1990).

From the perspective of institutional change, a recent study (Blanco et al., 2017) examines both unconditional and conditional transfers from outsiders to insiders. The results of this study present a rather dismal outlook in regard to how endogenous voluntary transfer options might improve cooperation. In particular, they find no evidence of a sustained positive reciprocal relationship between insiders and outsiders. In this study, however, decisions and outcomes were expressed in a neutral manner, with public good contributions and transfers expressed in experimental currency. Understanding how individuals might behave differently in richer contexts will be fundamental to understanding under what conditions a strong dynamic of reciprocity can be developed between insiders and outsiders.

Artifactual experiments and contextual variables

Over the last several decades, a growing number of studies utilize the methods of laboratory experimental methods but with subject pools drawn from field settings. In many cases, participants are from small rural communities whose daily lives depend on local CPRs (e.g. fisheries and forest products). Brent et al. (2016) provide an insightful discussion of the goals of field experiments as a complement to more standard laboratory experiments using university students. In addition to providing a discussion of the various types of field experiments, they provide a framework (with examples) for delineating key attributes of framed and artefactual experiments organized around: a) the institutional setting, b) intrinsic and extrinsic motivations, c) heterogeneity across the decision makers, d) uncertainty in relation to the productivity of the resource, and e) the context and framing of the decision setting.

Of particular interest is the recent research that develops experimental settings to examine the link between behavior and variation in ecological variables. For example, experimental research conducted in Africa shows that soil fertility and water availability affect risk aversion, loss aversion, discount rates, and anti-social behavior (Prediger et al., 2014; Tanaka and Munro, 2013). Tanaka and Munro (2013) show that farmers in the least favorable climatic zone in Uganda are the most risk averse, loss averse, and impatient. Similarly, Di Falco (2014) shows that more rainfall variability is associated with higher risk aversion in Ethiopia. These results are relevant insofar as they suggest that increased variability in the availability of resources may lower the propensity of individuals to undertake investments in physical and human capital, which in turn compromises economic development. Tanaka and Munro (2013) and Prediger et al. (2014) present results showing that in Namibian villages, where soil fertility is more limited due to geological characteristics, farmers display significantly higher levels of anti-social behavior. The authors attribute this behavior to increased competitive pressures among resource users that face greater scarcity. Similarly, Hendrix and Salehyan (2012) find a significant effect of rainfall conditions on an increased propensity for disruptive activities such as demonstrations, riots, strikes, and anti-government violence.

With the goal of studying how the natural environment impacts cooperative and competitive inclinations, Leibbrandt et al. (2013) and Gneezy et al. (2015) examine the behavior of fishers in a lake setting to that of fishers in an ocean setting. Interestingly, they find that fishers located by a lake tend to fish on their own, with less cooperation, while fishers located by the ocean tend to fish in groups and show higher cooperation levels. In a series of experimental games, the latter group is found to be more trusting and trustworthy. Cardenas et al. (2017) offers further insights into how context can impact behavior. Studies in villages in China, Colombia, Nepal, and Thailand show that greater integration within the global economy leads to lower cooperation within the community. Further, this effect is strengthened when subjects are confronted with risk on collective investments.

Finally, the research by Basurto et al. (2016), in Baja California, Mexico examines how pro-social and anti-social attitudes are influenced by outside regulation of natural resources. Specifically, they study behavior in communities that are in marine protected areas and those that are not. The research question is whether anti-social attitudes (related to a desire to be ahead, spite, envy, or conflict) interact with collective action among fishers and non-fishers under these two institutional contexts. Interestingly, they find in marine protected sites, pro-social and anti-social behavior is greater, and that in this setting the presence of anti-social behavior does not appear to negatively impact pro-social behavior. They speculate that market integration, economic diversification, and stronger group identity in the protected areas allow for the combination of the two types of behavior.

Concluding comments

The early literature on the commons (e.g. Gordon, 1954; Hardin, 1968; Ostrom, 1990) is a story of both tragedy and hope. The lessons learned are that resources without property rights or well-developed collective action institutions often lead to a path of suboptimal use of resources, resource degradation, and the potential for destruction of the resource. Yet, as Elinor Ostrom so eloquently argued, that path is not a necessary outcome. Individuals and communities can learn to create institutions that facilitate cooperation and long-lasting commons. In many ways, the now long history of experiments that focus on the commons tells a similar story.

In experimental settings with stark institutional environments that do not allow for communication, punishment, or other mechanisms to signal or enforce cooperation, subjects follow a path that leads to suboptimal outcomes (even destruction of the laboratory commons). Experimentalists, however, have provided ample evidence that institutional change can play an important role in promoting cooperation. Most of the experimental evidence comes from environments where the institutional change is created exogenously (introduced by the experimenter). Some evidence, however, comes from experimental environments where the institutional change arises endogenously, where there exists a second order public goods provision dilemma in providing the institution.

See Ostrom et al. (1994) and Ramalingam et al. (2016) for a discussion of this issue. Likewise, experimentalists have turned to field settings (often in areas of the world where individuals face issues of the commons that are vital to their livelihoods) attempting to better understand to what extent behavior is impacted by context, histories, and their reliance on resources.

In summary, employing a framework for analysis that begins with non-cooperative game theory and an assumption of self-regarding behavior provides a foundation from which one can examine the behavioral consequences of institutional change and context. As we argue, for completeness, this framework must include incentive structures that capture both production externalities related to appropriation and conservation externalities related to the public good nature of ecosystem services. Experimental economics, whether through the lens of university laboratories or field settings, provides methodological tools for gaining a better understanding of how human beings respond to institutional changes, changes in incentives, and ecological context.

Notes

1 As a starting point, see Ledyard (1995) and Chaudhuri (2011) for reviews of the literature on public good experiments. See Cherry et al. (2007) for a discussion of environmental economics and experimental methods and Cornes and Sandler (1986) for an extensive discussion of analytical models related to externalties and public goods.
2 This work has set the building ground for more recent contributions in this literature (Agrawal 2001; Anderies et al., 2004).
3 Also see Budescu et al. (1995) for an early experiment dealing with resource uncertainty.
4 www.teebtest.org/.
5 http://uknea.unep-wcmc.org/Resources/tabid/82/Default.aspx.
6 www.unep.org/maweb/en/Responses.aspx.

References

Agrawal, A. (2001) "Common property institutions and sustainable governance of resources" *World Development* 29: 1649–1672.

Anderies, J. M., Janssen, M. A., and Ostrom, E. (2004) "A framework to analyze the robustness of social-ecological systems from an institutional perspective" *Ecology and Society* 9(1): article 18. www.ecologyandsociety.org/vol9/iss1/art18/.

Barrett, S., and Dannenberg, A. (2012) "Climate negotiations under scientific uncertainty" *Proceedings of the National Academy of Sciences* 109(43): 17372–17376.

Basurto, X., Blanco, E., Nenadovic, M., and Vollan, B. (2016) "Integrating simultaneous prosocial and antisocial behavior into theories of collective action" *Science Advances* 2e1501220.

Berkes, F., Feeny, D., McCay, B. J., and Acheson, J. M. (1989) "The benefits of the commons" *Nature* 340: 91–93.

Bischoff, I. (2007) "Institutional choice versus communication in social dilemmas: an experimental approach" *Journal of Economic Behavior and Organization* 62: 20–36.

Blanco, E., Haller, T., Lopez, M. C., and Walker, J. M. (2016) "The opportunity costs of conservation with deterministic and probabilistic degradation externalities" *Environmental and Resource Economics* 64: 255–273.

Blanco, E., Haller, T., and Walker, J. (2017) "Externalities in appropriation: responses to probabilistic losses" *Experimental Economics* DOI 10.1007/s10683-017-9511.

Blanco, E., Lopez, M. C., and Villamyor-Tomas, S. (2015) "Exogenous degradation in the commons: field experimental evidence" *Ecological Economics* 120: 430–439.

Botelho, A., Dinar, A., Pinto, L., and Rapoport, A. (2013) "Time and uncertainty in resource dilemmas: equilibrium solutions and experimental results" *Experimental Economics* 17(4): 649–672.

Botelho, A., Dinar, A., Pinto, L. M. C., and Rapoport, A. (2015) "Promoting cooperation in resource dilemmas: theoretical predictions and experimental evidence" *Journal of Behavioral and Experimental Economics* 54: 40–49.

Brent, D., Friesen, L., Gangadharan, L., and Leibbrandt, A. (2016) "Behavioral insights from field experiments in environmental economics" *International Review of Environmental and Resource Economics* 10: 95–143.

Bromley, Daniel W., ed. (1992) *Making the commons work: theory, practice, and policy*. Institute for Contemporary Studies Press, San Francisco, CA.

Buckley, N., Mestelman, S., and Muller, R. (2008) "Baseline-and-credit emission permit trading: experimental evidence under variable output capacity" *Experimental methods in environmental economics*, Cherry T., Kroll S., and Shogren J. eds. Routledge, London and New York, 9–28.

Budescu, D. V., Rapoport, A., and Suleiman, R. (1995) "Common pool resource dilemmas under uncertainty: qualitative tests of equilibrium solutions" *Games and Economic Behavior* 10: 171–201.

Burton-Chellew, M., May, R., and West, S. (2013) "Combined inequality in wealth and risk leads to disaster in the climate change game" *Climatic Change* 120(4): 815–830.

Cardenas, J. C. (2009) "Experiments in environment and development" *Annual Review of Resource Economics* 1: 157–182.

Cardenas, J. C. (2011) "Social norms and behavior in the local commons as seen through the lens of field experiments" *Environmental and Resource Economics* 48: 451–485.

Cardenas, J. C., Janssen, M. A., Ale, M., Bastakoti, R., Bernal, A., Chalermphol, J., . . . and Anderies, J. M. (2017) "Fragility of the provision of local public goods to private and collective risks" *Proceedings of the National Academy of Sciences of the United States of America* 114: 921–925.

Cardenas, J. C., Stranlund, J. K., and Willis, C. E. (2000) "Local environmental control and institutional crowding-out" *World Development* 28: 1719–1733.

Chaudhuri, A. (2011) "Sustaining cooperation in laboratory public goods experiments: a selective survey of the literature" *Experimental Economics* 14(1): 47–83.

Cherry, T. L., Kroll, S., and Shogren, J. F. (2007) *Environmental economics, experimental methods*. Routledge, New York.

Cornes, R., and Sandler, T. (1986) *The theory of externalities, public goods, and club goods*. Cambridge University Press, Cambridge, UK.

Costanza, R., d'Arge, R., de Groot, R., Farber, S., Grasso, M., Hannon, B., . . . and van den Belt, M. (1997) "The value of the world's ecosystem services and natural capital" *Nature* 387: 253–260.

Costanza, R., de Groot, R., Sutton, P., van der Ploeg, S., Anderson, S. J., Kubiszewski, I., . . . Turner, R. K. (2014) "Changes in the global value of ecosystem services" *Global Environmental Change* 26: 152–158.

Delaney, J., and Jacobson, S. (2014) "Those outsiders: how downstream externalities affect public good provision" *Journal of Environmental Economics and Management* 67(3): 340–352.

Dickinson, D. L. (1998) "The voluntary contributions mechanism with uncertain group payoffs" *Journal of Economic Behavior & Organization* 35(4): 517–533.

Di Falco, S. (2014) "Adaptation to climate change in sub-Saharan agriculture: assessing the evidence and rethinking the drivers" *European Review of Agricultural Economics* 41: 405–430.

Engel, C., and Rockenbach, B. (2011) "We are not alone: the impact of externalities on public good provision" *MPI Collective Goods Preprint* 2009: 29.

Feeny, D., Berkes, F., McCay, B., and Acheson, J. (1990) "The tragedy of the commons: twenty-two years later" *Human Ecology* 18(1): 1–19.

Fehr, E., and Leibbrandt, A. (2011) "A field study on cooperativeness and impatience in the tragedy of the commons" *Journal of Public Economics* 95: 1144–1155.

Fischer, M. E., Irlenbusch, B., and Sadrieh, A. (2004) "An intergenerational common pool resource experiment" *Journal of Environmental Economics and Management* 48: 811–836.

Gangadharan, L., and Nemes, V. (2009) "Experimental analysis of risk and uncertainty in provisioning private and public goods" *Economic Inquiry* 47(1): 146–164.

Gneezy, U., Leibbrandt, L., and List, J. (2015) "Ode to the sea: workplace organizations and norms of Cooperation" *Economic Journal* 126: 1856–1883.

Gordon, H. S. (1954) "The economic theory of a common-property resource" *Journal of Political Economy* 62: 124–142.

Hackett, S., Schlager, E., and Walker, J. (1994) "The role of communication in resolving commons dilemmas: experimental evidence with heterogeneous appropriators" *Journal of Environmental Economics and Management* 27: 99–126.

Hardin, G. (1968) "The tragedy of the commons" *Science*, 162(3859): 1243–1248.

Hayo, B., and Vollan, B. (2012) "Group interaction, heterogeneity, rules and cooperative behaviour: evidence from a common-pool resource experiment in South Africa and Namibia" *Journal of Economic Behavior and Organization* 81: 9–28.

Hendrix, G. S., and Salehyan, I. (2012) "Climate change, rainfall, and social conflict in Africa" *Journal of Peace Research* 49: 35–50.

Herr, A., Gardner, R., and Walker, J. (1997) "An experimental study of time-independent and time-dependent externalities in the commons" *Games and Economic Behavior* 19: 77–96.

Janssen, M. A. and Rollins, R. (2012) "Evolution of cooperation in asymmetric commons dilemmas" *Journal of Economic Behavior and Organization* 81: 220–229.

Ledyard, J. (1995) "Public goods: a survey of experimental research," in *Handbook of experimental economics*, Kagel, J. and Roth, A., eds. Princeton University Press, Princeton, NJ.

Leibbrandt, A., Gneezy, U., and List, J.A. (2013) "Rise and fall of competitiveness in individualistic and collectivistic societies" *Proceedings of the National Academy of Sciences* 110(23): 9305–9308.

Margreiter, M., Sutter, M., and Dittrich, D. (2005) "Individual and collective choice and voting in common pool resource problem with heterogeneous actors" *Environmental and Resource Economics* 32: 241–271.

McCay, B., and Acheson, J. (1987) *The question of the commons: the culture and ecology of communal resources*. University of Arizona Press, Tucson, AZ.

Milinski, M., Röhl, T., and Marotzke, J. (2011) "Cooperative interaction of rich and poor can be catalyzed by intermediate climate targets" *Climatic Change* 109: 807–881.

Milinski, M., Sommerfeld, R. D., Krambeck, H. J., Reed, F. A., and Marotzke, J. (2008) "The collective-risk social dilemma and the prevention of simulated dangerous climate change" *Proceedings of the National Academy of Sciences* 105(2): 2291–2294.

Naidoo, R., Balmford, A., Costanza, R., Fisher, B., Green, R. E., Lehner, B., . . . and Ricketts, T. H. (2008) "Global mapping of ecosystem services and conservation priorities" *Proceedings of the National Academy of Sciences* 105: 9495–9500.

Netting, R. (1981) *Balancing on an Alp: ecological change and continuity in a Swiss mountain community*. Cambridge University Press, New York.

Olson, M. (1965) *The logic of collective action: public goods and the theory of groups*. Harvard University Press, Cambridge, MA.

Osés-Eraso, N., Udina, F., and Viladrich-Grau, M. (2008) "Environmental versus human-induced scarcity in the commons: do they trigger the same response?" *Environmental and Resource Economics* 40: 529–550.

Osés-Eraso, N., and Viladrich-Grau, M. (2007) "Appropriation and concern for resource scarcity in the commons: an experimental study" *Ecological Economics* 63(2–3): 435–445.

Ostrom, E. (1990) *Governing the commons*. Cambridge University Press, New York

Ostrom, E. (2006) "The value-added of laboratory experiments for the study of institutions and common-pool resources" *Journal of Economic Behavior and Organization* 61: 149–163.

Ostrom, E., Gardner, R., and Walker, J. (1994) *Rules, games, and common-pool resources*. University of Michigan Press, Ann Arbor, MI.

Ostrom, E., Walker, J., and Gardner, R. (1992) "Covenants with and without a sword: self-governance is possible" *The American Political Science Review* 86: 404–417.

Pfaff, A., Vélez, M. A., Ramos, P. A., and Molina, A. (2015) "Framed field experiment on resource scarcity & extraction: path-dependent generosity within sequential water appropriation" *Ecological Economics* 120: 416–429.

Poteete, A., Janssen, M., and Ostrom, E. (2010) *Working together: collective action, the commons, and multiple methods in practice*. Princeton University Press, Princeton, NJ.

Prediger, S., Vollan, B., and Frölich, M. (2014) "Resource scarcity and antisocial behavior" *Journal of Public Economics* 119: 1–9.

Ramalingam, A., Godoy, S., Morales, A., and Walker, J. (2016) "An individualistic approach to institution formation in public good games" *Journal of Economics Behavior and Organization* 129: 18–36.

Rodriguez-Sickert, C., Guzman, R. A., and Cardenas, J. C. (2008) "Institutions influence preferences: evidence from a common pool resource experiment" *Journal of Economic Behavior and Organization* 67: 215–227.

Rutte, C.G., Wilke, H.A.M., and Messick, D.M. (1987) "Scarcity or abundance caused by people or the environment as determinants of behavior in the resource dilemma" *Journal of Experimental Social Psychology* 23(3): 208–216.

Schmitt, P., Swope, K., and Walker, J. (2000) "Collective action with incomplete commitment: experimental evidence" *Southern Economic Journal* 66(4): 829–854.

Schott, S., Buckley, N., Mestelman, S., and Muller, R. (2007) "Output sharing in partnerships as a common pool resource management instrument" *Environmental and Resource Economics* 37(4): 697–711.

Schröter, D., Cramer, W., Leemans, R., Prentice, I. C., Araújo, M. B., Arnell, N. W., . . . and Zierl, B. (2005) "Ecosystem service supply and vulnerability to global change in Europe" *Science* 310: 1333–1337.

Schwartz-Shea, P., and Simmons, R. T. (1990) "The layered prisoners' dilemma: ingroup versus macro-efficiency" *Public Choice* 65(1): 61–83.

Sengupta, N. (1991) *Managing common property: irrigation in India and the Philippines*. Sage Publications, New Delhi.

Tanaka, Y., and Munro, A. (2013) "Regional variation in risk and time preferences: evidence from a large-scale field experiment in rural Uganda" *Journal of African Economies* 23: 151–187.

Tavoni, A., Dannenberg, A., Kallis, G., and Löschel, A. (2011) "Inequality, communication, and the avoidance of disastrous climate change in a public goods game" *Proceedings of the National Academy of Sciences* 108(29): 11825–11829.

Velez, M. A., Stranlund, J. K., and Murphy, J. J. (2009) "What motivates common pool resource users? Experimental evidence from the field" *Journal of Economic Behavior and Organization* 70: 485–497.

Vollan, B. (2008) "Socio-ecological explanations for crowding-out effects from economic field experiments in Southern Africa" *Ecological Economics* 67: 560–573.

Vollan, B., and Ostrom, E. (2010) "Cooperation and the commons" *Science* 330: 923–924.

Vyrastekova, J., and van Soest, D. (2003) "Centralized common-pool management and local community participation" *Land Economics* 79: 500–514.

Walker, J., and Gardner, R. (1992) "Probabilistic destruction of common-pool resources: experimental evidence" *Economic Journal* 102: 1149–1161.

Walker, J., Gardner, R., and Ostrom, E. (1990) "Rent dissipation in limited access common pool resource environments: experimental evidence" *Journal of Environmental Economics and Management* 19(3): 203–211.

Walker, J., Gardner, R., Herr, A., and Ostrom, E. (2000) "Collective choice in the commons: experimental results on proposed allocation rules and votes" *Economic Journal* 110: 212–234.

10

HUMANISTIC RATIONAL CHOICE

Understanding the fundamental motivations that drive self-organization and cooperation in commons dilemmas

Daniel A. DeCaro

Introduction

Human civilization is built on humanity's capacity for cooperation. Cooperation is necessary to provide the public good and overcome fundamental societal dilemmas, such as competition over limited resources and opportunities, which can contribute to war and environmental degradation (Parks et al. 2013). Broad-scale cooperation is supported by governance systems that promote rule-abiding behavior and direct members of society with diverse interests and beliefs towards more cooperative outcomes (Ostrom 1998; Tyler 2006). For example, in a common-pool resource (CPR) dilemma, stakeholders compete to use a shared resource (e.g., water) and are tempted to harvest more than their share (Hardin 1968). Without effective governance, such as conservation agreements and regulatory systems to limit consumption, smooth coordination, and build mutual trust, these stakeholders may collectively destroy the resource, destabilizing society and undermining human welfare (UNGA 2015).

Recent world events demonstrate that cooperation is not guaranteed. CPRs are collapsing worldwide, and there is heightened war and societal conflict (UNGA 2015). Furthermore, democratic societies are under great stress. Basic political civility seems to be unraveling. Fundamental institutions, such as democratic (shared) decision making and democratic pluralism (e.g., diversity, public discourse, polycentricity), which are necessary for robust societal cooperation, are being seriously challenged (e.g., Chokshi 2016; Norris 2016), jeopardizing societal sustainability and resilience (DeCaro et al. 2017a).

These cooperative failures arise in large part because contemporary governance systems are based on faulty behavioral assumptions (Poteete et al. 2010). Traditional rational choice theory, which assumes decision makers are narrowly self-interested—motivated primarily by personal costs/benefits—underestimates human capacity for voluntary cooperation (Bowles 2008); self-organization to collectively solve complex dilemmas (Ostrom 2010); concern for others' welfare (Fehr and Gächter 2002); social norms (Fehr and Fischbacher 2004); and preferences for equitable costs and benefits (Dawes et al. 2007). Social-psychological and justice perspectives (e.g., Deci and Ryan 2000), which emphasize more humanistic social motivations (e.g., self-determination, fairness) and intrinsic reasons for cooperation (e.g., moral responsibility, enjoyment) (e.g., Moller et al. 2006), alternatively underestimate the need for more stringent, rationally informed social safeguards, such as enforcement, to deter self-interested actors (Hardin 1968), uphold justice (Tyler 1990, 2006), and protect more altruistic, intrinsically motivated forms of cooperation, such as cooperative agreements (DeCaro et al. 2015).

As a result, the psychological processes involved in fundamental aspects of governance, such as shared decision making (Chess and Purcell 1999), enforcement (Bowles 2008), and communication (Balliet 2010; Pavitt 2011), are poorly understood (Anderies et al. 2013). This gap in scholars' and policymakers' understanding of the psychological effects of essential aspects of governance has generated both contradictory solutions and mixed outcomes (e.g., Bowles 2008; Agrawal and Ribot 2014), making it difficult to envision more effective governance systems (see DeCaro et al. 2017a for review).

Many explanatory frameworks of cooperative decision making have been proposed, and some focus on aspects of psychology (Parks et al. 2013). However, no existing frameworks or behavioral theories provide a unitary account of behavior or rival rational choice theory, with its emphasis on narrow self-interest, as the most informative behavioral theory of societal cooperation (see Ostrom 1998, 2010; Poteete et al. 2010 for discussion). Therefore, despite its failings, narrow self-interest is still assumed to be the best explanation for human behavior, and traditional rational choice theory remains the dominant public policy perspective.

This chapter seeks to advance behavioral theory by further developing Elinor Ostrom's (1998, 2010) theoretical account of democratic, citizen-led governance (see also Poteete et al. 2010). A deeper motivational foundation for cooperation is proposed by focusing on fundamental social-psychological motivations and cognitions, such as procedural justice, security, and legitimacy, which are just as fundamental as narrow self-interest. By focusing on fundamental psychological processes, it is possible to outline core principles of cooperative decision making that bridge disciplines and better explain several persistent questions about cooperation, improving public policy and accounting for a wider range of societal dilemmas.

Reframing the problem of self-interest

Self-interest is fundamental to behavioral theory of collective action in social dilemmas. Assumptions about self-interest influence how cooperative problems are framed, constraining research and public policy (Miller 1999; Ostrom 2010). These assumptions must be rethought to properly account for persistent questions about cooperation.

Rational self-interest

According to rational choice theory, people are too narrowly self-interested to solve important societal problems together. People are expected to act in their exclusive interest for strategic advantage, undermining mutual trust and any potential agreements that might support robust cooperation. Hence, people are assumed to be externally motivated to cooperate. They must be coerced into cooperating via outside intervention by a strong, centralized authority (Hobbes 1909 [1651]; Hardin 1968). Stakeholders will not self-organize, and external monitoring and enforcement are assumed necessary and sufficient to ensure compliance (Ostrom 1998, 2010).

Boundedly rational self-interest

Elinor Ostrom's (1990, 1998, 2010) account of self-governance systems where citizens manage complex societal problems together, demonstrates that traditional rational choice theory, for all its benefits, overgeneralizes narrow self-interest, overlooking humanity's capacity for cooperation. Self-organization is fundamental to human society, and humans are interested in broader goals than narrow self-interest (Poteete et al. 2010). Ordinary people can solve complex societal dilemmas under the right conditions; many do so voluntarily, with minimal outside intervention (Ostrom 1990).

Thus, the Ostrom School of Political Economy demonstrates that self-organization and cooperation are possible and identifies key factors (Cole and McGinnis 2015). However, Ostrom's perspective still adopts (boundedly) rational self-interest as its motivational starting point and assumes stakeholders are externally motivated (Ostrom 2010). Therefore it cannot explain why people decide to self-govern in the first place, or continue to voluntarily cooperate long term, especially when cooperation is risky or personally costly, as is often the case (Ostrom 1990; Poteete et al. 2010: appendix 9.1).

Boundedly rational humanistic self-interest

In this chapter, principles of human agency from humanistic psychology are adapted to develop a more fruitful motivational starting point for behavioral theory. Whereas traditional rational choice theory arguably represents the darker side of self-interest, assuming stakeholders must be coerced to cooperate despite their inner demons (Hobbes 1909 [1651]), humanistic psychology represents the brighter side of self-interest (Ryan and Deci 2000), assuming narrow selfishness is tempered by other, equally fundamental, motivations (e.g., DeCaro et al. 2015).

Considerable research in humanistic social psychology demonstrates that humans are neither inherently passive nor narrowly self-interested. People are agentic: they are motivated to solve problems that affect their well-being (Bandura 2001). They are also driven by fundamental social-psychological needs (Deci and Ryan 2000). These needs jointly guide self-interest. Narrow self-determination is tempered by needs like belonging and procedural justice (fair governance procedures), curtailing unrestrained self-interest. Social groups rarely tolerate outright selfishness (Fehr and Fischbacher 2004). Furthermore, societal systems (e.g., family, social norms, governance systems) are typically designed to socialize self-interest towards more acceptable forms of collective welfare, which take into account cultural preferences for important social needs and motivations (Ryan and Deci 2000; Chen et al. 2015). The brighter aspects of human nature can be cultivated to improve societal cooperation (Ostrom 1998).

This perspective reframes the problem of cooperation. Traditional rational choice theory assumes people are trapped in social dilemmas and must be forced to cooperate. Humanistic Rational Choice Theory (HRCT) acknowledges self-interest but also assumes people are motivated by fundamental needs. These needs drive self-organization and cooperation. Hence, people are inclined to solve important societal dilemmas to satisfy their fundamental needs, and they will cooperate best if those needs are satisfied. The problem is therefore one of finding proper societal supports to help stakeholders reach these goals, without getting in the way too much, or letting any need get out of balance. This implies that cooperation is poor when implementation of governance interventions, like shared decision making, communication, and enforcement, improperly address fundamental needs. An integrative interdisciplinary framework is developed next to clarify these psychological processes.

Theoretical framework

The currently proposed framework seeks to better explain why competitive stakeholders self-organize and how key aspects of governance exert their effects psychologically to influence cooperation. The framework is interdisciplinary, synthesizing multiple perspectives. It examines the involvement of several core social-psychological processes: legitimacy, fundamental needs, causal attributions (explanations for others' behavior), group merging (closeness and liking), rule acceptance and internalization, and trust. It is proposed that factors that promote cooperation do so in part by affecting fundamental needs, which then influence other important social cognitions, such as rule acceptance, group merging, and trust. These psychological processes influence behavior along with the perceived costs/benefits of cooperation.

Legitimacy, fundamental social-psychological needs, and motivation

Legitimacy plays a central role in governance, building trust, facilitating cooperation and rule compliance, and helping resolve contentious social dilemmas (e.g., Leach and Sabatier 2005) (see De Cremer and Tyler 2005; Tyler 2006 for review). Participatory and regulatory aspects of governance, such as shared decision making and enforcement, strongly determine legitimacy perceptions (Tyler 1990, 2006). Fundamental needs also influence legitimacy, because needs like procedural justice (Tyler 2006), self-determination (Brehm and Brehm 1981), and security (Hobbes 1909 [1651]) represent fundamental dimensions of governance.

The basic psychological processes that link fundamental needs to legitimacy, and subsequent trust and motivation (e.g., rule acceptance), have been described in social-psychology (Deci and Ryan 2000) and social justice research (Tyler 2006). Humans have fundamental needs, which are essential for well-being and fundamentally drive behavior (Baumeister and Leary 1995; Deci and Ryan 2000). Two of these needs are (a) procedural justice, fair decision-making procedures that support one's voice or decision-making influence (Tyler 2006) and (b) self-determination, pursuing goals in ways that align with one's core values and sense of agency (Ryan and Deci 2006). Security, belonging, competence (or efficacy), and material needs (e.g., money) are also important. Security refers to the predictability and orderliness of social interactions, governance systems, and outcomes. Security supports long-term planning and commitment (Hobbes 1909 [1651]). People also do not want to be taken advantage of by others, who may disobey rules or break their agreements (Ostrom 1990; Milinski et al. 2002). Finally, people need to feel like they belong in their social groups (Baumeister and Leary 1995), feel competent at solving important problems in their lives (Bandura 2001), and have the basic material resources to survive.

HRCT assumes people look to societal systems to fulfill their fundamental needs, asking themselves: Is this system procedurally fair? Does it support my self-determination, improve my competence, and support meaningful social relationships? Does it provide my material (economic) needs and security?

Systems that satisfy fundamental needs typically improve legitimacy, promoting acceptance and cooperation (Frey et al. 2004; Tyler 2006). Individuals are more cooperative and rule abiding when they accept the governance systems they encounter (institutional acceptance) and whole-heartedly endorse prescribed behaviors (internalized motivation) (Moller et al. 2006; Tyler 2006). Individuals who are internally motivated often cooperate more voluntarily (DeCaro and Stokes 2008, 2013). In contrast, individuals motivated primarily for external reasons, such as monetary reward or punishment, may not persist without continued external reinforcement (Bowles 2008).

When fundamental needs are not satisfied, people are motivated to correct the deficits (Vansteenkiste and Ryan 2013). It is therefore proposed that this motivates self-organization in many forms: activism and self-governance (Ostrom 2010), governmental reform (Tyler 1990), and even defiance and non-cooperation, to overturn illegitimate systems (McComas et al. 2011).

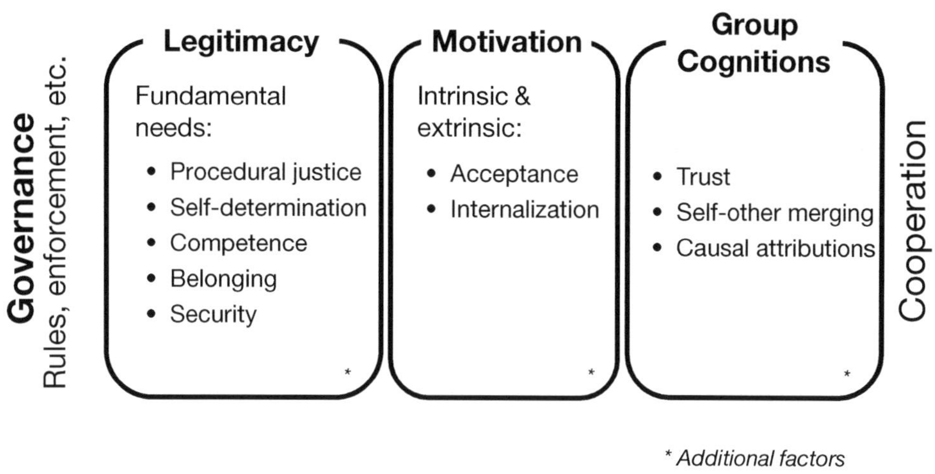

Figure 10.1 Framework for Humanistic Rational Choice Theory

In short, as illustrated in Figure 10.1, fundamental needs centrally influence cooperation. Subjective perceptions of the supportiveness of specific governance systems (e.g., enforcement), determine stakeholder motivation (e.g., rule acceptance), and subsequent cooperation. Cooperation is also affected by other factors, like group self-merging, causal attributions, and trust, as described next.

Self-other merging

Self-other merging refers to the process of transitioning from thinking of oneself as separate and emotionally detached (an "I"), to thinking of oneself as close and positively linked to a group ("We") (De Cremer et al. 2005). Cooperative groups typically develop a sense of cohesion, liking, and belonging, which supports trust and maintains group norms (McCay and Acheson 1987). When group members merge they tend to internalize the group's norms, and their individual self-interest more closely aligns with the group's (De Cremer and Van Vugt 1999), improving cooperation (Poteete et al. 2010).

Satisfaction of fundamental needs by governance systems facilitates self-other merging (De Cremer and Tyler 2005; DeCaro et al. 2015). Therefore, governance systems should strive to facilitate self-other merging, using methods that support both fundamental needs and group development (Figure 10.1).

Causal attributions and reputations

Beliefs about the causes of others' behavior (causal attributions) are also important (Dawes et al. 1977). For example, economic sanctions (monetary punishments) sometimes undermine long-term cooperation, by making group members think that no one wants to cooperate intrinsically (Mulder et al. 2006). Reputations influence cooperation similarly: Prior cooperation can improve group-liking, facilitate self-other merging, and bolster future cooperation (Milinski et al. 2002; Abele and Stasser 2008).

Group members' reputations and attributions, and their consequences for cooperation, are expected to depend on the broader social context and the governance system (Figure 10.1). In particular, different governance situations may satisfy different needs, yielding different attributions and outcomes.

Trust

Ostrom (1998, 2010) proposed that trust is central to societal cooperation. In social dilemmas, otherwise competitive stakeholders must take personal risks—contributing time, effort, and material goods—to provide public goods (e.g., irrigation systems). They must also put their best interests at stake in situations where they could be taken advantage of by others (e.g., leaving available water resources untouched so the pool can replenish). Factors that increase cooperation likely do so by fostering trust in these risky situations (e.g., Mulder et al. 2006).

Ostrom (1998: 9–15) described this basic process in terms of a reciprocal relationship among reciprocity, reputation, and trust. People are assumed to have a natural inclination towards reciprocity, responding in kind to others' behavior. Governance systems affect cooperation by establishing conditions that are more or less conducive to reciprocity and the formation of positive (or negative) reputations, which support long-term trust and cooperation. For example, communication among stakeholders sometimes facilitates cooperation. Communication may do so, in part, by giving dilemma stakeholders the opportunity to make initial cooperative agreements, build a positive reputation by honoring those agreements, and thereby begin a virtuous cycle of reciprocity, where others respond in kind, strengthening their trust and commitment over time (e.g., Ostrom et al. 1992). However, reciprocity does not provide a full account of cooperation, because cooperation does not always succeed (e.g., even with communication), people sometimes cooperate when others are uncooperative, and narrowly self-interested individuals must still somehow be motivated to take initial costly steps towards cooperation despite their self-interest (Ostrom 1998: 13; Poteete et al. 2010: 243).

Therefore, HRCT seeks to understand how fundamental needs are involved in this relationship, whereby governance systems facilitate development of trust and positive reputations (e.g., self-other merging, causal attributions) that motivate and sustain robust cooperation (Figure 10.1). These concepts are demonstrated in an empirical example next.

Demonstration: voting and enforcement

DeCaro et al. (2015) examined the psychological effects of shared decision making (voting) and enforcement (graduated economic sanctions) in a CPR experiment. Groups (N=40) harvested resources worth $0.01 each from a shared pool. During Phase 1 (baseline), there were no conservation rules, and the CPR was open access. During Phase 2 (treatment), voting and enforcement were introduced. Individuals in the Voted-Enforce condition voted on a predetermined list of conservation rules derived from real-world case studies and could use costly economic sanctions to enforce them, paying $0.02 whenever they wanted to deduct $0.04 from another group member. In other conditions, individuals could not vote (Imposed-Enforce condition), lacked enforcement (Voted condition), or both (Imposed condition). During Phase 3 (carryover), the ability to enforce was removed. Participants' fundamental motivations and perceptions were also measured, allowing the researchers to observe the immediate and long-term effects of voting and enforcement.

Groups in the Voted-Enforce condition showed the biggest increase in cooperation after voting and enforcement were introduced (Phase 2) and continued to cooperate voluntarily after enforcement was removed (Phase 3). Cooperation did not improve when enforcement was introduced in the Imposed-Enforce condition (Phase 2) and further declined when enforcement was removed (Phase 3). Thus, enforcement improved cooperation only when individuals voted. Cooperation was moderate and tied in the Voted and Imposed conditions. Thus, shared decision making was not sufficient by itself.

These outcomes were well accounted for by participants' perceptions. Specifically, in the Voted-Enforced condition, voting increased overall perceptions of procedural justice and self-determination; and the combination of voting with enforcement increased security. Fundamental needs (Phase 2) correlated with higher rule acceptance, internalized motivation, and self-other merging, which predicted subsequent voluntary cooperation (Phase 3) when controlling for anticipated earnings and other factors. Hence, voting on rules, and having the ability to initially enforce them, satisfied fundamental needs, thereby increasing rule acceptance and internalization, group self-other merging, and subsequent voluntary cooperation. The Imposed-Enforce condition did not have any of these beneficial psychological effects. The Voted condition and the Imposed condition each lacked one aspect of these benefits. For example, in the Voted condition, voting increased procedural justice and self-determination, but lacked security.

The Voted-Enforce condition was successful because voting and enforcement reinforced one another. Voting legitimized the rules and their enforcement. It made sense to protect rules chosen by the group, and having the ability to enforce them was empowering. Enforcement facilitated cooperation long enough (Phase 2) for the group to build positive reputations, build trust, and grow their support for the rule, thereby explaining why they continued to cooperate even after enforcement ceased (Phase 3). Enforcement was never possible in the Voted condition, so group members did not build positive reputations, develop robust trust, or cultivate support for the rule: The losing voters pillaged the resource while the winner voters complied, yielding moderate levels of cooperation overall. Finally, in the Imposed-Enforce condition, the rules were perceived as illegitimate, so enforcing them seemed unjustified. Enforcement felt purely coercive and crowded out internal motivation, giving little support for group members to build positive reputations or learn to trust one another.

Thus, differential satisfaction of fundamental needs, and associated effects on underlying motivations and perceptions, helped explain when and why two aspects of governance improve cooperation. It is important to ensure security without threatening other fundamental needs (e.g., procedural justice) or inadvertently undermining more intrinsic reasons for cooperating. Furthermore, enforcement systems should empower stakeholders by helping them reach shared goals and build positive reputations, not simply force compliance (Ostrom 2000). Voting with enforcement better addressed these principles in the current experiment, so humanistically self-interested agents responded with better cooperation.

Theoretical implications

DeCaro et al.'s (2015) research is informative, because social dilemma scholars often explain their results in terms of fundamental needs like security, procedural justice, and self-determination (e.g., Ostrom 2000; Bowles 2008; Vollan 2008), but few actually measure these psychological processes (Anderies et al. 2011). This kind of interdisciplinary, psychological research is necessary to understand cooperative behavior and reconcile disciplinary perspectives. DeCaro et al.'s (2015) project incorporated many disciplinary perspectives, but none fully explained the observed results. Enforcement without shared decision making (Imposed-Enforce condition) did not deter rule violators or improve cooperation, as commonly predicted by rational choice theory (e.g., Becker 1974; Bowles 2008). Voting (Voted condition) did not encourage high levels of voluntary cooperation by itself, as commonly assumed in social-psychological theories of compliance (e.g., Moller et al. 2006). Furthermore, voting with enforcement (Voted-Enforce condition) was not only beneficial but also increased voluntary cooperation, arguing against a strong crowding-out effect of enforcement as hypothesized in social psychology (e.g., Moller et al. 2006) and some interdisciplinary perspectives in behavioral economics (e.g., Frey et al. 2004; Bowles 2008). Finally, theoretical perspectives in social psychology and social justice research accurately predicted that fundamental needs influence rule acceptance and voluntary cooperation, but these perspectives did not predict the specific governance systems involved: Enforcement did not decrease perceptions of procedural justice or self-determination in the Voted-Enforce condition, or reduce voluntary cooperation and compliance, as commonly predicted. These findings also illustrate how fundamental needs may be involved in Ostrom's (1998) hypothesized relationship among trust, reciprocity, and reputations: Governance systems that satisfy fundamental needs create conditions which are more conducive to the formation of positive reciprocity and development of trust.

The current section demonstrates the value of HRCT. The next sections apply the theory to address major questions about cooperation, focusing on three essential aspects of governance: shared decision making, enforcement, and communication.

Shared decision making and enforcement

Governance mechanisms influence cooperation by altering structural components of the dilemma that determine perceived costs and benefits of cooperation (Messick and Brewer 1983). However, the effects of these mechanisms are often mixed. HRCT addresses several important questions, which have plagued shared decision making and enforcement.

Participatory democracy

- Why does shared decision making sometimes decrease cooperation?

Shared decision making is a type of participatory democracy ("participation"), where stakeholders contribute to governance feedback, design, decision making, or implementation. Participation is foundational to legitimacy (Tyler 2006), and it can promote cooperation, but not always (Arrow 1951).

DeCaro and Stokes (2008, 2013) reviewed case studies of CPR governance, demonstrating that experts often misunderstand participation's psychological complexities. Misconceptions lead to poor participatory design and implementation, which undermines legitimacy and decreases cooperation. For example, failure to recognize that different participatory methods yield different psychological effects in different contexts, has led to widespread use of simplistic, one-size-fits-all participatory processes, which actually undermine procedural justice and self-determination. Failure to grasp these nuances also allows participatory processes to be exploited by powerholders in society, through superficial or manipulative public engagement (e.g., Clement 2010; NCL 2013). Such misapplications create conflict and contribute to mixed outcomes of participation (e.g., McComas et al. 2011). The unreliability of participation has generated significant debate about its utility (Reed 2008).

To resolve these issues, scholars must ask how these participatory processes are perceived by various stakeholders in different contexts (DeCaro and Stokes 2013). Majority voting is a common form of shared decision making in Western societies, but it often yields mixed results. For example, Rauchdobler et al. (2010) found that voting for investments to a public good was counterproductive, because the losing voters would not comply. This fracturing effect of voting is common in research (e.g., Vollan 2008), and national politics (e.g., Brexit, U.S. election cycles; Arrow 1951). DeCaro et al. (2015) replicated and explained this effect: The losing voters felt that the majority vote limited their control over the chosen rules, undermining some of their procedural justice and self-determination. Hence, steps should be taken to mitigate this side-effect. Communication could help group members reach common ground when voting, allowing them to share ideas, explain costs/benefits of the options, and potentially reach a compromise (Ostrom et al. 1992), thereby supporting greater procedural justice and self-determination for everyone (Moller et al. 2006; Tyler 2006).

Legitimization of enforcement and empowerment

- How is enforcement legitimized, mitigating counterproductive effects?
- How do causal attributions influence enforcement?

Some voluntary cooperation is necessary for society. No social dilemma is perfectly enforceable (Frey et al. 2004). Moreover, real-world problems often require some self-organization (e.g., adaptive problem-solving) to be managed effectively (Ostrom 2000). Enforcement is important because it can create a sense of security, which ensures societal rules are obeyed (Hardin 1968; Rachlinski 1998). Effective enforcement can also improve credibility and build trust by protecting collective agreements and safeguarding cooperators (Ostrom 1998; Rustagi, et al. 2010). Research demonstrates that stakeholders to a cooperative dilemma are willing to bear the costs of enforcement if they perceive a need for it (Yamagishi 1986) and net benefits (Ostrom 1998: 7–13). However, benefits of enforcement are not always certain (Hilbe et al. 2014).

Several mechanisms may be involved when enforcement backfires (Bowles 2008; DeCaro et al. 2015). Enforcement can (a) signal distrust, encouraging others to reciprocate with distrust and uncooperative behavior (Mulder et al. 2006); (b) reframe moral obligations as economic transactions, so ethical behavior seems optional or like a purchasable "service" (Gneezy and Rustichini 2000); (c) create a sense of dependency, so that stakeholders lose faith in others' goodwill (Chen et al. 2009); and (d) undermine people's sense of self-determination, crowding out intrinsic motivation (Moller et al. 2006). The factors that determine when these effects occur are poorly understood. Therefore, real-world enforcement is unreliable (Agrawal and Ribot 2014).

DeCaro et al. (2015) demonstrated that enforcement was beneficial when combined with shared decision making, because it empowered groups, helping them to protect their collectively chosen agreements. Enforcement had the opposite effect when the rules were imposed. This finding implies that enforcement should be used to help stakeholders reach shared goals, not just force compliance. Enforcement systems may also need to be chosen or designed by the stakeholders whenever possible (Ostrom 2000) or legitimized in another way (Tyler 1990).

Hardin, whose seminal work inspired the widespread advocacy of enforcement within political science and economics, was also a proponent of this idea:

> To many, the word coercion implies arbitrary decisions of distant and irresponsible bureaucrats; but this is not a necessary part of its meaning. The only kind of coercion I recommend is mutual coercion, mutually agreed upon by the majority of the people affected.
>
> *(1968: 1248)*

Hardin's advocacy of democratic enforcement has been widely overlooked and misinterpreted. Many scholars assume enforcement is sufficient in itself, despite considerable evidence to the contrary.

Ostrom's (1990, 2010) investigation of self-governing systems demonstrates that enforcement systems that sustain cooperation are rarely imposed. Enforcement tends to be more effective when created collaboratively with shared decision making (Ostrom et al. 1992) and when elected by a majority than when exogenously imposed or voted against (Tyler 1990; Vollan 2008; Markussen et al. 2014). Enforcement systems also tend to be more effective when they use graduated sanctions that increase in severity with subsequent violations, along with deliberative democracy, designed to create a dialogue between enforcers and compliers, educate stakeholders, and rehabilitate defectors (Schlager et al. 2012).

Hence, according to Ostrom (1990, 2000), well-designed enforcement systems serve vital community-building functions, not only increasing security but also maintaining constructive relationships, helping stakeholders build positive reputations and trust. DeCaro et al. (2015) for instance found that, by satisfying fundamental needs, voting with enforcement (Voted-Enforce condition) resulted in very few punishments but increased security and group liking and closeness (self-other merging).

HRCT therefore implies that enforcement systems must be designed to support multiple psychological needs and social functions, not just security. Failure to recognize this principle often leads to overly punitive, narrowly conceived enforcement systems, which ultimately undermine cooperation.

Causal attributions

The mere presence of enforcement can alter how group members perceive one another, adversely affecting cooperation. For example, when enforcement is present, group members may attribute others' cooperation to narrow self-interest (e.g., fear of punishment), rather than altruism (Mulder et al. 2006). Hence, causal attributions may have also influenced the enforcement effects in DeCaro et al.'s (2015) study. For example, when rules were exogenously imposed (Imposed-Enforce condition), group members may have attributed others' cooperation primarily to fear of punishment (external). Therefore, when enforcement ended, cooperation declined. In contrast, when groups voted to decide rules (Voted-Enforce condition), they may have also attributed others' cooperation to internal causes, such as genuine endorsement of the rule and belief in their moral obligation to follow elected rules. These attributions may have safeguarded their trust and internal motivations, so that cooperation continued even when enforcement ended. It is important to study these effects to understand the decision-making calculus humanistically self-interested agents apply to cooperative decisions.

Communication

- How does communication improve cooperation?
- What motivates competitive stakeholders to initiate and sustain self-governance?

This section examines the motivations that compel otherwise competitive stakeholders to cooperatively self-organize, using communication as a test-bed. Traditional rational choice theory struggles to explain why people self-govern (Poteete et al. 2010) and why communication helps (Ostrom et al. 1992) if people are narrowly selfish and untrustworthy.

Why does communication help?

Despite traditional predictions to the contrary (Hobbes 1909 [1651]), communication among stakeholders often improves cooperation (Balliet 2010). However, the underlying mechanisms are unclear (Pavitt 2011). For example, Janssen (2010) coded group communication in a CPR experiment in terms of common topics, such as ecological dynamics and enforcement. Communication improved cooperation, but like many such studies, the specific topics groups discussed did not. Groups that exchanged more messages cooperated best, especially when each person contributed equally to the discussion. Given these results, HRCT would suggest that the benefits of communication are being driven by group identity and social justice effects (e.g., self-other merging, procedural justice). Being able to freely voice one's opinion and contribute equitably to group discussion typically increases perceptions of procedural justice, encourages self-other merging, and increases rule acceptance, improving cooperation (De Cremer and Tyler 2005). Communication style may also influence cooperation. Group members often shame defectors (Ostrom et al. 1992). When group members use controlling language, or fail to justify their requests, communication may be perceived as coercive and unfair, undermining cooperative motivation (Wilson et al. 2006). These predictions highlight the necessity of examining underlying motivations and perceptions, not just surface features of communication.

Stakeholders' goals for communication are also important (Wittenbaum et al. 2004). Pavitt et al. (2005) found that groups in CPR dilemmas spend a lot of time discussing how to make group decisions together, whether and why they accepted proposed strategies, and off-topic information like jokes (see also, Janssen 2010). These factors are often ignored in traditional rational choice theory but predicted cooperation. In contrast, HRCT expects these factors to improve cooperation, because they represent major aspects of procedural justice, self-determination, and institutional acceptance (DeCaro and Stokes 2008, 2013). Discussing off-topic information may also satisfy stakeholders' need to belong, increasing self-other merging (Buchan et al. 2006). Thus, when communication is possible, stakeholders use it as an opportunity to satisfy their fundamental needs; if they are successful, then cooperation is more likely.

Communication may also cultivate trust (Ostrom 1998). After communicating, group members expect others to cooperate more, especially if they promise to cooperate (Orbell et al. 1988). Communication can clarify misconceptions and reveal others' altruism. Prior to communication, people typically think others are uncooperative because of selfishness or bad character (Van Lange et al. 1990). However, people often defect for more benign reasons (e.g., accident). Communication can facilitate cooperation by revealing these benign reasons for non-cooperation, improving group members' impressions of one another (Tazelaar et al. 2004). Group members also tend to discuss how to achieve more equitable outcomes, make decisions fairly, and protect their conservation agreements (Pavitt 2011). These kinds of conversations further reveal stakeholders' intentions, showing that many stakeholders are open to cooperation and are humanistically self-interested, not narrowly self-interested. Each of these revelations potentially adds to individuals' positive reputations, triggering reciprocity and beginning the self-reinforcing cycle that sustains trust (Ostrom 1998: 12–14). Future research needs to study these social-psychological processes to fully understand their impacts.

Why does self-governing cooperation occur?

Humanistic psychology assumes that individuals are inherently self-motivated, energized to solve problems that affect their fundamental needs. HRCT further assumes that stakeholders are motivated to self-organize when a social dilemma threatens their fundamental needs.

For example, in the early stages of a CPR experiment, the dilemma is often managed poorly, the resource is depleted, and group members appear uncooperative (Tazelaar et al. 2004). Psychologically, this situation should feel out of control; security, belonging, and competence should be low; and legitimacy of the social interactions poor. Hence, individuals should feel deficits to their needs, motivating them to find a solution. All of the effects discussed in this chapter can be interpreted this way.

For example, after initial cooperative failure, groups are generally more supportive of enforcement to improve their security (Yamagishi 1986). When shared decision making seems unfair, groups generally try to improve procedural justice (De Cremer and Tyler 2005). When given the opportunity to communicate, groups usually try to solve the resource management problem. Doing so may improve feelings of competence. Belonging, trust, and other social needs and cognitions may also be addressed (Buchan et al. 2006). Thus fundamental needs help explain why people self-govern. Other factors further help explain when and why such cooperation is sustained long term, as previously discussed.

Synthesis and future directions

Cooperation is vital to a well-functioning society. The current chapter synthesized multiple perspectives on motivation and rational choice to develop a theoretical framework which helps to resolve several persistent questions about governance and cooperation. Fundamental needs and social cognitions drive human agency, determining the kinds of governance systems people desire and cooperate within.

There are many other frameworks of cooperative decision making (Weber et al. 2004; Parks et al. 2013). They generally focus on other aspects of social cognition (e.g., culture, identity, social, and cognitive biases), which are highly complementary to HRCT. The current perspective can and should incorporate other aspects of social cognition to more fully describe human behavior. For example, people care about fairness, but they are also prejudiced, form in-groups, and fear loss (loss aversion); these cognitions shape and constrain cooperative decisions in complex ways, which need further explanation from a humanistic perspective (DeCaro et al. 2017a).

It is also important to study the legal and institutional conditions which enable effective self-governance and democracy within complex societies (DeCaro et al. 2017b). Ostrom (1990, 2010) introduced several factors associated with successful self-governance at smaller scales, for example, shared decision making, enforcement, and communication. The current chapter essentially examined psychological mechanisms involved in those factors, and more can be done. Scholars have recently expanded Ostrom's work to address state-reinforced self-governance, situations where formal governments play a more direct role in supporting the self-organization of societal stakeholders at multiple scales, including large-scale systems parallel to local, state, and national government (Sarker 2013). For example, governments may provide financial support, grant stakeholders more authority to self-govern, and introduce policy sunsets to facilitate adaptation. The psychological effects of these factors should also be studied.

Conclusion

Societies rise and fall with their ability to encourage cooperation. Rational choice theory has taken humanity very far, allowing development of complex societies. However, by focusing on a narrow aspect of self-interest, traditional rational choice theory poses certain limitations to human welfare, doing a better job of characterizing the worst case scenario than solutions for it (Ostrom 1998, 2010). HRCT may take societies further by accounting for a larger set of the fundamental motivations and social cognitions that drive human behavior. This deeper level of understanding seems essential to resolve cooperative problems that continue to plague modern societies.

Acknowledgments

This material is based upon work supported by the National Science Foundation under Grant No. 1658608. Any opinions, findings, and conclusions or recommendations expressed in this material are those of the author and do not necessarily reflect the views of the National Science Foundation.

References

Abele S., and Stasser G. (2008) "Coordination success and interpersonal perceptions: matching versus mismatching" *Journal of Personality and Social Psychology*, 95(3): 576–592.

Agrawal A., and Ribot J. (2014) "Are Ostrom's design principles sufficient for design?" *Policy Matters*, 19: 111–115.

Anderies J. M., and Janssen M. A. (2013) "Robustness of social- ecological systems: implications for public policy" *Policy Studies Journal* 41(3): 513–536.

Anderies J. M., Janssen M. A., Bousquet F., Cardenas J.-C., Castillo D., Lopez M.-C., . . . and Wutich A. (2011) "The challenge of understanding decisions in experimental studies of common pool resource governance" *Ecological Economics*, 70: 1571–1579.

Arrow K. J. (1951) *Social Choice and Individual Values*. John Wiley & Sons, New York.

Balliet D. (2010) "Communication and cooperation in social dilemmas: a meta-analytic review" *Journal of Conflict Resolution*, 54(1): 39–57.

Bandura A. (2001) "Social cognitive theory: an agentic perspective" *Annual Review of Psychology*, 52(1): 1–26.

Baumeister R. F., and Leary M. R. (1995) "The need to belong: desire for interpersonal attachments as a fundamental human motivation" *Psychological Bulletin*, 117(3): 497–529.

Becker G. S. (1974) "Crime and punishments: an economic approach" *Journal of Political Economy*, 76(2): 169–217.

Bowles S. (2008) "Policies designed for self-interested citizens may undermine 'the moral sentiments': evidence from economic experiments" *Science*, 320: 1605–1609.

Brehm S. S., and Brehm J. W. (1981) *Psychological Reactance: A Theory of Freedom and Control*. Academic Press, New York.

Buchan N. R., Johnson E. J., and Croson R. T. (2006) "Let's get personal: an international examination of the influence of communication, culture and social distance on other regarding preferences" *Journal of Economic Behavior and Organization*, 60(3): 373–398.

Chen X. P., Pillutla M. M., and Yao X. (2009) "Unintended consequences of cooperation inducing and maintaining mechanisms in public goods dilemmas: Sanctions and moral appeals" *Group Processes & Intergroup Relations*, 12(2): 241–255.

Chen, B., Vansteenkiste, M., Beyers, W., Boone, L., Deci, E. L., Van der Kaap-Deeder, J., . . . and Ryan, R. M. (2015) "Basic psychological need satisfaction, need frustration, and need strength across four cultures" *Motivation and Emotion*, 39(2): 216–236.

Chess C., and Purcell K. (1999) "Public participation and the environment: do we know what works?" *Environmental Science and Technology*, 33(16): 2685–2692.

Chokshi N. (2016) "'Turn off the computer,' Oregon sheriff begs. Stop 'tearing each other apart'" *Post Nation*, 12 February. www.washingtonpost.com/news/post-nation/wp/2016/02/12/turn-off-the-computer-oregon-sheriff-begs-stop-tearing-each-otherapart/, accessed 19 July 2017.

Clement F. (2010) "Analysing decentralised natural resource governance: proposition for a 'politicised' institutional analysis and development framework" *Policy Sciences*, 43(2): 129–156.

Cole D. H., and McGinnis M. D. (2015) *Elinor Ostrom and the Bloomington School of Political Economy, Vol. 1: Polycentricity in Public Administration and Political Science*. Lexington Books, Idaho Falls, ID.

Dawes C. T., Fowler J. H., Johnson T., McElreath R., and Smirnov O. (2007) "Egalitarian motives in humans" *Nature*, 446(7137): 794–796.

Dawes R. M., McTavish J., and Shaklee H. (1977) "Behavior, communication, and assumptions about other people's behavior in a commons dilemma situation" *Journal of Personality and Social Psychology*, 35(1): 1–11.

DeCaro D. A., Arnold C. A., Frimpong-Boamah E., and Garmestani A. S. (2017a) "Understanding and applying principles of social cognition and decision making in adaptive environmental governance" *Ecology and Society*, 22(1): 33. https://doi.org/10.5751/ES-09154-220133.

DeCaro D. A., Chaffin B. C., Schlager E., Garmestani A. S., and Ruhl J. B. (2017b) "Legal and institutional foundations of adaptive environmental governance" *Ecology and Society*, 22(1): 32. https://doi.org/10.5751/ES-09036-220132.

DeCaro D. A., Janssen M. A., and Lee A. (2015) "Synergistic effects of voting and enforcement on internalized motivation to cooperate in a resource dilemma" *Judgment and Decision Making*, 10(6): 511–537.
DeCaro D. A., and Stokes M. (2008) "Social–psychological principles of community-based conservation and conservancy motivation: attaining goals within an autonomy-supportive environment" *Conservation Biology*, 22(6): 1443–1451.
DeCaro D. A., and Stokes M. (2013) "Public participation and institutional fit: a social–psychological perspective" *Ecology and Society*, 18(4): article 40. http://dx.doi.org/10.5751/ES-05837-180440.
Deci E. L., and Ryan R. M. (2000) "The 'what' and 'why' of goal pursuits: human needs and the self-determination of behavior" *Psychological Inquiry*, 11: 227–268.
De Cremer D., and Tyler T. R. (2005) "Managing group behavior: the interplay between procedural justice, sense of self, and cooperation" *Advances in Experimental Social Psychology*, 37: 151–218.
De Cremer D., Tyler T. R., and den Ouden N. (2005) "Managing cooperation via procedural fairness: the mediating influence of self-other merging" *Journal of Economic Psychology*, 26: 393–406.
De Cremer D., and van Vugt M. (1999) "Social identification effects in social dilemmas: a transformation of motives" *European Journal of Social Psychology*, 29: 871–893.
Fehr E., and Fischbacher U. (2004) "Social norms and human cooperation" *Trends in Cognitive Sciences*, 8(4): 185–190.
Fehr E., and Gächter S. (2002) "Altruistic punishment in humans" *Nature*, 415(10): 137–140.
Frey B. S., Benz M., and Stutzer A. (2004) "Introducing procedural utility. Not only what, but also how matters" *Journal of Institutional and Theoretical Economics*, 160: 377–401.
Gneezy U., and Rustichini A. (2000) "A fine is a price" *Journal of Legal Studies*, 29: 1–17.
Hardin G. (1968) "Tragedy of the commons" *Science*, 162(13): 1243–1248.
Hilbe, C., Traulsen, A., Röhl, T., and Milinski, M. (2014). "Democratic decisions establish stable authorities that overcome the paradox of second-order punishment" *Proceedings of the National Academies of Science*, 111(2): 752–756.
Hobbes T. (1909 [1651]) *Leviathan*. Clarendon, Oxford, UK.
Janssen M. A. (2010) "Introducing ecological dynamics into common-pool resource experiments" *Ecology and Society*, 15(2): 7. www.ecologyandsociety.org/vol15/iss2/art7/.
Leach W. D., and Sabatier P. A. (2005) "To trust an adversary: integrating rational and psychological models of collaborative policymaking" *American Political Science Review*, 99(4): 491–503.
Markussen T., Putterman L., and Tyran J. R. (2014) "Self-organization for collective action: an experimental study of voting on sanction regimes" *Review of Economic Studies*, 81(1): 301–324.
McCay B. J., and Acheson J. (1987) *The Question of the Commons: The Culture and Ecology of Communal Resources*. University of Arizona Press, Tucson, AZ.
McComas K. A., Stedman R., and Hart P. S. (2011) "Community support for campus approaches to sustainable energy use: the role of 'town–gown' relationships" *Energy Policy*, 39: 2310–2318.
Messick D. M., and Brewer M. B. (1983) "Solving social dilemmas: a review" in Wheeler L and Shaver O eds, *Review of Personality and Social Psychology* (4) Sage, Beverly Hills, CA, 11–44.
Milinski M., Semmann D., and Krambeck H.-J. (2002) "Reputation helps solve the 'tragedy of the commons'" *Nature*, 415(24): 424–426.
Miller D. T. (1999) "The norm of self-interest" *American Psychologist*, 54(12): 1053–1060.
Moller A. C., Ryan R. M., and Deci E. L. (2006) "Self-determination theory and public policy: improving the quality of consumer decisions without using coercion" *Journal of Public Policy and Marketing*, 25(1): 104–116.
Mulder L. B., van Dijk D., De Cremer E., and Wilke H. A. M. (2006) "Undermining trust and cooperation: the paradox of sanctioning systems in social dilemmas" *Journal of Experimental Social Psychology*, 42: 147–162.
National Civic League (NCL) (2013) "Making public participation legal" http://ncdd.org/rc/wp-content/uploads/MakingP2Legal.pdf.
Norris P. (2016) "It's not just Trump. Authoritarian populism is rising across the West. Here's why." *The Washington Post*, 3 March. www.washingtonpost.com/news/monkey-cage/wp/2016/03/11/its-not-just-trump-authoritarian-populism-is-rising-across-the-west-heres-why/?utm_term=.0b38f4a05fc6, accessed 19 July 2017.
Orbell J. M., Van de Kragt A. J., and Dawes R. M. (1988) "Explaining discussion-induced cooperation" *Journal of Personality and Social Psychology*, 54(5): 811–819.
Ostrom E. (1990) *Governing the Commons: Evolution of Institutions for Collective Action*. Cambridge University Press, Cambridge, UK.
Ostrom E. (1998) "A behavioral approach to the rational choice theory of collective action: Presidential address—American Political Science Association 1997" *American Political Science Review*, 91: 1–22.

Ostrom E. (2000) "Crowding out citizenship" *Scandinavian Political Studies*, 23(1): 3–16.
Ostrom E. (2010) "Beyond markets and states: polycentric governance of complex economic systems" *American Economic Review*, 100: 1–33.
Ostrom E., Walker J., and Gardner R. (1992) "Covenants with and without a sword: self-governance is possible" *American Political Science Review*, 86(2): 404–417.
Parks C., Joireman J., and Lange P. (2013) "Cooperation, trust, and antagonism: how public goods are promoted" *Psychological Science in the Public Interest*, 14(3): 119–165.
Pavitt C. (2011) "Communication, performance, and perceptions in experimental simulations of resource dilemmas" *Small Group Research*, 42(3): 283–308.
Pavitt C., McFeeters C., Towey E., and Zingerman V. (2005) "Communication during resource dilemmas: effects of different replenishment rates" *Communication Monographs*, 72(3): 345–363.
Poteete A. R., Janssen M. A., and Ostrom E. eds. (2010) "Pushing the frontiers of the theory of collective action and the commons" in *Working together: collective action, the commons, and multiple methods in practice.* Princeton University Press, Princeton, NJ, 215–247.
Rachlinski J. J. (1998) "Limits of social norms" *The Chicago-Kent Law Review*, 74: 1537–1567.
Rauchdobler J., Sausgruber R., and Tyran J. R. (2010) "Voting on thresholds for public goods: experimental evidence" *FinanzArchiv: Public Finance Analysis*, 66(1): 34–64.
Reed M. (2008) "Stakeholder participation for environmental management: a literature review" *Biological Conservation*, 141: 2417–2431.
Rustagi D., Engel E., and Kosfeld M. (2010) "Conditional cooperation and costly monitoring explain success in forest commons management" *Science*, 330(6006): 961–965.
Ryan R. M., and Deci E. L. (2000) "The darker and brighter sides of human existence: basic psychological needs as a unifying concept" *Psychological Inquiry*, 11(4): 319–338.
Ryan R. M., and Deci E. L. (2006) "Self-regulation and the problem of human autonomy: does psychology need choice, self-determination, and will?" *Journal of Personality*, 74(6): 1557–1586.
Sarker A. (2013) "The role of state-reinforced self-governance in averting the tragedy of the irrigation commons in Japan" *Public Administration*, 91(3): 727–743.
Schlager E., Heikkila T., and Case C. (2012) "The costs of compliance with interstate agreements: lessons from water compacts in the western United States" *Publius: The Journal of Federalism*, 42(3): 494–515.
Tazelaar M. J., Van Lange P. A., and Ouwerkerk J. W. (2004) "How to cope with 'noise' in social dilemmas: the benefits of communication" *Journal of Personality and Social Psychology*, 87(6): 845–859.
Tyler T. R. (1990) *Why people obey the law.* Yale University Press, New Haven, CT.
Tyler T. R. (2006) "Psychological perspectives on legitimacy and legitimization" *Annual Reviews of Psychology*, 57: 375–400.
UN General Assembly (2015) *Transforming our world: the 2030 agenda for sustainable development.* A/RES/70/1, 21 October 2015.
Van Lange P. A., Liebrand W. B., and Kuhlman D. M. (1990) "Causal attribution of choice behavior in three N-person prisoner's dilemmas" *Journal of Experimental Social Psychology*, 26(1): 34–48.
Vansteenkiste M., and Ryan R. M. (2013) "On psychological growth and vulnerability: basic psychological need satisfaction and need frustration as a unifying principle" *Journal of Psychotherapy Integration*, 23(3): 263–280.
Vollan B. (2008) "Socio-ecological explanations for crowding out effects from economic field experiments in southern Africa" *Ecological Economics*, 67: 560–573.
Weber J. M., Kopelman S., and Messick D. M. (2004) "A conceptual review of decision making in social dilemmas: applying a logic of appropriateness" *Personality and Social Psychology Review*, 8(3): 281–307.
Wilson J. M., Straus S. G., and McEvily B. (2006) "All in due time: the development of trust in computer-mediated and face-to-face teams" *Organizational Behavior and Human Decision Processes*, 99(1): 16–33.
Wittenbaum G. M., Hollingshead A. B., and Botero I. C. (2004) "From cooperative to motivated information sharing in groups: moving beyond the hidden profile paradigm" *Communication Monographs*, 71(3): 286–310.
Yamagishi T. (1986) "The provision of a sanctioning system as a public good" *Journal of Personality and Social Psychology*, 51(1): 110–116.

PART II

Commons interdisciplinary case studies

11
THE US PUBLIC LANDS AS COMMONS

Bruce Huber

For many US scholars, the federal public lands are (or once were) perhaps the archetypal commons. To visualize Hardin's pasture is, in all likelihood, to imagine a landscape somewhere in the American Mountain West, with herdsmen competing for valuable pasture in an otherwise arid territory. For Hardin, even the US National Parks—the crown jewels of the federal system of public lands—were paradigmatic commons, leading Hardin to conclude that "we must soon cease to treat the parks as commons or they will be of no value to anyone" (Hardin 1968, 1245).

It is a bit ironic, then, that scholars of the commons have largely bypassed study of the public lands of the United States. The bulk of contemporary research on federal public lands management applies modes of analysis other than the commons frameworks described in several chapters of this volume.

This is a mistake, but an understandable one. It is a mistake because the vast energies devoted to the elaboration of the Institutional Analysis and Development (IAD) framework and its kin have yielded an analytical toolkit that holds great promise for generating new insights into pressing resource management problems on US public lands.[1] At the same time, it is understandable that newcomers to these frameworks may assume an ill fit—perhaps because the highly regulated public lands no longer match the analyst's image of a commons, or perhaps because so much commons scholarship examines foreign resource governance systems in informal contexts and/or remote geographies (McGinnis and Walker 2010, 298). Any lack of fit is purely superficial, however. Although most public land in the United States today could not fairly be classified as common property, the usefulness of the IAD framework does not depend on the type of property system that governs a particular resource. Many public land issues present collective action problems involving common-pool resources that are rivalrous in consumption but from which exclusion is quite costly. Such scenarios are within the sweetspot of IAD analysis (Ostrom et al. 1994).

This chapter will begin with a brief history and overview of federal public land management. It will then describe several periods of public lands history that have been illuminated by study informed by commons approaches. Finally, it will identify gaps in contemporary public lands research where commons analysis could profitably supplement existing approaches.

A brief history of the US public lands

The public lands of the United States, as a legal category, have a lengthy and fascinating history.[2] When President Thomas Jefferson authorized the Louisiana Purchase, the fledgling federal government became, in a single stroke, the foremost public authority for the land of the North American continent. It would be decades before federal officials would establish even the most basic administrative presence in this newly acquired territory, but decisions made in Washington unmistakably set the direction for public land policy (Hibbard 1965, 56–143).[3]

During the first half of the 1800s, federal land policy represented a rough balance between two competing policy priorities: revenue generation and settlement (Robbins 1942, 3–116). The federal government's need for reliable sources of revenue was an acute concern given high levels of indebtedness. Federal lawmakers sought to encourage westward settlement for a variety of reasons, not least of which was domestic security. The interior lands west of the original states were regarded as a defensive vulnerability, not only on account of bellicose European powers but also because of unstable relationships with indigenous tribes whose land was being wrested away by settlers.

Although public land policy was hotly contested, nearly all agreed that federal land ought to be sold into private ownership. Thus, federal policy was unmistakably one of privatization of public lands. Disputes were over the terms of sale—the price; the size of land units; the pace and practice of land surveys; whether sale procedures should enable or discourage speculation in land—rather than whether federal policy objectives were best served by land disposal rather than retention. New US states, carved out of the federal public domain, bargained for massive federal land grants not only to establish schools and other public sites but also to finance state government via the sale of land (Gates 1979, 285–340). By the close of the Civil War, federal lawmakers foreswore revenue acquisition and enacted legislation (the Homestead Act, the General Mining Act) that incentivized land and resource development first and foremost. Hundreds of millions of acres passed from public to private ownership. Yet many millions of acres would remain unsettled and would come to form much of today's base of federal land.

Settlement did not lead straightforwardly to a stable system of private land ownership. Instead, land disposal proceeded awkwardly, in fits and starts (White 1991, 137–154). Land was claimed and then abandoned, passing in and out of public ownership. Federal administrative and law enforcement capabilities were miniscule (Feller 1984, 198). The boundaries of the public domain were poorly demarcated. Resource extraction often took place without any of the formalities of ownership or transfer of title. If a resource (timber, water, wildlife, minerals) was available and useful, settlers took it. In this sense, the public lands very much took on the character of a commons despite the enormous increase in private ownership across the West. Timber theft on public lands, for example, was rampant (Ise 1920, 62–118). Timber harvesters—both commercial producers and settlers that simply needed timber for construction—essentially moved their way from east to west, ravaging forests in their path and seldom pausing to establish land title.[4] The incipient cattle industry undertook enormous cattle drives across vast plains and subalpine pasture. These practices quickly reached scales that were unsustainable. As the 1800s came to a close, it became possible to speak not only of the end of the American Frontier (Turner 1961 [1893]) but also of timber famine, of species eradication (e.g., bison), and of the widespread degradation of a West that had seemed inexhaustible only a few decades prior.

Slowly but unmistakably, federal law evolved in response to these trends. The federal presence on the public lands grew, just as the scale and scope of the federal apparatus expanded in virtually all other respects. The first national parks took shape and evolved from protected enclaves into accessible and desirable public destinations (Keiter 2013, 13–63). Congress amended the federal code to change the terms of sale for lands rich in natural resources. Energy minerals such as coal were made subject to a competitive leasing system (Nelson 1983, 13–22). And finally, and perhaps most controversially, the range was closed with the passage of the Taylor Grazing Act (Peffer 1972 [1951], 214–231).

As land disposal slowed, the federal role on the public lands became distinctly managerial despite recurrent (if marginal) demands for continued disposal or allocation to the states. Most of the remaining federal lands are managed now for long-term retention under multiple-use statutes that blend conservation with other land use objectives (Coggins and Evans 1981). The federal land management agencies now have relatively large and thoroughly professionalized staffs. Bureaucratic culture and competency exert substantial influence on land management practices and at times yield policy inertia (Skillen 2009, 2015; Williams 2009). Federal environmental laws loom large: the environmental review demanded by the National Environmental Policy Act (NEPA) is a central part of planning and management decision making, and legal challenges stemming from NEPA review are common (see e.g., Broussard and Whittaker 2009). Many additional disputes arise in the shadow of the Endangered Species Act, although the Act's forceful language is seldom backed with maximal enforcement.

Commons research: moments in public lands history

Given this background, it is perhaps unsurprising that the period before the modern managerial structure took shape is the period of public land law that first attracted attention from scholars inspired and informed by commons approaches. Studies of US resource policy during the 1800s and early 1900s represent an effort to explain the official processes and social and economic forces that led to widespread resource depletion and deterioration across the American West during that time. Local, regional, and sometimes even national shortages in certain resources were standard. It is no wonder that resource policy then entered the US consciousness, or that resource conservation, as against the ever-present possibility of a tragedy of the commons, became a focal point of policy deliberation during the twentieth century.[5]

One productive line of scholarship has focused on the California Gold Rush. When the United States acquired the Mexican Session pursuant to the Treaty of Hidalgo in 1848, its military and administrative capacity in the Sierra Nevada range was effectively nil. Open access conditions characterized the territory of the gold rush of 1849–50, yet within a very short time, miners created—without the formal involvement of state or federal government—a relatively stable and effective system of property rights and administration. Although these rights and the associated norms were eventually codified and gained official imprimatur, the episode captures well many of the emergent themes of recent commons scholarship. Early analyses emphasized the more-or-less spontaneous inception of property institutions that facilitated cooperation and conflict resolution, reduced violence, and enabled orderly resource development (Umbeck 1981).[6] Subsequent work has both augmented and questioned these emphases. Zerbe and Anderson (2001) note that the establishment of effective social institutions was greatly aided by shared notions, such as fairness and "producerism," embedded in miners' cultural background. McDowell (2002) argues that commitments to egalitarianism were prior to and more important than commitments to specific property forms in explaining mining codes. Still other work has clarified that nascent resource systems took account of the interests of claim jumpers, not merely claim holders (Clay and Wright 2005).

Another historical strand subjected to thoughtful scholarly analysis is the degradation of the US range and the establishment of grazing rights on public lands. The use patterns associated with forage, unlike gold, did not lend themselves initially to the creation of anything resembling fee title in land. Meandering seasonal usage over vast distances made enforcement of boundaries exceedingly difficult. Federal law disallowed land claims sufficient to establish a year-round ranching operation.[7] Without a shared framework for the establishment of rights, violent conflicts emerged between competing users, especially between nomadic sheep herders and cattle ranchers. Herdsmen found themselves in an unstable business situation, and the forage production potential of the range diminished. The Public Lands Commission reported:

> There are, it is estimated, more than 300,000,000 acres of public grazing land, an area approximately equal to one-fifth the extent of the United States proper. . .
>
> At present the vacant public lands are theoretically open commons, free to all citizens; but as a matter of fact a large proportion have been parceled out by more or less definite compacts or agreements among the various interests. These tacit agreements are continually being violated. The sheepmen and cattlemen are in frequent collision because of incursions upon each other's domain. Land which for years has been regarded as exclusively cattle range may be infringed upon by large bands of sheep, forced by drought to migrate. Violence and homicide frequently follow, after which new adjustments are made and matters quiet down for a time. . .
>
> The general lack of control in the use of public grazing lands has resulted, naturally and inevitably, in overgrazing and the ruin of millions of acres of otherwise valuable grazing territory. Lands useful for grazing are losing their only capacity for productiveness, as, of course, they must when no legal control is exercised.[8]

The grazing situation at the turn of the century, then, had many of the markings of a genuine commons tragedy (Donahue 2001, 32). Resource depletion followed from the inability of stakeholders to coalesce around an effective strategy of resource management. Privatization was rendered almost impossible because of legal barriers and the physical characteristics of the forage resource (Libecap 2007, 271–275).

Assistance from the outside seemed necessary but would be slow in coming. The US Congress was deeply ambivalent over the proper treatment of grazing issues, as with broader questions of Western land use, and was reluctant to deviate from what it regarded as its central policy goal of settlement (Merrill 2002, 46–63). The Dust Bowl conditions of the 1930s, among other factors, finally galvanized national legislators into action, and in 1934 Congress passed the Taylor Grazing Act (Calef 1960, 49–62). This legislation came close to establishing fixed private rights in public rangelands. Although the allocation process was fraught (Raymond 2003, 109–152), the system of grazing permits that resulted became quite stable—so stable, in fact, that recent observers commonly decry land managers' reluctance to reduce grazing rights in the face of continued overgrazing (Donahue 2001).

This situation presents a bit of a puzzle. If grazing permits resemble property rights, why has overgrazing persisted? Commons theory would suggest, at least superficially, that private property rights, whether *de jure* or *de facto*, align the interest of the resource owner with the sustainability of the resource. There is a dearth of scholarship on this question—and so too is there a deficit in commons-based research devoted to other contemporary concerns in public land management.

Gaps in research and possible new directions

Part II of this volume provides an overview of various analytical frameworks that trace their genesis to problems of the commons. Despite these frameworks' longstanding association with commons problems, they have a far broader range of possible applications. Indeed, there is little in the world of political institutions, whether formal or informal, that could not be illuminated by work in the tradition of these commons-inspired paradigms. To date, however, there have been few efforts to apply these frameworks to contemporary issues involving the federal public lands.

Like many studies of US public policy, public lands research tends to focus on the formal outputs of the US Congress and federal agencies: legislation and regulation, that is, the rules on paper (e.g., Klyza 1996; Wilkinson 1992). There is a solid rationale for such a focus in that public lands management is today thoroughly professionalized; resource allocation involving state-owned property often turns in large part on managerial decisions made by public officials and the lawmakers in back of them. The large and sophisticated federal land management agencies are subject to extensive federal legislation and make decisions under the watchful eye of powerful and litigious environmental and industry groups. For good reason, then, scholars intent on understanding public land policy pay close attention to legislative and regulatory proceedings and the inevitable legal challenges that follow. The resulting literature generally explains these outputs as the result of contests among competing ideas and interest groups in standard venues of policy making (e.g., Culhane 1981), an approach that invites the application of public choice concepts in particular.[9] Other studies are more explicitly normative in that they describe a favored policy outcome and explain how and why current policy deviates from that outcome. Many of these studies, even those that are thinly-veiled advocacy pieces, draw primarily on general conventions of modern regulatory, legal, and policy analysis.[10] For example, the interested reader could review the numerous studies on issues such as hydraulic fracturing on federal land or federal wildlife management and would likely find in them much of the lexicon and conceptual toolkit popularized by some of the seminal policy studies of the last century.[11]

These sorts of studies, though important and useful, generally do not interact with the theoretical literature of the commons, or do so only superficially. Why does this matter, and what might we be missing?

First, scholarship may neglect important differences between formal enactments and informal practices. Analyses grounded in IAD and other commons-oriented approaches tend to focus on management rules in use rather than rules on paper. The distinction between rules in use and rules on paper was of critical importance to Elinor Ostrom and remains a central feature of the IAD framework that she and her collaborators developed (Ostrom 1990, 2005; see Chapter 5 of this volume). It is an especially significant distinction in the context of public property. On the one hand, public land ownership naturally draws attention to the legal and regulatory players and processes that generate formal land use decisions. Given the emergence and maturation of US land management agencies in the twentieth century, and the explosion of federal environmental law in the latter part of that century, the study of rules on paper is crucial. On the other hand, the US public lands have been profoundly affected by extralegal and informal activities over the years, and such activities still matter a great deal in many public lands contexts (Ellickson 1991; Fairfax et al. 1999).[12] Even now, depending on actual patterns of land use and law enforcement, state property may resemble private property, common property, or even nonproperty (Cole 2002, 9). The diversity of use patterns, and the corresponding complexity of *de facto* and *de jure* use rights in present-day public lands matters, suggest that there remain important areas of public land policy that could profit from rigorous study based on commons frameworks.

Second, public lands research could benefit from the rigorous and structured comparative study that IAD work permits. Much recent scholarship employs a case study approach to explore a single policy area. Yet as the previous section suggests, substantive resource management rules and policies involving public lands—rules that dictate whether to allow a particular resource to be used or consumed, and if so, by whom and in what quantities—have varied, and continue to vary, substantially across at least three dimensions. First, governance regimes have evolved over time, not only in terms of formal rules but also in terms of the enforcement of those rules. Importantly, evolution in rules on-the-ground may take place even during periods of apparent legal stability. Second, access rules vary according to the particular resource at issue. The same parcel of land may be subject to one regime for mineral development and another for grazing rights, and the administration of the relevant access rights may well differ markedly between the two (Fairfax et al. 1999, 634–641) such that land that bears the hallmarks of common property for the purposes of one resource may more closely resemble private property for the purposes of another. Finally, even resource-specific processes may vary by region, either because of differences in environmental conditions or because of regional administrative differences. This sort of variance, and the complex layering of land rights that it has produced, complicates the study of the public lands and makes comparative work ever more crucial. Even the most pressing contemporary public land issues are seldom analyzed alongside each other or subjected to the highly structured, comparative scrutiny that is the hallmark of the IAD approach.

There is, then, an opening for research that mixes political sophistication with careful attention to the many variables that shape and describe social-ecological systems, including those variables that cannot be pulled from the statute book or the Federal Register (e.g., Cole et al. 2014).

As an example, let us return for a moment to contemporary grazing policy. Critics contend that grazing allocations remain too high on federal public lands and that overgrazing continues to degrade Western rangelands. Most analyses make the case for administrative solutions such as an increase in the grazing fee, more stringent application of environmental standards, or reclassification of federal lands such that grazing is excluded. Whether or not such steps may be helpful, "brute force" recommendations for top-down policy change may not only strain political viability (let alone legal authority) but also neglect other variables that influence grazing levels on public lands. A complete analysis of the interactions between ranchers and public land managers would entail a deep examination of the biophysical and socio-legal factors that confront these actors in the course of their interactions. At present, there is little if any scholarship that describes comprehensively the decisional environment facing public land ranchers and managers. Such an inquiry could aid in the discovery of feasible and sustainable solutions to rangeland management problems.

Another example of an area in need of additional inquiry is the mineral leasing regime by which most energy commodities on the public lands are sold. Rights to mine coal, for example, are sold in a competitive bidding process—or, more accurately, a process that is supposed to be competitive according to federal law (Hein and Howard 2015). The actual process as it has played out over the last several decades is anything but competitive. In part, this appears to be a consequence of the fact that the law permits mining entities to nominate tracts for which competition is likely to be minimal (Huber 2015, 1522–1527). But the statute is flexible enough to allow for a more demanding review of nominated tracts, and in any event, reduced competition in coal production is itself the consequence of a long trend towards consolidation in the coal production industry. Neither of these aspects is fully exogenous to the problem of coal land management. Neither is a complete explanation; rather, each is an invitation to further inquiry: how is it that the coal industry's consolidation has come about, and how is it that administration officials have allowed this consolidation, through incremental decision making, to lead to a noncompetitive coal leasing process? These questions are raised rather than answered by existing scholarship, and the multi-level institutional analysis required to answer such questions could easily be facilitated with recourse to the IAD framework.

Finally, further analytical opportunities await at higher levels of abstraction. What lessons might be gained, for example, by comparing the case of grazing rights with that of timber harvests on public lands? Not long ago, overharvest of both timber and forage were considered quintessential "lords of yesterday"—stable policy pathologies rooted in century-old legal and cultural dynamics unique to the American West (Wilkinson 1992). Through the latter portion of the twentieth century, federal timber policy on the public lands was regularly criticized for despoiling virgin forest and prioritizing commercial timber harvesting at the expense of important forest habitat. The forest plans issued by the US Forest Service were ground zero for conflicts over forest policy, and the Service was thought incapable of moving beyond "clear cutting" and a "get out the cut" mentality (e.g., Clary 1986). But times have changed. Today, timber harvests are a fraction of previous levels, whereas grazing levels have declined only slightly from all-time highs. Why the different outcomes for timber and forage? Again, a structured, comparative inquiry of this question could bring needed attention to underappreciated variables in public lands matters.

Conclusion

The US public lands are vast. They span thousands of miles; they host countless forms of recreational and commercial activity; they surround and border myriad communities; and they contain resources in quantities sufficient to affect global markets. The federal land management agencies, persistently beset by fiscal woes and conflicts among their many stakeholders, cannot always impose a consistent federal policy across these lands. Even at their strongest, these agencies bring to bear the fragile agreement of disparate interests more often than the single-minded will of a resolute landowner. In such a context, where exclusion is difficult and conflict is certain, standard policy study can only go so far. Formal debates and official action represent only a fraction of the many facets of public land use. These facets are familiar and relatively easy to study, but they seldom provide a complete explanation for a state of affairs on the public lands. Fortunately, more comprehensive approaches are available. It remains only for scholars to pick up these tools and carry them, explorer-like, to the boundless public lands.

Notes

1 Late in her life, Ostrom and her collaborators proposed a new Social-Ecological Systems (SES) framework capable of incorporating a much broader set of social and ecological variables than contemplated with the IAD schema (Ostrom 2009; also Cole et al. 2014). I refer primarily to the IAD framework only due to its greater familiarity; the SES framework, with its more capacious view, may be as well or better suited to the study of some public lands conflicts.

2 This history is the subject of a rich and rewarding literature. For the uninitiated, this author suggests Gates (1979) for depth and detail and Wilkinson (1992) for readability.

3 A detailed account of the development of federal public lands administration during this period can be found in Rohrbough (1968) and is well supplemented by Feller (1984). For a legal historian's fascinating reinterpretation of the rise of federal land ownership, see Ablavsky (2017).

4 By far the most comprehensive account of nineteenth-century US timbering practices and law in any jurisdiction is Hurst (1964), focusing on the state of Wisconsin.

5 For an important and corrective account on conservation during the early twentieth century, see Hays (1959).

6 Umbeck's work accords well with a broader mode of inquiry initiated by Demsetz (1967), which explains the emergence of property institutions with reference to commodity values in relation to transaction costs. See also Ellickson (1991), emphasizing the importance of norms for the creation of property-based social equilibria.

7 Ranching operations often spanned tens of thousands of acres. The standard practice for cattle ranchers was to establish formal homestead claims for scarce verdant bottomlands with year-round access to water. Surrounding lands could be foraged as needed on a seasonal basis. Ranchers circumvented federal acreage restrictions by illegally fencing public lands, and eventually also formed armed claims associations to protect range land from intrusion (Peffer 1972 [1951], 72–89).

8 Report of the Public Lands Commission, Senate Document No. 154, 58th Congress, 3rd session, reprinted as Report of the Public Lands Commission, Gov't Printing Office, Washington DC. 1905, p. XXI.
9 I refer here to public choice in the casual sense, by which scholars apply the label to causal explanations involving strategic or self-interested official action.
10 "Convention" may be too formal a term, for I have in mind here the myriad studies (white papers, policy analyses, even many law review articles) that assume rather than defend the desirability of particular outcomes.
11 So, for example, the language of "regulatory capture" and "iron triangle" is regularly deployed, perhaps along with an emphasis on agenda control, windows of policy opportunity, and so forth. These ideas have their roots in dated but seminal works in political science such as Baumgartner and Jones (1993), Kingdon (2003), and Lowi (1969).
12 This reality, along with the fact of substantial regulation of private land use in the United States, leads some scholars to conclude that the distinction between public and private landownership is overblown or careless (Cole 2002, 8–14; Freyfogle 2007).

References

Ablavsky G. (2017) "The Rise of Federal Title" *California Law Review* 106: 631–631.
Baumgartner F. R. and Jones B. D. (1993) *Agendas and Instability in American Politics*. University of Chicago, Chicago, IL.
Broussard S. R. and Whittaker B. D. (2009) "The Magna Charta of Environmental Legislation: A Historical Look at 30 Years of NEPA-Forest Service Litigation" *Forest Policy and Economics* 11: 134–140.
Calef W. (1960) *Private Grazing and Public Lands*. Arno, New York.
Clary D. A. (1986) *Timber and the Forest Service*. University Press of Kansas, Lawrence, KS.
Clay K. and Wright G. (2005) "Order Without Law? Property Rights During the California Gold Rush" *Explorations in Economic History* 42: 155–183.
Coggins G. C. and Evans P. B. (1981) "Multiple Use, Sustained Yield Planning on the Public Lands" *University of Colorado Law Review* 53: 411–469.
Cole D. H. (2002) *Pollution & Property: Comparing Ownership Institutions for Environmental Protection*. Cambridge Press, Cambridge, UK.
Cole D. H., Epstein G., and McGinnis M. D. (2014) "Digging Deeper into Hardin's Pasture: The Complex Institutional Structure of the 'Tragedy of the Commons'" *Journal of Institutional Economics* 10: 353–369.
Culhane P. J. (1981) *Public Lands Politics: Interest Group Influence on the Forest Service and the Bureau of Land Management*. Johns Hopkins University (for RFF), Baltimore, MD.
Demsetz H. (1967) "Toward a Theory of Property Rights" *American Economic Review* 62: 347–349.
Donahue D. L. (2001) *The Western Range Revisited: Removing Livestock from Public Lands to Conserve Native Biodiversity*. University of Oklahoma, Norman, OK.
Ellickson R. C. (1991) *Order Without Law: How Neighbors Settle Disputes*. Harvard, Cambridge, MA.
Fairfax S. K. et al. (1999) "The Federal Forests Are Not What They Seem: Formal and Informal Claims to Federal Lands" *Ecology Law Quarterly* 25: 630–646.
Feller D. (1984) *The Public Lands in Jacksonian Politics*. University of Wisconsin, Madison, WI.
Freyfogle E. T. (2007) *On Private Property: Finding Common Ground on the Ownership of Land*. Beacon, Boston, MA.
Gates P. W. (1979) *History of Public Land Law Development*. Arno, New York
Hardin G. (1968) "The Tragedy of the Commons" *Science* 162: 1243–1248.
Hays S. P. (1959) *Conservation and the Gospel of Efficiency: The Progressive Conservation Movement, 1890–1920*. University of Pittsburgh, Pittsburgh, PA.
Hein J. F. and Howard P. (2015) *Reconsidering Coal's Fair Market Value: The Social Costs of Coal Production and the Need for Fiscal Reform*. NYU Institute for Policy Integrity, New York.
Hibbard B. H. (1965) *A History of the Public Land Policies*. University of Wisconsin, Madison, WI.
Huber B. R. (2015) "The Fair Market Value of Public Resources" *California Law Review* 103: 1515–1559.
Hurst J. W. (1964) *Law and Economic Growth: The Legal History of the Lumber Industry in Wisconsin, 1836–1915*. University of Wisconsin, Madison, WI.
Ise J. (1920) *The United States Forest Policy*. Yale University, New Haven, CT.
Keiter R. B. (2013) *To Conserve Unimpaired: The Evolution of the National Park Idea*. Island Press, Washington, DC.

Kingdon J. W. (2003) *Agendas, Alternatives, and Public Policies* (2nd ed.). Longman, New York.
Klyza C. M. (1996) *Who Controls Public Lands? Mining, Forestry, and Grazing Policies, 1870–1990*. University of North Carolina, Chapel Hill, NC.
Libecap G. D. (2007) "The Assignment of Property Rights on the Western Frontier: Lessons for Contemporary Environmental and Resource Policy" *Journal of Economic History* 67: 257–291.
Lowi T. J. (1969) *The End of Liberalism: The Second Republic of the United States*. Norton, New York.
McDowell A. G. (2002) "From Commons to Claims: Property Rights in the California Gold Rush" *Yale Journal of Law & Humanities* 14: 1–72.
McGinnis M. D. and Walker J. M. (2010) "Foundations of the Ostrom Workshop: Institutional Analysis, Polycentricity, and Self-Governance of the Commons" *Public Choice* 143: 293–301.
Merrill K. R. (2002) *Public Lands and Political Meaning: Ranchers, the Government, and the Property between Them*. University of California, Berkeley, CA.
Nelson R. H. (1983) *The Making of Federal Coal Policy*. Duke, Durham, NC.
Ostrom E. (1990) *Governing the Commons: The Evolution of Institutions for Collective Action*. Cambridge, New York.
Ostrom E. (2005) *Understanding Institutional Diversity*. Princeton University Press, Princeton, NJ.
Ostrom E. (2009) "A General Framework for Analyzing Sustainability of Social-Ecological Systems" *Science* 325: 419–422.
Ostrom E., Gardner R., and Walker J. (1994) *Rules, Games, and Common-Pool Resources*. University of Michigan, Ann Arbor, MI.
Peffer E. L. (1972 [1951]) *The Closing of the Public Domain: Disposal and Reservation Policies 1900–50*. Arno Press, New York.
Raymond L. (2003) *Private Rights in Public Resources: Equity and Property Allocation in Market-Based Environmental Policy*. Resources for the Future, Washington, DC.
Robbins R. M. (1942) *Our Landed Heritage: The Public Domain 1776–1936*. Princeton University Press, Princeton NJ.
Rohrbough M. (1968) *The Land Office Business: The Settlement and Administration of American Public Lands, 1789–1837*. Oxford University Press, New York.
Skillen J. R. (2009) *The Nation's Largest Landlord: The Bureau of Land Management in the American West*. University Press of Kansas, Lawrence, KS.
Skillen J. R. (2015) *Federal Ecosystem Management: Its Rise, Fall, and Afterlife*. University Press of Kansas, Lawrence, KS.
Turner F. J. (1961 [1893]) "The Significance of the Frontier in American History" in Billigton, R. A. ed., *Frontier and Section: Selected Essays of Frederick Jackson Turner*. Prentice Hall, New York, 37–62.
Umbeck J. (1981) *A Theory of Property Rights, With Application to the California Gold Rush*. Iowa State University Press, Ames, IA.
White R. (1991) *"It's Your Misfortune and None of My Own": A New History of the American West*. University of Oklahoma, Norman, OK.
Wilkinson C. F. (1992) *Crossing the Next Meridian: Land, Water, and the Future of the West*. Island Press, Washington, DC.
Williams G. W. (2009) *The U.S. Forest Service in the Pacific Northwest: A History*. Oregon State University, Corvallis, OR.
Zerbe R. O. and Anderson C. L. (2001) "Culture and fairness in the development of institutions in the California gold fields" *The Journal of Economic History* 61: 114–143.

12
WATER COMMONS
A critical appreciation and revisionist view

Eduardo Araral, Maitreyee Mukherjee, Faye Victoria Sit Ying Qi and Serik Orazgaliyev

Introduction

In this chapter, we address three questions. First, why and how has the commons analytic approach evolved in the study of water commons? Second, how might its application or results be different from other types of commons, due to the distinctiveness of water as a common pool resource (CPR)? Finally, what analytical frameworks, if any, lend themselves best to the study of the water commons and why? We conclude with the way forward.

The water commons literature has played an important role in shaping the scholarly discourse of the commons because it represents a natural experiment to study collective action problems such as cooperation, conflict, free riding and the classic tragedy of the commons. The water commons is not distinct from other types of commons such as forestry, fisheries, and watersheds and pasturelands because they all share the same characteristics in terms of excludability and rivalry.

The IAD framework offers a systematic approach to the study of the water commons. However, as this chapter shows, there remains some fundamental misunderstanding about the concepts of CPRs, common property and private property. Finally, the literature should move away from its predominant focus on small-scale commons and pay more attention to study of the global commons. The literature needs a critical appreciation and a revisionist view.

How has the analytic approach to the water commons evolved and why?

The core of the commons literature has long been informed by the study of three types of resources, namely water (especially irrigation and ground water), fisheries (coastal and open access) and forestry. For example, of the 15 case studies that Ostrom (1990) used in her classic book, *Governing the Commons*, seven were on water (irrigation and ground water), five on fisheries and two on forestry. The literature has also been extended to the urban commons, global commons, knowledge commons and to the study of anti-commons, but their influence in terms of theory building has been much less compared to the study of water resources, fisheries and forestry.

What is it about irrigation and ground water that makes them an attractive subject to study the commons? The short answer is that the irrigation setting represents a natural experiment to study collective action problems such as cooperation, conflict, free riding and the classic tragedy of the commons. For these reasons, Ostrom has likened the study of collective action in irrigation as the equivalent of the fruit fly to an evolutionary biologist.

For example, in an irrigation setting, there is a natural power asymmetry and hence inherent conflict between farmers in the head of the irrigation canal (who could always get water first) and farmers downstream. Non-cooperative game theory posits that those in the head of the canal will have a dominant strategy not to cooperate or share water downstream. The empirical evidence suggests that this is true in some cases but not in others. This raises an interesting question: under what circumstances would supposedly utility maximizing farmers in the headwaters cooperate or share waters with those downstream? More generally, under what circumstances do self-interested individuals cooperate when conventional game theory says they have no incentives to do so? This is one instance that makes irrigation systems an interesting unit of analysis to study collective action in the commons.

Second, an irrigation setting is also a natural experiment to study the classic collective action problem of free riding. The operation and maintenance of irrigation infrastructure requires the cooperation of a large number of individuals. For example, in large-scale irrigation settings, this could easily be thousands of farmers. In farmer-managed, ancient irrigation systems (those that Ostrom studied), the requirements for labor and materials to operate and maintain irrigation systems are substantial, for example see Araral (2003, 2009). Then there are problems of coordination as well as problems of enforcement, both of which are non-trivial problems to solve. Once the rules have been set, the problem of monitoring and enforcement is a non-trivial problem. How will the rules of water allocation and maintenance of canals and facilities be enforced when individual farmers have incentives not to comply?

All of these questions raise interesting possibilities for scholars of collective action in irrigation commons. For example, the theoretical literature suggests that free-riding and non-cooperation would be the equilibrium outcome in the irrigation setting. However, in the empirical literature, we see a lot of cooperation and contribution to the irrigation CPR. This leads to a more general interesting question: what explains successful collective action in the commons when players have incentives to free ride? This interesting question has been at the core of the irrigation commons research over the years.

Is the water commons distinct?

How might the application or results of water commons be different from other types of commons, due to the distinctiveness of water as a CPR? Although irrigation systems are widely regarded in the literature as CPRs, they are not, technically speaking, a CPR as defined in Ostrom and Ostrom's (1977) typology of goods (Table 12.1).

A CPR refers to a physical good, natural or man-made, large enough to make it costly to exclude potential users. Examples include large ground water reservoirs, lakes, open access fisheries, public forests, open grasslands, the atmosphere, among others, in which it is relatively difficult to exclude others from benefitting from the resource. These resources are subject to congestion,

Table 12.1 Typology of goods (adapted from Ostrom and Ostrom, 1977)

	Excludability	*Non-excludable*
Rivalrous consumption	Private goods (irrigation systems)	CPRs (open access fisheries, ground water)
Non-rivalrous consumption	Club goods	Public goods

pollution and overuse because consumption of their resource units (water, fish, wood, grass) is rivalrous. Because of this, CPRs need some kind of regulation, whether self-management by resource users, as documented by Ostrom, or externally imposed by government. CPRs can be owned by the state (and managed by its agents), collectively by resource users (referred to by Ostrom as common property), by private corporations, associations or individuals as private goods or resources, as open access goods (Hardin's open pasture; see Hardin 1968) or some combination thereof.

Whether water is a CPR or private good really depends on many factors. Water from a free flowing river is a public good (non-rivalrous/non-excludable). Ground water irrigation from public land is clearly a CPR but is a private good in a private land if the owner has rights to water. Once the river water enters an irrigation system, it can be a club good if 1) exclusion to its use is feasible and 2) there is more than enough supply to make consumption non-rivalrous. If enforcement of allocation rules is weak and there is scarce water, then it is a CPR.

Much of the confusion in the past (which still exists today) results from the conflation of the terms CPR (a physical good) with common property (a system of ownership by resource users).

From Table 12.1, irrigation systems (i.e. the infrastructure) should be regarded as a private good because it is rivalrous in consumption (water as a resource unit) and exclusion to the irrigation system is feasible, i.e. there are membership requirements. Those who do not abide by the rules can be effectively sanctioned. In contrast, ground water and offshore fisheries are the classic CPRs because of the difficulty of excluding resource users from harvesting the resource and because of rivalry in consumption. This distinctive feature of irrigation systems (private good, exclusion is feasible) makes it very different from other common pool goods. Despite this, most researchers still use irrigation systems as their unit of analysis. This is because the structure of the game in such a setting enables scholars to study collective action, especially the problems of free riding and the problem of cooperation when players have few incentives to cooperate.

Using the typology in Table 12.1, a critical appreciation of Ostrom's (1990) case studies would point to two important findings and conclusions (Table 12.2) that call for a rethink of the conventional wisdom about the commons. The first conclusion is that all of the successful or robust cases that Ostrom documented were in fact *privately owned* resource systems (i.e. exclusion to the resource system is feasible and consumption of resource units are rivalrous). In the meadows of Switzerland and Japan, exclusion is defined by membership in the village association or canton. In the ancient irrigation systems of Spain and the Philippines, exclusion is defined by rights to withdraw water from the system and membership in the irrigation association which owns the system.

The second conclusion is that, conversely, all cases that Ostrom classified as failed or fragile cases – ground water in California and fisheries in Sri Lanka, Canada and Turkey – were all in fact *open access resources* in which exclusion is difficult and consumption of resource units is rivalrous. These findings support the conventional wisdom that open access resources, what Hardin (1968) referred to as commons, would indeed lead to tragedy of the commons.

A reassessment of Ostrom's work in Table 12.2 suggests that the Nobel Prize Committee erred in their citation about Ostrom's work. According to the Prize citation:

> [Ostrom] has challenged the conventional wisdom that common property is poorly managed and should be either regulated by central authorities or privatized.
>
> *(Nobel Prize Press Release 2009)*

Table 12.2 shows that this conclusion is not entirely correct, especially the reference to common property and privatization. Privatized systems (irrigation, meadows) in which exclusion is feasible had robust outcomes, contrary to the claims of the Nobel Prize Committee. In contrast, open access goods (open fisheries, ground water) led to failed or non-robust outcomes.

Table 12.2 Reassessment/reinterpretation of the evidence from Ostrom's work

Examples of commons studied by Ostrom (1990)	*Feasibility of exclusion / clarity of property rights (based on author's reassessment of Ostrom 1990)*	*Institutional / resource outcomes (based on Ostrom's assessment*
Meadows / forests of Torbel Switzerland	Exclusion is feasible (access to commons defined by membership in local villages, corporations, cooperatives, cantons, associations which controls access (see pp. 61–65); in effect these are privately owned	Robust
Japanese mountain meadows	Exclusion is feasible / access to the commons defined by village membership, therefore privately owned; small group size makes exclusion feasible (see pp. 65–69)	Robust
Huerta irrigation in Spain	Exclusion is feasible / water rights are fixed and unitized, therefore privately owned; monitoring is credible and enforcement of private rights feasible (see pp. 69–81)	Robust
Zanjera irrigation in the Philippines	Exclusion is feasible; rights to the resource system (canals) and units (water) are unitized, quantified and are tradable, therefore privately owned; water control structures make enforcement credible; relatively small size and therefore exclusion is feasible (see pp. 82–88)	Robust
Mojave ground water basin	Exclusion is difficult because rights are unclear/ contested; therefore is open access/ common property; system is very large (15 interconnected basins) / no water associations were formed (see pp. 146–150)	Failure / conflict among water users / overdraft
(Raymond, West, Central basins) (earlier case)	Exclusion is difficult because rights are unclear and resource systems large (Raymund basin 277 square miles, West Basin 170) (see pp. 111–132)	Failure / conflict among water users / overdraft
Nova Scotia, Canada offshore fisheries	Exclusion is difficult (fishing ground covers 440 square km; considered open access; rights of local fishers to police not recognized) (see pp. 173–175)	Fragile institutions
Kirindi Oya irrigation, Sri Lanka	Exclusion is difficult (rights to irrigation unclear / no water control structures / irrigation association weak to make enforcement credible) (see pp. 157–172)	Failure of collective action
Alanya offshore fisheries, Turkey	Exclusion is difficult; unclear boundaries of resource and users; ineffective conflict resolution mechanisms (see pp. 18–21)	Fragile / conflict
Mawelle fisheries, Sri Lanka	Exclusion is difficult (for the number of nets that can be used and because ownership patterns constantly shifting while rights are not unitized, quantified and tradable) (see pp. 149–156)	Failure (overfishing)
Bodrum inshore fisheries, Turkey	Exclusion is difficult; three mile limit rarely enforced; no clear boundaries of resource and fishers; no monitoring, sanctioning, conflict resolution mechanisms (pp. 144–145)	Failure (overfishing / rent dissipation / sharp drop in catch per unit of effort)

(continued)

Table 12.2 (continued)

Examples of commons studied by Ostrom (1990)	*Feasibility of exclusion / clarity of property rights (based on author's reassessment of Ostrom 1990)*	*Institutional / resource outcomes (based on Ostrom's assessment*
Bay of Izmir fisheries, Turkey	Exclusion is difficult; too many fishers (1,800), large resource system, heterogeneous interests (see pp. 144–146)	Failure (overfishing)

While Ostrom has argued that exclusion is the key to the successful management of CPRs, she has a caveat. She argued that even if exclusion were feasible, there would still be considerable collective action problems, particularly the problems of supply of institutions, credible commitment and mutual monitoring.

This is true, but in privatized CPRs where resource units (water, woods, minerals) can be unitized, quantified and traded there will be an incentive for the holders of these rights to solve these institutional problems. As the value of the resource units increases, there will be a demand for institutions to monitor and protect them. Consequently, resource users will try to find ways to supply institutions (through self-regulation, private policing or by demanding external protection) if they are to capture the value of the resource. This is the mechanism that explains why privately owned resources are managed in a more sustainable manner than open access resources.

As for Hardin, he subsequently admitted that his conclusions have been over extended and should have been better framed as tragedy of the unregulated commons. By this he meant that the tragedy is the result of ill-defined and enforced property rights to these resources. The struggle to govern the commons persists especially in large scale commons, for example India and China's aquifers and their heavily polluted cities and rivers. Equally important, regional and global commons continue to deteriorate because of unclear property rights and effective regulations. We can see this is in global and regional commons, for example in Southeast Asia's Mekong River, Central Asia's Caspian and Aral Seas, Indonesia's forests and peatlands, the Amazon forest and the global atmosphere more generally.

While we know a lot about the commons from studies of small-scale, local commons in irrigation, forestry and fisheries, surprisingly we know little about regional and global commons. There are many unanswered questions in the literature. For example, why are some regional or global commons better managed compared to others? Under what circumstances do countries cooperate or not cooperate on regional or global commons?

This question has recently been studied by Orazgaliyev and Araral (working paper) in the case of the Caspian Sea, an oil rich and geopolitically important region dividing Europe and Asia. The Caspian Sea is interesting as it points to two different outcomes of a prisoner's dilemma game over a CPR. In the North Caspian Sea, claimant countries, Russia, Kazakhstan and Azerbaijan, have managed to cooperate and solve their conflicts over boundary and rights over oil resources. In contrast, claimants in the South Caspian Sea, Azerbaijan, Iran and Turkmenistan, have failed to do so despite decades of trying. The study shows that in the successful case of the North Caspian Sea, the following factors played a key role in conflict resolution – the presence of a hegemon (Russia), the personal chemistry among the political leaders (Yeltsin and Nazarbayev), social capital and historical ties, and domestic politics (i.e. role of powerful private sector players, bureaucratic politics). Similarly, the scholars speculate that countries in the South Caspian failed to cooperate because of the absence of these enabling factors. More studies of the global or regional commons are needed if we are to advance our understanding of how they can be better managed.

Analytical frameworks

What analytical frameworks, if any, lend themselves best to the study of water commons? Of the various analytical frameworks to study the commons, we find the Institutional Analysis and Development (IAD) Framework useful for the study of water commons in particular, and the commons in general, because it allows us to do a more systematic approach leading to the accumulation of knowledge over time.

First, it helps us to identify the structural variables that we need to focus on in our study of collective action. For instance, the outcomes of collective action in the irrigation commons depends on the structure of incentives which in turn is influenced by three variables, namely characteristics of the irrigation system (water scarcity, distance from markets, quality of the infrastructure, etc.), the attributes of the resource users (group size, social capital, etc.) and the governance or institutional context (the formal and informal rules and their enforcement mechanisms). From these three variables, we can identity many other specific variables to study.

Second, the IAD framework enables us to compare across different types of commons – water, fisheries, forestry, meadows, atmosphere, global commons, knowledge commons, etc. – leading to the accumulation of knowledge. Finally, the IAD framework allows us to do a meta-analysis of the literature and to systematically identify how different variables affect the incentives and outcomes in the commons. In the next section, we summarize what we know so far about the literature on the irrigation commons. We focus on three factors: characteristics of the resource, resource users and the institutional context.

Characteristics of the resource

Characteristics of the CPR matters to the incentive structure and outcomes of the commons. In irrigation, scholars have studied the effects of resource scarcity; external shocks such as droughts, typhoons or floods; proximity of the farms to markets; among others. These factors are posited to affect the incentives of farmers to cooperate or not.

Water scarcity

Scholars have mostly found that collective action among stakeholders is possible only when the resources are perceived to be scarce. In studies involving irrigation systems, Agrawal (2002), Bardhan (1993) and Uphoff et al. (1990) observed that cooperation in collective action is most difficult to achieve when water supply is abundant or becomes too scarce. Based on their findings, they proposed that the relation between resource scarcity and collective action is curvilinear, implying that collective action works efficiently when water is moderately scarce, while at the extremes stakeholders are motivated to exploit and profit most from available resources. If the irrigation system is so degraded that it is beyond any improvement effort, then members do not see any benefit in organizing a collective management group.

External shocks such as drought, flood or devastating typhoons also affect the ability of resource users to cooperate (Araral 2013). For instance, many irrigators' associations ceased to function in Bicol region, Philippines, after typhoons destroyed their infrastructure beyond any scope of repair. Araral (2009) also investigated the factors that facilitate collective action in irrigation commons by studying 1,958 irrigation associations in Philippines. His findings indicated that collective action was indeed related to water scarcity in curvilinear pattern, and the effect was further influenced by factors such as proximity to markets, governance structure, group and farm size.

Size of the resource

Smaller and clearly defined boundaries of CPRs require lower transaction costs in terms of monitoring or enforcement. Hence smaller resource size often leads to successful collective action groups (Wade 1988). On the other hand, a larger resource boundary constrains its management due to higher costs of monitoring and enforcement (Shah et al. 2002). However, Meinzen-Dick et al. (1997) point out that such a conclusion may be derived from flawed analysis, as often it is difficult to distinguish between size of an irrigation area and number of farmer-owners. In some settings, fragmentation of land leads to a number of farmers involved, while in others there may be a small number of farmers owning large land holdings.

Proximity to markets

Ostrom and Gardner (1993) hypothesized that integration with markets may result in enhanced market competition leading to anonymity among actors, reduction in mutual dependencies, loosening of traditional social bonds or less risk of penalty in case of defection, among collective action stakeholders. Simultaneously, Agrawal and Yadama (1997) also support the notion that proximity to markets may decrease chances of cooperation among actors managing CPRs.

On the contrary, Meinzen-Dick et al. (1997) propose that rather than level of commercialization, the structure of the labor market determines the kind of cooperation in collective action. Studying canal irrigation systems in Rajasthan and Karnataka, Meinzen-Dick et al. (2002) observe that farmers are likely to undertake collective irrigation operations if they have enough market incentives. In fact, farmers who are organized in larger commands, closer to market towns with religious centers, or under influential leadership are prone towards better collective action efforts to manage a large irrigation system.

Characteristics of resource users

Collective action in the water commons is also affected by the characteristics of resource users. Scholars have studied the effects of group size and its characteristics, the age and origin of resource users, wealth distribution among resource users, among others.

Group size and characteristics

Group size is a well-studied variable in the water commons and in the common literature in general (see Poteete and Ostrom 2004). While group theory and other conventional models advocate that collective action becomes difficult as group size increases, there is little consensus among scholars on what may be classified as a small or large group, or how context mediates the effects of group size in determining its social cohesion. Olson (1965) argued that as group size increases, significance of individual contribution towards collective actions diminishes. Rationally calculating individuals realize that and tend to break free from their group goals in order to improve their personal benefits. Following this line of argument, Tang (1992) as well as Fujie et al. (2005) show that controlling for all other factors, group size had inverse effects on level of cooperation, that is, it is more feasible to organize collective irrigation groups with fewer members. On the contrary, Ternström (2003), using an empirical study from Nepal, conclude that group size had no statistically significant effect on cooperation.

The disadvantages of larger group size have been explained by Meinzen-Dick et al. (1997) from the angle of tradeoff between transaction cost and potential economies of scale. Their paper details that higher group members increased transaction costs of supervision, monitoring, information gathering and associated asymmetries. Thus, marginal social cost of individual defection was insignificant when compared to marginal personal benefits. As a result, as groups tend to increase in size, their cohesion and accountability tend to decrease. On the other hand, a group that is too small is also not desirable, as farmers need to maintain an optimal size to enjoy substantial economies of scale from maintaining a common resource.

However, some scholars such as Frohlich and Oppenheimer (1970) and Sandler (1992) stipulate that group size effects on cooperation are conditional on other variables that are simultaneously affected by change in size of the group. These other variables might include heterogeneity of the group, conflicting interests, multiple leadership, face to face communication or unequal sharing among members. Agrawal and Gibson (1999) conclude that heterogeneity among group members leads to divergence of interests resulting in decreased cooperation. Vedeld (2000) also notes that homogeneity in interests among group elites improved collective cooperation, which ensured that the elites were a bit better off than the other subordinate members. Simultaneously Bardhan et al. (2007) and Rivera et al. (2017) demonstrate that inequality in distribution of benefits among group members destabilized its internal trust and degraded its efficiency. Lastly, Falk et al. (2002) and Ostrom et al. (1994) emphasize the importance of face to face communication among members to maintain group cohesion. As group size increases, such communication decreases leading to disconnection between members and feelings of less liability towards each other.

Age and origin of resource users

Scholars still debate if age and the origin of resource user groups affect their social capital capacity, in turn determining their collective action incentives. According to Meinzen-Dick et al. (1997), in older groups members have already accepted and adapted to its underlying rules, which kind of fixes their pattern of behavior. On the contrary, newer groups suffer from lack of trust, loyalty and reputation among members, resulting in chances for conflicts or defection (Fujie et al. 2005). However, studies like Ternström (2003) and Fujie et al. (2005) argue that age of irrigation system has no statistical significance on its relation to level of collective action behavior.

Apart from age of the group, its origin – that is whether it was self-organized within community members or founded by government agencies – also seems to influence its operational activities. For example, Ostrom and Shivakoti (2002) note that self-organized irrigation associations enjoyed a better sense of ownership, stronger cohesion and greater understanding among members than government arranged collective groups. In this line, Bardhan (2000) also argues that government involvement increases the probability of rule violation among members, along with lack of ownership (as groups can no longer appoint their own supervising agents), resulting in less cooperation. Further studies like Vedeld (2000) show that when groups were led by political elites or popular leadership, whose economic incentives are high if collective action is maintained, their cooperation level remained sustained. Alternately, if individual members in the community felt that they have the power to devise, or modify rules, or facilitate penalty systems for offenders, it led to internal trust and better compliance among members (Luintel et al. 2018).

Autonomy among members is yet another important determinant in establishing common understanding and consensus in the setting of rules. In addition to providing the sense of ownership, autonomy also decreases the cost of organization, management, decision making and implementation. Such autonomous systems are found in irrigation systems in West and Central Java, Indonesia, where water users' associations are involved in operations and maintenance but have no legal or political power to make decisions or settle disputes (Samad et al. 2000). The role of prior experience and locally influential leadership also influence the level of group cooperation among members. For example, Pradhan (2002) finds that in Nepal, groups with greater social capital are better equipped to deal with collective action problems.

Wealth distribution

Another theoretical variable in the irrigation commons literature is the effect of wealth distribution among farmers. Theory suggests that highly unequal wealth distribution has an adverse effect on collective action. The results from the literature are mixed. Based on his study of 23 cross-country community irrigation systems, Tang (1991) observed that a small variance in average annual family income among farmers was associated with greater rule conformance and effective maintenance of the resource. Baland and Platteau (1996, 1999) too assert that poverty (heterogeneity in social background and goals) poses a more serious threat to cooperation than heterogeneity in asset structure of members. Ternström (2003) examined empirical links between poverty and collective actions and identified inequality in wealth distribution as a potential barrier to cooperation. He showed that cooperation is more likely when economic conditions of members are similar. Further, his research showed that cooperation was highest among the poorest of communities who formed groups to survive the economic or ecological challenges of the CPR problem.

Moreover, if the CPR does not provide enough income to sustain their livelihood, farmers do not find it cost effective to organize themselves into functional groups, just as in the case of rich landlords in Andhra Pradesh and Gujarat (Koppen et al. 2002). Again, disparity in wealth or asset structure of farmers makes consensus in allocation of rules difficult, which complicates enforcement and maintenance of the resource (Bardhan and Dayton-Johnson, 2002), common in large-scale irrigation canals of Gujarat and Andhra Pradesh in India.

Bardhan (1993) provides plenty of evidence on how the disparity in individual gains from common property resource management can result in a situation in which certain participants may gain less or lose from cooperation, and this may cause collective action to fall apart. Intragroup heterogeneity specifically in payoffs between parties may thus have unfavorable effects on the sustainability of commons management. Internalization of cooperative norms, that is, the trust that others will play their part in the collective action, is tougher in such situations (Seabright 1990). Bardhan cites a few more examples, one of which is a study by Jayaraman (1981) noting how the egalitarian structure of a community was a critical factor in collective action in his case studies of farmers' organizations in surface irrigation projects in Gujarat.

Another study of ten irrigation cases in Tamil Nadu by Easter and Palanisami (1986) reveals that the smaller the variation in land holding among the farmers, the more likely they are to create water user organizations among themselves. Wade's (1987) research findings illustrate how rich farmers are able to appropriate sufficient water for their land without having to organize collectively and without having to incur large additional expenditures themselves. Hence, they even block the formation of a cooperative water control committee since it may hinder their irrigation freedom, at the expense of the poorer, smallholder farmers. However, land fragmentation may promote cooperation, since some of the big farmers may have pieces of land at different, disadvantaged locations and may thus be induced to participate in collective action.

Conversely, Boyce (1987) observes that in tubewell irrigation cooperatives in Bangladesh, when influential individuals manipulate the location of the tubewell and distribution of water in their favor, the resentment of those disadvantaged led them to sabotage the pump sets. Bandyopadhyay and von Eschen (1988) showed how villages in West Bengal experienced marked adverse effects of economic and social stratification on cooperation. An empirical analysis of 48 irrigation communities in South India by Bardhan (2000) found that factors such as inequality of landholding, as well as urban or market connections significantly decreased the level of cooperation among community members. On the contrary, cooperation was positively related to duration of access to water and resource monitoring facilities. Factors like social homogeneity, small group size, proportional cost-sharing rules, and competition over water resources with rival villages, also somewhat positively affected community cooperation.

From these studies, it is clear that wealth inequality matters to the outcomes in irrigation commons.

Institutional and historical context

Finally, the third set of variables in the IAD framework is the institutional context. The commons literature has studied how different sets of rules lead to different outcomes. In the water commons literature, scholars point to some key factors that affect outcomes – water rights, user associations, community management, decentralization, among others.

Water rights

Water rights is a core variable in the study of water commons, whether irrigation, riverbasin or ground water. If water rights are unclear or unenforceable, especially during periods of scarcity, we expect to see collective action problems and deterioration of the resource unit (overuse) and the resource system (breakdown of infrastructure). Water rights are of three types. The first is user-based allocation in which water infrastructure is commonly owned by the users but exclusion is feasible for non-members, making it a private good. This is typical of farmer-owned and ancient irrigation systems (see, for example, Araral 2013, 2014).

Long before formal water rights were defined by governments, most water users developed their own system of customary water rights which now exist alongside formal water rights. This literature has not been well studied. One such paper is by Meinzen-Dick and Bakker (1999) who provide a framework for analyzing the statutory and customary water rights of multiple users of water in the case of the Kirindi Oya irrigation system in Sri Lanka. The central argument in the paper is that stakeholders of an irrigation system go far beyond the owners and cultivators of irrigated fields, as its potential benefits extend to the irrigation of gardens, vegetation, watering animals, fish farms, and even supporting of different types of enterprises.

Thus, those beyond the irrigated fields that have a stake in these groups should also be considered claimants and need to be included in deliberations regarding transferring water from irrigation to other uses. The authors add that traditional approaches to water rights have tended to focus only on rights as defined in statutory law, often overlooking the significance of customary and religious law, as well as local norms and rules. In doing so, they have neglected the rights and claims of "secondary" water users and claimants. Therefore, instead of a singular approach to water rights, Meinzen-Dick and Bakker (1999) propose a framework for looking at pluralism in water rights, which implies looking for different uses of water, users of water, bases for claiming water rights, bundles of rights (rather than simple "ownership"), stronger and weaker rights, and institutions that ensure or mediate the extent to which rights are realized.

Water user associations

Water user associations have been well studied in the water commons literature (see, for example, Meinzen-Dick and Bakker (1999) and Araral (2005, 2009)). Most studies conclude that water user associations have a positive effect on the management of water commons. For example, Villamayor-Tomas and García-López (2017) use a large-N study obtained from water user associations and ejidos in Mexico, to provide evidence that the performance of community-based water management institutions has a positive and significant impact on the effectiveness of farmers' adaptation responses. Thus, they advocate that policies which strengthen the autonomy and capacity of water user associations' cooperation shall not only guarantee a long-advocated path for rural development but also help farmers deal with collective action problems.

Path dependency

Institutions do not simply arise out of a vacuum. There is a demand and supply of institutions and there is path dependency, i.e. how history matters to the choices and evolution of institutions. Heinmiller (2009) laments how scant attention has been given to the historical-institutional context of collective action and, accordingly, the significant limiting effects of path dependency. Path dependency in the commons context suggests that investments and adaptations in early resource management institutions make it very difficult for actors to abandon these institutions, thereby allowing them to influence subsequent collective action efforts. These effects are evident in Heinmiller's (2009) three transboundary water management cases, specifically in the Murray-Darling Basin of Australia, the Colorado Basin of the US and the Saskatchewan-Nelson Basin of Canada. In all three cases, early water institutions have proven to be strongly path dependent, significantly shaping subsequent collective action efforts. Thus, understanding institutional legacies may be as equally essential as the knowledge, preferences and mutual trust of actors in determining the outcomes of collective action efforts.

Conclusion

The water commons literature has played an important role in shaping the scholarly discourse of the commons. The IAD framework offers a systematic approach to the study of the commons. However, as this chapter has shown, there remains some fundamental misunderstanding about the concepts of CPRs, common property and private property. The assignment and enforcement of water rights would be a necessary though not sufficient condition for solving the problems associated with water commons. Finally, the literature should move away from its predominant focus on small scale commons and pay more attention to the study of the global commons. The literature needs a critical appreciation and a revisionist view.

References

Agrawal, A. (2002). Indigenous knowledge and the politics of classification. *International Social Science Journal*, 54(173), pp.287–297.

Agrawal, A. and Gibson, C. (1999). Enchantment and disenchantment: The role of community in natural resource conservation. *World Development*, 27(4), pp.629–649.

Agrawal, A. and Yadama, G. (1997). How do local institutions mediate market and population pressures on resources? Forest Panchayats in Kumaon, India. *Development and Change*, 28(3), pp.435–465.

Araral, E. (2003). Capacity assessment of irrigator's associations. Project preparation notes prepared for the World Bank-funded Philippines: Riverbasin and Watershed Management Project.

Araral, E. (2005). Bureaucratic incentives, path dependence, and foreign aid: An empirical institutional analysis of irrigation in the Philippines. *Policy Sciences*, 38(2–3), pp.131–157.

Araral, E. (2009). What explains collective action in the commons? Theory and evidence from the Philippines. *World Development*, 37(3), pp.687–697.

Araral, E. (2013). Does geography matter to institutional choice? A comparative study of ancient commons. *Geoforum*, 44, pp.224–231.

Araral, E. (2014). Ostrom, Hardin and the commons: A critical appreciation and a revisionist view. *Environmental Science & Policy*, 36, pp.11–23.

Baland, J. and Platteau, J. (1996). *Halting Degradation of Natural Resources*. New York: Food and Agriculture Organization of the United Nations.

Baland, J. M. and Platteau, J. P. (1999). The ambiguous impact of inequality on local resource management. *World Development*, 27(5), pp.773–788.

Bandyopadhyay, S. and von Eschen, D. (1988). Village failure to cooperate: Some evidence from West Bengal, India. In D. W. Attwood and B. S. Baviskar (eds.), *Who Shares? Cooperatives and Rural Development*. New Delhi: Oxford University Press, pp.112–145.

Bardhan, P. (1993). Analytics of the institutions of informal cooperation in rural development. *World Development*, 21(4), pp.633–639.

Bardhan, P. (2000). Irrigation and cooperation: An empirical analysis of 48 irrigation communities in South India. *Economic Development and Cultural Change*, 48(4), pp.847–865.

Bardhan, P. and Dayton-Johnson, J. (2002). Unequal irrigators: Heterogeneity and commons management in large-scale multivariate research. In E. E. Ostrom, T. E. Dietz, N. E. Dolšak, P. C. Stern, S. E. Stonich and E. U. Weber (eds.), *The Drama of the Commons*. Washington, DC: National Academy Press, pp.87–112.

Bardhan, P., Ghatak, M. and Karaivanov, A. (2007). Wealth inequality and collective action. *Journal of Public Economics*, 91(9), pp.1843–1874.

Boyce, J. (1987). *Agrarian Impasse in Bengal: Institutional Constraints to Technological Change*. Oxford, UK: Oxford University Press.

Easter, K. W. and Palanisami, K. (1986). Tank irrigation in India and Thailand: An example of common property resource management. Department of Agricultural and Applied Economics Staff Paper. St. Paul, MN: University of Minnesota.

Falk, A., Fehr, E. and Fischbacher, U. (2002). Appropriating the commons: A theoretical explanation. CESifo Working Paper Series, Munich.

Frohlich, N. and Oppenheimer, J. (1970). I get by with a little help from my friends. *World Politics*, 23(1), pp.104–120.

Fujie, M., Hayami, Y. and Kikuchi, M. (2005). The conditions of collective action for local commons management: The case of irrigation in the Philippines. *Agricultural Economics*, 33(2), pp.179–189.

Hardin, G. (1968). The tragedy of the commons. *Science*, 162, pp.1243–1248.

Heinmiller, T. (2009). Path dependency and collective action in common pool governance. *International Journal of the Commons*, 3(1), pp.131–147.

Jayaraman, T. K. (1981). Farmers' organizations in surface irrigation projects: Two empirical studies from Gujarat. *Economic and Political Weekly*, 16(39), pp.A89–A98.

Koppen, B., Parthasarathy, R. and Safiliou, C. (2002). *Poverty Dimensions of IMT in Large Scale Canal Irrigation in Andrah Pradesh and Gujarat, India*. Research Report 61. Colombo, Sri Lanka: IWMI.

Luintel, H., Bluffstone, R. A. and Scheller, R. M. (2018). An assessment of collective action drivers of carbon storage in Nepalese forest commons. *Forest Policy and Economics*, 90, pp.39–47.

Meinzen-Dick, R. and Bakker, M. (1999). Irrigation systems as multiple-use commons: Water use in Kirindi Oya, Sri Lanka. *Agriculture and Human Values*, 16, pp.281–293.

Meinzen-Dick, R., Mendoza, M. S., Saddoulet, L., Abiad-Shields, G. and Subramanian, A. (1997). Sustainable water user associations: Lessons from a literature review. In A. Subramanian, N. V. Jagannathan and R. S. Meinzen-Dick (eds.), *User Organizations for Sustainable Water Services*. World Bank Technical Paper 354. Washington, DC: World Bank.

Meinzen-Dick, R., Raju, K. and Gulati, A. (2002). What affects organization and collective action for managing resources? Evidence from canal irrigation systems in India. *World Development*, 30(4), pp.649–666.

Nobel Prize Press Release (2009). www.nobelprize.org/prizes/economic-sciences/2009/press-release/.

Olson, M. (1965). *The Logic of Collective Action: Public Goods and the Theory of Groups*. New York: Schocken Books.
Ostrom, E. (1990). *Governing the Commons: The Evolution of Institutions for Collective Action*. New York: Cambridge University Press.
Ostrom, E. and Gardner, R. (1993). Coping with asymmetries in the commons: Self-governing irrigation systems can work. *Journal of Economic Perspectives*, 7(4), pp.93–112.
Ostrom, E., Gardner, R., Walker, J. and Walker, J. (1994). *Rules, Games, and Common-Pool Resources*. Ann Arbor, MI: University of Michigan Press.
Ostrom, V. and Ostrom, E. (1977). Public goods and public choices. In Savas, E. D. (ed.), *Alternatives for Delivering Public Services: Toward Improved Performance*. Boulder, CO: Westview Press, pp.7–49.
Ostrom, E. and Shivakoti, G. P. (2002). *Improving Irrigation Governance and Management in Nepal*. Oakland, CA: ICS Press.
Poteete, A. and Ostrom, E. (2004). Heterogeneity, group size and collective action: The role of institutions in forest management. *Development and Change*, 35(3), pp.435–461.
Pradhan, J. (2002). Information technology in Nepal: What role for the government? *The Electronic Journal of Information Systems in Developing Countries*, 8(1), 1–13.
Rivera, J., Naranjo, M. A., Robalino, J., Alpizar, F. and Blackman, A. (2017). Local community characteristics and cooperation for shared green reputation. *Policy Studies Journal*, 45(4), pp.613–632.
Samad, M., Vermillion, D., Pusposotardjo, S., Arif, S. and Rochdyanto, S. (2000). An *Assessment of the Small Scale IMT Program in Indonesia*. IIMI Indonesia Country Paper. Colombo, Sri Lanka: IIMI.
Sandler, T. (1992). *Collective Action*. Ann Arbor, MI: University of Michigan Press.
Seabright. S. (1990). *Is Cooperation Habit-Forming?* Mimco. Cambridge, UK: Cambridge University.
Shah, T., Koppen, B., Merrey, D., Lange, M. and Samad, M. (2002). *Institutional Alternatives in African Small Holder Irrigation: Lessons from International Experience with IMT: Research Report 60*. Colombo, Sri Lanka: International Water Management Institute
Tang, S. (1991). Institutional arrangements and the management of common-pool resources. *Public Administration Review*, 51(1), pp.42–51.
Tang, S. Y. (1992). *Institutions and Collective Action: Self-Governance in Irrigation*. San Francisco, CA: ICS Press.
Ternström, I. (2003). *Causes for Conflicts in Irrigation Systems in Nepal: Using the IAD Framework to Find the Weakest Link*. Mini-Conference of the Workshop in Political Theory and Policy Analysis, Indiana University, Bloomington.
Uphoff, N., Wickramasinghe, M. L. and Wijayaratna, C. M. (1990). "Optimum" participation in irrigation management: Issues and evidence from Sri Lanka. *Human Organization*, 49(1), pp.26–40.
Vedeld, T. (2000). Village politics: Heterogeneity, leadership and collective action. *The Journal of Development Studies*, 36(5), pp.105–134.
Villamayor-Tomas, S. and García-López, G. (2017). The influence of community-based resource management institutions on adaptation capacity: A large-N study of farmer responses to climate and global market disturbances. *Global Environmental Change*, 47, pp.153–166.
Wade, R. (1987). *Village Republics: Economics & Conditions for Collective Action in Rural India*. Cambridge. MA: Cambridge University Press.
Wade, R. (1988). Village republics: Economic conditions for collective action in South India. *World Development*, 16(10), pp.1269–1270.

13
COMMONS ANALYSIS AND OCEAN FISHERIES

Bonnie J. McCay

Res nullius, public trust, and tragedies of the commons

The idea and metaphor of the commons has a long and complex history in marine affairs and scholarship. Hugo Grotius' 17th-century scholarship on behalf of Holland's merchants defined much of the sea as "res nullius," open to exploitation, and it helped establish in international law the idea that the seas beyond narrow territorial limits are owned by no one and under no rule (Burke 1994), the doctrine of "mare liberum," the "freedom of the seas." The idea of "the commons" as a condition of free access, beyond ordinary domains of property rights, is captured in the way that biologists and economists have more recently formulated problems in fisheries. The upside of this perspective is recognition that free and open access to the seas is important to transportation, trade, and resource extraction. The downside is that the "freedom" leaves the resources of the seas open to abuse, hence, efforts to come up with local, regional, and global systems for management of fishing, shipping, and other activities in the ocean commons.

A related notion that supports use of "the commons" idea for understanding oceans and their coasts is the idea of public rights to certain waterways and the sea as against private property claims. This is widely held to be based in Roman law. It has percolated through the millennia into other legal traditions, transmuted through the English experience into the notion of a "public trust" held by the royal sovereign that recognizes public rights of access for purposes of fishing, navigation, and, in recent decades, recreation (McCay 1993; Archer 1994; Slade et al. 1997). It can be argued that the public trust doctrine version of the commons not only protects public access rights but also establishes a duty on behalf of the sovereign to manage coastal seas and tidal waterways for public benefit (Osherenko 2006; Turnipseed et al. 2009). As was said in a late 19th-century court regarding legislation restricting certain kinds of fishing gear in New England waters, "what good is the freedom to fish when there are no fish to be found?" (McCay 1987).

In the fields of environmental science, conservation biology, and economics, people are viewed mainly as sources of perturbation to natural phenomena. Overfishing, dumping and polluting, draining the marshes, dredging the reefs, striking whales with ships are among the many anthropogenic activities threatening marine ecosystems and the populations dependent on them (Halpern et al. 2008). This perspective fits neatly with the explanation of environmental problems as "tragedies of the commons" (Hardin 1968, 1994). The villains are not just people, but people in a situation of free and open access. Either government must intervene or access should be closed.

Tragedy in the fisheries commons and bio-economics

Although Garret Hardin's famous 1968 essay, "Tragedy of the commons" (see also Hardin 1994) used livestock on a pasture as an example of how the exercise of individual self-interest can result in the deterioration of a common resource, marine fisheries are more popular as exemplars and teaching tools for commons problems.[1] The challenge of finding a sustainable solution to the commons problem is highlighted in marine fisheries, which have long histories of overexploitation and decline. In a very real sense the commons problem is "the fishermen's problem" (McEvoy 1986).

The simple version of the problem is "too many fishermen, too few fish." The more complex version is that open access to the common pool resources of the seas creates an incentive structure that leads people to overharvest and degrade the seas. Whether or not regulated by government, fisheries are often the sites of over-exploitation due in part to the relative freedom of enterprises to participate. Where large numbers compete for a shared but limited resource, it is in each individual's immediate interest to take what they can. Also, it is rational to avoid taking preventive or restorative measures. Because the resource is available to everyone, the results of an individual's sacrifice or investment will be shared with others. And the results of what others may do to help conserve the resource are free for one to take. Eventually, depending of course on the biology of the creatures being fished and the intensity of exploitation, the fisheries deteriorate, catches are low, and fishers are impoverished. Hence the "tragedy of the un-managed commons" (Hardin 1994).

Newfoundland cod: what kind of tragedy?

The northern population of Atlantic cod (*Gadus morhua L.*) of the east coast of Newfoundland and Labrador, Canada, suffered from a now infamous case of a tragedy of the (poorly managed) commons. Briefly, the cod population declined so drastically in the late 1980s that the government of Canada stopped the fishery in 1992. Over 18,000 people were affected by the closure, and the pathos of the story was further highlighted in a survey that showed the last significant spawning population (Rose 2007, p. 69). So few fish were left that one could make out individual survivors on a blurry sonogram.

Interpretations of the cause and consequences of this tragic event are many and diverse (Charles 1995; Hannesson 1996; Walters and Maguire 1996; McGuire 1997; Hutchings and Reynolds 2004; Rose 2007; May 2009). Briefly, the story of Newfoundland's northern cod fishery by the 1980s was that of a situation of relatively open access although not one lacking science-informed regulation. The Canadian Department of Fisheries and Oceans (DFO) had a long history of fisheries science and engagement in both domestic and international fisheries management. The regulations were mainly through annual catch limits that were established by the minister of DFO after consideration of scientific stock assessments and input from industry, unions, politicians, and others. However, there were no restrictions on how many boats or fishers could be licensed, nor on how much any one enterprise could take, a classic commons situation that fostered costly competition for shares of the quota as well as strong incentives to push for larger quotas despite scientists' recommendations. Both happened.

The process leading to disaster was complicated. It was marked by scientific uncertainty and hubris; interference by politicians; overreliance on data from industrial, offshore fleets; and missed cues from what was happening in the hard-to-measure inshore fisheries and from technological changes in offshore fleets (Anonymous n.d., Steele et al. 1992; Finlayson 1994; Martin 1995; Finlayson and McCay 1998). There were political trade-offs because jobs in fish plants were at stake, and environmental changes played a role too. Clearly the collapse of the great cod fishery was not just a "tragedy of the commons," but some of the classic dynamics of open access situations were at play.

Bio-economic modeling and enclosure of the commons

Much earlier, economists had explained problems of poor catches and poverty in Newfoundland's inshore fisheries in terms of the tragedy of the open access commons, working from the framework of bio-economic models. In 1922 Jens Warming provided the first economic analysis of the "commons problem" in fisheries (Andersen 1983). As further developed by resource economists for fisheries who built upon a biological model of fish populations subject to harvest (Gordon 1954; Scott 1955), the reigning paradigm for understanding fisheries became the notion of inevitable decline of the fish populations under conditions of open access, also cast as absence of exclusive property rights. The resulting theory about causes of fish scarcity served as support for government intervention in fisheries.

From the perspective of neo-classical economics, the situation was cast as one of market failure, and attention came to be focused on the issue of property rights. Without clearly defined, exclusive, and private property rights, market signals such as prices do not function to regulate relationships between people and scarce resources. Under such conditions, and depending on other factors, such as the rate of discounting the future and the reproductive rate of the exploited population, rational behavior by fishers, even by a sole owner of the resource, could destroy the resource, as was shown for whaling (Clark 1973). The general understanding that emerged among scholars of property rights and natural resources was enriched by experiments in game theory and the social sciences (Ostrom et al. 1994). Where the resource is shared and where the activities of one harvester can affect what is available to the other—i.e., a common pool resource—there are no incentives to restrict or cut back on harvesting, despite signs of lower abundance and smaller sizes. If one cut back, there was nothing to stop others from taking the fish, a classic commons dilemma. Similarly, there are major disincentives for being involved in coming up with rules for such a system, because all could share in the benefits of the rules and thus each would be better off letting others take on the costs of negotiation and enforcement.

The simpler bio-economic theory about causes of fish scarcity—rooted in open access—was a rationale for further government intervention in fisheries by restricting catches, seasons, gear, and so forth. It also became a stimulus for finding ways to create property rights that would restore market-based incentives for sustainable and profitable fisheries. It had become clear from experience with government regulation of fisheries that regulatory schemes based on biological issues often led to unwanted economic results such as over-capitalization, as fishers competed for limited quotas. For example, in February 2018, the fisheries management council with responsibility for Alaskan offshore fisheries was confronted with pressures to "rationalize" the Bering Sea trawl cod fishery because the fishing season was only 22 days.[2] The allowable catch, or quota, had been reduced in response to biological indicators of trouble, and the price had increased. Because there are no limits on entry (licenses are costly but not limited) and the fishery is competitive, this "derby" situation predictably leads to a sharp decline in the number of days before the season is closed.

Because of situations like this, which can be costly, endanger lives, and reduce product quality, economists have long argued for measures to restrict entry so that the fishery commons can be managed for both economic and biological objectives. This is what Canada did after 1992 in the Newfoundland case, pressuring many cod fishers to give up their licenses and creating a system that made it difficult for new entrants to gain access.

However, in other situations, such as Canada's west coast salmon fishery, it became evident that simply enclosing the commons by limiting entry could create new problems and replicate commons dilemmas at a smaller scale (Fraser 1979). Having diagnosed the fishery problem as one of a "lack of property rights," economists came up with the idea of combining limited entry with privileged and transferable rights to the resource, a kind of privatization (Christy 1973; Neher et al. 1989). Three decades have seen the development of individual transferable quota (ITQ) and related schemes for managing access to marine fisheries, a major policy innovation that is similar to tradeable allowances for control of greenhouse gas and pollutant emissions (Tietenberg 2002).

Now called "catch shares" in the United States, these systems began in the Netherlands, Iceland, and New Zealand and have been widely adopted in other nations (National Research Council 1999). In some cases, they have been very successful in forestalling tragedies of open access, such as where very limited quotas for valuable fish stocks led to "races for the fish," where huge numbers of boats competed and caught up the quota in very short times, even just one or two days, often dangerous situations where lives and boats were lost and the markets glutted with too much fish. Such was the case for Pacific halibut (Casey et al. 1995). In other cases, such as the Atlantic surfclam fishery, the existing and highly restrictive limited entry system created similar problems, and an ITQ system was implemented that provided a mechanism for reducing the number of vessels in an over-capitalized fishery (McCay et al. 2011).

Resistance to privatization

The call for "rationalizing" Alaska's Bering Sea trawl cod fishery noted earlier is currently being addressed through the North Pacific Fishery Management Council in terms of limiting entry, not ITQs. The voting majority of the Council, one of eight regional councils through which federal waters fisheries are managed in the United States and its territories, is opposed to them, however, and cast ITQs "as privatization adding another barrier to entry into the fishing world," although supporters "call it a reward for investment with benefits including substantial retirement income."

The effects of privatizing licenses or rights to shares of a quota can include better matching fishing effort and investments to the status of a resource, but it often means loss of work opportunities, high costs of entry, and dislocation of fishing enterprises (McCay 1995; Olson 2011). The "catch share" systems tend to result in concentration of ownership and participation in fewer enterprises, threatening fishery-dependent communities and making entry beyond the reach of most (Pálsson and Helgason 1995; Eythorsson 1998; General Accounting Office 2004; Pinkerton and Edwards 2009; Olson 2011). Consequently, there is a persistent debate among industry members and scholars about the use of ITQs in fisheries management. Some emphasize and show positive economic and conservation benefits (e.g., Costello et al. 2016); others question conservation benefits (Chu 2009; Acheson et al. 2015) and show negative economic and social consequences (Carothers and Chambers 2012). Some identify ways that secure ownership translates into stewardship (Costello et al. 2008), while others question that claim (Gilmour et al. 2012).

There has also been political resistance. For example, the US Congress made ITQs illegal for over a decade during which major studies were completed to determine their pros and cons, such as a National Academy of Science study (National Research Council 1999). In 2017–2018 the government of the Faroe Islands enacted a reform of its laws that prohibits allowing fishing licenses to be treated as private property.[3] Off the northeast coast of Newfoundland, Canada, ITQs in the crab fishery that came to replace the failed cod fishery were strongly resisted, in part because of fear that local fishing communities could lose their access to the bounties of the sea if ITQs were purchased by outsiders (McCay 1999). Such impacts on fishing communities happened in Iceland, where communities dependent on fish processing no longer had fish delivered to them because ITQ holders took the fish elsewhere (Pálsson and Helgason 1995), and ITQs became the focus of national political battles (Durrenberger and Pálsson 2014).

Considerable effort goes into finding ways to modify ITQ systems to protect communities and small-holders and reduce costs of entry for newcomers (McCay 2004). Examples include the Community Development Quota system used in Alaska to give native groups a stake in certain off-shore fisheries; requirements that owners of quota rights are bona fide fishers; special consideration for crew-members; and other measures (McCay 2004). U.S. fisheries legislation has been changed to allow for community-held shares of the catch, but thus far there has been little success in making this happen. The demand for modifications is a reminder that more than investments and profits are at stake. Put another way, property rights are not just the privileged, exclusive rights to licenses and quotas. The rights to share in a common pool resource are also property: common property, a kind of property that implies and ultimately requires the social relationships of community.

Common heritages and communities

An alternative analytical stance to the bio-economic model, one with fuzzier disciplinary boundaries, comes out of social science research with fishing families and communities. It tends to be critical of both government regulations and privatization policies for not adequately addressing social and equity issues. It has linkages with norms in international law of the sea as well. Complementing the doctrines of freedom of the high seas and public trust rights closer to shore is Arvid Pardo's formulation of the submarine resources of the high seas as a "common heritage of mankind" (Pardo 1975). In the context of law of the sea reforms at the United Nations, this was a way to bring all of the nations, rich and poor, coastal and landlocked, into discussions about exploitation of minerals found on the bottom of the high seas, with plans for a system for ensuring equitable sharing of the benefits from mining them. The notion of "common heritage" implies an international community with interests and rights in the sea, and some notion that responsibility for the healing of the oceans should be shared as well. It seems to be alive now in efforts underway within the United Nations to address threats to marine biodiversity in the open oceans.

The idea of "common heritage of mankind" resonates with the call for greater focus on and care for how common pool resource use and management interact with social relationships and community. Hardin, the widely cited author of the 1968 paper "Tragedy of the commons," 25 years later revised the title to "Tragedy of the unmanaged commons" (Hardin 1994) in response to critiques of his paper and how it was interpreted. Those critiques continue and are the basis of much scholarship about the marine commons today. A now well-established body of research, grounded in community-oriented case studies and social research, creates the potential for other more complex and nuanced understandings of the commons and common property, where access is not completely open, where common resource activities are in fact regulated, and where, in some cases, the commoners themselves are engaged in management. The bio-economic perspective and Hardin's tragedy scenario did not envision the possibility that communities of resource users—small, local; large, even international—may be engaged in managing common pool resources. Keeping the metaphor, one can talk about "tragedies of the commoners" and "comedies of the commons," too (Rose 1994).

In this regard, fisheries' research on the commons has fully melded with research on irrigation systems, pastoral ranching, urban settlements, and forest use. The ability of common resource users to come up with arrangements for management of those resources is classically exemplified in the Swiss alpine grazing lands (Stevenson 1991) and transparent in many other domains as well, including marine fisheries and even the provision of public services such as police protection (Ostrom et al. 1961). Anthropologists, sociologists, geographers, and others with extensive, detailed knowledge of the workings of fishing communities, as well as historians knowledgeable about the workings of ancient communities identified as having common properties, have shown the existence of local, often small-scale and informal systems for controlling access and regulating use of fish stocks and other common resources (McCay and Acheson 1987b; Berkes 1989; Cordell 1989; Acheson and Wilson 1996). Political scientist Elinor Ostrom compiled much of this as the basis for her ground-breaking analysis of conditions that seem to be required for successful small-scale community-based management of common property resources (Ostrom 1990). The most general message coming out of the critical enterprise is that commons management and mismanagement are possible with any kind of property regime, open-access, private, common, or government (Berkes et al. 1989; Feeny et al. 1990).

This broader critique envisions community as juxtaposed to (although not necessarily exclusive of) the market and the state as sources of governance for complex systems (Ostrom 1994; Apostle et al. 1998; McCay and Jentoft 1998). Community—constituted by networks of social relations, links to place, and heritage, histories, and political economies—is a valuable property of fisheries that influences how they work and what they are used for (Jentoft 2000; St. Martin 2006). Community can be a site and mechanism for learning about, experimentation with, and management of common resources (St. Martin et al. 2007).

The argument for community, coming largely out of sociology and anthropology, has roots in social theory, challenging an epistemology that sees the discrete individual as the key social actor, the basis of economics, with one that views individuals as inherently social (Jentoft et al. 1998; McCay and Jentoft 1998). In policy terms, the argument is one for more inclusive participation in resource monitoring and management (Béné and Neiland 2006). "Bottom-up" and participatory approaches are particularly salient for fishing and coastal communities, given the difficulties of obtaining knowledge about marine ecosystems and of enforcing regulations in remote areas, especially in developing countries and poor regions. Consequently, community-based management is widely embraced as a vehicle for marine management as it is for wildlife-management in general.

Community and tragedies of the "commoners:" social impacts

Understanding communities in relation to fisheries management requires investigating the politics of fisheries management as well as how local communities and fishing families are affected by marine policy, the field of "social impact" analysis. In the Newfoundland case as in most others, the metaphor of tragedy can be adapted to identify tragedies of "the commoners," of people dependent on use of commonly held resources. People are not only villains, playing their roles in different institutional and ecological settings, but they can also be cast as victims. Fishing families and businesses suffer the costs and dangers of troubles at sea, the brunt of regulations, and the consequences of the failure of the regulations to protect the resources.

The Newfoundland northern cod collapse presents a sad tale of the socio-economic impacts of the loss of a major fishery. The fisheries' closures that followed the collapse resulted in loss of work for at least 18,000 people, forcing many to accept modest government hand-outs and leading to mass out-migration. Rural depopulation quickly followed, with the challenges to communities and the quality of life that poses, such as schools shriveling in size and what they can offer. Significant social and health problems occurred (Arms 2004; Ommer 2008). There was also major decline and loss of the traditional small-boat, coastal fishery and its associated knowledge, skills, and culture.

Moreover, major changes in fisheries policy often have the unintended consequence of restructuring economies and communities. Governments can be pressured to come up with compensatory schemes, vessel and permit buy-backs, retraining, etc., but the restructuring that occurs can be profound and transformative, as has been the case in Newfoundland. The aftermath of the collapse of the cod fishery and subsequent closures was movement of people and capital into offshore crab and shrimp fisheries. These proved to be an economic boon because of fish abundance and strong markets (Arms 2004; Schrank 2005). But the costs of engagement in these fisheries were very high, requiring as they do larger, more powerful, and more seaworthy vessels and different kinds of gear, with the net effect of leaving most of the traditional cod fishermen with few options than to leave the fisheries or find positions on boats owned by someone else. Once egalitarian rural communities became fractured by a large economic gap between the large-boat owners and the rest.

Newfoundland's northern cod saga has a somewhat positive outcome. As of 2018, recovery of the cod population seems in sight. New spawning aggregations have been documented and there are other positive signs (Rose and Rowe 2015), not least of which is the ease of catching very large fish during the short seasons of the summer and fall when Newfoundlanders are allowed to fish for subsistence. However, one of the larger boats used in the offshore fishery has the following written on its bow: "She's Gone Boys, She's Gone," in reference to the cod fishery. It remains true even in the fact of healthier and growing cod populations, because the markets, skills, and licenses required for successful cod fishing are nearly gone from many Newfoundland fishing communities.

Community and comedies of the commons: TURFs

The interdisciplinary framework of "sustainability science" (Kates et al. 2001; McCay 2012) adds a perspective on people as active and innovative learners and responders, not just active villains or passive victims. People appear as problem-solvers as well as problem-creators, and when the systems involved are modeled, they should be viewed as part of the system, not exogenous to it. People are respected not just in terms of their exploitation of resources and uses of places but also in their roles in appreciating, understanding, managing, and conserving those resources and places. Returning to the metaphor of dramas of the commons, we may say that in this perspective, rather than cast as individuals with tragic flaws, as in the tragedy of the commons, they are cast as social beings, working with others to try to deal with and correct what seems out of kilter. In ancient Greek culture, this was a definition of comedy: "The drama of humans as social rather than private beings, a drama of social actions having a frankly corrective purpose" (Smith 1984; McCay and Acheson 1987a; Elliott 2001; see also Rose 1994).

This perspective leads to consideration of the roles of common resource users as managers, through their local institutions. There are many documented cases of community-based management in marine fisheries. Of current interest are those called TURFs, territorial use-right fisheries, where a local community or user-group has both rights and responsibilities for a defined fishing area and some or all of the resources in it. TURFs are among the many innovations in marine fisheries that have taken hold in small but growing ways, through local initiatives and partnerships among fishing groups, non-governmental organizations and benefactors, and government, as possible ways to avoid both tragedies of the commons and tragedies of the commoners. TURFs would seem to exemplify most of the features identified by Ostrom (1990) as important to the workings of small-scale, user-based systems for managing common pool resources, including well-defined boundaries (both spatial and social), the ability to monitor resources and members' behavior, user participation in rule-making and modifications, and matching the rules to local needs and conditions.

TURFs are found mainly in small-scale coastal fisheries for shellfish and relatively sedentary finfish (as in lagoons or coral reefs). They are often informal, in the sense that the territories used and defended are not recognized by government or formal law. In some cases territoriality is based on isolated circumstance, as in a Costa Rican community engaged in efforts to co-manage beach-nesting sea turtles (Campbell et al. 2007). The lobster "fiefs" of the Gulf of Maine are notorious examples of informal territoriality (Acheson 1988; Wilson 1996), although some aspects of the often fiercely defended territoriality of lobster catchers have been taken up into area-based co-management with the state (Acheson 2009). Another of Ostrom's "design principles" is ensuring that outside authorities respect the rule-making rights of community members, and the informal systems are often vulnerable when this is lacking.

TURFs are of particular interest as providing the basis for improved resource governance, particularly in the directions of effective community-based management and for co-management with government agencies. That has been a major focus of research in widely diverse areas, including the cases mentioned earlier, and in Japan, where cooperatives hold exclusive territory-based use rights to many coastal resources (Ruddle 1989; Makino and Matsuda 2005; Takahashi et al. 2006; Cancino et al. 2007). In Chile, exclusive territory-based use rights to fisheries for shellfish have been developed in recent decades (Castilla and Defeo 2001; Defeo and Castilla 2005; Gelcich et al. 2010).

An example I have studied comes from the fishing cooperatives of the Mexican North Pacific, along the coast of Baja California, which display remarkable independence and initiative in managing access to and use of abalone, lobster, and other fishery resources within territorial concessions that each cooperative holds. Over a fifty-year period of difficult transitions and learning experiences, they evolved to the point where they were able to show sustainable management of the lobster resources, gaining them certification from the Marine Stewardship Council (Ponce-Díaz et al. 2009; McCay et al. 2013). They exemplify most of the features identified by Ostrom (1990) as important to the workings of small-scale systems for managing common pool resources (McCay et al. 2013). This case is also a powerful lesson on how locally crafted and enforced rules can be effective, particularly where there is a high level of stakeholder participation.

The sustainability of community-based arrangements ultimately depends on their linkages with higher levels of governance, as also demonstrated in a study of community engagement in creating, maintaining, and enforcing a marine reserve network elsewhere in northern Mexico (Cudney Bueno and Basurto 2009). For one thing, local success can lead to poaching from outsiders, against which small communities need help. The cooperatives mentioned in the previous paragraph developed strong relationships with science institutions and government agencies, as well as the national system of cooperatives, which strengthened their ability to manage the fisheries.

"Co-management" is thus critical to the success of most small-scale systems for commons governance. It refers to arrangements linking local resource users and community-level institutions to government agencies, non-profit organizations, and others. The success of the Mexican North Pacific fisheries mentioned earlier is partly due to their cooperative arrangements with outsiders and government agencies (Ponce-Díaz et al. 2009). In principle co-management should be based on a formal sharing of authority, responsibility, and resources (Pinkerton 1989; Jentoft 2005). Large comparative studies show great variation in the performance of such arrangements and identify key factors such as leadership and social capital (Gutiérrez et al. 2011).

Constraints to community-based institutions

Small-scale, community-level fisheries management and co-management is inappropriate for many marine systems because of the much larger scale and transitory behavior of key components of the ecosystem. It tends to be relegated to benthic fisheries, such as abalone and lobster, as well as lagoon and other localized systems where the key goal of management may be mitigating competition and crowding rather than ensuring sustainable productivity. In contrast, the fishing communities of places like Fogo Island, Newfoundland, Canada, depend on access to fish populations that move many hundreds of miles in annual migrations, and thus must coordinate their actions with many other communities and interests, a situation that calls for some degree of centralized knowledge creation, negotiation, monitoring, and enforcement, that is, government. Related is the likelihood that where there are highly mobile fish populations there will be highly mobile fishing fleets, whose interests are opposed to those of more territorial coastal fishing fleets, a situation that has worsened with globalization of seafood economies (Berkes et al. 2006).

Yet another constraint is the existence of public trust doctrine, constitutions, and other institutions that make it difficult for local fishing communities to legitimately reserve access to community members. The public trust doctrine locates rights in the public at large, as against local communities, a matter that affects beach access in the United States (Potash 2016). The interstate commerce clause of the U.S. Constitution makes it difficult for state governments to reserve fishing to their waters to state residents (Schoenbaum and McDonald 1977).[4]

Consequently, the interests and claims of local communities must be voiced and addressed in larger political and economic arenas, where commoners and their political representatives, scientists, lawyers, and others become engaged in deliberations about allowable catches, fishing grounds, seasons, and the like. The highly participatory versions of fisheries (and other oceans) found in the United States, Canada, and other democratic regimes can be very quarrelsome and contentious and result in suboptimal policies. But they can and should be recognized as "comedies" or "dramas" of the commons, dramas of social action "having a frankly corrective purpose" (Smith 1984).

Concluding remarks

Commons analysis has a well-established and good fit with studies of the oceans and marine fisheries, given how much saltwater is public and common compared with private. Perhaps distinctive in applying commons frameworks of analysis is the difficulty establishing boundaries given the fluidity of marine ecosystems. The analytic frameworks used include both bio-economic modeling and classic institutional analysis; but also important are intensive case studies involving deep ethnography and long-term historical analysis. Normative assumptions vary greatly, mainly between those who hold economic views and those who are more grounded in the other social sciences and humanities. It is very simple: economists favor market-oriented solutions, and most other social scientists are chary of them because of implications for families and communities.

The marine fisheries field is marked by different approaches to and assumptions about how people fit in the marine commons. Traditional ecology and neo-classical economics emphasize the negative ("anthropogenic") effects of people on natural systems as major causes of environmental problems. Traditional social science emphasizes mainly the social and economic impacts of regulatory and other changes. These perspectives are exemplified in the recent history of Newfoundland northern cod and in fisheries experiencing policies such as ITQs. Increased credit given to communities and the commons in marine policy, as well as the interdisciplinary framework of sustainability science, add perspectives on the roles of people: as active "learners" and "responders," sometimes also as effective stewards and managers, participating in "comedies of the commons" (Rose 1994; McCay 2001). Co-management and territorial use-rights fisheries are examples of growing interest.

Notes

1 Marine fisheries may be chosen to exemplify commons problems so much more than livestock grazing because pasturage in common is now rare, but capture fisheries (as distinct from mariculture) always involve common pool resources with greater or lesser degrees of open access.

2 "Alaska: Bering Sea Trawl Cod Fishery May Have Been Shortest Ever, as High Prices Attract Effort," Jim Paulin, *Seafood News*, as reprinted with permission by SavingSeafood.Org, February 21, 2018.

3 According to the reform, all living marine resources in Faroese waters are the property of the people of the Faroe Islands, and as such, fishing licenses may never become private property, neither by law nor practice. Furthermore, fishing licenses cannot be traded directly between private buyers. To change hands, licenses must go through a public auction. See www.government.fo/news/news/the-faroese-parliament-passes-fisheries-reform/.

4 According to Blake Hudson, the test(s) are complicated: "Thus, in a dormant Commerce Clause case, a court is initially concerned with whether the law facially discriminates against out-of-state actors or has the effect of favoring in-state economic interests over out-of-state interests. Discriminatory laws motivated by 'simple economic protectionism' are subject to a 'virtually per se rule of invalidity,' City of Philadelphia v. New Jersey 437 U.S. 617 (1978), Dean Milk Co. v. City of Madison, Wisconsin, 340 U.S. 349 (1951), Hunt v. Washington State Apple Advertising Comm., 432 U.S. 333 (1977), which can only be overcome by a showing that the State has no other means to advance a legitimate local purpose, Maine v. Taylor, 477 U.S. 131(1986). See also Brown-Forman Distillers v. New York State Liquor Authority, 476 U.S. 573 (1986). On the other hand, when a law is 'directed to legitimate local concerns, with effects upon interstate commerce that are only incidental' (United Haulers Association, Inc.), that is, where other legislative objectives are credibly advanced and there is no patent discrimination against interstate trade, the Court has adopted a much more flexible approach, the general contours of which were outlined in Pike v. Bruce Church, Inc., 397 U.S. 137, 142 (1970) and City of Philadelphia v. New Jersey, 437 U.S. at 624. If the law is not outright or intentionally discriminatory or protectionist, but still has some impact on interstate commerce, the court will evaluate the law using a balancing test. The Court determines whether the interstate burden imposed by a law outweighs the local benefits. If such is the case, the law is usually deemed unconstitutional. See Pike v. Bruce Church, Inc., 397 U.S. 137 (1970)."

References

Acheson, J. M. (1988) *The Lobster Gangs of Maine*. Hanover, NH: University Press of New England.

Acheson, J. M. (2009) Co-management in the Maine lobster industry: Past successes, a questionable future. Unpublished Manuscript, 34.

Acheson, J., Apollonio, S., & Wilson, J. (2015) Individual transferable quotas and conservation: a critical assessment. *Ecology and Society*, 20(4): art. 7.

Acheson, J. M. & Wilson, J. A. (1996) Order out of chaos; The case for parametric fisheries management. *American Anthropologist*, 98, 579–594.

Andersen, P. (1983) On rent of fishing grounds: A translation of Jens Warmings' 1911 article, with an introduction. *History of Political Economy*, 15, 391–396.

Anonymous (n.d.) *History of the Northern Cod Fishery*. www.cdli.ca/cod/home1.htm.

Apostle, R., Barrett, G., Holm, P., Jentoft, S., Mazany, L., McCay, B., & Mikalsen, K. (1998) *Community, Market and State on the North Atlantic Rim: Challenges to Modernity in the Fisheries*. Toronto: University of Toronto Press.

Archer, J. H. (1994) *The Public Trust Doctrine and the Management of America's Coasts*. Amherst, MA: University of Massachusetts Press.

Arms, M. (2004) *Servants of the Fish; A Portrait of Newfoundland after the Great Cod Collapse*. Hinesburg, VT: Upper Access, Inc.

Béné, C. & Neiland, A. E. (2006) *From Participation to Governance: A Critical Review of the Concepts of Governance, Co-Management and Participation, and their Implementation in Small-Scale Inland Fisheries in Developing Countries. WorldFish Center Contribution No. 1750*. WorldFish Center, Penang, Malaysia, and CGIAR Challenge Program on Water and Food, Colombo, Sri Lanka.

Berkes, F. (1989) *Common Property Resources; Ecology and Community-Based Sustainable Development*. London: Belhaven Press.

Berkes, F., Feeny, D., McCay, B. F. & Acheson, J. M. (1989) The benefits of the commons. *Nature*, 340, 91–93.

Berkes, F., Hughes, T. P., Steneck, R. S., Wilson, J. A., Bellwood, D. R., Crona, B., . . . & Worm, B. (2006) Globalization, roving bandits, and marine resources. *Science*, 311 (17 March), 1557–1558.

Burke, W. T. (1994) *The New International Law of Fisheries: UNCLOS 1982 and Beyond*. Oxford, UK: Clarendon Press.

Campbell, L. M., Haalboom, B. J., & Trow, J. (2007) Sustainability of community-based conservation: Sea turtle egg harvesting in Ostional (Costa Rica) ten years later. *Environmental Conservation*, 34, 122–131.

Cancino, J. P., Uchida, H., & Wilen, J. (2007) TURFs and ITQs: Coordinated vs. decentralized decision making. *Marine Resource Economics*, 22, 391–406.

Carothers, C. & Chambers, C. (2012) Fisheries privatization and the remaking of fishery systems. *Environment and Society: Advances in Research*, 3: 39–59.
Casey, K. E., Dewees, C. M., Turris, B. R., & Wilen, J. E. (1995) The effects of individual vessel quotas in the British Columbia halibut fishery. *Marine Resource Economics*, 10, 211–230.
Castilla, J. C. & Defeo, O. (2001) Latin American benthic shellfisheries: Emphasis on co-management and experimental practices. *Reviews in Fish Biology and Fisheries*, 11, 1–30.
Charles, A. T. (1995) The Atlantic Canadian groundfishery: Roots of a collapse. *Dalhousie Law Journal*, 18, 65–83.
Christy Jr., F. T. (1973) Fishermen's quotas: A tentative suggestion for domestic management. In *Occasional Papers* 19, 7 pp. Kingston, RI: Law of the Sea Institute, University of Rhode Island.
Chu, C. (2009) Thirty years later: The global growth of ITQs and their influence on stock status in marine fisheries. *Fish and Fisheries*, 10: 217–230.
Clark, C. W. (1973) The economics of over-exploitation. *Science*, 181, 630–634.
Cordell, J. (1989) *A Sea of Small Boats*. Cambridge, MA: Cultural Survival.
Costello, C., Gaines, S., & Lynham, J. (2008) Can catch shares prevent fisheries collapse? *Science*, 321: 1678–1681.
Costello, C. et al. (2016) Global fishery prospects under contrasting management regimes. *Proceedings of the National Academy of Sciences USA*, 113: 1525–1529.
Cudney Bueno, R. & Basurto, X. (2009) Lack of cross-scale linkages reduces robustness of community-based fishery management. *PLoS ONE (Public Library of Science)*, 4, e6253.
Defeo, O. & Castilla, J. C. (2005) More than one bag for the world fishery crisis and keys for co-management successes in selected artisanal Latin American shellfisheries. *Reviews in Fish Biology and Fisheries*, 15, 265–283.
Durrenberger, E. P. & Pálsson, G. eds. (2014) *Gambling Debt: Iceland's Rise and Fall in the Global Economy*. Boulder, CO: University Press of Colorado.
Elliott, E. D. (2001) The tragi-comedy of the commons: Evolutionary biology, economics and environmental law. *Virginia Environmental Law Journal*, 20, 17–31.
Eythorsson, E. (1998) Metaphors of property: The commoditisation of fishing rights. In *Northern Waters: Management Issues and Practice*, ed. D. Symes. Oxford, UK: Fishing News Books, 42–51.
Feeny, D., Berkes, F., McCay, B. J., & Acheson, J. M. (1990) The tragedy of the commons: Twenty-two years later. *Human Ecology*, 18, 1–19.
Finlayson, A. C. (1994) *Fishing for Truth; A Sociological Analysis of Northern Cod Stock Assessment from 1977 to 1990*. St. John's, NL: Institute of Social and Economic Research, Memorial University of Newfoundland.
Finlayson, A. C. & McCay, B. J. (1998) Crossing the threshold of ecosystem resilience: The commercial extinction of northern cod. In *Linking Social and Ecological Systems: Institutional Learning for Resilience*, eds. C. Folke & F. Berkes. Cambridge, UK: Cambridge University Press, 311–337.
Fraser, G. A. (1979) Limited entry: Experiences of the British Columbia salmon fishery. *Journal of the Fisheries Research Board of Canada*, 36, 754–763.
Gelcich, S., Hughes, T. P., Olsson, P., Folke, C., Defeo, O., Fernandez, M., . . . & Castilla, J. C. (2010) Navigating transformations in governance of Chilean marine coastal resources. *Proceedings of the National Academy of Sciences*, 107, 16794–16799.
General Accounting Office (2004) *Individual Fishery Quotas: Methods of Community Protection and New Entry Require Periodic Evaluation*. www.gao.gov/cgi-bin/getrpt?GAO-04-277.
Gilmour, P. W., Day, R. W., & Dwyer, P. D. (2012) Using private rights to manage natural resources: Is stewardship linked to ownership? *Ecology and Society*, 17(3): art. 1.
Gordon, H. S. (1954) The economic theory of a common property resource: The fishery. *Journal of Political Economy*, 62, 124–142.
Gutiérrez, N. L., Hilborn, R. W., & Defeo, O. (2011) Leadership, social capital and incentives promote successful fisheries. *Nature*, 470, 386–389.
Halpern, B. S., Walbridge, S., Selkoe, K. A., Kappel, C. V., Micheli, F., D'Agrosa, C., . . . & Watson, R. (2008) A global map of human impact on marine ecosystems. *Science*, 319, 948–952.
Hannesson, R. (1996) *Fisheries Mismanagement: The Case of the Atlantic Cod*. London: Fishing News Books.
Hardin, G. (1968) The tragedy of the commons. *Science*, 162, 1243–1248.
Hardin, G. (1994) The tragedy of the unmanaged commons. *Trends in Ecology and Evolution*, 9, 199.
Hutchings, J. A. & Reynolds, J. D. (2004) Marine fish population collapses: Consequences for recovery and extinction risk. *BioScience*, 54, 297–309.

Jentoft, S. (2000) The community: A missing link of fisheries management. *Marine Policy*, 24, 53–59.
Jentoft, S. (2005) Fisheries co-management as empowerment. *Marine Policy*, 29, 1–7.
Jentoft, S., McCay, B. J., & Wilson, D. C. (1998) Social theory and fisheries co-management. *Marine Policy*, 22, 423–436.
Kates, R. W., Clark, W. C., Corell, R., Hall, J. M., Jaeger, C. C., Lowe, I., . . . & Svedin, U. (2001) Sustainability science. *Science*, 292, 641–642.
Makino, M. & Matsuda, K. (2005) Co-management in Japanese coastal fisheries: Institutional features and transaction costs. *Marine Policy*, 29, 441–450.
Martin, C. (1995) The collapse of the northern cod stocks. *Fisheries*, 20(5): 6–8.
May, A. (2009) The collapse of the northern cod. *Newfoundland Quarterly*, 102, www.mun.ca/harriscentre/reports/nlquarterly/NQ_article_Vol_102_No_2.pdf.
McCay, B. J. (1987) The culture of the commoners: Historical observations on Old World and East Coast U.S. fisheries. In *The Question of the Commons*, eds. B. J. McCay & J. M. Acheson. Tucson, AZ: University of Arizona Press, 195–216.
McCay, B. J. (1993) The making of an environmental doctrine: Public trust and American shellfishermen. In *Environmentalism: The View from Anthropology*, ed. K. Militon. London: Routledge, 85–96.
McCay, B. J. (1995) Social and ecological implications of ITQs: An overview. *Ocean and Coastal Management*, 28, 3–22.
McCay, B. J. (1999) "That's not right": Resistance to enclosure in a Newfoundland crab fishery. In *Fishing People, Fishing Places: Issues in Canadian Small-Scale Fisheries*, eds. D. Newell & R. Ommer. Toronto: University of Toronto Press, 301–320.
McCay, B. J. (2001) Community and the commons: Romantic and other views. In *Communities and the Environment: Ethnicity,Gender, and the State in Community-Based Conservation*, eds. A. Agrawal & C. C. Gibson. New Brunswick, NJ: Rutgers University Press, 180–191.
McCay, B. J. (2004) ITQs and community: An essay on environmental governance. *Review of Agricultural and Resource Economics*, 33, 162–170.
McCay, B. J. (2012) Peopling the marine ecosystem. In *Advancing an Ecosystem Approach in the Gulf of Maine*, eds. R. L. Stephenson, J. Annala, J. A. Runge & M. L. Hall-Arber. Bethesda, MD: American Fisheries Society.
McCay, B. J. & Acheson, J. M. (1987a) Human ecology of the commons. In *The Question of the Commons: The Culture and Ecology of Communal Resources*, eds. B. J. McCay & J. M. Acheson. Tucson, AZ: University of Arizona Press, 1–34.
McCay, B. J. & Acheson, J. M. (1987b) The question of the commons : The culture and ecology of communal resources. In *Arizona Studies in Human Ecology*. Tucson, AZ: University of Arizona Press, xvi.
McCay, B. J., Brandt, S., & Creed, C. F. (2011) Human dimensions of climate change and fisheries in a coupled system: The Atlantic surfclam case. *ICES Journal of Marine Science*, 68, 1354–1367.
McCay, B. J. & Jentoft, S. (1998) Market or community failure? Critical perspectives on common property research. *Human Organization*, 57, 21–29.
McCay, B. J., Micheli, F., Ponce-Diáz, G., Murray, G., Shester, G., Ramirez-Sanchez, S., & Weisman, W. (2013) Cooperatives, concessions, and co-management on the Pacific coast of Mexico. *Marine Policy*, 44, 49–59.
McEvoy, A. F. (1986) *The Fisherman's Problem; Ecology and Law in the California Fisheries, 1850–1980*. Cambridge, UK: Cambridge University Press.
McGuire, T. (1997) The last northern cod. *Journal of Political Ecology*, 4, 41–54.
National Research Council (1999) *Sharing the Fish: Toward a National Policy on Individual Fishing Quotas*. Washington, DC: National Academy Press.
Neher, P. A., Arnason, R., & Mollett, N. (1989) *Rights-Based Fishing*. Dordrecht, the Netherlands: Kluwer Academic Publishing.
Olson, J. (2011) Understanding and contextualizing social impacts from the privatization of fisheries: An overview. *Ocean & Coastal Management*, 54, 353–363.
Ommer, R. E. (2008) *Coasts Under Stress: Restructuring and Social-Ecological Health*. Montreal, Canada: McGill-Queen's University Press.
Osherenko, G. (2006) New discourses on ocean governance: Understanding property rights and the public trust. *Journal of Environmental Law and Litigation*, 21, 317–381.
Ostrom, E. (1990) *Governing the Commons : The Evolution of Institutions for Collective Action*. Cambridge, UK: Cambridge University Press.

Ostrom, E. (1994) *Neither Market Nor State: Governance of Common-Pool Resources in the Twenty-First Century.* Washington, DC: International Food Policy Research Institute.

Ostrom, E., Gardner, R., Walker, J., Agrawal, A., Blomquist, W., Schlager, E., & Tang, S. Y. (1994) *Rules, Games and Common-Pool Resources.* Ann Arbor, MI: University of Michigan Press.

Ostrom, V., Tiebout, C. M., & Warren, R. (1961) The organization of governments in metropolitan areas. *American Political Science Review*, 55, 831–842.

Pálsson, G. & Helgason, A. (1995) Figuring fish and measuring men: The individual transferable quota system in the Icelandic cod fishery. *Ocean and Coastal Management*, 28, 117–146.

Pardo, A. (1975) *The Common Heritage: Selected Papers on Oceans and World Order 1967–1974.* Valetta: Malta University Press.

Pinkerton, E. & Edwards, D. N. (2009) The elephant in the room: The hidden costs of leasing individual transferable fishing quotas. *Marine Policy*, 33, 707–713.

Pinkerton, E. W. (1989) *Cooperative Management of Local Fisheries.* Vancouver, BC: University of British Columbia Press.

Ponce-Díaz, G., Weisman, W., & McCay, B. J. (2009) Co-responsabilidad y participación en el manejo de pesquerías en México: lecciones de Baja California Sur. *Pesca y Conservación*, 1, 1–9.

Potash, J. (2016) Public trust doctrine and beach access: Comparing New Jersey to nearby states. *Law School Student Scholarship.* Paper 738. http://scholarship.shu.edu/student_scholarship/738.

Rose, C. M. (1994) The comedy of the commons: Custom, commerce, and inherently public property. In *Property & Persuasion; Essays on the History, Theory, and Rhetoric of Ownership*, ed. C. M. Rose. Boulder, CO: Westview Press, 105–162.

Rose, G. (2007) *Cod: The Ecological History of the North Atlantic Fisheries.* St. John's, NF: Breakwater Books.

Rose, G. A. & Rowe, S. (2015) Northern cod comeback. *Canadian Journal of Fisheries and Aquatic Science*, 72, 1789–1798.

Ruddle, K. (1989) Solving the common-property dilemma: Village fisheries rights in Japanese coastal waters. In *Common Property Resources*, ed. F. Berkes. London: Belhaven Press, 168–198.

Schoenbaum, T. J. & McDonald, P. E. (1977) State management of marine fisheries after the Fishery Conservation and Management Act of 1976 and Douglas V. Seacoast Products, Inc. *William & Mary Law Review*, 19, 17–42.

Schrank, W. E. (2005) The Newfoundland fishery: Ten years after the moratorium. *Marine Policy*, 29, 407–420.

Scott, A. (1955) The fishery: The objectives of sole ownership. *Journal of Political Economy*, 63, 116–124.

Slade, D. C., Kehoe, R. K., Stahl, J. K., & Coastal States Organization (U.S.) (1997) *Putting the Public Trust Doctrine to work.* Washington, DC: Coastal States Organization.

Smith, M. E. (1984) The Triage of the Commons. Paper presented at the Annual Meeting of the Society for Applied Anthropology, March 14–18, 1984, Toronto.

St. Martin, K. (2006) The impact of "community" on fisheries management in the US Northeast. *Geoforum*, 37, 169–184.

St. Martin, K., McCay, B. J., Murray, G., Johnson, T., & Oles, B. (2007) Communities, knowledge, and fisheries of the future. *International Journal of Global Environmental Issues*, 7, 221–239.

Steele, D. H., Andersen, R. R., & Green, J. M. (1992) The managed commercial annihilation of northern cod. *Newfoundland Studies*, 8, 34–68.

Stevenson, G. G. (1991) *Common Property Economics: A General Theory and Land Use Applications.* Cambridge, UK: Cambridge University Press.

Takahashi, S., McCay, B. J., & Baba, O. (2006) The good, the bad, or the ugly? Advantages and challenges of Japanese coastal fisheries management. *Bulletin of Marine Science*, 78, 575–591.

Tietenberg, T. (2002) The tradable permits approach to protecting the commons: What have we learned? In *The Drama of the Commons*, eds. E. Ostrom, T. Dietz, N. Dolsak, P. C. Stern, S. Stonich & E. U. Weber. Washington, DC: National Academy Press, 197–232.

Turnipseed, M., Crowder, L. B., Sagarin, R. D., & Roady, S. E. (2009) Legal bedrock for rebuilding America's ocean ecosystems. *Science*, 324, 183–184.

Walters, C. & Maguire, J.-J. (1996) Lessons for stock assessment from the northern cod collapse. *Reviews in Fish Biology and Fisheries*, 6, 125–137.

Wilson, J. A. (1996) Maine's lobster fishery: Managing a common property resource. In *Increasing Understanding of Public Problems and Policies*, 145–161. Oakbrook, IL: Farm Foundation.

14
COASTAL COMMONS AS SOCIAL-ECOLOGICAL SYSTEMS

Achim Schlüter, Stefan Partelow, Luz Elba Torres-Guevara and Tim C. Jennerjahn

Introduction

The coast is the interface between the land and the sea. It is characterized by a huge diversity of social-ecological systems (SES) and the interdependent interactions and outcomes between social and ecological subsystems (Ostrom, 2009). Many resources at the land-sea interface and property rights of those resources have characteristics of a commons, in the sense of being shared among many actors, ranging from a handful of fishers to the global community. A difference in the governance systems regulating the land and the sea is that the former have much more nuanced and diversified governance, where many aspects are institutionalized. Many more resources are becoming regulated by private property regimes on the continuum between open access and private property (Bromley, 1991; Schlager and Ostrom, 1992) or even anti-commons property (Schlüter, 2008), despite the fact that many ecosystem services produced by those resources are common, due to their characteristic of excludability (Woelfle-Erskine, 2017). This means that a lot of vested interests, particularly in relation to financial gains and livelihoods, are well established. Due to difficulties in accessibility, still relatively low scarcity, technological abilities and other factors, the number of institutions regulating the marine space have been lower in comparison to terrestrial commons so far. However, all of these factors are changing at an impressive speed.

Access to and therefore the usability of marine space is increasing substantially, including the types of resources and technologies used to harvest and monitor them. The desire to exploit the sea, might it be for providing the world with required protein through fisheries and aquaculture, or to put claims on resources believed to be valuable in the near future, is expanding at a great speed (Berkes et al., 2006; Halpern et al., 2009; McCauley et al., 2015). Terms like "blue growth", born at the Rio +20 conference in 2012, aim to ensure that this growth takes place in a sustainable way, but the mere existence of the term indicates that there is a huge drive to use the sea more intensively (Burgess et al., 2016; van den Burg et al., 2017).

Despite the recognition of the problems facing unregulated coastal and marine commons, there are still substantial difficulties in defining their property rights. This particularly holds true for the tropics, where a large proportion of coastal space is located. However, technological advancements and the capacity to build institutions for resource management are rising. This might not only lead to increases in the much needed regulation and enforcement of environmental rules (e.g. MPAs) (Halpern, 2014) but also to privatization processes. Fisheries, which have long been the iconic example of a common pool resource (CPR), are becoming "more private" (Foley and McCay, 2014). However, privatization is occurring in new forms. Individual transferable quotas do not yet privatize the fish, but the right to catch it. In contrast, the booming marine aquaculture sector, which will overtake the GDP of capture fisheries soon (FAO, 2016), privatizes the fish itself. Certification schemes like the Marine Stewardship Council have made sustainable fishing a privately traded commodity. What is currently happening on the coasts and the ocean could be described as a movement of enclosure of the marine and coastal commons (Ávila-García and Sánchez, 2012).

This private property in the sea has implications for the coastal commons. For example, the excrements and the surplus feed of aquaculture input waste into the sea water, leading to nitrogen and phosphorus enrichment and ultimately to eutrophication (Burford et al., 2003; Herbeck et al., 2013). These activities are likely to continue in coastal commons with substantial negative externalities, due to system characteristics such as the movement of water across boundaries. People who cannot pay the price premium of certified fish products will continue to "graze" with many others on the commons of the coast. Huge tanker fleets discharging their ballast water near the shore, or reducing air quality in harbor cities substantially (Wan et al., 2016), are examples of how sea-based activities have an influence on the coastal commons. Despite the fact that those processes might be more advanced on the land side, substantial pressures can also be observed on the coast. The coastal commons contain multiple overlapping forms of land and sea governance regimes relating to multiple goods with differing types of property rights within a relatively small area. On the coast manifold SES of the sea and the land amalgamate and interact with each other in unique ways. In the future this will require greater attention from governance practitioners and scholars due to its complexity and increased use (Portman, 2016).

In this chapter we use the Social Ecological System Framework (SESF) developed by Ostrom and colleagues (Ostrom, 2007, 2009) to structure the characterization of coastal commons as SES. It is a generic framework for studying many different commons and is therefore well suited to capture diverse SES occurring at the coast. The SESF was developed with the aim to understand collective action in the commons by identifying similarities and differences between similar system components. It was developed for being able to compare commons phenomena occurring in a broad range of resource use sectors. It provides a common language, which enables a joint understanding among scholars and practitioners specialized in the various individual SES interacting in the coastal commons, like forestry, irrigation, water catchment and fisheries, to name a few. However, we can also think of more modern problems related to commons, like (i) marine litter (born at the coast), (ii) air pollution, traffic congestion at harbor hubs (Halpern et al., 2008; Partelow et al., 2015), or (iii) tourism overuse resulting from large increases in sewage, light pollution or even conflicts among tourists for surfing spots (Kaffine, 2009). The diversity of commons and different SES involved make it clear that an understanding and analysis requires interdisciplinarity among and across the social and the natural sciences (Glaser et al., 2012; Markus et al., 2017; Partelow et al., 2017).

Before characterizing the coast from a commons perspective, we briefly look at the literature that is focusing on coastal governance and land-sea interactions. We then use the SESF to characterize the diversity of commons and collective action problems existing in the SES of the coast. This will allow us to grasp the distinctiveness of coastal commons in comparison to others, discussing the implications for studying and governing the coast.

Current coastal governance concepts

The scientific community is well aware of the particular role and complexity of the coastal realm and the land-sea interaction (Jennerjahn, 2012; Ramesh et al., 2015). Three approaches, characterized by their applied orientation towards the sustainable management of the coast, have mainly dealt with the interaction of the land and the sea (Pittman and Armitage, 2016).

First, Integrated Coastal Zone Management (ICZM) is a social science driven approach that also takes collective action, institutional (economics), public goods and CPR perspectives (Brown et al., 2012). Born (2012) argues that despite the very inclusive and holistic approach to coastal management, the interdependencies and multi-functionalities are not yet well considered. Also, recognizing the interactions between land and sea, ICZM still considers the land and the sea as separate units of the coast (Stoms et al., 2005).

Second, the Ecosystem Based Management (EBM) approach, somewhat addresses these shortcomings because it looks at the cumulative effects of various sectors in the marine realm (but not vice versa). However, in the center of interest are ecosystem health and functions, and the continued provision of ecosystem services. In the understanding of EBM humans are components of the ecosystem and the basic aim is to "manage people's influence on [those] ecosystems" (McLeod and Leslie, 2009). However, it could be argued that the EBM, whose development is basically led by natural scientists, could be understood as a response to the more social science oriented ICZM. It is much less anthropocentric and takes an ecology-oriented conservation perspective (Christie, 2011; Aswani et al., 2012).

And third, there are new approaches in conservation planning (Ban et al., 2013; Adams et al., 2014; Álvarez-Romero et al., 2015). These are aimed at integrating the interdependencies between the various SES determining land-sea interactions, with the goal to achieve efficient and effective conservation planning (Adams et al., 2014). Similar to EBM, it is well recognized that ecological and jurisdictional boundaries often do not coincide. It is particularly important to look at the interdependencies between the various ecosystems. The optimal protection of a coral reef might be the restriction of grazing for cattle near the coast as this might stop sediment and nutrient input (Adams et al., 2014). The focus is on the influence of the land on the sea. However, it is well recognized that those influences are bi-directional (Álvarez-Romero et al., 2011), with interacting drivers of change between them. Activities in the ocean could also have important influences on land-based ecosystems. For example, invasive species, either introduced by ballast water of ships or by an extension of their geographical distribution limit due to global warming, may substantially alter marine food webs and introduce pathogens into a specific region. This, in turn, can affect coastal communities by altering fisheries as a consumable resource and by having negative effects on human health. Also, global warming-driven sea level rise increases coastal erosion, which directly affects habitats of coastal flora and fauna and decreases living space for humans (Bax et al., 2003; Kopp et al., 2014).

Through the work of Ban et al. (2013), the approach of conservation planning is linked to the SESF. They argue that the SESF "is the most appropriate for use in conservation planning because it encompasses a comprehensive group of social and ecological factors, all of which are applicable to conservation" (Ban et al., 2013). According to them, the conservation community is lacking an understanding about which conservation practice is working in which particular context. As this is the goal of commons research (Ostrom, 1990; Agrawal, 2001), it shows how much a holistic commons perspective on the coast is needed.

Due to these reasons, and with the many applications of the SESF framework in the various land- and sea-based SES, the framework is a suitable tool for understanding collective action in diverse coastal settings. Here we just refer to SESF applications which have focused explicitly on coastal resources. Most of the studies focus on the use of a particular resource. Most prominent in this regard is fishing (Aswani et al., 2013; Cinner et al., 2013; Leslie et al., 2015; Partelow, 2015; Partelow and Boda, 2015; Torres-Guevara et al., 2016). Other resources considered are turtles (Schlüter and Madrigal, 2012; Madrigal-Ballestero et al., 2013) and aquaculture (Partelow et al., 2018).

The most comprehensive application of the SESF to coastal ecosystems having a focus on methodological issues, is the comparison of three case studies by Delgado-Serrano and Ramos (2015), dealing with a fresh water system, a forest and an estuary. However, the emphasis is not on the interactions between those systems. This focus on a single resource might be due to complexity but also to the initial setup of the SESF, which aimed at describing a particular resource system with its particular resource unit. However, in the reformulated and most current version of the SESF, this limitation has been addressed and the framework shall be adaptable for a multitude of interacting resource and governance systems, comprising various actors and resources (McGinnis and Ostrom, 2014). Ban et al. (2013) is to our knowledge the first paper that aims to make use of this introduced multidimensional perspective into the SESF.

Characterizing of coastal commons using the SES framework

Building an entire taxonomy of coastal commons is far beyond the scope of a single chapter. Therefore, we focus on exemplifying the diversity of various resource systems which form the coastal commons using the relevant SESF second tier variables. Coastal commons vary so much in their social and ecological configurations that for any example shown here one can find a counterexample.

Resource systems (RS)

In the coastal commons a huge variety of diverse resource systems are interacting. These include forests, water, pastures, beaches, estuaries and fish, just to name a few (see Figure 14.1).

Figure 14.1 Resource systems at the coastal interface

Many systems are in close proximity, have overlapping boundaries and closely connected to each other.

Marine and coastal commons are often characterized by unclear boundaries (RS2) (see Figure 14.2). In general, one might say that due to the movement of water the boundaries of the seascape are less clear than those of the landscape. The moving water might continuously change the boundaries of a system depending on weather conditions, tidal stage or season. But, this cannot be generalized for marine resource systems. Many migratory species might have a large, but stable resource system that is not greatly influenced by currents. In contrast, lobster eggs and juveniles are carried by open water currents which create unclear spatial (i.e., how far from the parent) and temporal (i.e., how long until settlement on a reef) boundaries in their species' reproduction and distribution. Nonetheless, once adult lobsters settle on a reef, they are largely non-mobile (Acheson, 2006), making a commons governance approach with spatially explicit property rights for adult lobsters an easier institutional arrangement to develop.

Also on land, resource system boundaries are often unclear. A forest has a clear boundary when referring to timber. Nevertheless, when looking at the watershed resources the forest depends on, such as water and nutrients, the boundary of the resource system is less clear, which affects the marine environment. For example, fresh water draining into a coastal lagoon strongly influences the salinity and thereby the abundance and community composition of fish (Torres-Guevara et al., 2016). Rivers carry the nutrients and sediments from forested watersheds into estuaries and lagoons, and eventually the open sea. However, nutrients on land have to a large extent the characteristics of a private good. Fertilizing a field is a targeted operation and will mainly improve the farmer's crop yield, but when excess nutrients enter the river, they exist in the commons of the watershed, which is carried to the sea. The nutrients which were added to the system under a private property regime of the farmer improving private yield, is then affecting the coastal commons, and thereby the resource users who depend on them such as coral reef fishers or mussel growers. For some resource systems, units and actors, those excess nutrient inputs entering the sea from upstream watersheds can create negative externalities, e.g., the reef fishers, and positive externalities for others, e.g., the mussel growers.

Mariculture as a resource system is another example where certain boundaries are clear and others are fluid. Fish in the cage are subject to clear boundaries set by the cage and to private property rights. However, the waste produced by aquaculture enters the water body and is distributed into the aquatic environment (Páez-Osuna, 2001; Biao et al., 2004). Therefore, growing private fish depends on the ecosystem conditions of the coastal commons, on which other resource users and systems also depend. This example demonstrates that there is large variability in the size of the relevant resource systems (RS3) and their equilibrium properties (RS6).[1] Aquaculture ponds, forests, beaches, a seagrass meadow or a coral reef, might only occupy a couple of square meters or many hectares. An estuary might be a middle-sized system, nevertheless, with open system boundaries to the land and to the sea. Different fish species have different resource system sizes. River catchment areas, which ultimately drain into the sea, are hundreds to thousands of square kilometers in area. Turtles that intend to nest at a particular beach must swim thousands of kilometers along the coast (Schlüter and Madrigal, 2012). The large variability implies that the optimal size of the governance systems varies substantially. There is more size variability than feasible governance levels.

The various interacting resource systems display high variability in the use and value of human constructed facilities (RS4). Some of those investments are spatially dependent, while others are not. This leads to different degrees of vested interests. A port owner, whose investment comprises a lot of asset specificity, might exert a substantial amount of power to secure a return on investment for her shareholders. The same might apply for big investors in the

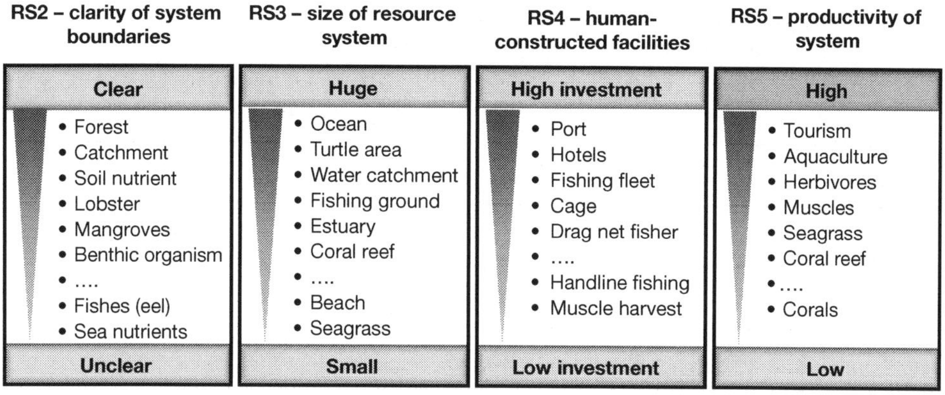

Figure 14.2 Diversity of characteristics of resource systems

tourism sector (Ávila-García and Sánchez, 2012). Large fishing fleets or container ships may have similarly high investments. However, they are mobile and can easily escape an exploited or degraded commons and move to the next. A fisher may just have invested into a handline he uses from the shore or may have bought a motorboat or dragnets. This investment is mobile, but only locally, and is less mobile than large fishing or shipping vessels which can move regionally or globally (Berkes et al., 2006). In contrast, a cage owner has invested into a stationary activity and may, therefore, heavily depend on finding a solution to the collectively caused problem of fish kills (caused by other owners, farmers in the catchment area, global CO_2 emitters, etc.). However, the amount of investment might be inversely related to the dependence on the resource and resource system.

The productivity of resource systems (RS5) in terms of creating direct (monetary) benefit to humans varies substantially. This leads to different potentials to establish property rights. Our understanding of many coastal commons is rather limited when we are dealing with the interaction effects among resource systems. This means that often equilibrium properties (RS6) are not known and predictions are difficult. According to Denzau and North (1994), the lack of knowledge on system properties leads to multiple mental models, which then become an important factor in understanding collective action, as uncertainty, belief systems, perceptions and a lack of understanding are challenges for governing the commons (RS7).

Governance systems (GS)

The sizes and boundaries of the resource systems interacting at the coast have been described as multidimensional, spatially overlapping and diverse in their characteristics. Therefore, it is impossible to create government organizations (GS1) corresponding to each resource system (see Figure 14.3). Also many existing government systems have not been created to govern those resource systems, but for other purposes, such as regulating a single resource unit within it. Often, when government agencies have been developed for a particular resource system, it was for the purpose of extracting particular resource units and only later with consideration for a sustainable use. For example, many different governance systems may exist such as a fishery authority, an agriculture extension service, a water board, a tourism regulator, or a protected area management structure. Nevertheless, these are neither designed considering the resource system boundaries, nor the interdependencies among the systems. They are designed for regulating a specific resource. Government bodies that cover land and sea issues are hardly existent (Adams et al., 2014). Considering the interaction effects of different resource units and resource systems in the coastal commons, it is difficult for sectoral organizations to jointly manage resource units and systems as no institutional structures exist. For example, governing the nutrient flows into a coral reef or a lagoon involves many government bodies dealing with forestry, agriculture, tourism and sewage disposal. Those different government bodies exist for historical reasons or due to their economic importance and are very different in size and their ability to help govern their singular commons. Government bodies regulating agriculture are often particularly strong because of the importance of the sector. Tourism developments are often of strategic importance for the development of a country's economy. Protection of landscapes or land ecosystems has already a longer tradition of well-established institutions for governance when compared to marine and coastal systems. Government expertise for marine protection often does not yet exist (Weber de Morais et al., 2015), leading to what has been termed "paper parks" (The World Bank, 2006), not having any practical implication.

What can be observed for the government is partly echoed by non-government organizations (GS2). The tourism sector, having strong economic interests and a relatively smaller number of influential players, might organize collective action rather easily (Olson, 1998 [1965]). The same might be true for the industrial fishing fleet of a country. Those characteristics might contribute to that sector's internal collective action problems being solved. Despite that, most of the coastal commons are used by various sectors. Collective choice rules (GS6) therefore might favor those well-organized sectors (Knight, 1992). Farmers of many well-established large-scale operations are typically well organized, compared to small-scale subsistence fishers. Less organized and less powerful fishers will have difficulties influencing well-established agriculture institutions when their interests are not aligned, costs and benefits are not proportional, and their governance systems are not designed to work together. Often international conservation organizations, like WWF, RARE or Conservation International, play an important role in governing coastal commons, at least within the Global South, where state government oversight of resource use may be less influential.

Property rights systems (GS4) vary substantially within the coastal commons. Ocean water might be the resource with the least amount of regulated property rights. Depending on the particular governance system for nutrient flows, water catchment areas, fisheries or beaches, property rights may or may not exist or be managed by a particular governance system. Marine Protected Areas exist in many places as de facto open access regimes. However, there are also de facto state run, common property regimes or an increasing number of private ones (Bennett et al., 2015). Although there may be common property regimes for agriculture and aquaculture, private property regimes are becoming increasingly dominant, entailing different types and degrees of externalities. If there are not joint incentives for users within different property rights regimes to institutionalize their cooperation in maintaining a common resource or the externalities which affect a common resource in another sector, the dilemmas in the coastal commons become increasingly difficult to address.

The existence of constitutional choice, collective choice and operational rules (GS5–7), varies substantially according to the commons issue we are looking at. Many fisheries around the world are managed by well-established common property regimes. The same applies for water catchments and forestry. However, these common regimes have been established to manage the use of a single resource like, e.g. water, fish or timber. They do not address the interdependencies between systems, units and actors. For example, an irrigation association might include all farmers using the water, but fishers may not be represented even though they also have an interest in how the freshwater extraction is regulated since they rely on the salinity of the lagoon for healthy

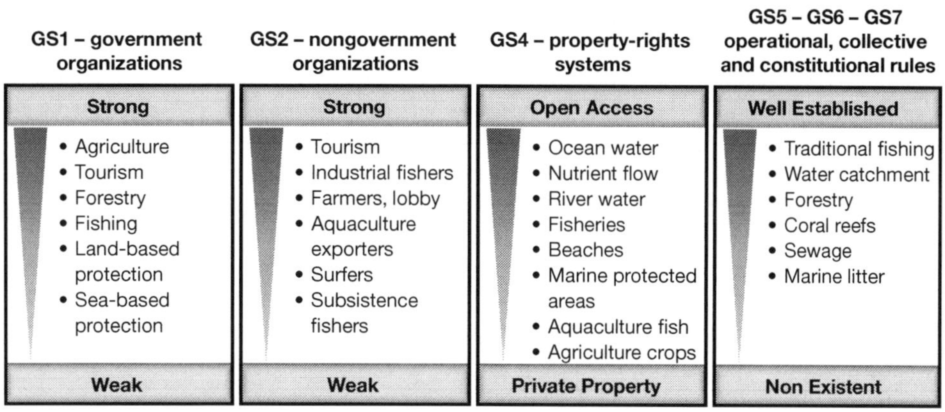

Figure 14.3 Diversity of characteristics of the governance system

fish populations. For other resources in the coastal commons, rules are often far less developed, such as for coral reefs or marine litter. The emergence of collective action and ultimately rules governing the commons might be affected by numerous issues. These include (i) the amount and type of SES or geographical levels involved, (ii) the observability, uncertainty and time horizon in which externalities occur, (iii) the economic situation of the actors, (iv) the costs associated with the establishment of those rules and (v) the pre-existing institutional structure those rules could be built on. For example, monitoring and sanctioning rules (GS8) might be less effective on the sea compared to terrestrial systems, as property rights enforcement is still more challenging in the open sea due to the large spatial area of the resource system, the high mobility of vessels and a lack of human presence which creates high costs for rule enforcement. Nevertheless, the effectiveness of rule enforcement also varies substantially between social systems, depending on the governance institutions which exist and the contextual issues described earlier.

Resource units (RU)

Defining what a resource unit is in the coastal commons is not as easy as in Hardin's example of the cow and meadow, which was additionally exemplified by Ostrom (2007). Many resource systems comprise various resource units, which have overlapping resource systems and interrelated externalities. Many resource units on the coast are similar to Hardin's example. However, most often the interdependence is more complex. The cows introducing nitrogen into the river and estuary are an example of a resource unit causing a commons problem. Important to consider is that this is not a commons problem among the various cow owners but between the cow owners and the users of other resource units downstream and connected aquatic systems, such as fishers or dive operators. Trees, fishes, mollusks, seagrass, turtles and turtle eggs, and water are all coastal resource units of this kind. But what about windmills or tourists? Windmills are installed in the North Sea by investors and planners, similar to a cow on a meadow. Ships on important sea routes and harbors are creating tragedies of the commons by using sea space and building infrastructure in common areas. Tourists are actors themselves using the commons. However, from the perspective of a hotel owner or a frequently visited national park, tourists are a resource unit, involved in creating a commons dilemma. Looking at those examples, one realizes that some resource units are immobile, like mangroves or windmills (RU1) (see Figure 14.4). Some are stable units, but in movement at the same time, like sand on the beach. Others are very mobile, like fish in the ocean, water in a catchment area or the cows on the meadow. The mobility patterns of those resources are substantially different. The sand does not move. It is moved by other forces, sometimes predictable, at other times not predictable (Mann and Westphal, 2016). When sand is removed by humans, it can cause serious commons problems for erosion and sedimentation of a coastal region (Boda, 2015). The fish swims wherever it wants, not considering any of our socially imposed jurisdictional boundaries. The water from the catchment area might also flow through various jurisdictions, though its ultimate destination is clear: it will end up in the sea. The mobility of a cow can be easily influenced by actors as long as fences are available to delineate a property rights system. Similar to mariculture cages in the open sea, it can be potentially put on the meadow or seagrass bed where the least commons problems are created (Adams et al., 2014).

Growth and replacement rates (RU2) vary substantially between the various resource units. As a non-renewable resource, at least on time scales of human lifespans, sand is not growing. It might only be replaced at a certain beach by sand coming from elsewhere. Mangroves reproduce, but slowly. Different fish species have different growth and reproduction rates. Even for individuals of a species this varies, depending on environmental conditions, like the availability of predators and prey, other food sources, temperature or the abundance of their species. This huge variability even within one CPR makes the predictability of resource availability rather difficult and therefore poses challenges for collective action and a governance system attempting to regulate their use. Often different resource units are heavily interdependent (RU3). Varying among species, the replacement rate of a fish depends heavily on the availability of other resource units such as food and habitat—which build the resource system for the species—like in mangrove forests, seagrass meadows or fresh water lagoons. This is again an indication of how resource units, resource systems and their corresponding governance systems and actors are intertwined on the coast.

The economic values (RU4) of the various resource units are different and are substantially fluctuating over time. Tourists have a rather high value, in terms of turnover per unit produced for the operator. This might mean that tourism operators often can out-compete more traditional uses of the coastal commons, like subsistence fishers. Demand for clean water is definitely on the rise in most coastal regions, due to a higher population pressure, changing lifestyle or increasing tourism business. Sand might be in high demand while the building industry is booming. The price for sea cucumbers might be high as long as Asian consumers are asking for them (Eriksson et al., 2015). The price obtained for selling a live reef fish might be high after the launch of *Finding Nemo* (McClenachan et al., 2012), thereby motivating many unsustainable extraction practices. High diversity and huge fluctuations in the economic value of a resource unit often make collective action for the commons more difficult. This holds particularly true if due to global economic interdependencies, prices do not reflect by any means local market scarcities (Gehrig, 2016), which could help manage the commons (Ferse et al., 2014). Rising prices will allow installing property systems which might solve certain commons problems (Barzel, 1989), though those forms of privatization often come along with problems of equity and justice (Foley and McCay, 2014).

Spatial and temporal distribution of resource units (RU7) varies substantially among the coastal commons. For example, fresh water availability in the tropics or the Mediterranean might vary substantially in the coastal commons over space and time. In temperate zones one might have a more constant rainfall pattern. Mangrove forests, for example, have their particular habitat in the intertidal zone, where they can thrive. They cannot move (Robertson et al., 1992). Once grown and managed in a sustainable way, they can create a constant benefit stream over time. The seasonal availability of various fish species is substantial and might require a diversification of fishing gear or even livelihood strategies of the fisher to diversify sources of income across seasonal change. However, restoring fish populations may not directly benefit local users if that fish species is highly mobile, and its distribution is spatially and temporally

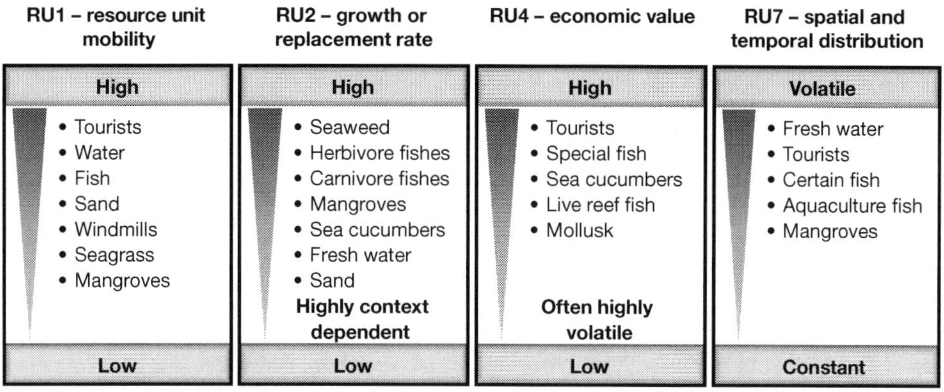

Figure 14.4 Diversity of characteristics of the resource units

heterogeneous. Although the spatial distribution of fish is not completely random, substantial variability does exist, which might favor common property regimes (Ostrom, 2000). It also implies that the rules governing this system might need to be more complex, such as comprising a rotation scheme among users.

Actors (A)

In the coastal commons a variety of heterogeneous actors are interacting, who might have to be considered depending on the particular collective action problem in focus (see Figure 14.5). The number of individuals in a group may vary substantially. There might be numerous fishers but only a few hotel owners. There might be few people living at the beach front but many in the entire catchment. Many small farmers may contribute to nutrient inflow into an estuary, but only a few companies might be responsible for the heavy metal content. The individuals within groups and between groups may vary substantially in their socioeconomic attributes (A2): they are financially wealthy and financially poor, old and young in age, educated and uneducated, male and female. Those attributes might vary within a group using the same resource unit or different resource units within the same system.

Coasts have always attracted a large diversity of people who make their livelihood from the abundance of resources available. However, coasts are currently the destination of many immigrants, who only recently started to interact with the SES (A3). Certain resource uses and activities have been well established for generations, like fishing or mangrove use, where the experience of using resources exists as part of the local culture. Others, like surfing, wind farming, coral reef restoration or whale watching are rather recent activities, where little experience about possible difficulties in managing the commons to avoid tragedies exists. Establishing sustainable governance rules in rather uncertain environments takes time. In modern times, the use of some resources happens too quickly for effective governance institutions to be developed before overuse tragedies occur or excessive externalities significantly affect other systems. Where actors are located (A4) in relation to the resource system, resource units and relevant governance systems can influence their interactions within the coastal commons. Farmers may be located far from the sea and may not consider the impacts of eutrophication when deciding agricultural policies for fertilizer use in their management committee. In addition, a fisher who must travel a long distance with a boat to a fishing area, which increases the economic costs of fishing, may assess the costs and benefits of applying for eco-certification which would bring a price premium on the market, and may be in favor of supporting governance arrangements which assist this process. Overall, a similarity between many coastal commons is that they are characterized by many diverse uses of their common resources by a diverse group of actors.

This diversity of actor groups also means that there is not necessarily a set of shared norms (A6) existing between those users. The social-ecological conditions an actor is interacting with leads to the emergence of norms, which seems appropriate in this particular system (Henrich et al., 2005). A fisher, a farmer, a tourist or the corresponding tourist operator are mainly engaging with diverse SES, or at least different parts of the same system, which often leads to different sets of rules and mental models. Leadership (A5) can play a key role in facilitating collective action for commons management among actor groups who may have differing norms and/or mental models. Social, economic and environmental conditions can influence the existence of leadership as well as many other factors such as social capital, which will likely influence management outcomes differently when they are strong or weak. Nonetheless, the coastal commons overlap with various sectors in the same or other resource systems, often with large geographical extensions. Therefore, it seems rather unlikely that a significant amount of social capital exists that could help in managing the coastal commons jointly. Strong leadership, which is able to create a strong sense of belonging to the particular coastal commons and which makes actors aware of their strong interdependence, might be of high importance (Gutiérrez et al., 2011).

The different resource systems interacting in the coastal commons vary in the amount of knowledge (A7) existing about the systems (Liquete et al., 2013; Partelow et al., 2017). Knowledge about land-based systems is probably more abundant than knowledge on coastal aquatic systems, due to much more use, interaction and research focused on them. Mangrove forests, water, or fish stock management might be areas where a lot of knowledge exists and thereby a certain joint understanding arises about how to manage the resource in a sustainable way. Still, this might just consider the individual resource, e.g. a particular fish species, but neither the entirety of trophic levels nor the interdependencies with other resources. For many other coastal resources the knowledge is even more limited and a diversity of mental models (A7) about the sustainable management of the resource exists, e.g. turtles, aquaculture, fishing or sewage (Fujitani et al., 2017). This also makes collective management difficult.

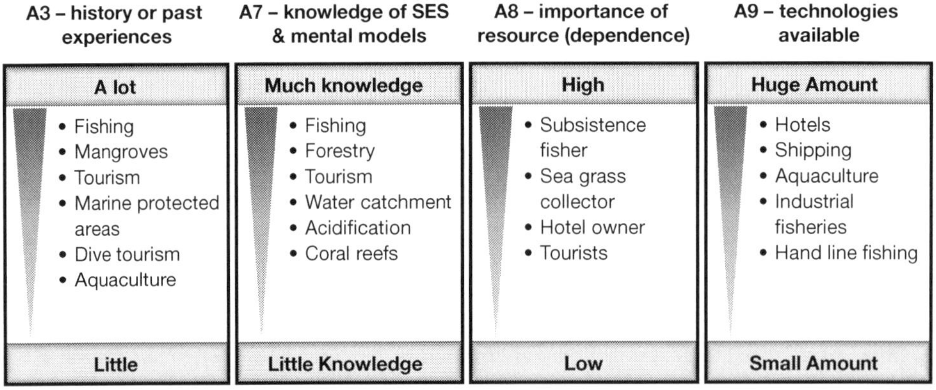

Figure 14.5 Diversity of characteristics of actors involved

Resource dependence (A8) on the coastal commons also varies substantially in various dimensions. Fishers or dive operators depend heavily on a healthy marine environment for their business. This is different for ship operators, who use the sea for different purposes. The farmer who uses the coastal commons might put a lot of nitrogen on her field. From her view this is seen as a necessity for the production process. Yet, she does not depend at all on the marine environment despite her strong influence on its health. Why should she care about solving commons problems? Some actors might earn their entire livelihood from the coastal commons; others just rely marginally on it. Some make a living from the commons which barely covers subsistence, while others make a fortune. This links to the variety of technologies at hand (A9). A hotel developer uses heavy machinery to work on the coastal commons. The subsistence farmer might use only an ox and plow. Nonetheless, he could also use heavy agrochemicals. A fisher might work with a handline, others with cyanide, gigantic nets or industrial size ships. All these potential differences between actors make collective action for the coastal commons rather difficult.

Action situations: interactions and outcomes (I – O)

"The action situation" in the sense of a physical or virtual space, where all actors of the coastal commons meet and interact, does not exist. The various resources have been managed by different entities and actors who are often not very well connected physically or socially via networks in disparate action situations. Network connections become more necessary when the overuse of a shared commons becomes apparent. New action situations have to emerge that bring actors together who use interlinked resource systems. This lack of connection often leads to independent streams of information sharing (I2), creating diverse mental models that make management difficult when consensus is required. Thus, management between different actor groups may require an investment in joint deliberation (I3), so that an understanding of the different values and uses of the commons can be developed to inform decisions which can reconcile individual and group interests. Successful collective action for the commons with little degree of conflicts (I4) is favored by an equal distribution of benefits and costs and by similar interests in the commons (Ostrom, 1990). But, the benefits and costs obtained from the coastal commons by various actors are largely heterogeneous in amount or in type. Fishers, for example, might agree on the optimal stock size, which leads to the social optimum that is economically viable for a fishing livelihood. However, the fisher's optimum harvest rate may be interpreted differently from the perspective of a dive operator whose optimum is a higher number of fish on a reef. Similarly, solving the problem of nitrogen input into the estuary does not benefit the farmer upstream but can reduce her benefits. It can cause difficulties for commons governance when the costs and benefits are asymmetric between actors. As indicated, some have made large investments (I5) in their use of the commons, and others not; regardless of the investment amount, their dependence on the commons may also vary. Some actor groups are better at getting organized, while others are too busy with sustaining their livelihoods, with no time left to think about lobbying or organizing the sector (I6, I7). This diversity shows how difficult it is to assess the social outcomes but also the reciprocal ecological outcomes of the diverse coastal commons (O1). Measuring outcomes would first of all require a deliberative process of actors involved, in which the characteristics of a good or bad performance are discussed. Outcomes will have different distributive effects on the various actors and actor groups involved (Ban et al., 2015) in asking the question about "good performance" and efficiency for whom (Bromley, 1989). Bearing this in mind and considering this question, might definitely be a need for managing the coastal commons in a socially resilient and sustainable way.

Related ecosystems (ECO)

Coastal SES will have to adapt to climate change impacts in numerous ways. Ocean acidification and increases in sea surface temperature will mainly affect marine organisms (Hoegh-Guldberg and Bruno, 2010). Sea level rise affects both sea and land: coral reefs or mangrove forests might have to move landward, if there is space available and sufficient time to adapt. Sea defenses for flood protection on land have to be built. Changes in the amount and seasonal distribution of rainfall in the catchment area have implications for the land and the sea. These changes will most likely require adaptations or more frequent responses from governance regimes. As mentioned, the boundaries of the coast as an ecological system are not clearly defined. Therefore, it is also not clear from which point onwards we should consider effects as an inflow and outflow into the focal SES. This must be defined for each individual case.

Strong interdependencies as the key characteristic of coastal commons

We have demonstrated that coastal commons are characterized by a huge diversity in resource systems and units, governance systems and even more actors, which are interacting in a countless variety of action situations. The features that best characterize the unique challenges facing the coastal commons are the interactions and the resulting interdependencies described earlier. Interdependencies exist within and between the various first tier variables. The conditions of a forest or a water catchment area have implications for a lagoon or a coral reef resource system. Certain fish species might heavily depend on mangrove or seagrass ecosystems or fresh water availability. With resource-related interdependencies come interdependencies among the various actor groups who work with each of the distinct resource systems or units. As many of these interdependencies are only now becoming more apparent and governing them is costly, there is often a link missing within and between the various governance systems. There is typically no fisher on the forest governance board. The interdependency within the various first tier variables (e.g. resource systems) leads to interdependencies across the various first tier variables of different systems comprising the coast (e.g. coastal farmland and a coral reef). The operational rules governing the allowed usage of nitrogen fertilizer in an agricultural area have implications for the growth and replacement rate of a fish population living in a coral reef. The collective choice rules determining the decision-making procedures for the overall extraction levels of water from an irrigation system, might have an equally important effect on the abundance of a fish population or the health of a coral reef as on the upstream managed resource unit, the agriculture field. How those collective choice and operational rules are changed depends substantially on the positive and normative mental models of the different actor groups involved. Being aware and having understood the causal mechanisms behind these interdependencies (positive mental model), might change the perceptions on what kind of collective choice rules are seen as appropriate. In the ideal case, a representative of the relevant fishers might be involved in the decision process of the irrigation association. As this interdependency is unidirectional, like many of them at the coast, one might expect that coming to an agreement is not easy and might require some compensation, as highlighted by the design principle of equivalence of benefit and costs (Ostrom, 1990). The conditions of these interactions, such as who compensates whom, are dependent, for example, on the normative models, or the economic conditions of the actors involved. Various institutional choices may exist, for example, an increase in water usage fees, any sort of compensation for not extracting water, or the simple presence of a fisher on the water board might all be solutions for managing interdependencies on the coast to avoid a tragedy of the commons.

There are various factors that make coastal interdependencies unique, particularly strong, and often difficult to manage. We now outline key features that characterize coastal commons:

- The coast is characterized by an extraordinary closeness of various systems, units and actors (see Figure 14.1). Due to the closeness of the various systems, commons or collective action problems on the coast are particularly multi-causal and involve many systems, actors and units. In other words, the extension of the SESF by McGinnis and Ostrom (2014) is a must if a holistic picture is the aim of the research or management.
- The movement of water, which usually crosses various resource and governance system boundaries, and is used by many resource units and actors, makes the interdependencies on the coast particularly complex.
- The coast is attracting more and more people due to its huge potential for increased use of the commons and our ability to cope with the associated risks. This creates an ever increasing amount of interdependencies and challenges for governance. These often arise in the absence of governance systems which could deal with those issues. Such governance systems still need to emerge.

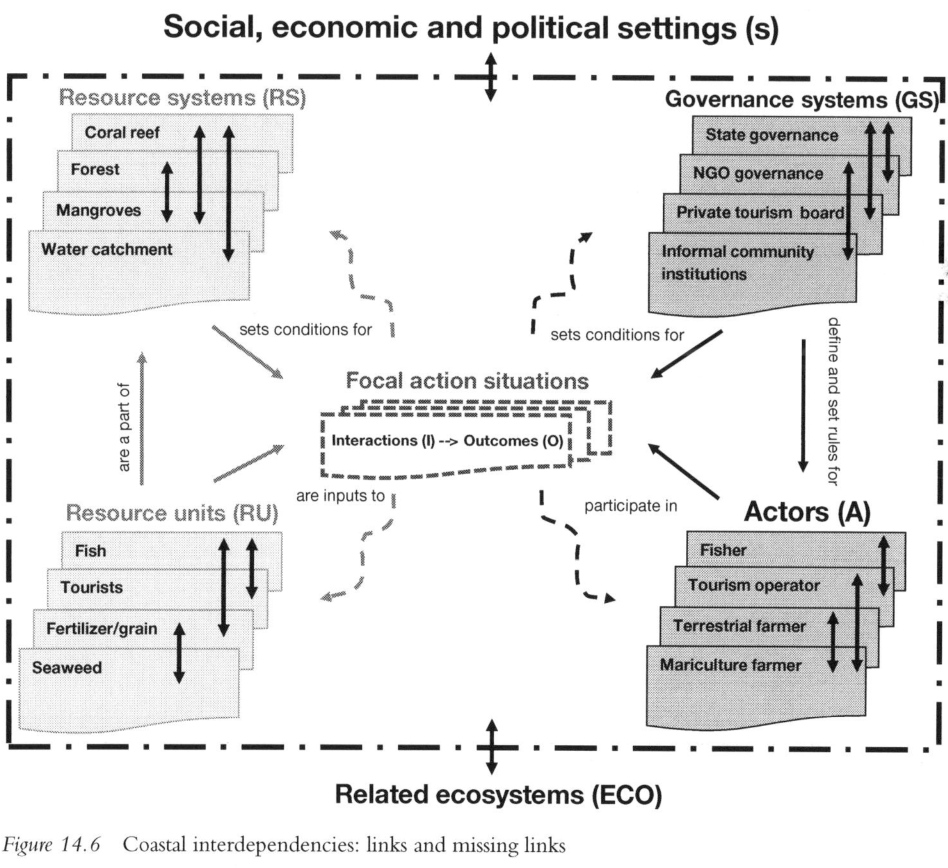

Figure 14.6 Coastal interdependencies: links and missing links

- Interconnectedness leads to complexity. As this is often combined with a lack of knowledge, uncertainty is a key feature. The mental models of different actors might be very diverse, which can impede the development of joint management solutions to interconnected commons problems.
- The ocean is one of the biggest commons and can have a huge buffering capacity. Interdependencies across such large spatial and temporal scales may be highly relevant from a long-term sustainability perspective but might not motivate collective action here and now.

Figure 14.6, built on the adapted multi system SESF of McGinnis and Ostrom (2014), exemplifies some of the interdependencies present on the coast.

Conclusion: analyzing coastal commons

We have demonstrated that coastal commons consist of diversity in all variables comprising the SESF. The conceptual expansion of the framework (McGinnis and Ostrom, 2014), which has enabled it to address interconnections and interdependencies, is definitely useful for obtaining a more comprehensive understanding of the coastal commons in its complexity (Ban et al., 2013; Partelow and Winkler, 2016). The conditions for collective action and thereby managing the various coastal commons in a sustainable way are different for each issue. The sheer dimensions of complexity and diversity preclude understanding or even describing the coastal commons in its entirety. Despite the need for integration and a holistic view, one needs to consider only parts of the system. However, this might lead us again to focus on a single CPR, a single resource and its associated collective action problems. From a scholarly perspective, which aims to understand where and why collective action emerges, this might be a justifiable reduction of complexity. However, from a governance, planning or sustainability perspective those partial views are not appropriate and might hinder the achievement of substantial sustainability gains (Adams et al., 2014).

Note

1 The abbreviations and numbers used correspond to the abbreviations used in the SESF. For a complete representation of the second tier variables of the SESF, see Chapter 3, this volume.

References

Acheson, J. M. 2006. Lobster and groundfish management in the gulf of Maine: A rational choice perspective. *Human Organization*, 65, 240–252.

Adams, V. M., Álvarez-Romero, J. G., Carwardine, J., Cattarino, L., Hermoso, V., Kennard, M. J., . . . & Stoeckl, N. 2014. Planning across freshwater and terrestrial realms: Cobenefits and tradeoffs between conservation actions. *Conservation Letters*, 7, 425–440.

Agrawal, A. 2001. Common property institutions and sustainable governance of resources. *World Development*, 29, 1623–1648.

Álvarez-Romero, J. G., Pressey, R. L., Ban, N. C. & Brodie, J. 2015. Advancing land-sea conservation planning: Integrating modelling of catchments, land-use change, and river plumes to prioritise catchment management and protection. *PloS one*, 10, e0145574.

Álvarez-Romero, J. G., Pressey, R. L., Ban, N. C., Vance-Borland, K., Willer, C., Klein, C. J. & Gaines, S. D. 2011. Integrated land-sea conservation planning: The missing links. *Annual Review of Ecology, Evolution, and Systematics*, 42, 381–409.

Aswani, S., Christie, P., Muthiga, N. A., Mahon, R., Primavera, J. H., Cramer, L. A., . . . & Hacker, S. 2012. The way forward with ecosystem-based management in tropical contexts: Reconciling with existing management systems. *Marine Policy*, 36, 1–10.

Aswani, S., Gurney, G. G., Mulville, S., Matera, J. & Gurven, M. 2013. Insights from experimental economics on local cooperation in a small-scale fishery management system. *Global Environmental Change*, 23, 1402–1409.

Ávila-García, P. & Sánchez, E. L. 2012. The environmentalism of the rich and the privatization of nature: High-end tourism on the Mexican coast. *Latin American Perspectives*, 39, 51–67.

Ban, N. C., Evans, L. S., Nenadovic, M. & Schoon, M. 2015. Interplay of multiple goods, ecosystem services, and property rights in large social-ecological marine protected areas. *Ecology and Society*, 20(4), article 2.

Ban, N. C., Mills, M., Tam, J., Hicks, C. C., Klain, S., Stoeckl, N., . . . & Chan, K. M. A. 2013. A social–ecological approach to conservation planning: Embedding social considerations. *Frontiers in Ecology and the Environment*, 11, 194–202.

Barzel, Y. 1989. *Economic Analysis of Property Rights*. Cambridge, UK: Cambridge University Press.

Bax, N., Williamson, A., Aguero, M., Gonzalez, E. & Geeves, W. 2003. Marine invasive alien species: A threat to global biodiversity. *Marine Policy*, 27, 313–323.

Bennett, N. J., Govan, H. & Satterfield, T. 2015. Ocean grabbing. *Marine Policy*, 57, 61–68.

Berkes, F., Hughes, T., Steneck, R., Wilson, J. A., Bellwood, D., Crona, B., . . . & Norberg, J. 2006. Globalization, roving bandits, and marine resources. *Science*, 311, 1557–1558.

Biao, X., Zhuhong, D. & Xiaorong, W. 2004. Impact of the intensive shrimp farming on the water quality of the adjacent coastal creeks from Eastern China. *Marine Pollution Bulletin*, 48, 543–553.

Boda, C. S. 2015. Power and rationality in coastal planning: Effects on participation and possibility in the management of barrier island dunes in Flagler Beach, Florida, USA. *Journal of Coastal Conservation*, 19, 561–576.

Born, K. M. 2012. Governance in rural landscapes. *Norsk Geografisk Tidsskrift [Norwegian Journal of Geography]*, 66, 76–83.

Bromley, D. 1991. *Environment and Economy; Property Rights and Public Policy*. Oxford, UK: Blackwell.

Bromley, D. W. 1989. *Economic Interests and Institutions: The Conceptual Foundations of Public Policy*. Oxford, UK: Basil Blackwell.

Brown, K., Tompkins, E. L. & Adger, N. W. 2012. *Making Waves: Integrating Coastal Conservation and Development*. Oxford, UK: Routledge.

Burford, M., Costanzo, S., Dennison, W., Jackson, C., Jones, A., McKinnon, A., . . . & Trott, L. 2003. A synthesis of dominant ecological processes in intensive shrimp ponds and adjacent coastal environments in NE Australia. *Marine Pollution Bulletin*, 46, 1456–1469.

Burgess, M. G., Clemence, M., McDermott, G. R., Costello, C. & Gaines, S. D. 2016. Five rules for pragmatic blue growth. *Marine Policy*, 87, 331–339.

Christie, P. 2011. Creating space for interdisciplinary marine and coastal research: five dilemmas and suggested resolutions. *Environmental Conservation*, 38, 172–186.

Cinner, J. E., Macneil, M. A., Basurto, X. & Gelcich, S. 2013. *Looking Beyond the Fisheries Crisis: Cumulative Learning From Small-Scale Fisheries through Diagnostic Approaches*. Oxford, UK: Elsevier.

Del Mar Delgado-Serrano, M. & Ramos, P. 2015. Making Ostrom's framework applicable to characterise social ecological systems at the local level. *International Journal of the Commons*, 9, 808–830.

Denzau, A. T. & North, D. C. 1994. Shared mental models, ideologies and institutions. *Kyklos*, 47, 1–31.

Eriksson, H., Österblom, H., Crona, B., Troell, M., Andrew, N., Wilen, J. & Folke, C. 2015. Contagious exploitation of marine resources. *Frontiers in Ecology and the Environment*, 13, 435–440.

FAO 2016. *The State of World Fisheries and Aquaculture 2016. Contributing to Food Security and Nutrition for All*. Rome.

Ferse, S. A., Glaser, M., Neil, M. & Schwerdtner Máñez, K. 2014. To cope or to sustain? Eroding long-term sustainability in an Indonesian coral reef fishery. *Regional Environmental Change*, 14(6), 2053–2065.

Foley, P. & McCay, B. 2014. Certifying the commons: Eco-certification, privatization, and collective action. *Ecology and Society*, 19(2), article 28.

Fujitani, M., McFall, A., Randler, C. & Arlinghaus, R. 2017. Participatory adaptive management leads to environmental learning outcomes extending beyond the sphere of science. *Science Advances*, 3, e1602516.

Gehrig, S. 2016. Social-cultural heterogeneity and scarcity in an artisanal fishery: Economic experiments in Chwaka Bay, Zanzibar, Tanzania. ISATEC Master's Thesis. Bremen University, Germany.

Glaser, M., Christie, P., Diele, K., Dsikowitzky, L., Ferse, S., Nordhaus, I., . . . & Wild, C. 2012. Measuring and understanding sustainability-enhancing processes in tropical coastal and marine social–ecological systems. *Current Opinion in Environmental Sustainability*, 4, 300–308.

Gutiérrez, N. L., Hilborn, R. & Defeo, O. 2011. Leadership, social capital and incentives promote successful fisheries. *Nature*, 470, 386–389.

Halpern, B. S. 2014. Conservation: Making marine protected areas work. *Nature*, 506, 167–168.
Halpern, B. S., Ebert, C. M., Kappel, C. V., Madin, E. M. P., Micheli, F., Perry, M., . . . & Walbridge, S. 2009. Global priority areas for incorporating land–sea connections in marine conservation. *Conservation Letters*, 2, 189–196.
Halpern, B. S., Walbridge, S., Selkoe, K. A., Kappel, C. V., Micheli, F., D'Agrosa, C., . . . & Fox, H. E. 2008. A global map of human impact on marine ecosystems. *Science*, 319, 948–952.
Henrich, J., Bowles, S., Boyd, R., Camerer, C., Fehr, E., Gintis, H., . . . & Tracer, D. 2005. "Economic man" in cross-cultural perspective: Behavioral experiments in 15 small-scale societies. *Behavioral and Brain Science*, 28, 1–61.
Herbeck, L. S., Unger, D., Wu, Y. & Jennerjahn, T. C. 2013. Effluent, nutrient and organic matter export from shrimp and fish ponds causing eutrophication in coastal and back-reef waters of NE Hainan, tropical China. *Continental Shelf Research*, 57, 92–104.
Hoegh-Guldberg, O. & Bruno, J. F. 2010. The impact of climate change on the world's marine ecosystems. *Science*, 328, 1523–1528.
Jennerjahn, T. C. 2012. Biogeochemical response of tropical coastal systems to present and past environmental change. *Earth-Science Reviews*, 114, 19–41.
Kaffine, D. T. 2009. Quality and the commons: The surf gangs of California. *Journal of Law and Economics*, 52, 727–743.
Knight, J. 1992. *Institutions and Social Conflict*. Cambridge, UK: Cambridge University Press.
Kopp, R. E., Horton, R. M., Little, C. M., Mitrovica, J. X., Oppenheimer, M., Rasmussen, D. J., . . . & Tebaldi, C. 2014. Probabilistic 21st and 22nd century sea-level projections at a global network of tide-gauge sites. *Earth's Future*, 2, 383–406.
Leslie, H. M., Basurto, X., Nenadovic, M., Sievanen, L., Cavanaugh, K. C., Cota-Nieto, J. J., . . . & Aburto-Oropeza, O. 2015. Operationalizing the social-ecological systems framework to assess sustainability. *Proceedings of the National Academy of Sciences*, 112, 5979–5984.
Liquete, C., Piroddi, C., Drakou, E. G., Gurney, L., Katsanevakis, S., Charef, A. & Egoh, B. 2013. Current status and future prospects for the assessment of marine and coastal ecosystem services: A systematic review. *PloS one*, 8, e67737.
Madrigal-Ballestero, R., Schlüter, A. & Lopez, M. C. 2013. What makes them follow the rules? Empirical evidence from turtle egg harvesters in Costa Rica. *Marine Policy*, 37, 270–277.
Mann, T. & Westphal, H. 2016. Multi-decadal shoreline changes on Takú Atoll, Papua New Guinea: Observational evidence of early reef island recovery after the impact of storm waves. *Geomorphology*, 257, 75–84.
Markus, T., Hillebrand, H., Hornidge, A.-K., Krause, G. & Schlüter, A. 2017. Disciplinary diversity in marine sciences: The urgent case for an integration of research. *ICES Journal of Marine Science*, 75(2), 502–509.
McCauley, D. J., Pinsky, M. L., Palumbi, S. R., Estes, J. A., Joyce, F. H. & Warner, R. R. 2015. Marine defaunation: Animal loss in the global ocean. *Science*, 347(6219), 1255641.
McClenachan, L., Cooper, A. B., Carpenter, K. E. & Dulvy, N. K. 2012. Extinction risk and bottlenecks in the conservation of charismatic marine species. *Conservation Letters*, 5, 73–80.
McGinnis, M. D. & Ostrom, E. 2014. Social-ecological system framework: Initial changes and continuing challenges. *Ecology and Society*, 19(2), article 30.
McLeod, K. & Leslie, H. M. 2009. *Ecosystem-Based Management for the Oceans*. Washington, DC: Island Press.
Olson, M. 1998 [1965]. *The Logic of Collective Action, Public Goods and the Theory of Groups*. Cambridge, MA: Harvard University Press.
Ostrom, E. 1990. *Governing the Commons. The Evolution of Institutions for Collective Action*. Cambridge, UK: Cambridge University Press.
Ostrom, E. 2000. Private and common property rights. In: Bouckaert, B. & De Geest, G. (eds.) *Encyclopedia of Law and Economics*. Cheltenham, UK: Edward Elgar.
Ostrom, E. 2007. A diagnostic approach for going beyond panaceas. *Proceedings of the National Academy of Sciences of the United States of America (PNAS)*, 104, 15181–15187.
Ostrom, E. 2009. A general framework for analyzing sustainability of social-ecological systems. *Science*, 325, 419–422.
Páez-Osuna, F. 2001. The environmental impact of shrimp aquaculture: Causes, effects, and mitigating alternatives. *Environmental Management*, 28, 131–140.
Partelow, S. 2015. Key steps for operationalizing social-ecological system framework research in small-scale fisheries: A heuristic conceptual approach. *Marine Policy*, 51, 507–511.

Partelow, S. & Boda, C. 2015. A modified diagnostic social-ecological system framework for lobster fisheries: Case implementation and sustainability assessment in Southern California. *Ocean & Coastal Management*, 114, 204–217.

Partelow, S., Schlüter, A., Von Wehrden, H., Jänig, M. & Senff, P. 2017. A sustainability agenda for tropical marine science. *Conservation Letters*, n/a-n/a.

Partelow, S., P. Senff, N. Buhari, and A. Schlüter. 2018. Operationalizing the social-ecological systems framework in pond aquaculture. *International Journal of the Commons* 12(1), 485–518.

Partelow, S., Von Wehrden, H. & Horn, O. 2015. Pollution exposure on marine protected areas: A global assessment. *Marine Pollution Bulletin*, 100, 352–358.

Partelow, S. & Winkler, K. 2016. Interlinking ecosystem services and Ostrom's framework through orientation in sustainability research. *Ecology and Society*, 21(3), article 27.

Pittman, J. & Armitage, D. 2016. Governance across the land-sea interface: A systematic review. *Environmental Science & Policy*, 64, 9–17.

Portman, M. E. 2016. *Environmental Planning for Oceans and Coasts: Methods, Tools, and Technologies*. Dordrecht, the Netherlands: Springer.

Ramesh, R., Chen, Z., Cummins, V., Day, J., D'elia, C., Dennison, B., . . . & Wolanski, E. 2015. Land-ocean interactions in the coastal zone: Past, present & future. *Anthropocene*, 12, 85–98.

Robertson, A. I., Alongi, D. M. & Boto, K. G. 1992. Food chains and carbon fluxes. In: A. I. Robertson & D. M. Alongi (eds.) *Tropical Mangrove Ecosystems*. 41st edition. Washington, DC: American Geophysical Union, pp. 293–326.

Schlager, E. & Ostrom, E. 1992. Property-rights regimes and natural resources: A conceptual analysis. *Land Economics*, 68, 249–262.

Schlüter, A. 2008. Small-scale European forestry: An anticommons? *International Journal of the Commons*, 2, 248–268.

Schlüter, A. & Madrigal, R. 2012. The SES framework in a marine setting: Turtle egg harvesting in Costa Rica. *Rationality, Moral and Markets*, 3, 148–167.

Stoms, D. M., Davis, F. W., Andelman, S. J., Carr, M. H., Gaines, S. D., Halpern, B. S., . . . & Warner, R. R. 2005. Integrated coastal reserve planning: Making the land–sea connection. *Frontiers in Ecology and the Environment*, 3, 429–436.

The World Bank 2006. *Scaling Up Marine Management : The Role of Marine Protected Areas*. Washington, DC, World Bank.

Torres-Guevara, L. E., Schlüter, A. & Lopez, M. C. 2016. Collective action in a tropical estuarine lagoon: Adapting Ostrom's SES framework to Ciénaga Grande de Santa Marta, Colombia. *International Journal of the Commons*, 10(1), 334–362.

Van den Burg, S. W. K., Stuiver, M., Bolman, B. C., Wijnen, R., Selnes, T. & Dalton, G. 2017. Mobilizing investors for blue growth. *Frontiers in Marine Science*, 3(291), 1–9.

Wan, Z., Zhu, M., Chen, S. & Sperling, D. 2016. Three steps to a green shipping industry. *Nature*, 530(7590), 275–277.

Weber de Morais, G., Schlüter, A. & Verweij, M. 2015. Can institutional change theories contribute to the understanding of marine protected areas? *Global Environmental Change*, 31, 154–162.

Woelfle-Erskine, C. 2017. Collaborative approaches to flow restoration in intermittent salmon-bearing streams: Salmon Creek, CA, USA. *Water (Switzerland)*, 9(217), 1–23.

15
CLIMATE AS A COMMONS

Jouni Paavola

Introduction

New institutional economics or "new institutionalism" has informed a significant body of research on local common property arrangements and international environmental conventions (e.g. Ostrom, 1990, 2005; Ostrom et al., 1994; Young, 2002). This interdisciplinary research encompassing economics, political science, sociology and anthropology has shed light on the conditions in which voluntary collective action can attain sustainable governance and use of environmental resources, and has identified design principles that characterize successful governance solutions.

The new institutional research on environmental governance has been phenomenally successful in terms of its volume growth and policy impact. Yet its potential is far from exhausted. Understanding the challenges and solutions of governing large and complex environmental resources such as atmospheric sinks were identified as key future tasks (Ostrom et al., 1999: 278) and some progress towards understanding their adaptive governance has been made (e.g. Dietz et al., 2003). However, much of the literature has examined single-level or uni-planar governance solutions. The governance of global environmental resources, however, is increasingly based on multi-level solutions operating at the local, national, international and intermediate levels simultaneously. This calls for finding ways to accommodate and deal with institutional diversity as part of the solution for adaptive governance (Ostrom et al., 1999: 278; Ostrom, 2005). In particular, there is a need to be able to deal with traditional national policies based on the enforcement power of the state in conjunction with solutions based on voluntary cooperation.

The greatest obstacle for the extension of the new institutional approach to new areas of research lies in its mostly implicit definition of "governance". The literature distinguishes between "governance" and "government" by considering the absence of coercive state power as the hallmark of "governance". Yet governance is what governments do. Sometimes – as when resource users govern themselves under customary institutions – environmental governance does not involve the *state*. Yet customary resource users perform the governmental functions of legislation, administration and adjudication, and therefore the *government* is involved. Rather than a monolithic external actor, the government and the state should be understood as arenas and instruments of collective action which are often pertinent in environmental governance.

The key implication of the involvement of the state is that it entails a different distribution of power than self-governance solutions. Otherwise, national environmental and natural resource use policies perform similar functions and rely on similar institutional solutions as customary common property arrangements, despite being formal, having larger jurisdictions and relying on the enforcement power of the state. Often these complex governance arrangements are polycentric (Ostrom, 2010, 2012).

This chapter suggests a broader definition of environmental governance: as the establishment, reaffirmation or change of institutions to resolve conflicts over environmental resources (Bromley, 1989, 1991; Knight, 1992). In this definition, *conflict* refers to a conflict of interest, not necessarily to an open conflict, between involved parties. This broader definition is applicable to the governance of all *environmental resources* from conventional renewable and non-renewable natural resources to biodiversity and atmospheric sinks, as well as to environmental safety and the quality of air and water. The definition does not limit the type or scale of environmental governance problems and solutions that can be examined. In what follows, the heuristics of commons will be applied to the problem of climate change (see also Paavola, 2008, 2011).

Atmospheric sinks of greenhouse gases as common-pool resources

The atmospheric sinks for greenhouse gases (GHGs) can be conceptualized as a common-pool resource not unlike a pasture or an aquifer. These sinks are stock resources, which have a limited capacity to provide a flow of sink services. Conventional common-pool resources such as aquifers and fisheries have a physical regeneration rate and thus a relatively well-defined capacity to generate a flow of resource units. Similarly, watercourses, air basins and global atmospheric sinks have a capacity to absorb pollutants and are replenished by natural processes at a certain pace. These sinks typically form a part of a larger resource system catering for multiple uses.

Therefore, the use of the units in the atmospheric GHG sink is always rival within the sink use (a unit of sink services used by one user is not available to others). It is possible that sink services can be jointly produced with other services. However, the sink service can become rival with other uses of the resource system if a threshold for multiple use is surpassed. Some resources have thresholds, which if surpassed, may lead to the collapse of the resource system. For example, the climate system may change its nature if the atmospheric CO_2 concentrations surpass 400–500 ppm.

The key challenge in governing atmospheric GHG sinks is the same as with all other common-pool resources: to constrain its use so as to prevent its destruction. A derivative task is to distribute the sustainable capacity of the atmospheric GHG sink to provide sink services among competing users. However, the challenges of governing the atmospheric GHG sink are also shaped by the difficulty of exclusion of potential users from the resource (Ostrom, 1990). The users of GHG sinks range from large coal and natural gas-powered electricity generation plants to families driving cars or keeping cattle. The size of the sink, the range and heterogeneity of activities that make use of it, and the large number of users make it difficult to monitor the use of sinks and to exclude unauthorized users. The absence of clear borderlines and the perfect mixing of GHG emissions in the atmosphere contribute to the difficulty of exclusion.

Because of the difficulty of exclusion, enforcement of entitlements to sinks is complicated. Users also have incentives to use the units of the GHG sink before other users make the units unavailable for them. Private ownership is not a feasible governance alternative when exclusion is difficult, but collective ownership and agreements to constrain resource use, and widely shared values, could help overcome the challenge of difficult exclusion.

There are also still further resource attributes influencing the governance challenges of atmospheric sinks. A consensus has emerged that the climate system is non-linear (Steffen et al., 2004). If the use of the sinks surpasses critical thresholds, the climate system may change towards a new equilibrium which may alter the conditions of life on Earth. There is uncertainty about what those thresholds are; the current estimates of safe CO_2 concentration levels vary between 400–500 ppm. There is also uncertainty regarding climate change impacts and their incidence. Therefore, governance solutions must facilitate the management of risks and uncertainty.

The challenges of governing the use of GHG sinks

The challenges of governing atmospheric GHG sinks are also shaped by the attributes of their users. User attributes determine the starting point for collective action aimed at establishing or modifying governance institutions, shape the costs and prospects of acting collectively, and influence what governance solutions can be agreed upon.

Collective action is shaped by political-economic factors as well as current patterns in the use of atmospheric sinks for GHGs. The most important aspect of the global political-economic order is the role of nation states as collective actors representing populations within their territories. The law on international relations treats nation states as formally equal, sovereign actors in international affairs. This formal equality contrasts with unequal developmental attainments. Industrialized developed countries have achieved high levels of per capita income and have strong, capable states. In the developing world, states are weak and at times dysfunctional, and they have been unable to promote income growth and wellbeing among their citizens. This also means that developing country states lack capacity to advance their (and their citizens') interests in international negotiations on the governance of atmospheric GHG sinks.

The economies of nation states also exhibit different degrees of complexity, which affects their vulnerability to climate change impacts. Complex economies of the developed countries offer numerous sources of income with different risk attributes: they are more resilient during periods of shocks and stresses. Economies of developing countries depend on primary production, agriculture in particular, and they are exposed to substantial climatic and economic risks. Because of underdeveloped financial and insurance sectors, people in developing countries cannot insure their assets and stand to lose them when tropical storms, floods or droughts occur (Paavola and Adger, 2006).

There are significant differences in the vulnerability of economies to weather-related disasters. In developed countries, per capita income and growth are not affected noticeably by extreme weather events such as the European drought and heat wave of 2003, even though the lost assets still measured up to a significant percentage of GDP. In contrast, extreme weather events such as Hurricane Mitch can tax over 10 percent of the GDP of a low-income country (see Linnerooth-Bayer et al., 2005).

The differences in vulnerability between developed and developing countries are even more significant in terms of loss of life. Disasters of comparable size claim a much higher magnitude of casualties in developing countries. For example, the magnitude 6.6–6.7 earthquakes in Northridge, California in 1994 and in Bam, Iran in 2003 killed 60 and 30,000 people, respectively. Hurricane Andrew killed 23 people in Florida in 1992 while a comparable typhoon killed over 100,000 people in Bangladesh in 1991 (see Adger et al., 2005). Brooks et al. (2005) have found that the level of educational attainment, level of health status and the quality of governance are important factors explaining the differences between countries in mortality due to natural disasters, alongside differences in the volume and quality of infrastructure and their effects on energy, water, transport and health care.

There are also other sources of heterogeneity that influence the ability of nation states to act collectively. These include political ideologies, such as beliefs in the ability of markets or states to generate desirable outcomes. They affect the range and assessment of governance alternatives which are perceived as feasible. Religious beliefs as well as secular beliefs in, for example, liberalism also situate nation states in international political arenas. Globalization is unlikely to reduce these heterogeneities. It is more likely to increase them because it will introduce heterogeneity to previously homogeneous societies and increase heterogeneity where it has already been present (Paavola, 2005).

Thus, the global community is divided by heterogeneities that make agreeing on a solution for governing the use of atmospheric sinks difficult. Developed countries have invested in energy-intensive lifestyles, technologies and infrastructure which make GHG reductions both expensive and time consuming. At the same time, developed countries have capacity to avoid adverse consequences of climate change, as well as to recover from them. Developed countries form a homogeneous and powerful negotiation block, which has significant experience from having acted collectively in other contexts. Developing countries – particularly the least developed countries – are in a different situation. They have contributed little to climate change because of their limited energy use and reliance on renewable sources of energy. But their economic development requires increasingly the use of energy and emissions of GHGs. At the same time, developing countries are highly vulnerable to adverse climate change impacts. Finally, developing countries make up a large and heterogeneous negotiation block, with members from oil producing countries to small island states that are threatened with inundation by rising sea levels.

There are, of course, more coalitions in climate change negotiations than just developed and developing countries, and the contours between the groupings are more complex than this discussion suggests. But even this limited account demonstrates that there are significant obstacles for acting collectively to govern atmospheric sinks. Actors start from uneven positions and their interests are different. Their views regarding feasible and acceptable solutions may also differ.

Progress in governing the use of global atmospheric sinks

The dominant view among scholars and policy makers has been that climate change governance should be based on international agreements, which involve most nations (see e.g. Hare et al., 2010). The United Nations Framework Convention on Climate Change (UNFCCC) and the Kyoto Protocol (KP) are cornerstones of this approach. These kinds of governance strategies face two key hurdles. First, wide participation has to be secured for any agreement to come into force. Second, all agreements need to be implemented through national policies. In light of the governance challenges discussed earlier, it is no surprise that a grand agreement on climate change involving the adoption of ambitious GHG mitigation commitments by a large number of states has been elusive.

One insight from the commons research is that collective action could more likely be successful if the number of involved actors could be reduced. In the case of atmospheric sinks, there are several ways to achieve this. For example, arguments have been made for the major polluters to negotiate their own agreement to overcome the large numbers' dilemma and heterogeneity (Naim, 2009; Victor, 2009). However, there are serious problems with agreements that omit the involvement and representation of affected parties for the sake of expediency (see Eckersley, 2012). Another proposal has argued for "a building blocks" approach: a strategy where sectoral or other partial agreements are negotiated incrementally among the involved and affected parties, with the expectation that over time they will together amount to a comprehensive approach to tackling global climate change (Falkner et al., 2010).

The Paris Agreement in 2015 in the end did not pursue a "mini-lateral" agreement or the building blocks approach. It instead avoided having to come up with a negotiated set of commitments by deferring it to the states to come up with their own "nationally determined contributions" (NDCs) to combat climate change. Recent analyses of the first round of these national pledges indicate that together they are not sufficient to limit global warming to below 2°C, not to speak of below 1.5°C warming (Robiou Du Pont et al., 2017). Most of the major polluters have pledged to reduce their emissions less than the alternative equitable and effective emission reduction strategies would entail (ibid.).

But more importantly, commons research has demonstrated that top-down governance solutions relying on the central role of the state have been a false panacea in the governance of many resources (see Ostrom et al., 2007). It has also highlighted the potential of polycentric governance arrangements to handle complex governance challenges. Polycentric order has been defined as "one where many elements are capable of making mutual adjustments for ordering their relationships with one another within a general system of rules where each element acts with independence of other elements" (V. Ostrom, 1999: 57). Polycentric order may emerge in a bottom-up way when diverse actors around a phenomenon like climate change seek to realize diverse benefits (or to avoid diverse costs) that accrue on different scales (see Ostrom, 2010). For example, mitigation actions do not only generate the global benefit of reduced GHG emissions and reduced rate of climate change; they also create co-benefits such as better air quality, reduced reliance on fossil fuels, reduced exposure to their price fluctuations and improved energy security. These benefits can be a sufficient motivation for mitigation actions, although perhaps not on a comprehensive scale.

A myriad of voluntary climate change initiatives already exist. For example, the Cities for Climate Protection (CCP) programme and the Sustainable Cement Sustainability Initiative (SCSI) represent attempts to address GHG emissions, comparable to those of major emitting states, as will be discussed later in greater detail. These initiatives have been successful in reducing GHG emissions or slowing their growth compared with business as usual. However, evidence suggests that voluntary initiatives may be at their best in realizing cost-saving emission reductions. Therefore, state-based and hybrid governance solutions may be needed to complement voluntary ones in order to stabilize atmospheric concentrations of GHGs at a safe level.

Examples of polycentric arrangements for climate governance

Polycentric climate change governance can involve a variety of actors, such as local governments and communities, non-governmental and church-based organizations, businesses, and governmental organizations in different combinations and roles. Some of the solutions are limited to one area of activity, such as local governmental activities or industry, while others can be more general in nature. Many of these solutions are voluntarily adopted and have voluntary membership, although the act of joining can create responsibilities. The CCP programme and the Cement Sustainability Initiative (CSI) will be discussed below as examples.

The CCP programme

Local governments have developed and implemented governance solutions for reducing the emissions of GHGs from their jurisdictions. The pioneer in this area has been the International Council for Local Environmental Initiatives (ICLEI) with its CCP programme. Others include Climate Alliance, C40, and the U.S. Mayors' Climate Protection Agreement (see Kern and Bulkeley, 2009; Gore, 2010; Román, 2010).

The ICLEI launched its CCP programme in 1993. It aimed to enlist 100 municipalities worldwide with joint emissions of 1 billion metric tonnes of CO_2 (ICLEI, 1993). The programme also sought to strengthen local commitments to GHG emission reduction; to develop and disseminate planning and management tools; to research and develop best practices; and to enhance the national and international ties between among municipalities (ibid.).

The CCP programme expects those joining it to develop a local action plan to reduce GHG emissions, to undertake measures to reduce emissions from municipal building stock and vehicle fleets, to institute public awareness campaigns on climate change, and to join procurement initiatives that seek to create demand for climate-friendly products and services. Those joining are also expected to link with local governments in developing country and emerging-economy countries' local governments to foster technological and financial transfers (see ICLEI, 1993).

The CCP progress report published in 2006 (ICLEI Local Governments for Sustainability, 2006) highlighted that 550 local governments had joined the programme since 1993. Their combined population was a quarter of a billion, or more than 4 percent of the global total. The combined GHG emissions from participating local governments were 1.85 billion tons of eCO_2, or more than 6 percent of the global total (excluding emissions from land use and land use change). That is, GHG emissions of the CCP members are comparable to those of large Annex 1 countries, such as Germany, Japan and Russia. The participants reduced their joint emissions by 3 percent or 60 million tonnes of CO_2 between 1990 and 2006. These emission reductions brought substantial savings to participating cities amounting to about $35 per reduced ton of CO_2 emissions (ICLEI Local Governments for Sustainability, 2006: 2).

Cement Sustainability Initiative (CSI)

Another example of climate change governance is provided by the CSI, a programme of the World Business Council for Sustainable Development (CSI, 2002) that has been considered a model for the "sectoral" approach to climate change mitigation (Schmidt et al., 2008; Meckling and Chung, 2009). The cement industry is a significant GHG emitter: its worldwide CO_2 emissions are about 5 percent of the global total, making them comparable to those of Germany, Japan and Russia in 2004 (CSI, 2002; UNDP, 2007).

Ten large cement manufacturers formed the CSI in 2002. Its members represent nearly two-thirds of the global cement manufacturing capacity outside China (CSI, 2009). The CSI aims to increase the cement industry's contribution to sustainable development and the public understanding of that contribution. The agenda for action adopted in 2002 contained six key areas of work which were: (1) climate protection; (2) fuels and raw materials; (3) employee health and safety; (4) emissions reduction; (5) local impacts; and (6) international business processes (CSI, 2002: 5). The agenda invited other cement producers to join and and committed producers to report on their progress three years after the date from which they joined (ibid.).

GHG emissions of the cement industry originate from the chemical reactions of the key raw material, limestone (50 percent of the total), fuel used in the manufacturing processes (40 percent of the total), and electricity consumption, transport and other sources (10 percent of the total). Thus, the initiative encompasses raw material considerations (which influence half of emissions); fuel mix (the use of renewable sources of energy or energy derived from waste); process technology and its efficiency; product quality (which influences the use of cement per output unit); logistics; and other factors.

The CSI developed a CO_2 protocol for use in defining and making publicizing baseline emissions of involved companies. It facilitated the setting of targets by involved companies against their baseline emissions and annual reporting of CO_2 emissions (CSI, 2002: 19–20).

The data suggests that CO_2 emissions per produced tonne of clinker have decreased 6 percent between 1990 and 2006. Thermal energy efficiency has improved by 14 percent over the same period. But the emissions of CSI members increased by 35 percent because their output grew by 50 percent in the same period.

The CSI data suggests that operational optimization has limited scope to influence CO_2 emissions because it is tied to the technological design of plants. Industry performance improves through the addition of new, efficient plants and decommissioning of old, inefficient plants. Alternative fossil fuels, waste and biomass contribute to the fuel mix in different ways in different regions (CSI, 2009). Raw material and fuel mix and product choices have substantial potential to reduce sectoral CO_2 emissions over the long run.

Key observations from the examples

Climate change governance initiatives such as the CCP and the CSI can cover GHG emissions comparable to those of major Annex 1 countries. The CCP has also achieved GHG emission reductions comparable to those of major Annex 1 countries, and it has done so by providing cost savings to the participants. The CSI has improved performance compared with business as usual in a period when the cement industry's output grew by 50 percent (CSI, 2009). But voluntary initiatives such as the CCP and the CSI are most likely to realize cost-saving emission reductions. These are not insignificant – as Enkvist et al. (2007), suggest, nearly a third of the global emission reductions needed by 2030 would actually provide a net benefit.

New forms of climate change governance may also have other, less tangible implications. The CCP and the CSI have established processes for assessing current performance and for setting targets and planning for their attainment. These processes make performance transparent and can create stakeholder pressure for further improvement. The CCP and the CSI have also identified and disseminated best practices and have pursued market creation for new climate friendly products and services. So, over time, they may help to bring down the marginal abatement costs of carbon and thus to create new cost-effective measures for reduction of GHG emissions.

But because two-thirds of the GHG emission reductions needed by 2030 entail economic sacrifices, there clearly remains a role for conventional state-based solutions as part of a wider polycentric governance strategy. This raises the question: what should the division of labor be among state-based, hybrid and voluntary governance solutions, and how do they interact? Voluntary industry initiatives such as the CSI are likely to benefit from the existence of political commitments because those commitments provide a basis for longer-term planning and investment. State-based governance solutions can also foster and facilitate the functioning of hybrid and voluntary climate change governance initiatives. For example, markets need backing by the states, such as legal recognition and enforceability of contracts in courts, to be credible and to function.

From the other viewpoint, hybrid and voluntary forms of climate change governance may play an important role in legitimizing and mainstreaming climate change to actors participating in them and to external political and economic decision-makers (see Gillard et al., 2017). They may lower the threshold of participating in mitigation activities and increase pressure to make progress in conventional state-based forms of climate change governance. At the same time, voluntary and hybrid forms of climate governance as part of a wider polycentric governance strategy offer a decentralized, flexible and incentivized way of learning, innovating and experimenting with promising ways of reducing GHG emissions and targeting R&D investments (ibid.).

Although the discussion here has focused on the potential and promises of hybrid and voluntary forms of climate change governance, they can also have problematic implications. Collaborative industry initiatives may not in reality be open to all, and they may result in restraints on competition. Voluntary initiatives in general are not representative, and their accountability remains unclear. These issues are increasingly drawing attention in research (see Unerman and O'Dwyer, 2006; Bäckstrand, 2008).

Conclusions and future directions

The governance framework for climate change is still largely in the making, but both new institutional arguments about polycentricity and the emerging empirical evidence suggest that institutional diversity will characterize it. The governance framework will partly be based on the UN Framework Convention for Climate Change and the protocols and decisions of parties made under it. However, national policies and regulations, sub-national and local policies and plans, and a variety of hybrid and voluntary initiatives will also play a role in climate change governance. Together, these institutional responses will create a wider polycentric governance strategy for climate change that will disperse authority and responsibility.

Although the dynamics of different kinds of institutional solutions as part of a wider polycentric governance strategy largely remain to be studied, something can be said about it. Voluntary and hybrid governance initiatives can clearly be comparable to major Annex 1 countries in terms of GHG emissions and emission-reduction achievements. While these initiatives will be at their best in realizing emission reductions that save money, they can also help to create markets for carbon-friendly products and abatement technologies, and help to bring down the marginal abatement cost of carbon over time. However, climate stabilization will also require emission reductions that will entail economic sacrifices. This means that state-based governance solutions will remain a part of the wider polycentric governance strategy.

The question is: how will different governance solutions within the wider polycentric strategy interact? Voluntary solutions may benefit from political commitment, which can provide a basis for longer-term planning and investment. State-based governance solutions can also foster hybrid solutions involving markets. Voluntary initiatives may in turn play a role in mainstreaming and legitimizing climate change to actors participating in them and to external political and economic decision-makers. They can lower the threshold of participating in voluntary climate change measures and create pressure for making progress in state-based forms of climate change governance. Voluntary and hybrid forms of climate change governance also offer a decentralized, flexible and incentivized way of learning about low-cost and promising ways of reducing GHG emissions and targeting R&D investments effectively.

There clearly is a need to improve the evidence base on the performance of non-conventional forms of climate change governance and the interaction of different types of governance solutions that form parts of a wider polycentric governance strategy. The scholarship on common-pool resources and polycentricity is well placed to make a contribution, because it can draw on both a conceptual apparatus and comparable empirical evidence.

But climate change is not only a problem of managing and reducing the level of GHG emissions to avoid dangerous climate change. The atmospheric concentration of GHGs has already increased to a level that will entail long-lasting impacts on a variety of socio-ecological systems. Even ambitious reductions of GHG emissions in the near future cannot anymore prevent these climate change impacts that are already occurring and intensifying. In the short run, some of the impacts of climate change such as an extended growing season in the far north may be beneficial, but in the long run many impacts of climate change will be adverse.

In high latitudes such as Northern Europe and the northern part of North America, mean annual temperatures will increase much more than elsewhere and annual rainfall will increase in many places. At the same time, snow and ice cover will reduce in extent and duration.

Seasons will change because of a longer growing season, and shifts in ecological zones of flora and fauna may also occur. Flood risk will increase, as will the likelihood and impacts of extreme weather in general. At the lower latitudes, which already often experience water stress or scarcity, rainfall will further decrease, which will lead to challenges in agriculture, public water supply and ecosystem management. Yet more extreme weather patterns will still pose increased risk of flooding alongside droughts and increased risk of wildfires. Sea level rise will impact coastal communities everywhere, for example by increasing the impact of given levels of sea surges alongside creeping coastal erosion.

Climate change adaptation can include public policies and plans, public or private investments, provision of new private or public goods or services, burden sharing arrangements, or changes in behaviors or practices. For example, national adaptation strategies and plans can establish adaptation priorities and identify the key structures and processes for adaptation. Investments for adaptation can include climate proofing of critical infrastructure for energy, communications or transport. Weather forecasts of different durations and advance warning systems are examples of climate services that can be provided publicly or privately. Insurance and compensation schemes are examples of burden-sharing arrangements. Behavioral and practice changes include, for example, farmers altering their crop choices to better handle climate risks.

Many adaptation actions seek to provide either a common-pool resource or a public good. In particular, there is an interest in what is called ecosystem-based climate change adaptation, which can, for example, involve the use of land cover or wetlands to manage flood risks, or the use of urban blue and green infrastructure to alleviate the impacts of heat waves. The provision of these more localized benefits of ecosystems in terms of climate change adaptation is the traditional terrain of commons research, with perhaps a new twist. Ecosystem-based adaptation is also embraced by western developed countries within which provision will take place in the complex institutional setting involving both top-down and bottom-up processes as well as both state and non-state actors.

References

Adger, W. N., Hughes, T.P., Folke, C., Carpenter, S.R., and Rockström J. (2005) Social-Ecological Resilience to Coastal Disasters. *Science* 309: 1036–1039.

Bäckstrand, K. (2008) Accountability of networked climate governance: Rise of transnational climate partnerships. *Global Environmental Politics* 8: 74–102.

Bromley, D.W. (1989) *Economic Interests and Institutions: The Conceptual Foundations of Public Policy*. Blackwell, Oxford, UK.

Bromley, D.W. (1991) *Environment and Economy: Property Rights and Public Policy*. Blackwell, Cambridge, MA.

Brooks, N., Adger, W.N., and Kelly, P.M (2005) The determinants of vulnerability and adaptive capacity at the national level and the implications for adaptation. *Global Environmental Change* 15: 151–163.

CSI, Cement Sustainability Initiative (2002) *The Cement Sustainability Initiative: Our Agenda for Action*. Geneva: World Business Council for Sustainable Development.

CSI, Cement Sustainability Initiative (2009) *Cement Industry Energy and CO_2 Performance: Getting the Numbers Right*. Geneva: World Business Council for Sustainable Development.

Dietz, T., Ostrom, E., and Stern, P.C. (2003) The struggle to govern commons. *Science* 302: 1907–1912.

Eckersley, R. (2012) Moving forward in the climate negotiations: Multilateralism or minilateralism? *Global Environmental Politics* 12: 24–42.

Enkvist, P.-A., Nauclér, T., and Rosander, J. (2007) A cost curve for greenhouse gas reduction. *McKinsey Quarterly* 1: 35–45.

Falkner, R., Stephan, H., and Vogler, J. (2010) International climate policy after Copenhagen: Towards a "building blocks" approach. *Global Policy* 1: 252–262.

Gillard, R., Gouldson, A., Paavola, J., and Van Alstine, J. (2017). Can national policy blockages accelerate the development of polycentric governance? Evidence from climate change policy in the United Kingdom. *Global Environmental Change* 45: 174–182.

Gore, C.D. (2010) The limits and opportunities of networks: Municipalities and Canadian climate change policy. *Review of Policy Research* 27: 27–46.
Hare, W.C., Stockwell, C., Flachsland, C., and Oberthür, S. (2010) The architecture of the global climate regime: A top-down perspective. *Climate Policy* 10: 600–614.
ICLEI, International Council for Local Environmental Initiatives (1993) *Cities for Climate Protection: An international campaign to reduce urban emissions of greenhouse gases.* Available online at http://www.iclei.org/index.php?id=10829.
ICLEI, International Council for Local Environmental Initiatives (2006) *ICLEI International Progress Report: Cities for Climate Protection.* www.iclei.org/index.php?id=10829.
Kern, F. and Bulkeley, H. (2009) Cities, Europeanization and multi-level governance: Governing climate change through transnational municipal networks. *Journal of Common Market Studies* 47: 309–332.
Knight, J. (1992) *Institutions and Social Conflict.* Cambridge University Press, Cambridge, UK.
Linnerooth-Bayer, J., Mechler, R., and Pflug, G. (2005) Refocusing disaster aid. *Science* 309: 1044–1046.
Meckling, J.O., and Chung, G.Y. (2009) Sectoral approaches for a post-2012 climate regime: A taxonomy. *Climate Policy* 9: 652–668.
Naim, M. (2009) Minilateralism: The magic number to get real international action. *Foreign Policy* 173: 135–136.
Ostrom, E. (1990) *Governing the Commons: The Evolution of Institutions for Collective Action.* Cambridge University Press, Cambridge, UK.
Ostrom, E. (2005) *Understanding Institutional Diversity.* Princeton University Press, Princeton, NJ.
Ostrom, E. (2010) Polycentric systems for coping with collective action and global environmental change. *Global Environmental Change* 20: 550–557.
Ostrom, E. (2012) Nested externalities and polycentric institutions: Must we wait for global solutions to climate change before taking actions at other scales? *Economic Theory* 49: 353–369.
Ostrom, E., Burger, J., Field, C.B., Norgaard, R.B., and Policansky, D. (1999) Revisiting the commons: Local lessons, global challenges. *Science* 284: 278–282.
Ostrom, E., Gardner, R., and Walker, J. (1994) *Rules, Games and Common-Pool Resources.* University of Michigan Press, Ann Arbor, MI.
Ostrom, E., Janssen, M.A., and Anderies, J.M. (2007) Going beyond panaceas. *Proceedings of the National Academies of Sciences of the United States (PNAS)* 104: 15176–15178.
Ostrom, V. (1999) Polycentricity: Part 1. In M. McGinnis (ed.) *Polycentricity and Local Public Economies.* University of Michigan Press, Ann Arbor, MI, pp. 52–74.
Paavola, J. (2005) Interdependence, pluralism and globalisation: Implications for environmental governance. In J. Paavola and I. Lowe (eds) *Environmental Values in a Globalising World: Nature, Justice and Governance.* Routledge, London, pp. 143–158.
Paavola, J. (2008) Governing atmospheric sinks: The architecture of entitlements in the global commons. *International Journal of the Commons* 2: 313–336.
Paavola, J. (2011) Climate change: The ultimate "tragedy of the commons"? In D. Cole and E. Ostrom (eds) *Property in Land and Other Resources.* Lincoln Institute for Land Policy, Cambridge, MA, pp. 417–433.
Paavola, J. and Adger, W.N. (2006) Fair adaptation to climate change. *Ecological Economics* 56: 594–609.
Robiou Du Pont, Y., Jeffery, M.L., Gütschow, J., Rogelj, J., Christoff, P., and Meinshausen, M. (2017) Equitable mitigation to achieve the Paris Agreement goals. *Nature Climate Change* 7: 38–43.
Román, M. (2010) Governing from the middle: The C40 cities leadership group. *Corporate Governance* 10: 73–84.
Schmidt, J., Helme, N., Lee, J., Houdashelt, M., and Höhne, N. (2008) Sector-based approach to the post-2012 climate change policy architecture. *Climate Policy* 8: 494–515.
Steffen, W., Sanderson, A., Jäger, J., Tyson, P.D., Moore III, B., Matson, P.A., and Wasson, R.J. (2004) *Global Change and the Earth System: A Planet Under Pressure.* Springer Verlag, Heidelberg, Germany.
UNDP, United Nations Development Programme (2007) *Human Development Report 2007/2008: Fighting Climate Change: Human Solidarity in a Divided World.* Palgrave Macmillan, Basingstoke, UK.
Unerman, J. and O'Dwyer, B. (2006) Theorising accountability for NGO advocacy. *Accounting, Auditing and Accountability Journal* 19: 349–376.
Victor, D. (2009) Plan B for Copenhagen. *Nature* 461: 342–344.
Young, O.R. (2002) *The Institutional Dimensions of Environmental Change: Fit, Interplay and Change.* MIT Press, Cambridge, MA.

16
GOVERNING WILDLIFE COMMONS
Wild boars, wolves, and red kites

Christian Schleyer, Nina Hagemann and Katharina Rauchenecker

Introduction

Wildlife can be analyzed as part of a social-ecological system (SES), where forests and agricultural lands constitute the primary natural resource systems and various forms of wildlife are the resource units. Excluding someone from using or appropriating wildlife is often difficult or induces high transaction costs. At the same time, wildlife has the characteristic of being rival in consumption and can, therefore, be characterized as a common pool resource. Wildlife are important for humans, providing food, conserving biodiversity, and supporting various forms of recreation. Human–wildlife conflicts can arise when some actors benefit from wildlife (e.g., through hunting wild boars) while others bear the costs (e.g., wild boar trampling on arable land) (Milner et al. 2014). Such human–wildlife conflicts may range from actually endangering lives to concerns over wildlife entering urban areas and making residents unfamiliar with such wildlife feel uneasy or insecure. Diffuse anxieties are linked to wildlife as well; an example is the 'reappearance' of wolves in Western Europe (Thiel et al. 2012; Vitali 2014).

In the scientific literature, including the commons literature, wildlife is covered extensively. The focus is often on case studies showing and analyzing the variety of wildlife, wildlife management, and human–wildlife conflicts: from Africa with its wildlife-rich savannahs (Ramutsindela and Noe 2012) to North America (Henri 2012) and Europe with its fish and otter (Klenke et al. 2013), wolves (Thiel et al. 2012; Hiedanpää 2013), moose, and reindeer (Helle et al. 1990). In Manfredo et al. (2009), the great variety of human dimensions of wildlife management is presented.

In this chapter, we follow the definition of wildlife management by Messmer (2000, 97) "as the application of ecological knowledge to balance wildlife populations with human needs." Messmer (2000) further differentiates between direct and indirect management: the former includes activities with a direct effect on the birth or death rates of wildlife populations; the latter has only indirect effects on birth and death rates as it focuses on improving or otherwise modifying or affecting habitat conditions.

In many developing countries, wildlife are a crucial source of food and income, especially for poor people (Ntuli and Muchapondwa 2017). In contrast, in industrialized countries, this dependence is not so strong; there are alternative sources for satisfying, for example, the demand for meat, and income from hunting domestic wildlife such as wild boar or game is low. Still,

Table 16.1 Examples of benefits and impairments and adverse effects of human–wildlife interactions

Benefits	*Impairments and adverse effects*
Food	(Human) casualties
Fur and other materials	Habitat loss of rival species
Wildlife products trade (meat or materials)	Economic costs of crop and livestock damages as well as management (e.g., fencing)
Services in the tourism sector	Opportunity and transaction costs of wildlife management
Biodiversity conservation	Social costs of human wildlife conflicts
Pharmacy / healing	Political conflicts
Existence value	Mental health
Spiritual and aesthetical values	
Cultural identity (rituals/customs)	
Leisure/recreation	
Education	

Sources: Based on Freeze (1998); Messmer (2000); Bisi et al. (2007); Aiyadurai et al. (2010); Barua et al. (2013).

wildlife management has an economic dimension which leads to the definition and assignment of property rights on wildlife. In both cases, wildlife are closely linked to biodiversity conservation, and significant changes in wildlife population can trigger trophic cascades. Further, ecosystem services provided by wildlife are important factors for the economy and for human-wellbeing in general, including tourism and recreational services. However, wildlife often cause damage. Especially in densely populated areas such as Europe, human–wildlife conflicts are numerous and prominent. Table 16.1 outlines examples of benefits as well as impairments and adverse effects from human–wildlife interactions.

Securing the benefits of human–wildlife interactions and managing externalities such as damage to property (land, flocks) require specific management and governance approaches. Social-ecological wildlife systems differ across the globe and cannot be exhaustively analyzed in this chapter. Therefore, this chapter uses various examples of wildlife management in Germany to highlight and analyze typical human–wildlife interactions, in particular focusing on actual conflicts and governance and management strategies to cope with these conflicts. Here, we look at both managed and unmanaged ('wild') forms of wildlife.

Wildlife management

Property rights are usually assigned to resource system land, such as forested or agricultural land, yet, the various resource units of 'wildlife' are often not included and wildlife property rights are absent (Naughton-Treves and Sanderson 1995). On the one hand, this is due to the high mobility of wildlife, which makes it difficult to assign specific property rights. Ecological habitat seldom corresponds with the spatial boundaries of ownership rights on land and administrative boundaries. Thus, since wildlife are "fugitive resources . . . property rights are only invested in the individual [animal] when the resource is taken [i.e. the animal is killed/hunted successfully]" (Naughton-Treves and Sanderson 1995, 1269). In addition, since wildlife can be characterized as common pool resource "harvesters are induced to harvest as quickly as they can in order to establish their proprietary rights" (Naughton-Treves and Sanderson 1995, 1269). Implementing institutions to overcome this potential—and often actual—problem of 'overharvesting' is difficult due to high transaction costs for making these institutions effective (Hanley and Sumner 1995; Schlager et al. 2016).

On the other hand, rights and duties with respect to wildlife are often defined in separate legal frameworks like hunting laws and nature conservation policies. However, designing and enforcing rights and duties on wildlife use and protection raises questions about the right scale of setting and implementing norms and determining those with legitimate rights to the resource (Naughton-Treves and Sanderson 1995). In many countries, property rights related to wildlife, such as hunting rights, are assigned to a small group of people, usually hunters that have certified knowledge about wildlife characteristics and the complexity of ecosystems and their management (Naughton-Treves and Sanderson 1995; Nahrath 2000). The following subsections provide examples of how these rights and duties are distributed in Germany: for managed wildlife, we use the example of wild boars, and for unmanaged (or 'wild') wildlife, we investigate the examples of wolves and the red kite.

Managed wildlife

In European (national) hunting laws, various property rights regimes are embedded in different hunting systems. For example, private property regimes on managed wildlife exist in Germany and France where hunting rights are linked to property rights on land. Nahrath (2000) shows that for Switzerland numerous property rights regimes are implemented for wildlife management: a license-based, a renting-based, and a state monopoly system. The monopoly system is exercised in a canton where hunting is prohibited and state authorities are responsible for wildlife management. In some other cantons, local hunting associations can rent territories and manage the wildlife based on quotas issued by the state. The license-based system is the most common one. The state sets hunting quotas, and based on these quotas, hunters are permitted to hunt wherever they want within the assigned territory.

The German renting system differs from the Swiss: in Germany, hunting rights are connected to land ownership. If the land owned by an individual person or a single organization exceeds 185 acres (ca. 75 ha), the land owner herself is entitled to execute the hunting right, provided that she also holds a shooting license. However, due to the specifics of German inheritance law, parcels owned by individual owners are often small and, thus, land ownership is fragmented. Here, the federal hunting law denotes reasonably large hunting territories of at least 371 acres (ca. 150 ha) and establishes hunting associations with compulsory membership of all (small) land owners within a particular hunting territory. The hunting associations, de facto government-owned corporations, then transfer hunting rights to an individual person, usually the tenant, who is now authorized to hunt but also needs to pay for this hunting right (thus, the renting system) (Rauchenecker 2010).

Hence, hunting rights are used by either land owners themselves or by their tenants. In both cases, hunting licenses are required that are issued by local state authorities only after successful examination, including weapon handling and practical exercises to determine knowledge of wildlife ecology and the hunting law, and a personal suitability test (*Eignungsprüfung*). Hunting rights are linked with several obligations, such as a minimum number of shoots, which is defined in shooting plans issued by the respective regional hunting authority. This becomes part of the contract between land owner and hunting tenant. As a consequence, shooting tenants are also held responsible for damage by wildlife if the population gets too large, i.e. if the shooting quota is not met: "Hunting must respect the intentions of game-preservation, and game-preservation must rely on hunting as one method to achieve its intentions" (Hasenkamp 1995, 453). For the individual hunter, this results in a permanent trade-off "between the relinquished (private) property right to use the land owned [or the hunting tenancy for this land] also for (unregulated) hunting and a 'right' to exist granted to wild animals and game" (Hasenkamp 1995, 465).

The wild boar is a prominent example of a human-(managed) wildlife conflict. The number of wild boars has increased across Europe since the 1980s (Massei et al. 2015) and is a typical wild animal in German forests. Wild boars are omnivorous animals, which allows them to spread and adapt in many different places (Schley and Roper 2003). Although wild boars with newborns may at times feverously defend their young, the human–wildlife conflict in this case is not about fears but rather about the damages caused by the increasing number of wild boar in Germany. Damages by wild boar are mainly caused by eating up seeds and fruits, by ploughing up grassland and cultivated land in their search for food, trampling on plants and soils, and causing traffic accidents (Schley and Roper 2003).

Management options for damage prevention include hunting, fencing, aversive conditioning, and feeding (Geisser and Reyer 1998; Schley and Roper 2003; Honda et al. 2009). In Germany, hunters are responsible for hunting and feeding, whereas fencing and aversive conditioning are usually carried out by farmers or other land users. The German Federal Hunting Law determines rights and duties of hunters in general, and also provides guidance for damage compensation processes (see also Löppmann 2017). To receive compensation, the damaged area has to be a 'huntable area', and not all crops and plants are covered by compensation schemes. In general, compensation payments must be paid by the local hunting association (*Jagdgenossenschaft*) which is subject to the respective federal state hunting authority (§ 29(1) German Federal Hunting Law). However, details of these compensation processes are specified in the individual hunting laws of the German federal states and, thus, differ considerably. In some cases, compensation agreements are also part of lease contracts between land owner and farmer (if not the same party). There is also a great variety of compensation agreements between land owners and hunting tenants (Rauchenecker 2010).

'Wild' wildlife

Wolves

Wolves were almost extinct in Western Europe before they returned at the end of the 20th century as a result of new regulations and resettlement efforts. As in many other European countries (see Bisi et al. 2007; Kojola et al. 2009 for Finland), wolves also returned to Germany at the end of the 1990s (Ronnenberg et al. 2017). In 2017, about 60 packs with about 150 to 160 wolves were living in Germany.[1] Wolves require large habitats and spread over large territories, which usually do not coincide with political-administrative boundaries, thus making wolf management a challenging endeavor (Gehring and Potter 2005).

The return of wolves had diverse effects on biodiversity and ecosystem service provision in these regions: they are competing with hunters for wild animals, but as they go for weak and ill game, they also provide an important regulating service (Thiel et al. 2012). When wolves returned to Germany different mind-sets existed. Farmers were afraid that the wolves would go for their livestock, and hunters were afraid of the general development and competition for wildlife (Ansorge et al. 2010), whereas the general public was rather curious (Kaczensky 2006). Some stakeholder groups, such as the tourism sector and scientists, can even draw advantages from the return of wolves. With the increasing number of wolf packs, especially in the Eastern part of Germany and the federal state of Lower Saxony, more and more attacks on livestock have occurred. Whereas the society as a whole benefits from the return of wolves through, for example, an increase in biodiversity and preservation of cultural heritage, the costs, such as livestock damages, often lie with individuals in the areas with wolf population (Thiel et al. 2012).

Wolves are strictly protected under the EU Habitats Directive, categorized as a critically endangered species in many EU member states. The protection prohibits killing and translocation of wolves. Exceptions exist only in cases where wolves cause severe damage or physically threaten people; and usually only individual wolves are taken off the packs. Non-compliance is severely punished, usually with fines. Wildlife such as wolves that receive greater protection require specific management and governance approaches because many management options, such as hunting, are excluded. Remaining options for avoiding damage, especially for livestock herders, include fencing or using herding dogs.

As in many other countries (for the US see Treves et al. 2009), in Germany damage (ex-post) is compensated and protection measures (ex-ante) are subsidized. The specific design of the compensation measures are subject to federal state regulation. Thiel et al. (2012) describe the rules that apply in the federal state of Saxony: the Nature Protection Law of Saxony compensates each (commercial or non-commercial) livestock holder if wolves prey upon livestock. However, the state requests livestock holders to also contribute to wolf protection. In demarcated wolf areas, i.e. areas where wolves are actually present, livestock herders are required to fence their livestock and/or keep herding dogs to protect their livestock within a 30 km buffer zone around the wolf area in order to receive compensation payments. These areas are adapted whenever wolf packs move. Ex-ante protection measures are supported by the state with 60% of the costs (SMUL 2009). Herders outside the 30 km buffer zone around wolf areas are not obliged to take preventive measures to be compensated if wolves attack their livestock. Other measures to reduce human-wolf conflicts are the state-led Saxon wolf management system that includes research to better understand wolf habits and characteristics as well as wolf-related education of the general public, technical advice, and financial support. The example of wolves shows the broad range of instruments applied to reduce the human–wildlife conflict in case of a fugitive and strongly protected wild animal: individual preventive actions are requested from potentially affected livestock owners; however, these actions can be financially supported. Moreover, data and information are gathered to better understand wolf movements and habitat requirements as well as to understand what kind of precautionary measures are required to reduce the amount of livestock damage.

Red kites

Threatened bird species are protected or even reintroduced for reasons of biodiversity conservation; the black and the red kite are examples of such birds. The European red kite (*milvus milvus*) population is small and accounts for only about 21,000 breeding pairs living in a few areas across Europe (Nicolai et al. 2009). However, 'managing' a wild animal such as the red kite is complicated (see, for example, Sergio et al. 2005 for Spain). Their habitats and populations are endangered by many factors: the intensification of grasslands, reduction of landscape elements, and the establishment of wind power plants as well as inter-species competition with other large birds such as the buzzard (Nicolai et al. 2009). The red kite is also quite sensitive to crop changes. For example, increasing maize cultivation affects red kite populations negatively (Sauerbrei et al. 2014).

The diet of the red kite requires open landscapes with a variety of elements. These elements are habitats of insects, worms, mice, and field mice, the main food sources of the carnivore red kite. However, with structural changes in agriculture, landscape features change and elements are removed (Plieninger et al. 2006). Therefore, in several German federal states agri-environment schemes have been introduced financially supporting farmers who either protect habitats for the red kite or create new habitats. One example is the program on perennial protective strips in Lower Saxony for breeding, feeding, or retreat areas for the red kite. The preparation of these strips requires a minimum width as well as specific cultivation of perennial crops and so-called rest time, when the field is left without any disturbance.

Poisoning of the birds is another reason for the decline in the population, whether intended or unintended, as examples from Spain, France, and Scotland show (Sergio et al. 2005; Berny and Gaillet 2008; Smart et al. 2010). The red kite as a scavenger can, for example, suffer from the application of pesticides on agricultural land, especially rodenticide that is aimed at poising mice. Also, establishing physical new technology barriers in the habitats of red kites, such as wind turbines, can cause a decline in their numbers. In Germany, the Renewable Energy Act was implemented with the aim of increasing the percentage of energy produced by renewable sources such as wind. Subsidies, so-called feed-in-tariffs, are paid to increase the number of wind turbines. While wind turbines provide many benefits, they also result in harm, like habitat disruption, soil sealing, and bat and bird casualties (such as of the red kite). Schaub (2012) shows that human population growth rates are linked with an increase in wind turbines in the area of the red kites' home range. Moreover, Rasran and Mammen (2013) found that populations outside wind turbine areas are larger than within those areas.

That does not mean that red kites and wind turbines cannot coexist, but the discourse shows that environmental and nature protection efforts can have adverse effects on other conservation aims and thus create trade-offs. Resolving or reducing these trade-offs requires societal discourse, research, and sound planning efforts. Schaub (2012) argues that construction and operation schedules positively influence red kite populations. As foraging flights in the breeding season lead to most collisions with wind turbines (Eichhorn and Drechsler 2010), operating hours during breeding season could be limited or suspended to protect populations.

Aspects of selecting governance and management options

With the wild boar, wolves, and the red kite we presented only a few examples of social-ecological systems and wildlife commons that can be found in Europe and, to some extent, also in North America. Moreover, the governance approaches and specific management options presented are only examples of how potential or actual human-environment conflicts are addressed; there are many more that could not be covered in this chapter. Preferences of how to manage wildlife properly can differ widely and always depend on the specific cultural context, the type of wildlife, the threat wildlife pose to humans, domestic animals, livestock, and property, and other factors. One particular reason why wildlife management differs are divergent interests of livestock owners aiming at protecting their livestock from carnivores, and nature-conservationists who see wildlife as a natural part of fauna and cultural heritage that has to be protected (Lundmark and Matti 2015). To regulate these diverse interests, in Sweden, a system of regional Wildlife Management Delegations was introduced as a new structure for decision-making in relation to the management of large carnivores (Lundmark and Matti 2015). The idea is to increase the legitimacy of the management system by capturing the interests and beliefs of the actors involved and to allow an exchange of reasonable and informed arguments between stakeholders. Integrating citizens in, for example, collecting wildlife information (appearance, crossing, or mortality), as in Canada (Lee et al. 2006), could also increase the exchange between different interest groups who focus on wildlife. Such organizations also exist in Germany: for example, the contact office 'Wolves in Saxony' where citizens receive information about wolves and where they can attend guided tours through the area and learn all about wolves. But citizens can also provide information if they make some observations about wolves in the area.

The choice of wildlife management options is dependent on the characteristics of the social-ecological system (including species and habitat characteristics) and the governance system.

Moreover, cultural background and economic incentives determine the acceptance of certain management options (see, for example, Hanley et al. 2010; Liordos et al. 2017). Traditions and culture hugely influence how wildlife are managed: in Sweden, hunting is a social event aimed at hunting game for food extraction, whereas US hunters prefer hunting alone without any support by, for example, hounds, and also aim to hunt for trophies (Heberlein 2000).

Governance options can range from the relocation of animals (Benjamin-Fink and Reilly 2017); mobility restrictions such as fencing, deterrence (smell, noise, etc.) or feeding; lethal control, especially in cases of overpopulation or if people are at risk; habitat manipulation such as the grizzly bear habitat extension in Canada (Braid et al. 2016); and education. Education as one of the soft approaches to ease human–wildlife conflicts is relevant for the acceptance of any management option. Of importance is objective fact-based information that does not only cover the risks and behavioral norms but outlines the benefits of wildlife (Bruskotter and Wilson 2014). Trust in information holders, for example state authorities, is important for successful—meaning behavior-changing—information campaigns (Bruskotter and Shelby 2010).

Human preferences regarding wildlife are an important factor of wildlife management. These preferences and perceptions on wildlife and wildlife conflicts differ, depending on culture, severity of the conflicts, and experiences with wildlife. The distribution of property rights (especially management rights) allows some actors to prevail in their preferences; others without any property rights can only verbally articulate their preferences and perceptions. Therefore, management and governance options need to be well tailored to the resource conflict and also to the human dimension of the conflict.

An important aspect in easing human–wildlife conflicts is to identify all costs of human–wildlife conflicts, including hidden costs. Barua et al. (2013) identified several of these hidden and usually uncompensated costs, such as decreased psycho-social wellbeing in case of fatalities, disruption of families, poor health status, and also transaction costs for requesting compensation payments. All these factors are of higher relevance in countries with large wild mammals such as elephants, human casualties, and poor families with high dependence on livestock. However, they apply to a less severe extent also to the cases here, such as wolves. As a minimum, disagreement between family members or friends, reduction in wellbeing due to awareness of endangered livestock, and transactions costs of applying all measures to be eligible for compensation influence the perspective of people on wildlife and the acceptance of protection measures.

Conclusion

The three examples of managing wild boar, wolves, and red kite as parts of a social-ecological system with the characteristics of a common pool resource showed different governance and management options for governing these commons. It was shown that many management options exist that are dependent on the conservation aim as well as trade-offs with other goods that are societally relevant. Wolves, red kite, and other species that are characterized by their high mobility also raise the question of how to set up monitoring systems and integrate different actor groups such as hunters, biologists, and others to cooperate on monitoring and establishing integrated rules for wildlife management.

The examples also illustrated the complex interrelationship between animals and humans that can be traced back to a lack of information, economic motives, and cultural backgrounds. The transferability of the sketched examples to other world regions is rather limited because they are context specific. Even within Europe, different options are possible given the acceptance of the respective measures, which are crucial and require ongoing educational work and extension services. Moreover, the spatial scales are relevant. In Central Europe, habitats of wild animals are much smaller and more scattered than in regions of the US, Africa, or Sweden. These scales also limit or extend governance and management options.

Note

1 www.zeit.de/wissen/2017-11/woelfe-population-deutschland-anstieg-debatte-politik.

References

Aiyadurai, A., Singh, N.J., and Milner-Gulland, E.J. (2010) "Wildlife hunting by indigenous tribes: A case study from Arunachal Pradesh, north-east India" *Oryx*, 44(4): 564–572.

Ansorge, H., Holzapfel, M., Kluth, G., Reinhardt, I., and Wagner, C. (2010) "Die Rückkehr der Wölfe. Das erste Jahrzehnt" *Biologie in Unserer Zeit*, 40(4): 244–253.

Barua, M., Bhagwat, S.A., and Jadhav, S. (2013) "The hidden dimensions of human–wildlife conflict: Health impacts, opportunity and transaction costs" *Biological Conservation*, 157: 309–316.

Benjamin-Fink, N., and Reilly, B.K. (2017) "Conservation implications of wildlife translocations: The state's ability to act as conservation units for wildebeest populations in South Africa" *Global Ecology and Conservation*, 12: 46–58.

Berny, P., and Gaillet, J.-R. (2008) "Acute poisoning of red kites (*milvus milvus*) in France: Data from the Sagir network" *Journal of Wildlife Diseases*, 44(2): 417–426.

Bisi, J., Kurki, S.P., and Svensberg, M. (2007) "Human dimensions of wolf (*canis lupus*) conflicts in Finland" *European Journal of Wildlife Research*, 53(4): 304–314.

Braid, A.C.R., Manzer, D., and Nielsen, S. E. (2016) "Wildlife habitat enhancements for grizzly bears: Survival rates of planted fruiting shrubs in forest harvests" *Forest Ecology and Management*, 369: 144–154.

Bruskotter, J.T., and Shelby, L.B. (2010) "Human dimensions of large carnivore conservation and management: Introduction to the special issue" *Human Dimensions of Wildlife*, 15(5): 311–314.

Bruskotter, J.T., and Wilson, R.S. (2014) "Determining where the wild things will be: Using psychological theory to find tolerance for large carnivores" *Conservation Letters*, 7(3): 158–165.

Eichhorn, M., and Drechsler, M. (2010) "Spatial trade-offs between wind power production and bird collision avoidance in agricultural landscapes" *Ecology and Society*, 15(2): article 10.

Freeze, C.H. (1998) *Wild Species as Commodities. Managing Markets and Ecosystems for Sustainability*. Washington, DC: Islands Press.

Gehring, T.M., and Potter, B.A. (2005) "Wolf habitat analysis in Michigan: An example of the need for proactive land management for carnivore species" *Wildlife Society Bulletin*, 33(4): 1237–1244.

Geisser, H., and Reyer, H. (1998) "Efficacy of hunting, feeding, and fencing to reduce crop damage by wild boars" *Journal of Wildlife Management*, 68(4): 939–946.

Hanley, N., Czajkowski, M., Hanley-Nickolls, R., and Redpath, S. (2010) "Economic values of species management options in human–wildlife conflicts: Hen harriers in Scotland" *Ecological Economics*, 10(1): 107–113.

Hanley, N., and Sumner, C. (1995) "Bargaining over common property resources: Applying the Coase theorem to red deer in the Scottish Highlands" *Journal of Environmental Management*, 43: 87–95.

Hasenkamp, G. (1995) "The economics of hunting, game-preservation, and their legal setting" *European Journal of Political Economy*, 11(3): 453–468.

Heberlein, T. (2000) "The gun, the dog and the thermos" *Sweden & America*, 13: 24–29.

Helle, T., Aspi, J., and Kilpelä, S.-S. (1990) "The effects of stand characteristics on reindeer lichens and range use by semi-domesticated reindeer" *Rangifer*, 10(3): 107–114.

Henri, D. (2012) *Managing Nature, Producing Cultures: Inuit Participation, Science and Policy in Wildlife Governance in the Nunavut Territory, Canada*. PhD Thesis, Linacre College, Oxford, UK.

Hiedanpää, J. (2013) "Institutional misfits: Law and habits in Finnish wolf policy" *Ecology and Society*, 18(1). https://pdfs.semanticscholar.org/ecbc/6dc48995e9c8128eebeeb93e9378ab090470.pdf?_ga=2.236539906.247743770.1540383357-2043343316.1540383357.

Honda, T., Miyagawa, Y., Ueda, H., and Inoue, M. (2009) "Effectiveness of newly-designed electric fences in reducing crop damage by medium and large mammals" *Mammal Study*, 34(1): 13–17.

Kaczensky, P. (2006). *Medienpräsenz und Akzeptanzstudie Wölfe in Deutschland*. Report. www.kora.ch/malme/05_library/5_1_publications/K/Kaczensky_2006_Akzeptanzstudie_Woelfe_in_Deutschland.pdf.

Klenke, R.A., Ring, I., Kranz, A., Jepsen, N., Rauschmayer, F., and Henle, K. (eds.) (2013) *Human-Wildlife Conflicts in Europe. Fisheries and Fish-Eating Vertebrates as a Model Case*. Berlin, Heidelberg: Springer.

Kojola, I., Kaartinen, S., Hakala, A., Heikkinen, S., and Voipio, H.-M. (2009) "Dispersal behavior and the connectivity between wolf populations in Northern Europe" *Journal of Wildlife Management*, 73(3): 309–313.
Lee, T., Quinn, M.S., and Duke, D. (2006) "Citizen, science, highways, and wildlife: Using a web-based GIS to engage citizens in collecting wildlife information" *Ecology and Society*, 11(1): article 11.
Liordos, V., Kontsiotis, V.J., Georgari, M., Baltzi, K., and Baltzi, I. (2017) "Public acceptance of management methods under different human–wildlife conflict scenarios" *Science of the Total Environment*, 579: 685–693.
Löppmann, S. (2017) Wildschäden auf landwirtschaftlichen Flächen - Durch Wildschweine (sus scrofa) verursachte Wildschäden in Deutschland - Maßnahmen der Bundesländer Hessen und Niedersachsen im Vergleich. Bachelor Thesis. Universität Kassel, Germany.
Lundmark, C., and Matti, S. (2015) "Exploring the prospects for deliberative practices as a conflict-reducing and legitimacy-enhancing tool: The case of Swedish carnivore management" *Wildlife Biology*, 21(3): 147–156.
Manfredo, M.J., Vaske, J.J., Brown, P.J., Decker, D.J., and Duke, E.A. (eds.) (2009) *Wildlife and Society. The Science of Human Dimensions*. Washington, DC: Island Press.
Massei, G., Kindberg, J., Licoppe, A., Gačić, D., Šprem, N., Kamler, J., . . . and Náhlik, A. (2015) "Wild boar populations up, numbers of hunters down? A review of trends and implications for Europe" *Pest Management Science*, 71(4): 492–500.
Messmer, T. (2000) "The emergence of human wildlife conflict management: Turning challenges into opportunities" *International Biodeterioration and Biodegration*, 45: 97–102.
Milner, J. M., van Beest, F.M., Schmidt, K.T., Brook, R.K., and Storaas, T. (2014) "To feed or not to feed? Evidence of the intended and unintended effects of feeding wild ungulates" *Journal of Wildlife Management*, 78(8): 1322–1334.
Nahrath, S. (2000) "Governing Wildlife Commons ?" A Comparative Analysis of Switzerland's Three Hunting Systems. The 8th Biennial Conference of the International Association for the Study of Common Property (IASCP), *Crafting Sustainable Commons in the New Millennium*. Bloomington, IN, 1–30.
Naughton-Treves, L., and Sanderson, S. (1995) "Property, politics and wildlife conservation" *World Development*, 23(8): 1265–1275.
Nicolai, V.B., Günther, E., and Hellmann, M. (2009) "Artenschutz beim Rotmilan" *Naturschutz und Landschaftsplanung*, 41(3): 69–77.
Ntuli, H., and Muchapondwa, E. (2017) "Effects of wildlife resources on community welfare in Southern Africa" *Ecological Economics*, 131: 572–583.
Plieninger, T., Höchtl, F., and Spek, T. (2006) "Traditional land-use and nature conservation in European rural landscapes" *Environmental Science & Policy*, 9: 317–321.
Ramutsindela, M., and Noe, C. (2012) "Scalar thickening: Wildlife management areas and conservation scales in southeast Tanzania" *Singapore Journal of Tropical Geography*, 33(1): 137–151.
Rasran, L., and Mammen, U. (2013) "Populationsentwicklung und Bruterfolg von Greifvögeln in Beziehung zur Entwicklung der Windkraftnutzung in Deutschland" in Hötker, H., Krone, O., Nehls, G. (Hrsg.), Greifvögel und Windkraftanlagen: Problemanalyse und Lösungsvorschläge, 314–328; Bundesministerium für Umwelt, Naturschutz und Reaktorsicherheit; NABU, Leibniz-Institut für Zoo- und Wildtierforschung, BioConsult SH, Bergenhusen, Berlin, Husum https://forschung.boku.ac.at/fis/suchen.person_publikationen?sprache_in=de&menue_id_in=102&id_in=132713.
Rauchenecker, K. (2010) *Institutioneller Wandel im Bereich Jagd und Wildtiermanagement. Das Beispiel der Jagdgenossenschaften*. Aachen: Shaker.
Ronnenberg, K., Habbe, B., Gräber, R., Strauß, E., and Siebert, U. (2017) "Coexistence of wolves and humans in a densely populated region (Lower Saxony, Germany)" *Basic and Applied Ecology*, 25: 1–14.
Sauerbrei, R., Ekschmitt, K., Wolters, V., and Gottschalk, T.K. (2014) "Increased energy maize production reduces farmland bird diversity" *GCB Bioenergy*, 6(3): 265–274.
Schaub, M. (2012) "Spatial distribution of wind turbines is crucial for the survival of red kite populations" *Biological Conservation*, 155: 111–118.
Schlager, E., Blomquist, W., and Tang, S.Y. (2016) "Mobile flows, storage, and self-organized institutions for governing common-pool resources" *Land Economics*, 70(3): 294–317.
Schley, L., and Roper, T.J. (2003) "Diet of wild boar *sus scrofa* in Western Europe, with particular reference to consumption of agricultural crops" *Mammal Review*, 33(1): 43–56.
Sergio, F., Blas, J., Forero, M., Fernández, N., Donázar, J.A., and Hiraldo, F. (2005) "Preservation of wide-ranging top predators by site-protection: Black and red kites in Doñana National Park" *Biological Conservation*, 125(1): 11–21.

Smart, J., Amar, A., Simm, I.M.W., Etheridge, B., Cameron, D., Christie, G., and Wilson, J.D. (2010) "Illegal killing slows population recovery of a re-introduced raptor of high conservation concern: The red kite *milvus milvus*" *Biological Conservation*, 143(5): 1278–1286.

SMUL (Sächsisches Staatsministerium für Umwelt und Landwirtschaft) (2009) *Managementplan für den Wolf in Sachsen*. Dresden: SMUL.

Thiel, A., Schleyer, C., and Plieninger, T. (2012) "Wolves are mobile, while fruit trees are not! How characteristics of resources and supranational regulatory frameworks shape the provision of biodiversity and ecosystem services in Germany" *Environmental Policy and Governance*, 22(3): 189–204.

Treves, A., Jurewicz, R.L., Naughton-Teves, L., and Wilcove, D.S. (2009) "The price of tolerance: Wolf damage payments after recovery" *Biodiversity and Conservation*, 18(14): 4003–4021.

Vitali, C. (2014) "A frame-analytical perspective on conflict between people and an expanding wolf *canis lupus* population in central Italy" *Oryx*, 48(4): 575–583.

17

ECOSYSTEM SERVICES AS COMMONS?

Tatiana Kluvankova, Stanislava Brnkalakova, Veronika Gezik and Michal Maco

Ecosystem services, resource overuse and social dilemma?

Approaches to the management and protection of natural resources and the landscape often overlook the functioning of ecosystems and the resilience of complex biophysical systems. Loss of biodiversity is caused by drivers such as land use change, deforestation and habitat fragmentation or environmental pollution resulting from climate change. Moreover, environmental externalities, a lack of adequate property rights, market power and asymmetrical information are the indirect causes of ecosystem degradation. In spite of the fact that European policies are progressing, they are still not effective in addressing societal challenges. Moreover, land users' lack of awareness of ecosystem service value and unsustainable management has also had adverse effects on ecosystems' resilience (European Commission, 2011; Andrés et al., 2012).

The concept of ecosystem services was introduced in the 1970s as an analytical concept as well as a policy tool for conservation management to address the relationship between human societies and the natural environment. As declared by the Millennium Ecosystem Assessment (MEA, 2005), the aim of the ecosystem service concept has been to overcome traditional approaches to the management of natural resources. Such approaches have been based on the static conservation of nature and landscape, which often overlook the functioning of ecosystems and the resilience of complex biophysical systems.

The main expectation of MEA is based on the introduction of a paradigm shift in natural resource management by targeting ecosystems, which has the aim of providing essential values for human well-being through service provisions (Amsworth et al., 2007; Skroch & López-Hoffman, 2009). Four main ecosystem service categories are recognized by MEA (2005): (i) provisioning ecosystem services as an object or resource that humans can take from the environment (food, timber, fossil fuels, drinking water); (ii) regulating ecosystem services as a natural process within an ecosystem that benefits human existence (water purification, pollination, carbon storage); (iii) cultural ecosystem services as a non-material benefit that contributes to an area's culture and experience (recreation, national parks, art, relaxation, visual display); and (iv) supporting ecosystem services as a natural process that supports all other ecosystem services (water cycle, photosynthesis, biomass production, soil formation). The meaning of ecosystem services for society's economic benefit was expanded by TEEB (2010) and subsequently included society's role in coordinating individual and collective interests through institutions and governance (Gómez-Baggethun & Kelemen, 2008; Muradian & Rival, 2012; Primmer et al., 2015) also adopted and modified by CICES (2010).

Understanding the potential benefits of ecosystem services is a pre-condition for the development of effective integrated policies for resource management and addressing the well-being of local communities. Regulatory and market-based instruments are traditional approaches in pursuing environmental conservation goals. Nevertheless, such instruments' capacity to operate across scale and with imperfect information is limited. The state – as the main regulator – rather than correcting market inefficiencies instead established restrictive measures which had a questionable impact on environmental effectiveness (Vatn, 2010; Kluvánková-Oravská et al., 2013).

In many cases, the state failed to manage ecosystems in an effective manner and thus created de jure state property but de facto open access. The absence of effective management and enforcement could simply turn valuable ecosystems into a resource for exploitation on a first-come-first-served basis (Bromley, 1992).

The market in particular has difficulties (i) in capturing ecosystem dynamics and long-term time periods (Estes et al., 2011) and (ii) with asymmetric imperfect information and targeting the trade-off of selected services that may result in the degradation of other services due to a drip-down effect and negatively affecting the resilience of ecosystems (Muradian & Rival, 2012).

Ecosystem services as common goods thus face the traditional social dilemma of individual and collective interests. Potential beneficiaries cannot be easily excluded from utilizing ecosystem services, and the subtractability of use is very high (Ostrom, 2010). Distant users operate across governance scales and with diverse interpersonal and social interests, often ignoring the sustainability and carrying-capacity of local ecosystems. The producers and users of ecosystem services rely on different information sources that may in certain cases contribute to the overuse and depletion of some ecosystem services, on the one hand, yet support the protection of ecosystems (and related services), on the other. We addressed this key question: in which ways can common-pool resource regimes (CPRs) contribute to the sustainable management of ecosystem services over spatial scales?

Examples of ecosystem services as common goods include among others carbon storage, pollination, and urban climate. We argue that a cooperative approach is required to support the decision-making process and navigating the behavioural change of ecosystem producers and users to deal with asymmetric and imperfect information and complexity for the sustainable governance of ecosystem services. As such, markets (Williamson, 1991) and governments (Panayotou, 1993) tend to fail to address this task. Panayotou (1993) stresses the importance of self-organization, and the essential shift of the state's role in environmental decision-making from command and control to coordination. Hybrid regimes are more suitable for coping with the challenges that accompany ecosystem service provision across governance scales (Muradian & Rival, 2012). In our chapter, we argue that the common pool resource regime (Ostrom, 1990, 2009) can be seen as (i) an institution for ecosystem service governance in natural and urban areas under conditions of asymmetric and imperfect information; and (ii) that self-organization has a basis for behavioural change from sectoral to ecosystem service governance.

In the next section, we introduce ecosystem service governance in comparison to sectoral approaches. Then we review analytical frameworks to determine the institutional robustness of self-organization in CPR regimes and behavioural change to study the social dilemma of ecosystem services in forest and urban areas are explored. The final section analyses the motivation factors that trigger behavioural change from sectoral to ecosystem service governance.

From sectoral approach to ecosystem service governance

Sectoral policies and approaches to the management and protection of natural resources and the landscape often fragment ecosystem services, and are primarily concerned with market interests that result in overuse and constitute a barrier to achieving the sustainable use and management of such ecosystem services. In contrast, national biodiversity conservation policies isolate ecosystem services from human use by focusing on outcomes and impacts rather than on how governance turns policies into practice (Doremus, 2003; Primmer et al., 2015). Several authors (Kluvánková-Oravská et al., 2009; Andrés et al., 2012; Sarvašová et al., 2013) claim that nature protection as a policy sector is not isolated but is rather directly or indirectly influenced by many other sectors (e.g. forestry, agriculture, water management, rural development, energy, etc.). Global processes such urbanization, privatization, large-scale forestry and agriculture, and diverging interests of multiple actors across scales have significantly increased the isolation of existing sectoral and nature conservation policies, which have negatively affected ecosystem service providers and increased ecosystem vulnerability and fragmentation (EEA, 2010; De Groot et al., 2012; Primmer et al., 2015).

The quality of ecosystem service provision largely results from individual and group decisions, and the regulation of such decisions by decentralized norms and formal centralized governance schemes across decision-making levels (Parker & Meretsky, 2004) and various financial incentives. The scale at which decisions are made often differs from the scale where such decisions finally benefit different actors, leading to rivalry in the consumption of goods and services. The resolution of conflicts between the various beneficiaries of ecosystem goods and services typically involves collectively beneficial but individually costly actions (Gibson et al., 2000; Ostrom et al., 2002). Such weakness often results in costly regulations, ineffective or unfair market allocation, mismatch between ecosystems and the institutions meant to manage them, as well as the erosion of intrinsic motivations for conservation (Corbera et al., 2007; Chobotova, 2013; Midler et al., 2015; Rode et al., 2015). The vulnerability of ecosystem services is thus aggravated by multilevel factors, as well as the public or commons character of those goods.

The concept of ecosystem services is thus seen as promising "therapy" to bridge sectoral policies into the integrative governance of ecosystem services and shifting the motivation of resource users toward sustainability. The ecosystem services governance approach offers a novel governance model whereby community well-being, economic prosperity, and environmental effectiveness can be sustained via a hybrid governance model that combines market and self-governance components. Aspects such as social, economic, and environmental factors can be integrated into the management of natural landscapes and broadening biodiversity conservation practices, with the aim to strengthen arguments and tools for protecting ecosystems (Constanza, 2008; De Groot et al., 2012; Schroter et al., 2014). Ecosystem services governance views ecosystem services as common goods, facilitates cooperation between groups, disciplines, or sectors with different paradigms or interests, and can also foster interdisciplinary research processes. However, many issues still need to be resolved in order to fully integrate ecosystem services governance into everyday planning, management and decision-making (Jahn et al., 2012).

Analytical frameworks to study the social dilemma of ecosystem services in forest and urban commons

Ecosystem services face a situation whereby, similar to other common goods, property rights fail to adequately address the social dilemma (Demsetz, 1967). As Ostrom et al. (1994) highlight, "understanding the conditions under which the users of CPRs successfully develop and maintain effective institutions is critical to facilitating improved resource policies".

Numerous studies such as Berkes and Folke (1985), Ostrom (1990), Henrich et al. (2001), Cárdenas and Carpenter (2008), Poteete et al. (2010), and others, have claimed that local users in CPR regimes are capable of crafting their own rules that allow for the sustainable and equitable management of ecosystem services, and which can solve resource management problems without external authorities and even without any external financial incentives. In such a case, the market approach would be implemented only as "support" for performed activities, even if the market instruments are absent. These regimes are typified by a transfer of knowledge, resources and institutions across the scales, which may potentially form a set of independent self-governed systems. Due to their institutional maturity and robustness, CPR regimes can be viewed as capable of addressing the social dilemma of ecosystem service governance across scales.

The primary conditions to undertake these assumptions is compliance with the design conditions of Ostrom's (1990, 2005, 2008, 2010) robust governance. In order to identify which particular aspects contribute to the increased quality of ecosystem service governance, we applied eight design principle analyses to determine the conditions for ecosystem service governance and the effectiveness of the regime that reduces the costs of central monitoring in two different types of institutional settings: carbon sequestration in forest common pool resource regimes, and microclimate in urban commons. If ecosystem services governance is successfully implemented, and innovative incentives for the climate regulation ecosystem service are beneficially set for local CPR regimes (as ecosystem service providers), then these regimes could ensure the balanced use and protection of natural resources, and thus contribute to scaling down strategic global environmental policy objectives, as well as increasing the well-being of local communities (Mace et al., 2012; Kluvankova & Jilkova, 2013; Brnkalakova, 2016).

Ecosystem services examples of forest and urban commons

Traditional forest CPRs established in medieval Europe are mostly private holdings owned in common and managed jointly by shareholders. This type of forest ownership regime faces diverging interests and values of multiple shareholders, as well as forces of globalization that increase the challenge of managing common forests. The management of carbon via CPR regimes has been identified as a potential business strategy to increase the competitiveness of the regime on the global market.

Evidence of forest CPR regimes seen as effective regimes for the provision of climate regulation ecosystem services was documented in Kluvankova et al. (forthcoming a). The institutional robustness of CPR regimes in adapting to innovations in forest management has been found essential, in particular, to integrate carbon forestry management practices that result in increased carbon sequestration compared to state or private forest regimes.

CPRs in urban spaces may be of public, common-property or private ownership, yet all are characterized by multiple users of the benefits of climate regulation ecosystem services and collective action over resource management. As Hardin (1968) noted, the pool of multiple users' demands often exceeds the capacity of the resource, and the combination of private and shared property limits collective action in overcoming the social dilemma. The empirical application of CPR regimes to foster the sustainable provision of microclimate regulation ecosystem service in urban areas will concern semi-public spaces in Bratislava (Kluvankova et al., forthcoming a). Semi-public spaces understood as dynamic spatial units are defined by three basic attributes: physical boundaries, property relations, and institutions (Maco, 2015) and have the capacity to engage local actors, create relations between users and space, and accommodate community-based management arrangements. These are, for example, courtyards, community gardens, and vacant lots.

In both examples, the CPR regime is seen as creating and maintaining the collaboration of ecosystem service providers and users across the scale, thus enhancing the ecosystem service governance model.

We firstly highlight the importance of rights. To prevent the overuse of CPRs, a limitation of access to beneficiaries is often required, a process of exclusion through physical and institutional means that is repeatedly difficult and costly. Following Schlager and Ostrom (1992), property ownership does not play a significant role in effective management, but the distribution of property rights does. Besides collective ownership, a robust regime can also be developed on private or public properties. More important is user access to rules-in-use, which define users' competences towards the space – hence access, withdrawal and management rights are essential attributes for the sustainable use of ecosystem services. In this context, it is crucial to determine which incentives support the provision of the first three property rights (see Boxes 17.1 and 17.2).

Box 17.1 Carbon sequestration

Forest CPRs, if based on shared property and resource management, offer opportunities to learn how to manage common forests more effectively and sustainably from issues coming from participation, collective decision-making and shared responsibility, and traditional knowledge.

Box 17.2 Urban climate

Users of green infrastructure in urban areas often face open access issues. If these three rights are provided to all members of an urban community in the CPR regime created in a semi-public space, the willingness to participate on green infrastructure management increases due to the fact that individual interests become collective interests – with an incentive to promote the continuity of collective action.

In the next section, we provide an empirical analysis of conditions under which the users of CPRs successfully develop and maintain effective institutions for the self-governance of ecosystem services in our two climate regulation examples. The applicability of eight design conditions is presented in Table 17.1, followed by arguments as to how historical and human-constructed CPR regimes can address the social dilemma of ecosystem service governance across spatial scales.

Rules-in-use (Principles 1–6, Ostrom, 1990) embedded in management regimes have the role of controlling and navigating the behaviour of users, as well as empowering them and promoting self-management. European forest commons constitute historical CPR regimes, where rules-in-use evolved as tradition along with the institutional maturity of self-governing communities following traditional knowledge and collective action. These were found to be crucial for the effective management of natural resources and ecosystem services provision

Table 17.1 Analysis of CPR regimes according to the eight design principles

Eight design principles	*Forest carbon sequestration*	*Urban climate*
1. ***Boundaries***: *Resource and group boundaries are clearly defined.*	Clear physical boundaries, but fuzzy users' boundaries because of unknown, inactive, non-resident or new shareholders.	Clear physical and user borders by connecting ES providers and users.
2. ***Local needs and conditions***: *Rules governing the use of collective goods are well matched to local needs and conditions.*	Customary rules and traditional knowledge of natural resources cause unintentional protection of ecosystems. The vulnerability of forests increases because of extreme weather events.	Mixed balance between costs and benefits (source of conflict between individual and collective interests).
3. ***Management and collective action***: *Most individuals affected by these rules can participate in modifying the rules.*	Self-management and collective decision-making is historically determined – although today facing the pressures of global market and diverging interests.	Self-management and collective decision-making introduced as new commons.
4. ***Monitoring***: *A system for monitoring members' behaviour exists.*	Shareholders are monitored internally by assembly and externally by governmental regulations.	Informal monitoring by residents and users.
5. ***Sanctions***: *A graduated system of sanctions is used.*	Conflicts are mostly solved internally by common agreement. If not, a national court is held.	Internal sanction mechanism based on custom and social exclusion in development.
6. ***Low-cost conflict resolution***: *Community members have access to low-cost conflict resolution mechanisms.*	Conflict resolution, as well as sanctioning, is embedded in national customary law where rules are already provided.	Conflict resolution mechanism is not part of national customary law but can be subject to informal customs.
7. ***Respected by external authorities***: *The rights of community members to devise their own rules are respected by external authorities.*	The number of forest CPRs (e.g. Slovakia, Sweden, Slovenia) is defined in national law as private holdings owned in common and managed jointly by shareholders.	Self-organizing rights are created by public or private contract.
8. ***Nested enterprises***: *CPRs are parts of larger systems.*	Nested in institutional systems, parts of protected areas, evidence of the increasing role of forest CPRs in global climate change mitigation and integrated in the global market, recently also by green economy activities.	Nested in supranational legal and institutional systems such as connectivity to municipal CO_2 mitigation and adaptation plans and policies.

Sources: Holmgren et al. (2010), Bassi (2012), Feliciano et al. (2015), Prempl et al. (2015), Kluvankova and Gezik (2016).

(Holmgren et al., 2010; Bassi, 2012; Feliciano et al., 2015; Prempl et al., 2015; Brnkalakova, 2016; Kluvankova & Gezik, 2016).

In contrast, the clear rights and responsibilities of CPR regime users in urban areas are problematic as they often face an open access situation. The most successful examples of transformations to the CPR regime can thus be seen in semi-public spaces, where users and providers are identical and locally-determined. Community empowerment is the basic incentive that promotes participation. However, actors' motivation to participate must be cooperative rather than competitive. Secondly, the separation of individual and collective rights and interests in semi-public housing units is critical – such as individual ownership rights and collective interests in mutual conflict. Hence the establishment of an informal or formal institution to separate and enforce collective interests only in the shared area has been found an essential solution to the social dilemma (Kluvankova et al., forthcoming b).

Despite ongoing natural and social disturbances, forest commons have been struggling to adapt to traditional management practices and customary uses of natural resources according to global market demands. As a result, a number of forest CPR regimes in Europe are transforming solely from timber cutting to multiple forest use or green economy activities. In this way, the provision of multiple non-market public and common goods is increasing, which legitimizes the regime in the global arena (Principles 7 and 8, Ostrom, 1990).

The legitimacy and acceptance of CPR regimes in urban semi-public spaces is questionable, unless such is nested in legal and institutional systems at the municipal or supranational level. Successful cooperation with higher authorities is essential in such as connectivity to municipal CO_2 mitigation and adaptation plans and policies.

In addition, the importance of traditional CPR forest management practices has been recognized as key for the implementation of carbon smart forestry. The reconfiguration of forest management practice with the self-organization of forest management and governance has been found to be a social innovation for increasing the potential of climate regulation ecosystem services. If new policy tools and incentives are created with the aim of supporting carbon management in forest CPRs, such could play a significant role not only in climate change mitigation, but also in contributing to the higher economic performance and well-being of forest CPRs typically in marginalized mountain areas.

Ecosystem Services as motivation for behavioural change?

In order to support better environmental practices and prevent the depletion of a common pool resource, direct payment schemes (also called payments for ecosystem services – PES) have been proposed as a possibly cheaper and more effective alternative vis-à-vis the regulatory approach (Bräuer et al., 2006; McNeely, 2009).

Payment schemes may contribute to changing practices and behaviour, which are normally assumed to be critical features of collective action problem-solving. However, ending a payment often results in a return to the initial state, since behavioural changes are highly conditional on monetary transfer (Muradian & Rival, 2012; Kluvánková-Oravská et al., 2013–2014). Payment schemes are not a panacea for the protection of ecosystem services. Many examples show that PES should complement rather than substitute regulatory approaches (Nunes & Riyanto, 2005; Pascual & Perrings, 2009; Muradian et al., 2010; Pascual et al., 2010).

As the regulatory approach ensures that an upper limit of biodiversity damage is set at the regional or national level, the market instrument approach should assure flexibility and efficiency, leading to the equal distribution of costs and benefits of biodiversity conservation (Nunes & Riyanto, 2005; Pascual & Perrings, 2009; Muradian et al., 2010; Pascual et al., 2010). We argue that the hybrid approach avoids the weaknesses and inefficiencies that may occur when adopting either the hierarchy or market instrument approach.

Although evidence to support our argument is scarce, such can be documented for example in the Lion Guardians community conservation model (Hazzah et al., 2014). That model has developed as a novel institution based on interpersonal trust, leadership, and collective action of the Maasai community. The idea was to combine with The Predator Compensation Fund's incentive-based conservation of predators to foster traditional indigenous values to implement effective lion protection policy in Kenya and Tanzania. Maasai warriors – traditional lion killers – have been transformed into lion protectors. They contribute with unique knowledge of local ecosystems, and learn the skills needed to effectively mitigate conflicts between people and wildlife, monitor lion populations, and help their own communities live with lions in coexistence. The Predator Compensation Fund effectively reduces lion killings by 87% to 91%, and in combination with Lion Guardians incidents of lion killings have almost been eliminated (Hazzah, et al., 2014). More details can be found in Table 17.2.

Table 17.2 Example of robust institutions and payment for ES incentive

	Predator Compensation Fund (PCF)	*Lion Guardians (LG)*	*PCF + LG*
Program aim	Increase tolerance of carnivores and change behavioural outcomes.		
Program characteristics	Complex system of incident verification, payments, and penalties for violating program rules. The program pays local people market price for depredated livestock, and imposes penalties for lion killing. Community pressure is a major incentive to conserve lions.	Conservation-related employment, training in literacy and scientific monitoring (lion numbers and movements), and community assistance. The program employs well-respected ilmurran, many of whom gained respect by killing lions. It can utilize their traditional leadership roles to prevent lion killings.	Compensation does not diminish the frequency of depredation, but is a response mechanism to depredation, with effort applied to preventing hunts through peer pressure with the threat of loss of payments. The LG program specifically aims to reduce depredation incidents, and attempt to intervene when hunts are planned.
Incentive	Based primarily on an economic model of behaviour.	Draws on community and traditional Maasai self-organization.	
Results	Compensation is associated with a marked reduction in lion killing, implying that at least some Maasai pastoralists value monetary compensation for livestock losses enough to change their behaviour. However, there were still incidents of lion killing.	Associated with the near-total cessation of lion killings in areas where implemented. The LG model appears more cost-effective.	Both PCF and LG reduced lion killing. However, the PCF + LG combination eliminated lion killing to almost zero (99%).

Source: Hazzah et al. (2014).

Despite the mixed evidence of the role of PES in the long-term behavioural changes of consumers and producers towards the more environmentally-friendly use of CPRs, significant interest can be seen in PES when incorporated into hybrid regimes such as those documented in Table 17.2. The positive effect can be explained in part by encouraging higher transparency, and more flexibility and cooperative capacity by actors in selecting how to achieve a certain goal. The hybrid regime with the introduction of PES also helps to establish a dialogue between various interest groups and creates community trust that has hitherto been lacking. In conjunction with traditional regulation, such a model can become critical in achieving biodiversity conservation objectives (Chobotova, 2013).

Conclusion

Our chapter primarily aims to address the social dilemma of ecosystem services as common goods. By building on analyses of institutional robustness, we generated arguments as to how CPRs can foster ecosystem service governance under the asymmetric and imperfect information of the global arena. Empirically applied in the context of climate regulation ecosystem services in traditional forest commons and new commons in urban semi-public spaces, key factors were identified for the robust long-term institution of the ecosystem service governance model. In particular, traditional CPR regimes were found sufficiently robust to address the ecosystem service dilemma via mature self-governance regimes, which are adaptable to scale-down global CO_2 targets to the local level, and the innovation of forest management and governance practices to include carbon sequestration. In new CPR regimes, the incentive for a change in behaviour from individual to collective actions will require a local institution to bridge the border between ES providers and users.

In contrast to traditional approaches to conserving resources and managing ecosystem services derived from imposing regulatory measures or market incentives alone, we demonstrate the potential of a hybrid model whereby the capacity of a mature community and self-organized governance structures enhance behavioural change from sectoral to ecosystem service governance. A combination of PES and a robust self-governed institution provides reasoning for behavioural change from sectoral to ecosystem service governance in the context of the global world's uncertainty and complexity. Such a mechanism has the potential not only to increase the well-being and competitiveness of local communities in the threatened CPRs in marginalized regions but also to contribute to climate change mitigation, which is a global problem. This poses challenging options for the manner in which commons and community resource management become central to ecosystem service policies.

References

Amsworth, P., Chan, K., Daily, G., Ehrlich, P., Kremen, C., Ricketts, T. & Sanjayan, M. (2007). 'Ecosystem-service science and the way forward for conservation'. *Conservation Biology*, 21(6), pp. 1383–1384.

Andrés, S.M., Mir, L.C., van den Bergh, J.C.J.M., Ring, I. & Verburg, P.H. (2012). 'Ineffective biodiversity policy due to five rebound effects'. *Ecosystem Services*, 1, pp. 101–110.

Bassi, M. (2012). 'Recognition and support of ICCAs in Italy'. *Technical Series*, 64.

Berkes, F. & Folke, C. (1985). *Linking Social and Ecological Systems: Management Practices and Social Mechanisms for Building Resilience*. Cambridge, UK: Cambridge University Press.

Bräuer, I., Mussner, R., Marsden, K., Oosterhuis, F., Rayment, M., Miller, C. & Dodokova, A. (2006). *The Use of Market Incentives to Preserve Biodiversity. Final Report*. A project under the Framework contract for economic analysis. ENV.G.1/FRA/2004/0081.

Brnkalakova, S. (2016). Adaptive management of mountain ecosystem services. n.p.: Doctoral thesis.

Bromley, D.W. (1992). 'The commons, common property, and environmental policy'. *Environmental and Resource Economics*, 2, pp. 1–17.

Cárdenas J.C. & Carpenter J., 2008. 'Behavioural development economics: Lessons from field labs in the developing world'. *Journal of Development Studies*, 44(3), pp. 337–364.

Chobotova, V. (2013). 'The role of market-based instruments for biodiversity conservation in central and eastern Europe'. *Ecological Economics*, 95, pp. 41–50.

CICES (2010). *Proposal for a Common International Classification of Ecosystem Goods and Services (CICES) for Integrated Environmental and Economic Accounting. European Environment Agency.* New York.

Constanza, R. (2008). 'Ecosystem services: Multiple classification systems are needed'. *Biological Conservation*, 141, pp. 350–352.

Corbera, E., Brown, K. & Adger, W.N. (2007). 'The equity and legitimacy of markets for ecosystem services'. *Development and Change*, 38(4), pp. 587–613.

De Groot, R., Brander, L. & van der Ploeg, S. (2012). 'Global estimates of the value of ecosystems and their services in monetary units'. *Ecosystem Services*, 1, pp. 50–61.

Demsetz, H. (1967). 'Toward a theory of property rights'. *The American Economic Review*, 57(2), pp. 347–357.

Doremus, H. (2003). 'A policy portfolio approach to biodiversity protection on private lands'. *Environmental Science & Policy*, 6, pp. 217–232.

EEA (2010). *European Environment Agency Report. Europe's Ecological Backbone: Recognising the True Value of Our Mountains*. n.p.: n.n.

Estes et al. (2011). 'Trophic downgrading of planet earth'. *Science*, 333, pp. 301–306.

European Commission (2011). *Forestry in the EU and the World. A Statistical Portrait*. Luxembourg: Publications Office of the European Union.

Feliciano, D., Alves, R., Mendes, A., Robeiro, M. & Sottomayor, M. (2015). 'Forest land change ownership in Portugal'. In: *COST Action FP1201 FACESMAP Country Report*. Vienna: European Forest Institute Central-East and South-East European Regional Office, p. 50.

Gibson, C., McKean, M.A. & Ostrom, E. (2000). *Communities, Institutions and Governance*. Cambridge, MA: MIT Press.

Gómez-Baggethun, E. & Kelemen, E. (2008). *Participatory Methods for Valuing Ecosystem Services. Methods and Tools for Environmental Appraisal and Policy Formulation*. Lisbon: THEMES Summer School.

Hardin, G. (1968). 'The tragedy of the commons'. *Science*, 3859(162), pp. 1243–1248.

Hazzah, L., Dolrenry, S., Naughton, L., Edwards, C.T.T., Mwebi, O., Kearney, F. & Frank, L. (2014). 'Efficacy of two lion conservation programs in Maasailand, Kenya'. *Conservation Biology*, 28(3), pp. 851–860.

Henrich, J., Boyd, R., Bowles, S., Camerer, C., Fehr, E., Gintis, H. & McElreath, R. (2001). 'In search of Homo economicus: Behavioral experiments in 15 small-scale societies'. *American Economic Review*, 91(2), pp. 73–78.

Holmgren, L., Holmgren, E., Fridman, J. & Lidestav, G. (2010). 'Biological diversity indicators: A comparison of northern Swedish forest commons and other forest ownership categories'. *Forest Research*, 25, pp. 61–68.

Jahn, T., Bergmann, M. & Keil, F. (2012). 'Transdisciplinarity: Between mainstreaming and marginalization'. *Ecological Economics*, 79, pp. 1–10.

Kluvankova, T., Brnkalakova, S., Marek, M.V., Valatin, G., Hopkins, J., Kovac, U. & Nijnik, M. (forthcoming a). 'Carbon sequestration for the well-being of European mountain regions'. *Ecological Economics*.

Kluvankova, T. & Gezik, V. (2016). 'Survival of commons? Institutions for robust forest social-ecological systems'. *Journal of Forest Economics*, 24, pp. 175–185.

Kluvankova, T. & Jilkova, J. (2013). 'From government to governance'. In: *From Governing to Governance Reconsidered*. Ruzomberok, Slovakia: VERBUM, pp. 11–19.

Kluvankova, T., Kaiser, A. & Maco, M. (forthcoming b). 'Smart governance'. *The International Journal of the Commons*.

Kluvánková-Oravská, T., Chobotová, V., Banaszak, I., Slavikova, L. & Trifunovova, S. (2009). 'From government to governance for biodiversity: The perspective of central and eastern European transition countries'. *Environmental Policy and Governance*, 19(3), pp. 186–196.

Kluvánková-Oravská, T., Chobotová, V. & Smolková, E. (2013). 'The challenges of policy convergence: The Europeanization of biodiversity governance in an enlarging EU'. *Environment and Planning, Government and Policy*, 31, pp. 401–413.

Kluvánková-Oravská, T., Chobotova, V., Udovč, A. & Brnkalakova, S. (2013–2014). *Deliverable D.5.5. Stakeholders' Analysis Regarding the Acceptability of Alternative Soil Policy Options*. s.l.: ECOFINFERS project.
Mace, G.M., Norris, K. & Fitter, A.H. (2012). 'Biodiversity and ecosystem services: A multilayered relationship'. *Trend in Ecology and Evolution*, 27(1), pp. 19–26.
Maco, M., 2015. Theory of Commons in Urban Governance : Application to Semi-public Spaces Theory of Commons in Urban Governance : Application to Semi-public Spaces. Slovak University of Technology, pp. 156.
McNeely, J.A. (2009). 'Payments for ecosystem services: An international perspective'. In: K. N. Ninan (ed.) *Conserving and Valuing Ecosystem and Services and Biodiversity. Economic, Institutional and Social Challenges*. London: Earthscan.
MEA (2005). *Ecosystems and Human Well-Being: Synthesis*. Washington, DC: Island Press.
Midler, E., Pascual, U., Druckerd, A. & Narloch, U. (2015). 'Unraveling the effects of payments for ecosystem services on motivations for collective action'. *Ecological Economics*, 120, pp. 394–405.
Muradian, R., Corbera, E., Pascual, U., Kosoy, N. & May, P.H. (2010). 'Reconciling theory and practice: An alternative conceptual framework for understanding payments for environmental services'. *Ecological Economics*, 69, pp. 1202–1208.
Muradian, R. & Rival, L. (2012). 'Between markets and hierarchies: The challenge of governing ecosystem services'. *Ecosystem Services*, 1(1), pp. 93–100.
Nunes, P.A.L.D. & Riyanto, Y.E. (2005). 'Information as a regulatory instrument to price biodiversity benefits: Certification and eco-labeling policy practices'. *Biodiversity and Conservation*, 14, pp. 2009–2027.
Ostrom, E. (1990). *Governing the Commons: The Evolution of Institutions for Collective Action*. New York: Cambridge University Press.
Ostrom, E. (2005). 'Understanding the diversity of structured human interactions'. In: *Understanding Institutional Diversity*. Princeton, NJ: Princeton University Press.
Ostrom, E. (2008). *Polycentric Systems as One Approach for Solving Collective-Action Problems*. Available at SSRN: http://ssrn.com/abstract=1304697.
Ostrom, E. (2009). 'A general framework for analyzing sustainability of social-ecological systems'. *Science*, 5939(325), pp. 419–422.
Ostrom, E. (2010). 'A long polycentric journey'. *Annual Review of Political Science*, 13, pp. 1–23.
Ostrom, E., Dietz, T., Dolsak, N., Stern, P., Stovich, S. & Weber, E. (2002). *The Drama of the Commons*. Washington, DC: National Academy Press.
Ostrom, E., Gardner, R. & Walker, J. (1994). *Rules, Games, & Common-Pool Resources*. Ann Arbor, MI: The University of Michigan Press.
Panayotou, T. (1993). Empirical tests and policy analysis of environmental degradation at different stages of economic development. Working Paper WP238, Technology and Employment Programme. Geneva: International Labour Office.
Parker, D. & Meretsky, V. (2004). 'Measuring pattern outcomes in an agent-based model of edge-effect externalities using spatial metrics'. *Agriculture, Ecosystems, and Environment*, 101, pp. 233–250.
Pascual, U., Muradian, R., Rodriguez, L.C. & Duraiappah, A. (2010). 'Exploring the links between equity and efficiency in payments for environmental services: A co nceptual framework'. *Ecological Economics*, 69, pp. 1237–1244.
Pascual, U. & Perrings, C. (2009). 'Developing mechanisms for in situ biodiversity conservation'. In: Ninan, K.N. (ed.) *Conserving and Valuing Ecosystem Services and Biodiversity: Economic, Institutional and Social Challenges*. London: Earthscan, pp. 151–174.
Poteete, A., Janssen, M. & Ostrom, E. (2010). *Working Together: Collective Action, the Commons, and Multiple Methods in Practice*. Princeton, NJ: Princeton University Press.
Prempl, T., Udovč, A., Bogotaj, N. & Krč, J. (2015). 'From restitution to revival: A case of common re-establishment and restitution in Slovenia'. *Forest Policy and Economics*, 59, pp. 19–26.
Primmer, E., Jokinen, P., Blicharska, M., Barton, D.N., Bugter, R., Potschin, M. & Grandin, U. (2015). 'A framework for empirical analysis of ecosystem services governance'. *Ecosystem Services*, 16, pp. 158–166.
Rode, J., Wittmer, H., Emerton, L. & Schröter-Schlaack, C. (2015). ''Ecosystem service opportunities': A practice-oriented framework for identifying economic instruments to enhance biodiversity and human livelihoods'. *Journal for Nature Conservation*, 33, pp. 35–47.
Sarvašová, Z., Šálka, J. & Dobšinská, Z. (2013). 'Mechanism of cross-sectoral coordination between nature protection and forestry in the Natura 2000 formulation process in Slovakia'. *Journal of Environmental Management*, 123, pp. 65–72.

Schlager, E. & Ostrom, E. (1992). 'Property-rights regimes and natural resources: A conceptual analysis'. *Land Economics*, 68(3), pp. 249–262.

Schroter, M., van der Zanden, E.H., van Oudenhoven, A.P.E., Remme, R.P., Serna-Chavez, H.M., de Groot, R.S. & Opdam, P. (2014). 'Ecosystem services as a contested concept: A synthesis of critique and counter-arguments'. *Conservation Letters*, 7(6), pp. 514–523.

Skroch, M. & López-Hoffman, L. (2009). 'Saving nature under the big tent of ecosystem services: A response to Adams and Redford'. *Conservation Biology*, 24(1), pp. 325–327.

TEEB (2010). *The Economics of Ecosystems and Biodiversity*. Birkirkara, Malta: Progress Press.

Vatn, A. (2010). 'An institutional analysis of payments for environmental services'. *Ecological Economics*, 69, pp. 1245–1252.

Williamson, O. (1991). 'Comparative economic organization: The analysis of discrete structural alternatives'. *Administrative Science Quarterly*, 36(2), pp. 269–296.

18

URBAN COMMONS OF THE GLOBAL SOUTH

Using multiple frames to illuminate complexity

Seema Mundoli, Hita Unnikrishnan and Harini Nagendra

Introduction

For that which is common to the greatest number has the least care bestowed upon it.
(The politics of Aristotle, *translated by Jowett 1885, 30)*

Aristotle's concern about the nature of commons indicates that popular perceptions about commons, as a resource open to use by all and cared for by none, date back to ancient times. Similarly, starting as far back as the 13th century, with the enclosure of commons in England, the management, customs and contestations around commons have also been documented (Weston and Bollier 2013; Linebaugh 2014; Kratzwald 2015). While the growth of scholarship on natural resource commons has been substantial in recent decades, much of the research has predominantly focused on forest, fisheries and irrigation systems occurring in rural landscapes (van Laerhoven and Ostrom 2007). However, in recent years, the rapid urbanization across the globe, particularly intense in the Global South, (UN-Habitat 2016), has brought the discussion on commons into the urban sphere as well.

Urban common pool natural resources (henceforth urban commons) are under threat across the world (Elmqvist et al. 2013). When compared to rural commons, urban commons are a more contested resource. The extremely heterogeneous demography of urban areas results in the same commons being prioritized by different groups for different reasons (Kip 2015), and simultaneously inhibits social cohesiveness, making collective action and commons management a huge challenge in the urban context (Foster 2011; Garnett 2012). Urban commons are also especially challenging to categorize for purposes of management and protection. The categories of public, private, toll goods and common pool resources can be very fuzzy (Helfrich 2012) when categorizing urban natural common pool resources. Urban commons are likely to change to public, private or toll goods, and in urbanizing cities that witness rapid land use transformations, these shifts across categories can happen very swiftly and often go unnoticed. As a result, natural spaces lose their status as commons, with adverse implications for those dependent on commons and the environmental sustainability of cities.

Urban commons occur in different social and cultural contexts and have a variety of institutional arrangements for governance. Unraveling the unruly complexity of urban commons requires the use of multiple analytical lenses and research methods. Urban commons are complex social-ecological systems (SES), and their management often presents a "wicked problem." Rapid urbanization is often associated with outcomes that include rapid changes to land use and customary land rights, generation of urban waste, and pollution and depletion of local resources such as groundwater, which severely impact a variety of urban ecosystems (including urban commons) and reduce their capacity to provide important services. There is no tame solution for addressing problems of complex SES; but one need not be overwhelmed by difficulties in the management of ecosystems (DeFries and Nagendra 2017) such as urban commons either. Incorporating analytical frameworks that recognize complexities but that are also directed towards finding positive outcomes are important to formulating solutions for the "wicked problems" of managing urban commons.

Since the late 2000s, we have assembled an interdisciplinary team of ecologists and social scientists to conduct extensive research on the urban commons in Bengaluru, one of India's largest and fastest growing cities. In our research urban commons are the ecosystems found in the city core and peri-urban areas—these include lakes, ponds, wells, remnant patches of groves, grazing lands, and tree cover along roads and in cemeteries. These commons continue to be accessed—as was done traditionally—for livelihood and subsistence use, and for religious and cultural purposes. In the past there were also different arrangements for management by the communities. Some aspects of informal management persist even today. Ownership was also varied—commons could be the property of the village, State or private individuals, as was the case with wooded groves (Mundoli et al. 2017a). Today these commons are primarily owned by the State. There are exceptions; for example lakes leased to private entities for a specified period. Additionally the conversion in land use of commons such as ponds, grazing lands and wooded groves have today made them private or public goods. There is also a wide range of actors who access, interact with and manage these urban commons—including long-term residents and recent migrants from different economic strata, members of NGOs and civil society groups, and officials of local, State and central administration.

In our research we have conceptualized urban commons as closely linked social-ecological systems (SES) associated with a range of actors and institutions for management (Nagendra 2016). Our normative assumption is that the ecosystems mentioned above need to be recognized as urban commons. Moreover, we assume the positive potential for commons as a governance regime for better management of natural resources (De Moor 2012; Weston and Bollier 2013), especially in cities of the Global South where urbanization poses a double challenge to environmental sustainability and social equity. We believe that the collective management of urban ecosystems as commons, with equitable rights to access, withdrawal and management, will advance progress towards an environmentally sustainable and socially just urban future. Thus, our research argues for a reimagining of cities, to go beyond the traditional approach of using hierarchical master plans developed using the "best science" alone and envisage planning as a collaborative and negotiated approach by urban commoners working alongside planners.

We have used multiple analytical frameworks to understand the transformation of the urban commons of Bengaluru, and the consequent impact on ecological sustainability, and human well-being. SES frameworks unpack the different components of the urban commons, providing a clearer picture of the system in its totality and the interactions between different components of the system over different temporal and spatial scales. Ostrom's (2007, 2009; McGinnis and Ostrom 2014) SES framework urges researchers to go "beyond panaceas," to embrace complexity and develop solutions that are socially and ecologically context specific. The structure of the SES, in this case the urban commons, needs unpacking by defining the attributes of the resource system, the resource units that it generates, the users of that system, and the governance system.

The interactions and outcomes between these attributes are nested in larger social, political, and economic settings and ecological systems, and the SES as an interdisciplinary framework helps to work within this complexity (Nagendra and Ostrom 2014). Scale and time also become critical in analysing and understanding urban commons as multi-tiered urban SES. By acknowledging the complexity of urban commons as an SES, not only is Hardin's theory of the "tragedy of commons" challenged as an impoverished understanding of the commons, but the framework urges for working *with* complexity to address concerns of environmental sustainability (Ostrom 2007, 2009). The SES framework helps advance this goal by integrating social and ecological components and combining the advantages of both case-based and rule-based reasoning to build knowledge (Epstein et al. 2013).

We also expand on the property rights system using the framework proposed by Schlager and Ostrom (1992). Here, bundles of rights are associated with different users or actors. These property right bundles are the right to access and withdrawal, to manage, to exclude, and to alienate. Schlager and Ostrom (1992) also draw distinctions between formal rights provided by the State and informal rights used in practice. In the management of urban commons, a mix of both formal and informal property rights are usually present. Testing the property rights framework in a range of empirical situations can advance our understanding of messy institutional arrangements for urban commons management, though this has not been applied to urban contexts so far.

Further, long standing commons have histories: an ecological history and a history of human-resource interaction, that also includes an understanding of how rules for use of a resource have been adapted to changing conditions over time (Hess 2008). Although urban commons may be relatively new in rapidly urbanizing countries like India, commons in cities are not without their histories. Understanding the nature of contemporary urban commons requires consideration of "legacy effects" that have contributed to determining the urban landscape of present times (Larondelle and Strohbach 2016). Nature is an integral part of human history (Cronon 1992, 1993). Urban commons as ecological spaces are shaped by and in turn shape the lives of urban residents. Residents have built relationships with urban commons that extend across generations, both at a personal level and as part of a community. Interactions with commons for livelihood, subsistence, and for social and cultural reasons have not only shaped these ecological spaces but have contributed to shaping the life of the community as well (Nagendra 2016). Thus, the discipline of environmental history becomes important in the study of contemporary commons that have been in existence, often for centuries.

The environmental history of commons, and even more so of urban commons, is still an emerging area of study. An important gap in our historical understanding of commons is to know how institutions, norms and rules, vital for managing and preserving the commons, have emerged and changed over time (De Moor 2011, 2012; De Moor et al. 2016; Laborda-Pemán and De Moor 2016). The role of path dependency in enabling or constraining collective action becomes fundamental in this regard (Heinmiller 2009). Tracing environmental history aids not just in attempting to understand the past but also helping take decisions for the future. And while it may not provide solutions to problems, environmental history is important for the kind of questions it asks.

Further, history helps in effectively communicating the complex transformations to urban commons across a variety of audiences: from academics and policy makers to urban residents. These narratives of environmental change need to be seen as "our chief moral compass for the world" (Cronon 1992, 1375). Thus, for instance, tracing the history of urban commons can serve as a warning against social injustice witnessed in the past (Unnikrishnan et al. 2016a). Studying the past as well as the contemporary history of urban commons highlights the potential of commons as a governance regime for a wide range of circumstances over centuries (Weston and Bollier 2013). Tracing the environmental history of urban commons is also important to provide a continuity of narrative about change and its present day implications, especially in the analysis of institutions (Laborda-Pemán and De Moor 2016). This is essential, as a suggested solution at one point in time can become a cause for concern at another point.

Environmental history and human culture are intimately interwoven, and one cannot ignore the latter in tracing the former. Fernandez (1987), in laying out the principles of a new social contract, specifies the importance of communities in reproducing culture for the protection of their commons. Culture is built around local ecologies, and land management practices that connect people to their landscapes. Thus cultural relations with urban commons are shaped by long-term livelihood and subsistence dependencies on commons. These dependencies in turn create spiritual connections and/or a sense of place. However, cultural severance breaks down these relationships, bringing with it ecological, economic and social problems. This severance can be physical, where subsistence and livelihood use is discontinued, or psychological, when humans are distanced from their environment. The loss of commons as a result of conversion, enclosure, privatization or degradation thus results in a loss of sense of place and erosion of spiritual and symbolic connections with the commons (Rotherham 2013; Baindur 2014). Culture's role in the protection of commons is contested—it can protect or destroy commons—but undeniably culture shapes practices and can play an important role in promoting sustainable use of commons (Wall 2014). Redefining cultural connections to urban commons does not necessarily mean recreating the past in its entirety (Rotherham 2013) and will need reconceptualization keeping the modern context of cities and urban populations in mind.

Along with the discipline of history and its close links with culture, integrating urban ecology into urban commons research is vital (Wall 2014). Urban commons are social-ecological systems characterized by extensive and intensive interactions between humans and nature. Trying to understand these complex, linked systems within the disciplinary silos of natural and social sciences is limiting (Alberti et al. 2003). Urban ecology, with its interdisciplinary nature, is able to overcome disciplinary limitations by enquiring into the variety of relationships between ecosystems and society in human-dominated urban landscapes (Marzluff et al. 2008). Preservation of the ecological diversity of the commons is important for the protection of social and cultural dependence: urban ecology helps to understand these links and can contribute to addressing the challenges of strengthening them.

The ecosystem services framework presents another, related frame to examine the benefits from natural resource commons in cities. Ecosystem services include provisioning services such as food, medicinal resources, water and raw materials that support subsistence and livelihoods needs; cultural services that include the aesthetic, recreational and spiritual needs; supporting services that enable biodiversity to flourish and enhance genetic diversity; and regulating services such as water purification, micro-climate regulation, and pollination (MEA 2003). The actors and beneficiaries can vary depending on the services, from local residents who access water from a lake, or fuelwood from a tree, to residents across the city who contribute to the pollution that impacts trees, and benefit from the pollution-combating services provided by trees in urban commons such as city parks. The tangible and intangible ecosystem services derived from urban commons can be either nurtured or compromised based on how we understand and undertake to manage commons as ecosystems.

Context and objectives

Bengaluru is an old city with a history of settlement dating back at least 6000 years, with a long history of dependence on commons (Nagendra 2016). Since the city is situated in a rain-shadow region, with no access to a perennial water source, a series of interconnected man-made tanks and lakes were built over centuries and dedicated to the larger community, by powerful chieftains and commoners alike. Contiguous with the lakes, the landscape has included other common pool resources such as wells, ponds, wooded groves, cemeteries and grazing lands. These urban commons supported a variety of essential provisioning and cultural ecosystem services. Livelihoods such as agriculture, horticulture, grazing, fishing, brick making and washing clothes have been supported by these commons, along with subsistence use such as collecting fuelwood, foraging for leafy vegetables and fruits, and water for domestic use. Local residents have developed close cultural and spiritual ties with their commons via specific place-based rituals and practices (Unnikrishnan et al. 2016a, 2016b; Mundoli et al. 2017b, 2017c). As the city has grown, the urban commons have also been accessed by migrant labour residing in hutments adjacent to these commons who collect firewood for cooking, and (in the absence of alternatives), use these spaces as public toilets. Like other ecological spaces in cities, urban commons have also provided supporting and regulating ecosystem services, including biodiversity support, flood control, micro-climate regulation, and air and water purification (Nagendra 2016).

The growth of Bengaluru has transformed land use and altered the governance of the urban commons. Urban commons have been encroached upon and degraded. They have also been converted to other forms of land use and are no longer commons but private or public goods. These transformations have adversely impacted both the environment and the lives of urban residents (especially already marginalized groups). The periphery of Bengaluru is the location where many commons are still visible, albeit under constant pressure of conversions and degradation. Our research analyses the type and extent of land use change taking place in the urban commons, within the backdrop of larger urban changes experienced by the city. This is accompanied by questions of changes in perceptions of these commons, among long-term local residents, recent migrants, and administrators in charge of formal management. We also document existing and past dependence (both material and cultural) on commons by different users over time. Keeping in mind the socio-economic inequalities and disparities that increasingly characterize urban India today, we also seek to illustrate a growing trend towards the prioritization of commons for recreational and aesthetic purposes. This trend limits the capacity of commons to provide provisioning ecosystem services (such as food, fodder and fuelwood) that have supported local livelihoods and subsistence. The resulting alienations and exclusions, in contemporary and historical times, provide glimpses into what the city and its residents can expect in future if urban natural resource spaces are not valued as commons.

Governing the commons is also a challenge in the context of urbanization. Different perceptions held by stakeholders (users, residents, administrators and builders, for instance) manifest as conflicts, undermining efforts to protect these commons. India is at the cusp of urbanization, and the coming years will see more than 50 percent of its population residing in cities. The State's vision for accommodating this shift has been to promote new models for cities such as "smart cities," with an increasing trend towards privatization of commons via the favoring of public-private partnerships (Unnikrishnan and Nagendra 2015; Mundoli et al. 2017c). In this vision, the assessment of the implications on the environment and socio-economic vulnerability is inadequate. Therefore, a focus of our research is to critically examine these impacts through the transformations to commons in the city of Bengaluru.

Diverse methods of enquiry

We adopt a mixed methods approach, keeping in mind the complex nature of urban commons, which also allows for triangulation of our findings. We collected primary data by conducting individual interviews and group discussions using detailed questionnaires. Local residents, commons users, NGOs, elected leaders and administrators/government officials were sources of information. Questions covered the past and present uses and extraction from the commons, conflicts around the commons, perceptions of changes in the resource and users' relationship with the resource over time. Exclusions (for example, limiting access by way of fencing, specifying timings of entry, or charging user fees) were also a focus of our research. We sought to understand the interactions between users, residents and government officials and their perceptions about how the commons should be governed. Apart from questionnaires, open ended discussions and oral histories provided a rich narrative of change that helped in uncovering the sense of identity and place among the local residents. Additionally, we used transect walks in some instances on specific days and times in order to observe activities around the commons, classifying these activities into provisioning, cultural, regulating or supporting ecosystem services. We also recorded biodiversity data of species found and used in the commons. For example, plant species recorded on site, and observations on current use, were contrasted with oral histories of the past.

Understanding land use change was an important component in our analysis of change. We used mapping and spatial analysis tools (GIS and Remote Sensing) to determine boundaries of commons, built up features such as roads and settlements, vegetation, and waterbodies. Maps of Bengaluru's landscape from different points in time—1791, 1884–85, 1935–36, and the current date (Nagendra 2016)—were useful in tracing historical change. These maps were overlaid and registered, and land use around these commons was digitized using ArcGIS. We also used village revenue maps of survey numbers acquired from local administrative offices to identify locations of commons, along with lists relating survey numbers to specific commons. Field visits with a GPS were used to record the current location and condition of commons. We recorded encroachments, conversions, and activities causing degradation such as pollution and dumping of urban waste.

We used comparative and gradient analysis to enable comparison of contemporary challenges. The study sites fell on a gradient from the city core to the periphery. This enabled an analysis of how aspects such as types of land use and institutions of local government, along this spectrum from urban core to periphery, influence the preservation or degradation of commons (Mundoli et al. 2015, 2017a). In comparative analyses, lakes governed under public-private partnerships were compared with city-managed "public" lakes to compare commons uses and to determine what type of management aided preservation of a wider range of ecosystem services (Unnikrishnan and Nagendra 2015).

Archival material provided an important source of data to examine the discourses around historical transformations of commons. Public archives within and outside India as well as libraries of institutions were the sources of our data. These archival materials included administrative records from the late 18th century, gazetteers, travelogues, paintings, maps, photographs and travel notes (Nagendra 2016). The documents provided insights into the perspectives of the government regarding the utility of commons, conflicts in some instances and the management decisions taken with regard to commons.

Using archival data has helped us in the research, but importantly it has enabled us to narrate the ecological story of change and challenges faced by the city to a wider non-academic audience. This research method helps to develop a narrative of change (for example to understand how a lake commons was converted into a polo ground), and additionally helps to build an idea of the social world of those times. This is done by asking questions about who wrote archival documents, whom these documents were intended for, and the stakeholders mentioned; but more importantly, *those not mentioned as well* (Lerner 2010), bringing into question the distribution of power and its implications. In incorporating archival data as a research method we have tried to find a balance by presenting the information as a mix of chronicle (reporting a sequence of events) and a narrative (developing a story) (Cronon 1992). We focus on a range of actors (citizens, planners, and others), and of users (largely local residents), within the constraints of the archival and oral history material available.

Using mixed methods also helped us to account for the limitations of using a single method. Thus, GIS/RS tools that enable a broad categorisation of land use were coupled with collection of ground truth data. Archival material provides mainly the views of the literate, most often State views, and could result in a skewed representation. Maps of past land use are likely to represent the political orientations of that period. To address this, we incorporated oral interviews at study sites, focusing on vulnerable communities who were on the losing side of conflicts around the commons. By adopting mixed methods, we attempted to triangulate our findings to get a multi-dimensional perspective of the complexity of issues around the use and management of the urban commons over time.

Research studies

In this section, we describe three research studies, showing how we have used the frameworks and methods described above for a better understanding of the complex transformations of the urban commons of Bengaluru.

Of lakes lost: the transformation of Sampangi lake to Sri Kanteerava Sports Stadium

Located in the bustling core of the megapolis of Bengaluru is a small water body, the Sampangi Tank. Hemmed in by roads, residences, offices and an indoor stadium, this tiny tank today represents the only remnant of what was once a large lake covering several acres. Our research examined the transformation of this lake to a tank, tracing the origins of the change from as far back as 1885 (Unnikrishnan et al. 2016a) using multiple frameworks and methods (see Table 18.1).

During the 1880s, conflicts around the lake emerge in the archives. Maps and British administrative records show that the lake formed a transitional area governed by the colonial British Government as well as the native Mysore monarchy. The lake had supported a variety of provisioning ecosystem services, including agriculture and horticulture. Oral histories tell us that paddy and finger millet (*Eleusine coracana*) were cultivated in the adjacent wetland and surroundings. Flowers and vegetables were grown by *Vannhikula Kshatriyas*, members of a horticultural community who had migrated to Bengaluru centuries before. The lake provided a number of ecosystem services for local subsistence and livelihoods. Water from the lake, as well as the surrounding wells recharged by the lake, supplied the needs of the *Vannhikula Kshatriyas* and other long-term residents such as the *Vokkaligas* (agriculturists), *Vaishyas* (merchants) and *Devangas* (weavers). Water from the lake also supplied the British Cantonment. Fish were found in abundance in the lake, the mud from the lake bed was used in brick making and the area around was used as a site for grazing cattle. Thus prior to the 1900s, even as land use changes such as roads, a brewery, tennis courts and other built-up areas were present, and some restrictions on drawing water and making bricks were being introduced, the lake continued to be an important source of essential provisioning services such as food, fodder and water. It was also an important cultural site for the *Karaga*, a festival celebrated by the *Vannhikula Kshatriya* community, centred around the Sampangi Lake and other water bodies in the vicinity.

Between 1900 and 1935, maps demonstrate drastic changes in land use around the lake. Increased urban construction included schools, churches, hospitals and even a golf course. In order to prevent the flooding of colonial properties (including a brewery), and to create a polo ground for the use of a British regiment, the lake was drained in this period. Conflicts with regard to the use of lake began to emerge. There were restrictions imposed by the British administration on cattle grazing and sinking of new wells. Archival correspondence includes a petition submitted by horticulturists against the draining of the lake, concerned that their crops would be affected. However, their appeals seemed to have been disregarded as indicated by the events between 1935 and 1973. The lake was reduced to a tiny rectangular tank during this time period. While the *Vannhikula Kshatriyas* recall that paddy, finger millet, areca (*Areca catechu*) and coconut (*Cocos nucifera*) were cultivated until the 1940s, the lake was increasingly being degraded. Mention of the lake as a malodorous and malarial swamp was followed by the use of the lake bed for carnivals, cattle fairs, and eventually its conversion to a stadium. The water in the lake became polluted and unfit for human consumption, though adjacent slums still depended on the lake for domestic use. The lake bed and surrounding area was also used for open defecation.

The time period between 1974 and to 2014 was marked by widespread changes in land use. There were more roads with heavy vehicular traffic, an increase in offices and residential pockets, and a large indoor stadium was built. Along with this there were changes in the social demography. Older residents moved out or shifted to other livelihoods, while new migrants into the city made their home in the neighbourhood. Now, only the tiny rectangular tank remains as a memory of the once wetland-lake landscape in the minds of older residents, gaining prominence once a year during the *Karaga* festival.

We applied the Schlager-Ostrom framework to examine changes in property rights and the associated impacts on social vulnerability. The lake changed hands in terms of formal management, from the British Government and Mysore rulers in the 19th and early 20th century, to local bodies charged with administration. The users of the lake, including agriculturists, horticulturists, grazers, fishers, and brick makers, had different property rights, related to the distribution of power. While all four rights of access, appropriation, alienation and management were vested with the powerful British and Mysore rulers, subaltern groups such as agriculturists and fishers who depended on the lake were only given permissions to access and withdraw: permissions that were in constant danger of being revoked as the priorities of those in power changed. Shifts in ecosystem services were also perceptible: from its early beginnings. The lake, prized for its supporting, provisioning and spiritual ecosystem services, was prioritized for recreational use during the colonial period. Beginning with the draining of the lake for a polo ground in the late 19th century, the process of transformation to a recreational space was completed with the draining of the lake and construction of an indoor stadium post-Independence. Provisioning ecosystem services were thus completely lost, and the social demography of the neighbourhood altered as those who accessed these services left the area or shifted to other livelihoods (Unnikrishnan et al. 2016a).

The topography and ecology of Sampangi Lake and its surrounding area has thus been irrevocably altered to the detriment of the city's environment. During periods of heavy rain, the stadium built on the lake bed is submerged, necessitating huge expenditure to pump out water and rebuild infrastructure. The lessons from this narrative of change of Sampangi Lake serve as a warning against the relentless focus on "development" as an imperative, which has led to widespread conversions of natural common pool resources such as Sampangi Lake across the city. The city has become more vulnerable in terms of its ecology and hydrology, as well as in the lives of already disadvantaged and marginalized urban residents.

Of lakes restored: the importance of polycentricity

The pressures of urbanization present a continuous challenge to the status and governance of lakes in Bengaluru. In a collaboration with Elinor Ostrom, Harini Nagendra used the SES framework to understand the importance of different social and ecological variables in determining collective action and ecological restoration of lakes in Bengaluru (Nagendra and Ostrom 2014). This research also incorporated multiple methods (Table 18.1), and was focused on seven lakes in the peripheral areas of Bengaluru city that were in different conditions. Some were restored and being managed by local residents in collaboration with the administration, others were polluted, dried up, or encroached.

We applied the SES framework to determine the factors that contributed to successful collective action. The lakes not only had divergent ecological attributes but also presented a range of social attributes including differences in the number of actors, level of trust and types of community rules. A defined subset of ten variables included seven second-tier variables, and three third-tier variables, and assessed for their potential as barriers or facilitators of collective action. Overall, collective action was low in only one lake, clearly challenging the notion that urban residents will not come together to resolve problems of urban commons. However, there were differences in the success of collective action. The more successful lakes were of small size, had lower levels of pollution, with greater collaboration between residents and government officials.

The results of this study clearly illustrated the usefulness of the SES framework in identifying the different combinations of variables that contribute to effective management of these urban commons. The research also highlighted the need for collaboration between citizen groups and the government to tackle the challenges of restorations of urban commons. The SES framework demonstrated the importance of polycentric arrangements for governance of the urban commons, such that local residents can organize based on their local presence, knowledge and on-ground monitoring capacity, while linking with government administrators to address technical and policy challenges that require infrastructure, high expense, and legal mechanisms of enforcement.

Of the culture and nurture of urban commons

Apart from lakes, the mosaic of urban commons in Bengaluru includes cemeteries, ponds, remnant grazing lands and wooded groves (locally known as *gunda thopes*). These wooded groves have formed a part of the landscape of Bengaluru since historical times. The groves were planted with native fruiting trees: mango (*Mangifera indica*), jamun (*Syzygium cumini*), Mowra butter tree (*Madhuca longifolia*), jackfruit (*Artocarpus heterophyllus*), tamarind (*Tamarindus indica*), and different species of *Ficus*, mainly cluster fig (*Ficus racemosa*), Mysore fig (*Ficus mysorensis*), banyan (*Ficus benghalensis*) and sacred fig (*Ficus religiosa*), to meet local needs for fruit and wood. While these groves seem to have been important local commons, there is little knowledge about their past use, how the use has changed and what the impacts of the transformations were for different urban groups.

Our research on these groves, conducted along a rural-urban gradient of 18 peri-urban settlements, covering 43 groves in the periphery of Bengaluru (Mundoli et al. 2017a) focused on understanding the contribution of these groves (as social-ecological commons) to the resilience of the city and its residents. The various methods used in this research (Table 18.1), include archival work, oral histories, and socio-economic and livelihood interviews. We also recorded tree species and numbers to document the biodiversity in the groves. During the field visits, we marked the location of the groves using GPS points and checked these against village maps and Google Earth to ensure accuracy of sites visited. The ecosystem services framework provided a useful conceptual lens to categorize the multiple benefits derived from these groves, and the changed use from the past.

Oral histories and archival documents indicated groves were accessed extensively for provisioning and cultural ecosystem services. The provisioning services supported by these groves such as a supply of wood, fodder, fruit and fuelwood were closely linked to the cultural life of the local communities, which formed an important core for the motivation of collective action for commons management. Temples formed the centre of the socio-cultural life of the communities in the past, and wood from these groves, especially of the mango tree, was often used to build or renovate temples. Individuals used wood for household purposes such as for fuel and construction, and for ceremonies such as marriages and cremations. This was done especially by marginalized landless residents who lacked access to their own trees. The entire community used the groves as a meeting site during festivals; they accessed fuelwood from the groves in preparing feasts. Some of the groves also had small shrines for regular prayer, while families worshipped their deities by consecrating stones at the foot of trees.

Groves provided fodder and shelter to grazers. In addition, nomadic communities of astrologers and fortune tellers, whose services were used by local residents, took temporary residence in these groves. Adults and children ate fruit from the mango trees, and mango leaves decorated homes and temples during festivals. The residents protected the fruit bearing trees of mango, jamun and tamarind. In one studied site, arrangements for sharing the fruit indicate a sense of community. In this site, during the fruiting season, local residents would appoint a watchman to protect the mango trees. Households in the community shared the fruit harvested. It was evident from our interviews that these groves were an important urban commons that not only supported livelihood and subsistence but were central to the cultural lives of the residents.

However, the spatial and demographic expansion of Bengaluru has placed immense pressure on these groves, resulting in a decline in ecosystem services they provided. Of the 43 groves, 20 have been converted to other forms of land use such as schools, government offices, bus stops and so on, while 18 were in a degraded state. These conversions and degradations have meant a loss of provisioning services that has had adverse impacts on the urban poor who continue to depend on commons for livelihoods and subsistence. Perceptions about these groves are also changing among the local residents. From being a collectively managed commons, residents perceive the groves today as State-owned, resulting in a reduced interest in their protection. As a result, their attitudes indicate a preference for prioritising the groves for recreational use. In one of the sites, local residents had come together to convert a grove to a park with a play area for children and paths for elders to walk. Administrative decisions impacted access to the groves, further distancing local residents. These included restrictions on accessing the wood and even the groves itself by way of fencing. The groves are today fading from the memory of residents, both elderly and young. This was evident during the course of our field work: we were unable to trace the location of some groves marked in government records, as residents had forgotten that they even existed.

The importance of extra-economic values such as culture in building urban resilience is under-researched (Gómez-Baggethun and Barton 2013). Our research indicates that wooded groves were important socio-cultural nodes, whose linked cultural and provisioning ecosystem services contributed to building resilience of local residents. The conversion, degradation and encroachments of groves, the associated loss of multiple ecosystem services, and the consequent disconnect from the commons has had adverse impacts on the social and ecological resilience of the city and its residents (Gómez-Baggethun and Barton 2013). Protecting ecosystem services thus becomes imperative for the continued preservation of the commons. Leveraging the social and cultural memories that people have in order to build resilience is crucial in this regard. Cultural ecosystem services that extend beyond the purely recreational can play a critical role in reviving eroding links to commons. These services can also address issues of conflict in the urban landscape by promoting social cohesion (Andersson et al. 2015; Plieninger et al. 2015).

Table 18.1 Frameworks, disciplines and methods used in our research

Paper	*Frameworks*	*Disciplines*	*Methods*
Unnikrishnan et al. 2016a	Schlager-Ostrom, Ecosystem services	Environmental history Urban ecology Cultural geography	GIS/RS Archival Oral histories
Nagendra and Ostrom 2014	SES	Urban ecology	GIS/RS Field interviews Observations on biodiversity and land use
Mundoli et al. 2017a	Ecosystem services	Environmental history Urban ecology Cultural geography	Field interviews Observations on biodiversity and land use

Conclusion: a commons ethic as the norm for governance

The urbanization challenges witnessed by Bengaluru are not unique. Across India and in most cities of the Global South, urbanization has led to challenges both for environmental sustainability and socio-economic equity (UN-Habitat 2016). But what needs to be acknowledged is that each city is unique in its ecological base, as well as the place-based interactions developed between urban ecosystems and city residents extending back hundreds of years (Nagendra 2016). To envision an environmentally sustainable urban future, we need to understand the lessons from the past, and to explicitly outline the normative positions we adopt to guide management of resources in the urban future.

Adopting a commons outlook in the management of natural resources can help address the joint urban challenges of environmental sustainability and social equity (Weston and Bollier 2013). But while there is considerable scholarly work based on empirical research on the collective management of resources, not much of this is in the urban context. Even today, ecological resources in the urban context are predominantly viewed as public goods, owned and managed by the State, rather than as commons (Garnett 2012). Urban residents are also often unaware that an urban space such as an ecosystem can even be a commons (Kip 2015). Thus, more research on the multiple dimensions of urban commons is necessary to argue and present a case for the idea of managing resources as commons in cities (Foster 2011; Garnett 2012). For this, first and foremost, we need to acknowledge that human dominated ecosystems, including commons in cities, emerge necessarily from local, community and place based interactions between nature and society (Alberti et al. 2003).

Interdisciplinary frameworks that can bridge natural and social sciences become imperative to understand the interactions between humans and ecological systems across different locations. The SES framework is a diagnostic tool that aids in unpacking the components of closely inter-linked social and ecological systems, allows for comparison and enables communication of research across groups (McGinnis and Ostrom 2014). Similarly, the Schlager-Ostrom framework with its emphasis on nested attributes too is a widely applied framework for understanding complex systems, especially in the context of institutional changes (Galik and Jagger 2015). However, while both frameworks have been applied in the context of fisheries, forests and rural water management scenarios (Epstein et al. 2013; Galik and Jagger 2015), our research on the commons of Bengaluru provides perhaps the first applications of these frameworks in an urban context.

The range of ecosystem services provided by urban commons contributes to enhancing the resilience of cities (Gómez-Baggethon and Barton 2013; Hasse et al. 2014; McPhearson et al. 2014). The ecosystem services framework, despite its limitations, has considerable scope for aiding urban policy and planning as it presents a more complete picture of an ecological space in cities, and should be incorporated into decision-making (Fish 2011). However, empirical studies on urban ecosystem services are themselves limited especially in countries of the Global South. Among different categories of ecosystem services, moreover, provisioning services (which assume especial significance in the context of the commons, as these are extractive services providing food and fodder which can be subtractive, and therefore act as defining characteristics of common pool resources) are the least studied (Hasse et al. 2014). In our research we have focused on provisioning services in the context of Bengaluru that are important for building resilience of the poorest and most marginalized of urban residents.

In addition to provisioning ecosystem services, we need to value cultural ecosystem services as cultural interactions of urban residents with the commons play a major role in developing a strong sense of place and a motivation for collective action. However, research on cultural ecosystem services has been limited. In terms of planning, decision-makers rarely pay attention to cultural ecosystem services. There is a tendency to prioritize only certain types of cultural services such as recreation, and to marginalize other, more nebulous and difficult to define (but equally important) attributes that people might tend to place greater value on, such as a sense of place (Daniel et al. 2012; Milcu et al. 2013; Hasse et al. 2014). Our research in Bengaluru is aimed at identifying the cultural ecosystem services of urban commons, going beyond the purely recreational (Mundoli et al. 2017c).

Long-term analysis is imperative if one is to understand what can happen to urban commons if they are not governed properly. Historical analysis can reveal potentially useful institutional arrangements, or even arrangements to avoid in governing commons (Laborda-Pemán and De Moor 2016). Path dependency shapes collective action for the future, and it is therefore important to look back at past institutions and perceptions about commons (Heinmiller 2009). For instance, our research on the historical transformation of Sampangi Lake (Unnikrishnan et al. 2016a) is aimed to caution planners against perceiving commons as merely sites for recreation as was done in colonial times; a trend being replicated today, to the detriment of the city and its residents. There is a need for similar research across different cultures and geographies that can contribute to the creation of an "institutional toolbox" to guide what kinds of governance arrangements could work in different contexts (De Moor 2011). Historical research and commons studies can benefit tremendously from exchanges, especially in the long-term analysis of institutional changes to commons (Laborda-Pemán and De Moor 2016)—learnings which are useful for future decision-making.

As Helfrich (2012) says "Commons do not simply exist; they are created", and it is their collective care and use that makes as well as retains the essence of what a commons is. It is often when a public good is threatened in the city and when people come together to protect it that it becomes a commons (Kratzwald 2015). Urban commons can even provide a way of reimagining the management of not just natural resource commons in cities as described in this chapter, but also of other types of commons, such as built infrastructure including streets, squares and so on (De Moor 2011: 431). While we need to adopt a commons outlook in management, this is not to say that the role of the State is irrelevant, or that it necessarily has to be seen in parallel with that of the market. The State has an important, even essential, role to play in enabling collective management and in minimising the risks of decentralized management (Ostrom 1999; Foster 2011; Weston and Bollier 2013). The State needs to also function as an implementer of what the public wants, by taking citizen partnership seriously and providing recognition to positive actions of citizens (Garnett 2012; Kratzwald 2015). This enabling role of the State should, as Foster puts it (2011: 92–93) lie on a spectrum from a *de minimis* government role to one where the presence of the government is necessary for management of a resource. Citizens need to participate in building policies and plans for governing commons from the ground, and not leave it only to far off decision-makers of the State (Wall 2014). Ostrom (1999) argues for a polycentric governance system for managing natural resources such as urban commons. Such a governance system is nested to incorporate different actors and is also adaptive.

These are ecologically, economically and socially challenging times that pose a threat to sustainability of cities and well-being of urban residents (UN-Habitat 2016). Changing the way we manage our resources becomes imperative if we need to address concerns for environment and equity in access to resources. Adopting a new paradigm of governance, i.e. a commons-based governance that empowers communities in managing ecological resources, is imperative. A commons approach is a rights-based approach that ensures participation of vulnerable populations in decision-making where natural resource management is concerned. At the same time it can also help reinvent peoples' relationships with nature and contribute to the protection of ecosystems (Weston and Bollier 2013). Managing space as commons in cities thus not only benefits the environment and urban residents but is normatively indicative of the kind of society that we would like to build: one that values the principles of collaboration, sustainability and equity.

Acknowledgements

Harini Nagendra would like to acknowledge a deep intellectual debt to Elinor Ostrom. The early ideas of study of the urban commons of Bengaluru were developed during the course of weekly Tuesday conversations with Lin between 2009 and 2012; she is sorely missed. The authors gratefully acknowledge funding from a USAID PEER Grant to Harini Nagendra at the Ashoka Trust for Research in Ecology and the Environment (ATREE), and research funding from the Centre for Urban Ecological Sustainability, Azim Premji University. Hita Unnikrishnan also acknowledges the Newton International Fellowship for their support. We also thank B. Manjunatha for his invaluable assistance with field and archival research.

References

Alberti, M., Marzluff, J.H., Schulenberger, E., Bradley, G., Ryan, C. and Zumbrunnen, C. (2003) "Integrating humans into ecology: Opportunities and challenges for studying urban ecosystems," *BioScience*, 53(12): 1169–1179.

Andersson, E., Tengö, M., McPhearson, T. and Kremer, P. (2015) "Cultural ecosystem services as a gateway for improving urban sustainability," *Ecosystem Services*, 12: 165–168.

Baindur, M. (2014) "Bangalore lake story: Reflections on the spirit of a place," *Journal of Cultural Geography*, 31(1): 32–56.

Cronon, W. (1992) "A place for stories: Nature, history and narrative," *The Journal of American History*, 78(4): 1347–1376.

Cronon, W. (1993) "The uses of environmental history," *Environmental History Review*, 17(3): 1–22.

Daniel, C.T., Muhar, A., Arneberger, A., Aznar, O., Boyde, J.W., Chan, K.M.A., Constanza, R., Elmqvist, T., Flint, C.G., Gobster, P.H., Grêt-Regameyj, A., Lave, R., Muhan, S., Penker, M., Ribe, R.G., Schauppenlehner, T., Sikor, T., Soloviy, I., Spierenburg, M., Taczanowska, K., Tam, J. and von der Dunk, A. (2012) "Contribution of cultural services to the ecosystem services agenda." *PNAS*, 109(23): 8812–8819.

DeFries, R. and Nagendra, H. (2017). "Ecosystem management as a wicked problem," *Science*, 356: 265–270.

De Moor, T. (2011) "From common pastures to global commons: A historical perspective on interdisciplinary approaches to commons," *Natures Sciences Sociétés*, 4(19): 422–431.

De Moor, T. (2012) "What do we have in common? A comparative framework for the old and new literature on the commons," *IRISH*, 57: 269–290.

De Moor, T., Laborda-Peman, M., van Weeren, R. and Winchester, A. (2016) "Ruling the Commons. Introducing a new methodology for the analysis of historical commons," *International Journal of the Commons*, 10(2): 529–588.

Elmqvist, T., Fragkias, M., Goodness, J., Güneralp, B., Marcotullio, P.J., McDonald, R.I., Parnell, S., Schewenius, M., Sendstad, M., Seto, K.C. and Wilkinson, C. (2013) *Urbanization, biodiversity and ecosystem services: Challenges and opportunities*. Dordrecht, the Netherlands; Heidelberg, Germany; New York; London: Springer.

Epstein, G., Vogt, J.M., Mincey, S.K., Cox, M. and Fischer, B. (2013) "Missing ecology: Integrating ecological perspectives with the social-ecological system framework," *International Journal of the Commons*, 7(2): 432–453.

Fernandez, J.W. (1987) "The call to the commons: Decline and recommitment in Austrias, Spain," in McCay, J.B. and Acheson, J.M. eds, *The question of the commons: The culture and ecology of communal resources*. Tucson, AZ: The University of Arizona Press, 266–289.

Fish, R.D. (2011) "Environmental decision making and an ecosystems approach: Some challenges from the perspective of social science," *Progress in Physical Geography*, 35(5): 671–680.

Foster, S.R. (2011) "Collective action and the urban commons," *Notre Dame Law Review*, 87(1): 57–134.

Galik, C.S. and Jagger, P. (2015) "Bundles, duties and rights: A revised framework for analysis of natural resource property rights regimes," *Land Economics*, 91(1): 76–90.

Garnett, N.S. (2012) "Managing the urban commons," *University of Pennsylvania Law Review*, 160: 1995–2027.

Gómez-Baggethun, E. and Barton, D.N.(2013) "Classifying and valuing ecosystem services for urban planning," *Ecological Economics*, 86: 235–245.

Hasse, D., Larondelle, N., Andersson, E., Artmann, M., Borgström, S., Breuste, J., Gomez-Baggethun, E., Gren, A., Hamstead, Z., Hansen, R, Kabisch, N., Kremer, P., Langemeyer, J., Rall, E.L., McPhearson, T., Pauleit, S., Quershi, S., Schwarz, N., Voight, A., Wurster, D. and Elmqvist, D. (2014) "A quantitative review of urban ecosystem service assessments: Concepts, models and implementation," *Ambio*, 43: 413–433.

Heinmiller, T.B. (2009) "Path dependency and collective action in common pool governance," *International Journal of the Commons*, 3(1): 131–147.

Helfrich, S. (2012) "Common goods don't simply exist—they are created," in Bollier, B. and Helfrich, S. eds, *The wealth of the commons: A world beyond market and state*. Amherst, MA: The Commons Strategies Group Levellers Press.

Hess, C. (2008) "Mapping the new commons," Presented at The Twelfth Biennial Conference of the International Association for the Study of the Commons, Cheltenham, UK, 14–18 July 2008.

Jowett, B. (1885). *The politics of Aristotle (translated into English with introduction, marginal analysis, essays, notes and indices*. Oxford, UK: Clarendon Press.

Kip, M. (2015) "Moving beyond the city: Conceptualising urban commons from a critical urban studies perspective," in Dellenbaugh, M., Kip, M., Bieniok, M., Müeller, A.K. and Schwegmann, M. eds, *Urban commons: Moving beyond state and market*. Basel, Switzerland: Birkhaüser Verlag GmbH, 42–59.

Kratzwald, B. (2015) "Urban commons: Dissident practices in emancipatory spaces," in Dellenbaugh, M., Kip, M., Bieniok, M., Müeller, A.K. and Schwegmann, M. eds, *Urban commons: Moving beyond state and market*. Basel, Switzerland: Birkhaüser Verlag GmbH, 26–41.

Laborda-Pemán, M. and De Moor, T. (2016) "History and the commons: A necessary conversation," *International Journal of the Commons*, 10(2): 517–528.

Larondelle, N. and Strohbach, M.W. (2016) "A murmur in the trees to note: Urban legacy effects on fruit trees in Berlin, Germany," *Urban Forestry and Urban Greening*, 17: 11–15.

Lerner, N. (2010) "Archival research as a social process," in Ramsay, A.E., Sharer, W.B., L'Epplattenier, B. and Mastrangelo, L.S. eds, *Working in the archives: Practical research methods for rhetoric and composition*. Carbondale, IL: Southern Illinois University Press, 195–205.

Linebaugh, B. (2014) *Stop, thief! The commons, enclosures and resistance*. Oakland, CA: PM Press.

Marzluff, J., Shulenberger, E., Endlicher, W., Alberti, M., Bradley, G., Ryan, C., ZumBrennen, C. and Simon, U. (2008) *Urban ecology: An international perspective on the interaction between humans and nature*. New York: Springer.

McGinnis, M.D. and Ostrom, E. (2014) "Social-ecological system framework: Initial changes and continuing challenges," *Ecology and Society*, 19(2): 30–41.

McPhearson, T., Andersson, E., Elmqvist, T. and Frantzeskaki, N. (2014) "Resilience of and through urban ecosystem services," *Ecosystem Services*, 12: 152–156.

MEA (2003) *Ecosystems and wellbeing: A framework for assessment*. London; Washington, DC; Covelo, CA: Island Press.

Milcu, A.I., Hanspach, J., Abson, D. and Fischer, J. (2013) "Cultural ecosystem services: A literature review and prospects for future research," *Ecology and Society*, 18(3): article 44.

Mundoli, S., Manjunatha, B. and Nagendra, H. (2015) "Effects of urbanisation in the use of lakes as commons in the peri-urban interface of Bengaluru," *India International Journal of Urban Sustainable Development*, 7(1): 89–108.
Mundoli, S., Manjunatha, B. and Nagendra, H. (2017a). "Commons that provide: The importance of Bengaluru's wooded groves for urban resilience," *International Journal of Urban Sustainable Development*, 9(2): 184–206.
Mundoli, S., Manjunatha, B. and Nagendra, H. (2017b) "Nurturing urban commons for sustainable urbanization," *IIC Quarterly*, 43(3&4): 258–270.
Mundoli, S., Unnikrishnan, H. and Nagendra, H. (2017c) "The "sustainable" in smart cities: Ignoring the importance of urban ecosystems," *Decision (Journal of the Indian Institute of Management—Calcutta)*, 44(3): 103–120.
Nagendra, H. (2016) *Nature in the city: Bengaluru in the past, present and future*. New Delhi: Oxford University Press.
Nagendra, H. and Ostrom, E. (2014) "Applying the social-ecological system framework to the diagnosis of urban lake commons in Bangalore, India," *Ecology and Society*, 19(2): 67. http://dx.doi.org/10.5751/ES-06582-190267.
Ostrom, E. (1999) "Polycentricity, complexity, and the commons," *The Good Society*, 9(2): 37–41.
Ostrom, E. (2007) "A diagnostic approach for going beyond panaceas," *PNAS*, 104(39): 15181–15187.
Ostrom, E. (2009) "A general framework for analysing sustainability of social-ecological systems," *Science*, 325: 419–421.
Plieninger, T., Bieling, C., Fagerholm, N., Byg, A., Hartel, T., Hurley, P., López-Santiago, C., Nagabhatla, N., Oteros-Rozas, E., Raymond, C.M., van der Horst, D. and Huntsinger, L. (2015) "The role of cultural ecosystem services in landscape management and planning," *Current Opinion in Environmental Sustainability*, 14: 28–33.
Rotherham, I.D. ed. (2013) *Cultural severance and the environment: The ending of traditional and customary practice on commons and landscapes managed in common*. Dordrecht, the Netherlands: Springer.
Schlager, E. and Ostrom, E. (1992) "Property-rights regimes and natural resources: A conceptual analysis," *Land Economics*, 68(3): 249–262.
UN-Habitat (2016). *Urbanization and development: Emerging futures (World Cities Report 2016)*. United Nations Human Settlements Programme (UN-Habitat).
Unnikrishnan, H., Manjunatha, B. and Nagendra, H. (2016a) "Contested urban commons mapping the transition of a lake to a sports stadium in Bangalore," *International Journal of Commons*, 10(1): 265–293.
Unnikrishnan, H., Mundoli, S. and Nagendra, H. (2016b) "Down the drain: The tragedy of the disappearing urban commons of Bengaluru," *SAWAS*, 5(3): 7–11.
Unnikrishnan, H. and Nagendra, H. (2015) "Privatizing the commons: Impacts on ecosystem services in Bangalore's lakes," *Urban Ecosystem*, 18(2): 613–632.
Van Laerhoven, F. and Ostrom, E. (2007) "Traditions and trends in the study of the commons," *International Journal of the Commons*, 1(1): 3–28.
Wall, D. (2014) *The commons in history: Culture, conflict and ecology*. Cambridge, MA: MIT Press.
Weston, H.B. and Bollier, D. (2013) *Green governance: Ecological survival, human rights and the law of the commons*. New York: Cambridge University Press.

19

OSTROM IN THE CITY

Design principles and practices for the urban commons

Sheila R. Foster and Christian Iaione

Introduction

If cities are the places where most of the world's population will be living in the next century, as is predicted, it is not surprising that they have become sites of contestation over use and access to urban land, open space, infrastructure, and culture. The question posed by Saskia Sassen in a recent essay—*Who Owns the City?*—is arguably at the root of these contestations and of social movements that resist the enclosure of cities by economic elites (Sassen 2015). One answer to the question of *who owns the city* is that we all do. In our work we argue that the city is a common good or a "commons"—a shared resource that belongs to all of its inhabitants, and to the public more generally.

We have been writing about the urban commons for the last decade, very much inspired by the work of Jane Jacobs and Elinor Ostrom. The idea of the urban commons captures the ecological view of the city that characterizes Jane Jacobs' classic work, *The Death and Life of Great American Cities* (Foster 2006). It also builds on Elinor Ostrom's finding that common resources are capable of being collectively managed by users in ways that support their needs yet sustains the resource over the long run (Ostrom 1990).

Jacobs analyzed cities as complex, organic systems and observed the activity within them at the neighborhood and street level, much like an ecologist would study natural habitats and the species interacting within them. She emphasized the diversity of land use, of people and neighborhoods, and the interaction among them as important to maintaining the ecological balance of urban life in great cities like New York. Jacobs' critique of the urban renewal slum clearance programs of the 1940s and 1950s in the United States was focused not just on the destruction of physical neighborhoods, but also on the destruction of the "irreplaceable social capital"—the networks of residents who build and strengthen working relationships over time through trust and voluntary cooperation—necessary for "self-governance" of urban neighborhoods (Jacobs 1961). As political scientist Douglas Rae has written, this social capital is the "civic fauna" of urbanism (Rae 2003).

This social capital—the norms and networks of trust and voluntary cooperation—is also at the core of urban "commoning." The term commoning, popularized by historian Peter Linebaugh, captures the relationship between physical resources and the communities that live near them, utilize and depend upon them for essential human needs (Linebaugh 2008). In other words, much of what gives a particular urban resource its value, and normative valence, is the function of the human activity and social network in which the resource is situated. As such, disputes over the destruction or loss of community gardens, of open and green spaces, and of spaces for small scale commercial and artistic activity are really disputes about the right to access and use (or share) urban resources to provide goods necessary for human flourishing (Foster & Iaione 2016).

The urban commons framework thus raises the question to which Elinor Ostrom's groundbreaking work provides an intriguing answer. Ostrom demonstrated that there are options for managing shared, common goods which are neither exclusively public nor private. She found examples all over the world of resource users cooperatively managing a range of natural resources—land, fisheries, and forests—using "rich mixtures of public and private instrumentalities" (Ostrom 1990). Ostrom identified the conditions and "design principles" which increase the likelihood of long-term, collective governance of shared resources (Ostrom 1990). In many of these examples, users work with government agencies and public officials to design, enforce and monitor the rules for using and managing the resource.

Building in part on the insights of Vincent Ostrom, and others, she referred to this kind of decision making as "polycentric" to capture the idea that while the government remains an essential player in facilitating, supporting, and even supplying the necessary tools to govern shared resources, the government is not the monopoly decision maker (Ostrom et al. 1961). Polycentric systems have multiple governing entities or authorities at different scales and each governing unit has a high degree of independence to make norms and rules within its own domain (Ostrom 2010b). Polycentric systems also can unlock what she called "public entrepreneurship"—opening the public sector to innovation in providing, producing, and encouraging the co-production of essential goods and services at the local level. Public entrepreneurship often involves putting heterogenous processes together in complementary and effective ways (Ostrom 2005).

As such, our work has explored whether the commons can be a framework for addressing a host of internal and external resource challenges facing cities, and specifically to rethinking how city space and shared goods are used, who has access to them, and how their resources are allocated and distributed. Recognizing that there are many tangible and intangible urban resources on which differently situated individuals and communities depend to meet a variety of human needs, what might it look like to bring more polycentric tools to govern the city, or parts of the city, as a "commons?" Is it possible to effectively manage shared urban resources without privatizing them or exercising monopolistic public regulatory control over them, especially given that regulators tend to be captured by economic elites? Can the Ostrom design principles be applied to cities to rethink the governance of cities and the management of their resources? We think they cannot be simply adapted to the city context without significant modification.

Cities and many kinds of urban commons are different from natural resources and more traditional commons in important ways. This is why, starting ten years ago, we both began to explore the governance of the urban commons as a separate body of study first investigating individually how different kinds of urban assets such as community gardens, parks, neighborhoods (Foster 2006, 2011) and urban infrastructure such as urban roads (Iaione 2010) could be reconceived as urban commons, and later jointly to conceive the whole city as a commons (Foster & Iaione 2016). We realized that we needed a different approach to bridge urban studies and commons studies and therefore to pose a slightly different set of questions for the governance of the urban commons (Iaione 2015). We also needed to define a different set of design principles for the management of urban commons in the city and the city itself as a commons.

For this reason, we have been surveying and mapping 100+ cities around the world and 200+ examples of urban commons within them (www.commoning.city). The goal of this research project is to enhance our collective knowledge about the various ways to govern urban commons, and the city itself as a commons, in different geographic, social and economic contexts.

From this study, we have extracted a set of design principles that are distinctively different from those offered by Elinor Ostrom and which can be applied to govern different kinds of urban commons, and cities as commons. Specifically, we investigate whether these design principles can help cities transition to fairer, inclusive, sustainable, resilient futures given existing patterns of urbanization and the contested nature of urban resources such as public spaces, open or vacant land, abandoned and underutilized structures, and aging infrastructure. In this chapter, we see examples of how these resources can be governed as a commons in cities around the world.

The city as a commons

To say that the city is a commons is to suggest that the city is a shared resource—open to, shared with, and belonging to many types of people. In this sense, the city shares some of the classic problems of a common pool resource—the difficulty of excluding people and the need to design effective rules, norms and institutions for resource stewardship and governance. Indeed, "the city analog to placing an additional cow on the commons is the decision to locate one's firm or household, along with the privately-owned structure that contains it, in a particular position within an urban area" (Fennell 2014: 1382). Congestion and overconsumption of city space can quickly result in rivalrous conditions in which one person's use of space subtracts from the benefits of that space for others. For instance, different kinds of urban infrastructure (roads, telecommunications systems, water systems, parks) otherwise considered to be a nonrivalrous public good can become rivalrous either through increased demand (Frischmann 2012) or because of regulatory slippage (Foster 2011).

In addition to more traditional concerns about congestion and rivalry, the openness of cities also raises the question of distribution in the commons. Many contestations of city space and resources revolve around the question of how best to "share" the finite resources of the city among a variety of users and uses (Foster and Iaione 2016). To be sure, distributive concerns fall outside of the considerations that motivated Garret Hardin's *Tragedy of the Commons*—i.e. consuming resources beyond the point where they benefit anyone and in fact reduce the overall benefit of the resource for everyone (Fennell 2004). But Ostrom's institutional approach to managing shared resources applies to a much broader range of human behavior and social dilemmas than avoiding suboptimal results from the cumulative actions of rational actors. (Ostrom 2010a). Ostrom's work generated an approach that can be used in the analysis and design of effective institutions (or instruments) to manage not just common pool resources but many different types of shared resources.

The "commons," as defined by scholars who build on Ostrom's institutional analysis and development (IAD) approach, is as much a reference to community management or governance of shared resources as it is to the nature of the resource itself. "The basic characteristic that distinguishes commons from noncommons is institutionalized sharing of resources among members of a community" (Madison, Frischmann & Strandburg 2014: 2). As such, it is not surprising to see the emergence of "new" commons—or nontraditional common pool resources—such as knowledge commons, cultural commons, infrastructure commons, neighborhood commons, among others (Hess 2008). These new commons seek to provide an alternative to the public-private (government) binary of governance solutions. These new kinds of commons focus on "communities working together in self-governing ways to protect resources from enclosure or to build newly open-shared resources" (Hess 2008: 40).

It is tempting, in asking whether shared *urban* resources (including the city itself) can be governed by local communities working together, to apply Ostrom's design principles to the city and to apply them to the management of many kinds of public and shared resources in the city. For many reasons, however, Ostrom's ideas cannot be wholly adapted to the city the way they have been used to understand the management and governance of natural resources. Ostrom's framework needs to be adapted to the reality of urban environments, which are already congested, heavily regulated and socially and economically complex. Without such adaptation, Ostrom's design principles will be lost in translation.

Ostrom's study focused mainly on close knit communities in which it was clear who was from the place and who was not (principle 1). For these communities, social control/monitoring and social sanctioning were two central pillars of Ostrom's design principles for the governance structure that communities often put in place to manage a common pool resource (principles 5 and 6). For this reason, she observed that rules of cooperation among users were written or modified by those who would be entrusted with both the duty to obey them and the responsibility to enforce them (principle 3). The fact that these rules were written by the same community of users that applied them suggested the need to leave some room for adaptation of such rules to local needs and conditions (principle 2). Of course, these structures and rules were premised on the assumption that communities' rights to self-govern the resource would be recognized by outside authorities (principle 4).

Ostrom found, however, that for more complex resources this governance responsibility or power was shared with other actors to form nested enterprises (principle 8). Notwithstanding this, she observed that conflicts might arise because even the most united communities have internal fractions and therefore require accessible, low cost tools to solve their own disputes (principle 7). These are the basic design principles which for years have been driving the study and observation of common, shared resources—namely scarce, congestible, renewable natural resources such as rivers, lakes, fisheries, and forests.

Cities and many kinds of urban resources are different from natural resources and more traditional commons in ways that render necessary adjustments to some of Ostrom's principles. First, cities are typically not exhaustible nor nonrenewable, although they can become quite fragile over time due to internal and external threats. There are, of course, natural resources such as lakes, rivers, trees, and wetlands in urban environments that can be rendered quite vulnerable by rapid urbanization, migration, and landscape change (Unnikrishnan et al. 2016). Because they resemble in most ways traditional common pool resources, researchers have approached the possibility for collective governance and polycentric management of these "urban commons" in a similar fashion (Nagendra & Ostrom 2014). However, much of the city consists of built urban infrastructure—open squares, parks, buildings, land, streets, roads and highways—which can be purposed and repurposed for different uses and users. In this way, these resources—the kind of "urban commons" to which we refer—are quite distinct in character and design from the forests, underwater basins and irrigation systems that were the subject of Elinor Ostrom's study of common pool resource governance.

Second, cities and many of their resources are what we might call "constructed" commons, the result of emergent social processes and institutional design (Madison, Frischmann and Strandburg 2010) As with knowledge commons, the urban commons often require the creation of governance or management structures that allow for not only the sharing of existing resources but also the production of new resources which will be shared by a group or community of actors. (Madison et al. 2010) The process of constructing a commons—what some refer to as "commoning"—involves a collaborative process of bringing together a wide spectrum of actors that work together to co-design and co-produce shared, common goods and services at different scales (Bollier & Helfrich 2015; Bresnihan & Byrne 2015). They can be created at the scale of the city, the district, the neighborhood, or the block level.

Third, cities do not exist in a pre-political space. Rather, cities are heavily regulated environments and thus any attempt to bring the commons to the city must confront the law and politics of the city (Foster 2011). Managing and creating urban common resources most often requires changing or tweaking (or even hacking, in a sense) the regulation of public and private property and working through the administrative branches of local government to enable and/or protect collaborative forms of resource management. Legal and property experimentation is thus a core feature of constructing different kinds of urban commons (Ela 2016).

Fourth, cities are incredibly complex and socially diverse systems which bring together not only many different types of resources but also many types of people (Portugali et al. 2012). Because of this diversity and the presence of often thick local (and sublocal) politics, social and economic tensions and conflicts occur at a much higher rate and pace than many natural environments. The economic and political complexity of cities also means that governance of urban commons cannot be just about communities governing themselves. Rather, collective governance of urban commons almost always involves some forms of nested governance—perhaps involving other levels of government (Hudson & Rosenbloom 2013)—and in most cases cooperation with other urban actors and sectors.

Design principles for the urban commons

Based on these differences, we began to think anew about design principles for the urban commons, taking into account what Ostrom learned about successful governance of natural resources commons. While many of her principles have clear applicability to constructed urban commons—such as recognition by higher authorities (principle 7), the importance of nestedness for complex resources (principles 8), the existence of collective governance arrangements (principle 3), and resource adaptation to local conditions (principle 2)—others are of limited utility or need to be adapted to the urban context.

For instance, communities should drive, manage, and own the process of governing shared urban resources, but we have seen time and time again that they can rarely avoid dealing with the state and the market. While this can be true of natural commons, and rural communities, we think both the state and the market are even more omnipresent in cities, making it difficult to side step them over the long run. As such, we observe that many types of urban commons tend to benefit from cooperation with other than internal community members and resource users. Rather, they need to collaborate and manage resources with other commons-minded actors, such as those constituting knowledge institutions and civil society organizations.

We have observed that in contexts where the State is the strongest, and markets are not as strong, local and provincial government actors can lend assistance to, and form a solid alliance with, communities to advance collective governance of urban resources. In this sense, the State generally acts as an enabler of cooperation and pooling of resources with other actors.

On the other hand, where the State is weak or weaker, either because of corruption or lack of resources, the market seems to be the only answer to enable the pooling of resources (i.e. human, economic, cognitive, etc.) needed for collective action and collaborative management of urban resources. The market could subsidize the commons if proper legal structures and participatory processes are put in place and there is sufficient social and political capital among resource users to negotiate with market actors.

In both cases, the concept of "pooling" seems to capture the true essence of commons-based projects and policies in the urban environment. For these reasons, we have identified in our work two core principles underlying many kinds of urban commons as an enabling state (Foster 2011) and pooling economies (Iaione & De Nictolis 2017).

We also observed for instance that technology in cities plays a key role in enabling collaboration and sustainability, as well as pooling users of urban assets, shared infrastructure, and open data management. Further, urban commons-based governance solutions are cutting-edge prototypes and therefore often require careful research and implementation. In other words, they are experimental: new approaches and new methodologies are constantly being developed and require prototyping, monitoring and evaluation.

These basic empirical observations are now the cornerstone of a much larger and scientifically driven research project that we established and call the "Co-Cities Project." The idea of the "Co-City" (Iaione 2016) is based on five basic design principles, or dimensions, extracted from our practice in the field and the cases that we identified as sharing similar approaches, values and methodologies. While some of these design principles resonate with Ostrom's principles, they are each adapted to the context of the urban commons and the realities of *constructing* common resources in the city. We have distilled five key design principles for the urban commons:

- **Principle 1**: *Collective governance* (or co-governance) refers to the presence of a multi-stakeholder governance scheme whereby the community emerges as an actor and partners (through sharing, collaboration, cooperation, and coordination) with four other possible categories of urban actors in a loosely coupled system;
- **Principle 2**: *Enabling State* expresses the role of the State (usually local public authorities) in facilitating the creation of urban commons and supporting collective governance arrangements for the management and sustainability of the urban commons;
- **Principle 3**: *Social and economic pooling* refers to the presence of autonomous institutions (e.g. civic, financial, social, economic, etc.) that are open, participatory, and managed or owned by local communities operating within non-mainstream economic systems (e.g. cooperative, social and solidarity, circular, cultural, or collaborative economies, etc.) that pool resources and stakeholders often resulting in the creation of new opportunities (e.g. jobs, skills, education, etc.) and services (e.g. housing, care, utilities, etc.) in underserved areas of the city or for vulnerable inhabitants;
- **Principle 4**: *Experimentalism* is the presence of an adaptive, place-based and iterative approach to design legal and policy innovations that enable the urban commons;
- **Principle 5**: *Tech justice* highlights access, participation, co-management and/or co-ownership of technological and digital urban infrastructure and data as an enabling driver of cooperation and co-creation of urban commons.

These design principles articulate the types of conditions and factors that we observe are present and that instantiate the city as a cooperative space in which various forms of urban commons not only emerge but are sustainable. These conditions shape and define what we call a "co-city." The concept of the co-city imagines the city as an infrastructure on which participants can share resources, engage in collective decision-making and co-production of shared urban resources and services, supported by open data and technology, guided by principles of distributive justice. A co-city is based on polycentric governance of a variety of urban resources such as environmental, cultural, knowledge and digital goods that are co-managed through contractual or institutionalized public-community or public-private-community partnerships.

Polycentric urban governance involves resource pooling and cooperation between five possible categories of actors—social innovators or the unorganized public, public authorities, businesses, civil society organizations, and knowledge institutions—the so-called "quintuple helix governance" approach. (Iaione & Cannavo 2015). These co-governance arrangements have three main aims: fostering social innovation in urban welfare provision, spurring collaborative economies as a driver of local economic development, and promoting inclusive urban regeneration of blighted areas. Public authorities play an important enabling role in creating and sustaining the co-city.

The ultimate goal of a co-city, we believe, is the creation of a more just and democratic city, consistent with the Lefebvrian approach of the right to the city (Foster & Iaione 2016; Iaione 2017a). In the next section, we will turn to looking at an array of mechanisms, or tools, that constitute the commons-based approach to urban resources and facilitate the transition to a co-city. Few cities have adopted a comprehensive and self-conscious approach to transition from a city in which there are urban commons present to the city itself as a commons. In Europe, three cities exemplify this transition in our view: the city of Barcelona (Ajuntament Barcelona 2017), the city of Turin through the UIA Co-City Turin Project (Iaione 2017b) and the city of Ghent through its commons transition plan (Bauwens & Onzia 2017).

The design principles in action

Having identified design principles for urban commons, we now turn to some of the recurring institutional, financial, and legal mechanisms or tools that are employed to construct, govern, and sustain a variety of shared urban resources consistent with these principles. The design principles can lead to the production of very different urban commons governance devices which need to be adapted to the local context and the needs of local communities. We have grouped these forms of urban commons governance mechanisms and tools into four main categories: institutional, legal, financial, and digital. For each, we offer a brief description and an example or two from our case studies and field observations and explain how they manifest different design principles.

Institutional mechanisms

In defining the various forms that the concept of commons has taken in history, Elinor Ostrom and Charlotte Hess underscore that the commons requires an institutional approach to governing the production, use, management, and/or preservation of various kinds of type of resources (Hess & Ostrom 2007). These resources include both tangible and intangible shared resources, such as knowledge or cultural commons, as well as infrastructure resources. The term commons thus denotes more than a shared resource, or the community that benefits from that resource. The commons is the institutional arrangement that allows the coordination and sharing of those resources, and helps to ensure their accessibility and sustainability for a wide variety of users (Madison et al. 2016).

In this section, we introduce a set of institutional mechanisms for implementing co-governance of different kinds of shared urban resources. This is a tentative, and not exhaustive, list. From our empirical research as well as our own experience, we observe that the institutional ecosystem of a commons-infused city involves several closely related forms of collective governance at different scales. These include: Policy Innovation Labs, collaborative districts, collaborative working hubs, and collaborative housing, among others. We give a few examples in this section and how they relate to the design principles discussed earlier.

Policy Innovation Labs

Policy Innovation Labs are government sponsored facilities, entities or working groups devoted to the collective co-design, experimentation, governance and monitoring of urban policies. An emerging administrative innovation in many cities around the world, Policy Innovation Labs exist outside of the often-ossified environment of city hall and employ a deliberative and co-working style of interaction to engage a variety of urban stakeholders. Policy Innovation

Labs situate urban inhabitants and local communities at the center of the policy design process, including the experimentation and evaluation of proposals, enabled and implemented by the local government. These labs are experimental initiatives, often undertaken by public administration members with the support of social innovators and public innovation experts. They are also places where new instruments of collaborative governance, often supported by digital technologies and data analytical tools, can emerge and be supported before being embedded deeply in the local administrative state.

One such Lab grew out of the city of Bologna's effort to become a "Collaborative City," inspired by the concept of the urban commons. The role of the Policy Innovation Lab is carried out by Bologna's newly formed Office for Civic Imagination and operationalized through "collective civic enterprises," which are enabled by the public administration through co-design rather than ordinary participatory processes (Iaione 2016). Civic imagination is the process through which civic actors, individually and collectively, "envision . . . better political, social, and civic environments and work towards achieving those futures" (Baiocchi et al. 2014). Bologna's Office for Civic Imagination is conceived as a permanent co-design laboratory that can manage and monitor all collective action processes, collect and disseminate public data, aggregate skills and tools, and support collective practices of city inhabitants including experimenting with new forms of urban cooperativism.

Similar laboratories in cities serve the same function of pairing civic innovation and collaboration between government and residents throughout the city. The MindLab in Denmark is another example of a policy innovation unit whose aim is to enable public authorities to manage collaborative policies implementing co-working/service design processes that involve citizens. It involves a cross-governmental collaboration (Ministry of Industry, Business and Financial Affairs, the Ministry of Employment, the Ministry of Education and municipal: city of Odensa) to distribute among civil servants the methods and tools for facilitating the exchange of knowledge/experiences and mutual learning mechanisms through an experimental approach (Cartensen & Bason 2012).

New York City's recently launched "NYCx Co-Lab" is focused at the neighborhood level and is designed to propose and test new solutions to modernize public infrastructure, support community driven development, and bridge the digital divide in low-income and ethnic minority areas of the city. The city touts it as a program "to transform urban spaces into hubs for tech collaboration, research, testing and development" toward helping the City "advance its goals to be the most fair, equitable and sustainable city in the world." (New York City, Office of the Mayor 2017). Mexico City's Lab for the City (Laboratorio Para La Ciudad) is another standout among the many emerging city labs, in that it has created a platform for experimental policies that can transform the urban environment through community-engaged processes and using digital technologies to enhance that engagement.

Collaborative districts or neighborhood agencies

Collective action in cities and urban environments most often occurs at the neighborhood level. Although the form of these efforts varies, for shorthand we will refer to institutional arrangements that represent collective, co-management at the neighborhood level as "collaborative districts" or neighborhoods. The idea is that community residents work together in a deliberative and collaborative manner with other stakeholders, and at a defined geographic scale, to maintain or regenerate the assets and services within their neighborhood or district. These assets are then transformed into a collaboratively managed resource. These districts and neighborhoods represent a form of polycentricism in the way that Ostrom discussed in her work, meaning that they are nested in (or networked with) other government processes or governance structures. They also can be one way of applying the governance devices conceived and prototyped through public deliberation processes and co-design labs to the neighborhood/district level.

Neighborhood cooperative development agencies are one example of this kind of institutional tool. These agencies are inspired by *regies des quartier*, or neighborhood management, widely implemented in France (Regies des quartier 2017) and with ancient roots in Italy and currently manifesting through "community foundations" throughout the country (Laino 2012). The French *regies des quartier* are non-profit agencies coordinated at the national level but governed at the local level through an association "de loi 1901," the traditional model of volunteer non-profit civic association composed by a partnership of urban inhabitants, the city authorities, neighborhood level associations, social workers and other actors relevant for the inclusive economic development of the area. The agencies are distributed all over French cities, although the higher concentration is in the city of Paris, in the North-East Area (Regies des quartier 2017). They specifically address local development and community development issues and support residents finding jobs by investing in neighborhood infrastructure (Eme 2004).

Two notable examples in Italy include the Pilastro Development Non-profit Agency in Bologna and the Neighborhood Houses in Turin. Both follow the "quintuple helix" approach by bringing together at least five possible urban actors or stakeholders in a collaborative institutional format to instigate peer-to-peer production of neighborhood goods and services. The Pilastro Development agency was created in a deprived neighborhood and constituted by the city government, neighborhood associations or collectives, the local housing agency, a civic organization formed by unemployed neighborhood residents to offer a range of community services, small neighborhood business and economic actors (e.g. agri-food center), the neighborhood's mall, a prominent regional banking institution, and a philanthropic foundation (Pilastro 2016).

Similarly, the Neighborhood Houses (NHs)—or *Case del Quartiere*—in Turin is a network of eight neighborhood community centers located in formerly distressed or abandoned spaces that were regenerated and made available to the local community for civic, cultural, and educational uses that meet the needs of residents (Caponio & Donatiello 2017). The networking of these houses is promoted and supported by the municipality and other public and private stakeholders to foster cooperation and free exchanges between the houses and to innovate neighborhood level responses to the needs of vulnerable individuals, specific groups and the community.

In China, some urban villages are partly governed by village "collective economy organizations" which survive the transition from rural to urban land (Tang 2015). Village leaders convert the village's collective landholdings, along with the land expropriation compensation funds given by the city government, into a village shareholding company which then offers shares to the villagers and rents out portions of collective land and buildings to the local city government and investors for industrial projects. Although urban public services are typically provided through government established urban residents' committees, successful village shareholding companies partially usurp those committee functions and play a leading role providing villagers with economic and social welfare, subsidizing community administration services, and mobilizing residents.

"Collaboratories" and urban living labs

Collaborative working and living are often entry points into establishing polycentric governance of a variety of urban resources. The collaborative hub is a co-working space through which the relationship between district or neighborhood level institutions and community residents is mediated. These hubs can also interact with Policy Innovation Labs, described earlier, to coordinate and monitor the path of experimental civic collaboration policies. The main difference between Policy Innovation Labs and collaborative hubs is where they originate from. Most often collaborative hubs emerge from and exist outside of the local government structure and administrative apparatus, although they work with public actors. In this way, they are part of an ecosystem of distributed, polycentric systems of decision making within a city or metropolitan area. They are also a form of knowledge and resource pooling, as well as collective governance.

One example of how these institutional forms manifest is the "living lab." Living labs are user-centered, co-working open innovation spaces where different categories of actors, or stakeholders, can integrate their skills and energies and make them available to the city. At the heart of these living labs is a methodological protocol known as "collaboratory," conceived in the late 1980s in the field of computer science and later applied to fields such as environmental or energy research (Grimes 2016). The key idea behind the development of collaboratories in scientific research is that knowledge is an activity that is inherently collaborative (Finholt 2002). They operate as a catalyst to foster mutual learning and co-creation (Hess & Ostrom 2007: 13, 327).

In the urban commons context, the collaboratory takes the form of a physical or virtual setting where innovative and cultural forces of the city converge, share resources and knowledge and join efforts for generating new forms of shared resources (including knowledge). In these spaces, collaboration teams can coordinate project planning steps in the city, trainings on innovative skills such as digital manufacturing, and more generally be a feeder into the process of generating and disseminating social innovation (Cossetta & Palumbo 2014). They function as a living lab for innovation in the design of urban policy solutions by aggregating different actors (social innovations, enterprises, public institutions, knowledge institutions) in a co-design process where new public policies can be generated outside of the local government. These labs can be structured as public-private-community institutional partnerships involving many actors and they can be established at the city, neighborhood, or institutional level. They can also be connected to each other, as in the European Network of Living Labs.

The city of Reggio Emilia in Italy has recently launched a process to co-design with many stakeholders a city-level collaboratory. The diverse and historic Hillman City neighborhood in Seattle, Washington established a community-based collaboratory composed of a co-working space, a community garden, a multipurpose space, and a learning kitchen. Universities can also become urban living labs, or collaboratories, as evidenced by our own LABoratory for the GOVernance of the Commons (LabGov), established at LUISS Guido Carli University of Rome. LabGov promotes co-governance experiments on the urban commons and policies which situate cities as facilitators of the urban commons in different contexts. It played a co-leading role in the Bologna Regulation (see later) and is now working on the establishment of a collaboratory in the neighborhoods of the city of Rome and through other projects throughout the country. Such university labs provide a platform for a rigorous and interdisciplinary approach to urban policy and democratic experimentalism (Sabel & Dorf 1998). University labs can also be an instrument available to different urban stakeholders to guide field trials, city to city experiential exchanges, and empirical observation (Biber 2016).

Collaborative housing

Co-housing and housing cooperatives have a long history in many American and European cities, and not all are examples of collaborative or commons-based housing governance. Some are what property scholar Carol Rose has described as "commons on the inside and private property on the outside" (Rose 1988). In other words, not all are accessible or are collaboratively managed toward the goal of providing affordable or collective goods. The goal of those living in collaborative housing spaces today is not just that of sharing common spaces or saving money, but also having a lower impact on the environment, building relationships and satisfying socialization needs, as well as gaining access to social services or creating collaborative economic activities. Additionally, some emerging forms of co-operative and co-managed housing reflect institutional, technological or design innovations that address social inequalities, greater ecological sensitivity, and the shortage of affordable housing.

Collaborative or co-housing already constitutes a successful policy agenda for inclusive urban development in Germany. After a period of strong state support to forms of self-organization for housing (1984–2000), which led to the blossoming of about 350 experiences, state support was interrupted. Over the past six years, however, Germany has seen a revival of self-organized co-housing, partly due to the exponential increase in housing costs. In 2008, the city of Berlin set up an agency to promote the networking of existing co-housing projects and to facilitate the birth of new co-housing projects (Droste 2015).

A similar transformation is happening in Italy, where the housing sector is moving beyond forms of social housing based on public management or public-private partnerships towards public-community and public-private-community partnerships for the supply of affordable housing (Federcasa-Nomisma 2017; Federcasa-LabGov 2018). For example, the Abit@giovani project, in Milan, is formed by a partnership comprising regional authorities, philanthropic foundations, and a real estate fund. It addresses the housing needs of young residents by offering them access to recently regenerated houses throughout the city and creates a collaborative network of apartments through which renters are connected in co-working labs and a digital platform (Abita Giovani 2017).

In Barcelona, Spain, the non-profit housing cooperative La Borda also enables a high degree of self-management and relies on the self-construction of buildings by its members. A novelty of the La Borda model is that it uses a "grant of use" tenancy for the allocation of housing units (Cabrè & Andrés 2017). The land belongs to the city of Barcelona and the cooperative pays a lease for 75 years. The funding for the project is provided by a hybrid regime of social economy, ethical banking, and voluntary contributions by individuals and groups (La Borda 2017).

Beyond co-housing, much innovation in the area of collaborative housing arises from legal innovations such as community land trusts (CLTs), described in more detail below. A CLT is different from the traditional housing cooperative in that it separates land *ownership* from land use and transforms what might otherwise be a collection of individuals owning property (in the typical housing cooperative ownership model) into a collaboratively governed institution which manages collectively shared goods in a way that ensures their long-term affordability. Residents in the CLT lease the land but own the building and their housing units. The affordability of the housing (or other uses) is secured through lease covenants (contractual promises) which restrict the resale price of the building (or housing units) and thus the amount of profit that can be extracted from the building or unit by any individual owner.

Legal tools

There are a range of legal instruments available to implement an urban co-governance scheme and to support the kinds of institutional mechanisms offered earlier. Legal tools instantiate forms of urban collective governance and resource pooling. They often result in the formation of partnerships representing the quintuple helix of urban governance mentioned earlier. This set of legal devices is applied across a range of urban resources—ranging from urban heritage to public services—transforming them into shared assets and shared governance between at least three to as many as five different actors in the urban environment. They include specific legal instruments such as land trusts and collaboration pacts that can either be contractual or institutionalized. The tools are also designed to ensure that the resources meet the needs of local communities, and often are focused on making those resources more available, accessible and affordable to a broader range of urban residents.

Public-community pacts

Legal tools for the urban commons most prominently came into focus with the Bologna Regulation on Collaboration for the Care and Regeneration of the Urban Commons. The Regulation established "pacts of collaboration," a co-designed partnership between the public administration and local communities to share the regeneration and/or management of a variety of urban resources. These resources can range from public spaces and urban parks to abandoned buildings and urban services. Since that adoption, many cities in Italy have uncritically adopted the regulations, without attention to local context. This kind of cut and pasting of laws, without attention to local context, is not consistent with the design principles we have articulated earlier. The only city that has invested in the adaptation of this kind of regulation to its local context, and the creation of an experimentation process and an institutional ecosystem to implement it, is the city of Turin through its "Co-City" project.

Urban civic uses

Other examples of national or local regulations that can reflect enabling state and collective governance, as well as experimentalism, include the UK's Localism Act which gives local communities the right to keep buildings or other assets that have use value for the community. In a similar vein, the city of Naples in a 2016 Resolution formally recognizes as "civic uses" abandoned or underutilized public properties that are being used and transformed by residents into collective uses or "common goods." The city allows their use, without transferring ownership or even leases to the residents, and has appointed a multimember advisory body to study, analyze, oversee, and control the management and protection of these collectively utilized spaces and places.

Land trusts

One legal mechanism that is often employed to support commons-based institutions in the urban environment, particularly where there is a speculative real estate market, is the community land trust (CLT). Land owned by a CLT is removed from the real estate market and put into a legal structure that is democratically governed by a diverse membership of public and private actors and inclusive of community residents where it is based. The governing board of a CLT is most often "tripartite"—an equal number of seats represented by users or people who lease the land from the CLT, residents from the surrounding community who do not lease land from the CLT, and the public and private sector (usually public officials, local funders, nonprofit providers of housing or social services, and others). CLTs are also membership organizations, made of voting members from the surrounding community served by the CLT.

Although community land trusts are often used to protect and sustain the affordability of housing in a neighborhood, the Dudley Street Neighborhood Initiative (DSNI) example from the United States demonstrates that the land trust is much more malleable than its most common use in the housing context. DSNI, borne out of efforts by residents in a poor Boston neighborhood, formed an urban redevelopment corporation called Dudley Neighbors, Inc. (DNI) that worked with state and local government and private financers to acquire acres of vacant lots and to construct an "urban village" consisting of hundreds of affordable housing units, six public green common spaces, two community centers, an urban farm, refurbished schoolyards, and numerous playgrounds (Medoff & Sklar 1994).

On a similar scale, the Granby Four Streets CLT, a four-streets triangle located in a working class neighborhood in the Toxteth area of Liverpool, is endeavoring to create a vibrant mixed income community consisting of low-cost affordable housing, green public/social spaces, a retail and enterprise hub for both commercial and creative activities, as well as street markets and event spaces. After two decades of struggle against property degradation and redevelopment policies by the city council, Granby Four Streets CLT was able to create a partnership with the local city council and other housing associations, secure funding from private local investors, and maintain continuous citizens' involvement to restore ten properties (Thompson 2015).

The CLT model is also increasingly being adopted as official city policy as in New York City, which is promoting the use of community-driven CLTs throughout the city to meet affordable housing development and neighborhood revitalization goals. The city recently pledged over $1.5 million to create and expand CLTs, as well as to fund a CLT "Learning Exchange" for community-based organizations and affordable housing developers that want to form CLTs in their neighborhoods. The CLT model is also being used by cities to protect the affordability of informal settlements against the immense pressure of forced evictions and gentrification in those areas brought on by the real estate speculation affecting the entire city. Recently, a community in San Juan, Puerto Rico became the first informal settlement to be placed in a community land trust, managed through an independent government agency with a governing board made up of representatives from the public and private sectors and community residents (Bernardi 2017).

Financial mechanisms

One critical element for the development of the urban co-governance, or polycentric governance at the local or district level, is the development of social project financing for collaborative forms of urban infrastructure and urban services. Commons institutions and co-governance in the urban environment must be financed somehow given the pressures on public financing and the lack of access to private capital by many poor populations. This kind of financing is one example of "pooling economies" (Iaione & De Nictolis 2017). Pooling economies are instigated or created by the attraction of resources, in this case funding, from different actors or segments of society. These economies form around the collaborative economy, supporting efforts of residents and others to pool their efforts and cooperate around new forms of infrastructure and services.

Here we identify three financial tools that reflect pooling economies and which support production and co-creation of urban commons throughout a city. This is not to say that traditional forms of funding from philanthropic foundations, private investment entities, and institutional or individual investing are not important funding tools. Indeed, they can be part of a pooled strategy to co-create or co-produce common goods and services. The following financial tools reflect, in our view, the principles set out earlier: crowdfunding, solidarity funding, and social project finance.

Crowdfunding and civic financing

Crowdfunding (through online platform or person to person fundraising) can raise awareness and mobilize the local community as a first step leading toward the regeneration of blighted urban areas (Patti & Polyak 2017). In addition to pooling monetary resources, crowdfunding can also be the means to allow different actors to pool their efforts in a campaign to activate urban commons projects. For instance, the city of Bologna, in collaboration with a social enterprise and a local civil committee, launched a crowdfunding platform and campaign to mobilize local inhabitants to participate in the restoration of the "Portico di San Luca," an historical site in the city Center (Un Passo per San Luca 2014). Over 300,000 euros were raised from over 7,000 donors, and the platform has remained open for other collaborative projects in the city.

Another example is the "commons buy-out" (i.e. civic acquisition by a local collective) of the Macao building in a suburb of Milan by a group of users occupying the city-owned building in which they held social and cultural activities. The group launched a campaign that attracted over 2,000 current users and external members to purchase the building and place it into some form of a cooperatively owned and managed space (Macao 2017). Consider also the recently formed "real estate investment cooperative" (REIC) in New York City, composed of hundreds of city residents in New York who pool their money and seek other contributions in order to acquire vacant or underutilized city-owned buildings and transform them into affordable spaces for civic, cultural and cooperative uses (REIC 2018). Anyone can become a member of the REIC, paying a small membership fee of $10, and become active in fundraising and governance of the non-profit organization. To date, the REIC has raised over $1 million.

The Macao model is inspired by the Miethauser Syndicat model in German cities with around 123 properties purchased through a traditional bank loan and in which ownership and use mimics the form of a community land trust mentioned earlier. Unlike the CLT, however, ownership and use are not divided between the building and the land. Rather, the building structure is owned by two entities. The title to the property belongs to a limited liability company (LLC), established specifically for this purpose and composed of the Association of Tenants living in the building (the Miethauser mainly deal with housing projects), and another LLC represents the overall tenement trust comprised of all Tenant Associations (Horlitz 2012).

Solidarity funding

A similar funding mechanism that also relies on pooling economies is social solidarity funding, which is fueling the boom of cooperative housing in some regions of the world. As an example, two Swiss cooperative housing federations are maintaining solidarity funds, constituted by taking sums from individual housing cooperatives and creating a larger, common pool of revolving funding (Scruggs 2017). Cooperatives can then access this funding to finance new housing and land acquisition or repairs to existing housing. This concept has fueled, for example, Switzerland's housing co-op boom during the global financial crisis and has motivated the Geneva non-profit housing developer Codha to raise millions from co-op residents to finance hundreds of new affordable housing units in one of the most expensive cities in the world (Scruggs 2017).

Social project financing: social bonds and impact investing

The development of a social and community-based economy is distinct from the traditional charity model based on grants. Instead, it tends to blend the profit and social motives, through social *impact investing* (Bugg-Levine & Emerson 2011). This model of investment is widely adopted and utilized in the United States, the UK and the EU either through "social corporate responsibility" practices by private firms or through Community Interest Companies (CIC) (2013), a creature of UK law. The CIC is mainly focused on providing benefits for the local communities and is designed to be flexible (Cho 2016). What differentiates the CIC from a company is the statutory clauses of the asset lock which ensures the assets are used for the benefit of the community (CIC Regulator case study series 2017). The CIC, for instance, is set up as a social enterprise whose investments are directed toward its social purposes and whose profits are then reinvested exclusively for this purpose rather than to maximize profit for shareholders. They can also serve several social purposes fitting with the urban commons including: community development and inclusion, regenerating disadvantaged areas, and delivering new and innovative services at the neighborhood level (UK Government 2017).

According to the UK government, CICs are diverse and can include local community enterprises, social firms, co-operatives, or national or international organizations (UK Government 2017). One example of a UK CIC is the Social Finance Group, which carries out social finance projects. A leading case is the social project finance that led to the development of the very first experimentation of social impact bonds in the UK in 2007, the Peterborough Project (Social Finance Group, 2017a). The project was aimed at the production of services for vulnerable populations, with the support of volunteers and organized communities (NGOs, foundations). Another example is the Bright Futures Fund, launched in November 2014 by the Social Finance and Kin Capital. It is the first national fund to take advantage of the new Social Investment Tax Relief (SITR). The Fund aims to reach £3 million and will invest in charities and social enterprises working to improve the lives of children, young people and vulnerable groups across the UK (Social Finance Group 2017b).

A similar corporate model is embodied by the US model of Benefit Corporation. These corporations are business entities that pursue the general interest as a statutory requirement. It is the most popular corporate form of social enterprise (Murray 2012). A cooperative alternative to the corporate model would be the adoption of the Italian "community cooperatives" model (Mori 2014) or the French "collective interest cooperative" model (Mystica 2016).

Because urban commons can constitute some forms of social infrastructure—places for the care, shelter and education of local inhabitants—social bonds (SBs) and social impact bonds (SIBs) can also play an important role in financing schemes for these resources. SBs and SIBs are designed to attract private capital to finance innovative solutions in welfare services (Fransen et al. 2018). SBs or SIBs are used to implement social project finance schemes and through public-private-community partnerships regenerate urban spaces or run some community-based services, which are then able to generate the social value needed to be eligible for an SB (International Capital Market Association 2017) or support the repayment of the SIB (Zheng Lu et al. 2015).

Digital and technological tools

Technology can provide crucial tools for communication, and the process of commoning. Digital tools are often employed to connect actors and to facilitate their involvement, and collaboration. Communicating the possibility of collaboration is an essential step, as it facilitates the pooling of assets and actors. Communication and connectivity as it relates to commoning is more than the simple transmission of a message from a sender to a recipient, but rather is an act of social participation—something that is connected to the etymological root of the verb communicate (κοινω and κοινονεο: I make common sense, I join and participate).

As important, digital tools and technology can help to address urban inequality and the inaccessibility of a range of commons goods and services. This is what we refer to when we suggest that "tech justice" is a critical design principle. Without connectivity and the ability to communicate, it is impossible to realize one's goals, to flourish, and to connect to others and build social capital across economic and cultural lines. In this last section we will focus on some of the more important digital and technological tools that we have observed can enable commoning and that also reflect the principle of tech justice.

Digital platforms

A digital platform can activate a dissemination process, enabling all urban inhabitants to access transparent and comprehensive information on the projects launched in a territory and ongoing decision-making processes within the administration. Through the platform, citizens are made aware of the possibilities of participating in such processes and of contributing to the creation of public policies on issues that concern them closely. In addition, they come to know the tools provided to them by the administration to facilitate the activation of collaboration and participation processes. The dissemination of information is fundamental, but it is not in itself sufficient to ensure the development of new forms of urban co-governance and is often complemented by two other functions.

The first function is the platform's ability to enable processes whereby citizens can play an active role in shaping public policies, sometimes by voting or otherwise registering their preferences. These platforms can empower early and meaningful participation in the co-creation process by allowing participants to propose ideas and begin working collaboratively on the development of solutions. The second function, on the other hand, is to enable the introduction of learning pathways. These platforms are accessible to citizens who, even if they do not have the means to undertake specialized education, are able to increase their knowledge and acquire skills which can then be employed within collaborative decision-making processes.

The local governments of the recent EU-level legislatures have recognized the relevance of digital democracy and application of the principles of Open Government through the implementation of institutional platforms of transparency, participation and collaboration. For example, in Barcelona the government of Ada Colau has created a powerful digital infrastructure for public consultations, resulting in increased transparency and participation in the activities of the city (Decidim Barcelona 2018). Another example is the city of Paris, which has chosen to give great weight to online deliberation and voting as part of its participatory budgeting process (Paris Mayor 2014).

The city of Athens has also developed a platform, SynAthina, to facilitate urban co-governance partnerships. The platform acts as a network coordinator: citizens, NGOs or informal groups can submit ideas for voluntary activities in public spaces or ways to utilize other urban assets in collaboration with relevant government representatives, NGOs, and private actors that might support their projects. Online applications and discussions continue with offline meetings organized in a physical space, the "SynAthina Kiosk." The platform and the meetings currently host almost 3,000 users, organized by over 300 groups of urban inhabitants in cooperation with over 100 sponsors (SynAthina 2013).

Increasing the digitization and accessibility of democratic, collaborative processes at the local (regional and municipal) level is also exemplified by the process in the city of Bologna to become a Collaborative City. The Bologna *Iperbole* digital platform has functioned as a dissemination platform, allowing the collaborative and participatory processes to be widely known. Through the "Collaborare è Bologna" campaign the city shed light on the implementation of its now more than 300 pacts of collaboration. The Co-Bologna platform also allows users to observe and participate in the experimentation processes developed in urban co-governance field labs in three neighborhoods. The platform's section on "experiment with us" enables interested parties to pursue new projects or experiments in the city and to share their experiences doing so. The city of Turin is implementing a similar, even more innovative, platform applying blockchain through the First Life platform.

Local networks and connectivity

The production and governance of infrastructure commons at the neighborhood/district level could be performed through several structures. Community created and user-managed "mesh" networks—decentralized wireless access points connected to each other—are one example. Mesh networking relies on existing urban infrastructure and the combined efforts of many local actors—including public and private property owners who grant access to buildings and other structures to mount the access points—to create a solution to the "last mile" connectivity gap. The goal of these networks is to bring internet service to communities and populations that lack high speed wireless or broadband access, increasingly seen as a necessary public good or "fourth utility." In this model, no one owns the entire infrastructure (open and free access), but everyone who wants access can contribute with their own resources to run the network which, in turn, is managed and governed by the community.

Mesh networks have been established in many European and US cities, including the famed community mesh network in Red Hook Brooklyn designed to overcome the digital divide, the Detroit, Michigan network which provides connectivity to the 40 percent of its residents without internet access, and the mesh network in Berlin, Germany designed to provide vital internet service to newly arrived migrants living in refugee shelters. Among the well-known and celebrated examples of wireless or metropolitan area wireless networks in the EU are the Spanish Guifi, the Greek AWMN, the Italian projects Neco, and Ninux and Freifunk in Germany. Many of those are considered as a democratic re-appropriation of technology (De Filippi & Treguer 2016).

Many of the design principles applied by these community networks mirror the design principles for urban commons. As adopted in the Declaration of Community Connectivity, these include: a) collective ownership (the network infrastructure is owned by the community where it is deployed), b) social management (the network infrastructure is governed and operated by the community), c) open design (the network implementation details are public and accessible to everyone), d) open participation (anyone is allowed to extend the network, as long as they abide by the network principles and design), e) free peering and transit (community networks offer free peering agreements to every network offering reciprocity and allow their free peering partners free transit to destination networks with which they also have free peering agreements), and f) the consideration of security and privacy concerns while designing and operating the network. The Declaration was facilitated by the UN Internet Governance Forum (IGF) Dynamic Coalition on Community Connectivity (DC3) (Belli 2016).

Community mesh and broadband networks promote what legal scholar Olivier Sylvain calls "broadband localism," an approach that seeks to overcome broadband infrastructure and service disparities by race, ethnicity and income (Sylvain 2012). It is also what he is getting at when he pushes regulators to go beyond "network neutrality" to "network equality" (Sylvain 2016). Network connectivity is thus an important element of "tech justice," as it can generate economic opportunity and facilitate access to essential goods and services, such as job opportunities, as well as to provide resiliency in network access in the face of climate and weather events.

In a similar vein, we also see emerging locally networked energy production within a community through the establishment of "micro-grids" to become more energy self-sufficient and resilient (Duda et al. 2016). Communities such as West Harlem in New York City, for example, are exploring cooperatively owned micro-grids as a means of helping vulnerable populations withstand climate change impacts and reduce energy costs in the transition from a fossil fuel to a renewable energy economy. Like wireless and broadband community networks, micro-grids involve pooling through different forms of financing, accessing excess energy from public buildings, and combining national and international incentives for energy efficiency. Through incentives, therefore, the community will be able to install photovoltaic systems that will be owned and managed by the community itself. The energy produced might also be able to support the creation of a community network so that the local community can self-manage and share the wireless mesh network.

Concluding remarks

The design principles and practice are based on our observation and study of the ways that a variety of resources in cities, both existing and created, are being managed or governed by local communities in a cooperative fashion with other actors and often enabled by government bodies and officials. The five design principles, and some of the mechanisms through which they manifest, together with the co-city policy cycle/process (Iaione 2016), compose the beta version of what we call "the co-city protocol." We interpret such protocol as a language that could guide collaboration among urban communities experimenting with the governance of the urban commons, as well as the exchange of ideas and practices on the commons at the urban level without impairing institutional diversity and adaptiveness. Much like in the digital and open source world, this protocol would allow local communities to build a shared language that could be iteratively updated and could increase shared knowledge around the city, ultimately contributing to the construction of an urban methodological approach to commons in the city and to governing the city itself as a commons.

References

Abita Giovani (2017) Available at www.abitagiovani.it/index.php?lang=it. (last accessed October 9, 2018).

Ajuntament Barcelona (2017) "Comuns Urbans: cap a l'ús i la gestió comunitària d'allò public," available at http://ajuntament.barcelona.cat/participaciociutadana/ca/comuns-urbans-cap-lus-i-la-gestio-comunitaria-dallo-public-0 (last accessed October 8, 2018).

Baiocchi, G., Bennett, E., Cordner, A., Klein, P. & Savell, S. (2014) *Civic Imagination Making a Difference in American Political Life*. New York: Routledge.

Bauwens, M. & Onzia, Y. (2017) *Commons Transition Plan for the City of Ghent*, available at https://stad.gent/sites/default/files/page/documents/Commons%20Transition%20Plan%20-%20under%20revision.pdf (last accessed October 8, 2018).

Belli, L. (2016) *Community Connectivity: Building the Internet from Scratch*, Annual Report of the UN IGF Dynamic Coalition on Community Connectivity, available at http://internet-governance.fgv.br/sites/internet-governance.fgv.br/files/publicacoes/community_connectivity_-_building_the_internet_from_scratch_0.pdf (last accessed October 20, 2017).

Bernardi, M. (2017) "An informal settlement as a community land trust. The case of San Juan, Puerto Rico," *LabGov.org*, available at http://labgov.city/thecommonspost/the-favela-as-a-community-land-trust-the-case-of-san-juan-puerto-rico/ (last accessed October 6, 2018).

Biber, E. (2016) "Law in the Anthropocene epoch," *UC Berkeley Public Law Research Paper*.

Bollier, D. & Helfrich, S. (2015) *Patterns of Commoning*. Amherst, MA: Levellers Press.

Bresnihan, P. & Byrne, M. (2015) "Escape into the city: Everyday practices of commoning and the production of urban space in Dublin," *Antipode* 47: 36–54.

Bugg-Levine, A. & Emerson, J. (2011) "Impact investing: Transforming how we make money while making a difference," *The MIT Press Journal*, 6(3): 9–18.

Cabrè, E. & Andrés, A. (2017) "La Borda: A case study on the implementation of cooperative housing in Catalonia", *International Journal of Housing Policy*, 1(21): 412–432.

Caponio, T. & Donatiello, D. (2017) "Intercultural policy in times of crisis: Theory and practice in the case of Turin, Italy," *Comparative Migration Studies*, 5(13), available at www.ncbi.nlm.nih.gov/pmc/articles/PMC5581365/ (last accessed October 8, 2018).

Cartensen, V. & Bason, C. (2012) "Powering collaborative policy innovation: Can innovation labs help?" *Innovation Journal*, 17(1): 2–26.

Cho, M. (2016) "Benefit corporations in the United States and community interest companies in the United Kingdom: Does social enterprise actually work," *Northwestern Journal of International Law & Business*, 37: 149–172.

Cossetta, A. & Palumbo, M. (2014) "The co-production of social innovation: The case of living lab," in Dameri, R.P. & Rosenthal-Sabroux, C. (eds.) *Smart City. How To Create Public And Economic Value With High Technology in Urban Space*, Cham, Switzerland: Springer.

Decidim Barcelona (2018) *La plataforma de participació de Barcelona*, available at http://decidim.barcelona (last accessed January 7, 2018).

De Filippi, P. & Treguer, F. (2016) "Wireless community networks: Towards a public policy for the network commons?" in Belli, L. & De Filippi, P. (eds.) *Net Neutrality Compendium*. Cham, Switzerland: Springer.

Droste, C. (2015) "German co-housing: An opportunity for municipalities to foster socially inclusive urban development?" *Urban Research & Practice*, 8(1): 79–92.

Duda, J. Hannah, T. & Burke, M. (2016) "Building community capacity for energy democracy: A deck of strategies," *Democracy Collaborative*, available at http://prototypes.democracycollaborative.org/energy democracy/fullscreen.html (last accessed October 8, 2018).

Ela, N. (2016) "Urban commons as property experiment: Mapping Chicago's farms and gardens," *Fordham Urban Law Journal*, 43: 247–294.

Eme, B. (2004) "Généalogie Des Régies De Quartier," *Ecologies et Politique*, 1(28): 45–65.

Fennell, L.A. (2004) "Common interest tragedies," *Northwestern Law Review*, 98: 907–990.

Fennell, L.A. (2014) "Agglomerama," *Brigham Young University Law Review*, 2014: 1373–1414.

Federcasa-Nomisma (2017) "Report on the housing emerging in Italy," available at www.federcasa.it (last accessed October 8, 2018).

Federcasa-LabGov (2018) "Report on collaborative housing," available at www.federcasa.it (last accessed October 8, 2018).

Finholt, T. (2002) "Collaboratories," *Annual Review of Information Science and Technology Banner*, 36, 1: 73–107.

Foster, S.R. (2006) "The city as an ecological space: Social capital and urban land use," *Notre Dame Law Review*, 82: 527–582.

Foster, S.R. (2011) "Collective action and the urban commons," *Notre Dame Law Review*, 87: 57–134.

Foster, S.R. & Iaione, C. (2016) "The city as a commons," *Yale Law and Policy Review*, 34: 281–349.

Fransen, L., del Bufalo, G. & Reviglio, E. (2018) "Boosting investment in social infrastructure in Europe." Report by the High-Level Task Force on investing in social infrastructure in Europe chaired by Romano Prodi and Christian Sautter. Available at www.sipotra.it/wp-content/uploads/2018/01/Boosting-investment-in-social-infrastructure-in-Europe.-Report-by-the-High-Level-Task-Force-on-investing-in-social-infrastructure-in-Europe-chaired-by-Romano-Prodi-and-C.pdf (last accessed October 15, 2018).

Frischmann, B.M. (2012) *Infrastructure: The Social Value of Shared Resources*. Oxford, UK: Oxford University Press.

Grimes, H. (2016) "Creating a 'collaboratory' environment to transcend traditional research barriers: Insights from the United States," *Energy Research & Social Science*, 19: 37–38.

Hess, C. (2008) "Mapping the new commons," Working Paper, available at https://ssrn.com/abstract=1356835 (last accessed October 8, 2018).

Hess C. & Ostrom E. (2007) *Understanding Knowledge as a Commons*. Cambridge, MA: MIT Press.

Horlitz, S. (2012) "Housing beyond profit: A comparison of U.S. and German alternative ownership models," *American Institute for Contemporary German Studies*, available at www.aicgs.org (last accessed October 17, 2017).

Hudson, B. & Rosenbloom, J.D. (2013) "Uncommon approaches to commons problems: Nested governance commons and climate change," *Hastings Law Journal*, 64: 1273–1342.

Iaione, C. (2010) "The tragedy of urban roads: Saving cities from choking, calling on citizens to combat climate change," *Fordham Urban Law Journal*, 37(3): 889–929.

Iaione, C. (2015) "Governing the urban commons," *Italian Journal of Public Law*, 7: 170–221.

Iaione, C. (2016) "The CO-city," *American Journal of Economics and Sociology*, 75: 415–455.

Iaione, C. (2017a) "The right to the co-city," *Italian Journal of Public Law*, 80: 170–217.

Iaione, C. (2017b) "The co-city project journal № 1," available at www.uia-initiative.eu/en/uia-knowledge-first-uia-expert-journals-published (last accessed October 8, 2018).

Iaione, C. & Cannavo, P. (2015) "The collaborative and polycentric governance of the urban and local commons," *Urban Pamphleteer* 5: 29. www.ucl.ac.uk/urbanlab/research/urban-pamphleteer/UrbanPamphleteer-5 (last accessed October 8, 2018).

Iaione, C. & De Nictolis, E. (2017) "Urban pooling," *Fordham Urban Law Journal*, 44: 665–701.

International Capital Market Association (2017) "The social bond principles," available at www.icmagroup.org/assets/documents/Regulatory/Green-Bonds/SocialBondsBrochure-JUNE2017.pdf (last accessed 21 January 2018).

Jacobs, J. (1961) *The Death and Life of Great American Cities*. New York: Vintage Books. Reissue edition (December 1, 1992).

La Borda (2017) Available at www.laborda.coop/en/ (last accessed November 2, 2017).

Laino, G. (2012) *Il fuoco nel cuore e il diavolo in corpo*. Milan, Italy: Franco Angeli.

Linebaugh, P. (2008) *The Magna Carta Manifesto: Liberties and Commons For All*. Berkeley, CA: University of California Press, p. 59.

Macao (2017) Available at: www.macaomilano.org/ (last accessed October 9, 2018).

Madison, M.J., Frischmann, B.M. & Strandburg, K.J. (2010) "Constructing commons in the cultural environment," *Cornell Law Review*, 95: 657–710.

Madison, M.J., Frischmann, B.M. & Strandburg, K.J. (eds.) (2014) "Governing knowledge commons," in *Governing Knowledge Commons*. Oxford, UK: Oxford University Press.

Madison, M.J., Strandburg, K.J. & Frischmann, B.M. (2016) "Knowledge commons," in Menell, P. & Schwartz, D. (eds.) *Research Handbook on the Economics of Intellectual Property Law* (Vol. II – Analytical Methods). Cheltenham, UK: Edward Elgar Publishing.

Medoff, P. and Sklar, H. (1994) *Streets Of Hope: The Fall And Rise Of An Urban Neighborhood*. New York: South End Press.

Mori, P.A. (2014) "Community and cooperation: The evolution of cooperatives towards new models of citizens' democratic participation in public services provision," *Annals of Public and Cooperative Economics*, 85: 327–352.

New York City, Office of the Mayor (2017) "De Blasio administration and Brownsville community leaders announce NYCx Co-Lab challenges in Brownsville," available at www1.nyc.gov/office-of-the-mayor/news/679-17/de-blasio-administration-brownsville-community-leaders-nycx-co-lab-challenges-in (last accessed January 7, 2018).

Mystica, A. (2016) "A comparative look at international approaches to social enterprise: public policy, investment structure, and tax incentives," *William & Mary Policy Review*, 7(2): 1–34.

Murray, H.J. (2012) "Choose your own master: social enterprise, certifications, and benefit corporation statutes," *American University Business Law Review*, 2: 1–54.

Nagendra, H. & Ostrom, E. (2014) "Applying the social-ecological system framework to the diagnosis of urban lake commons in Bangalore, India," *Ecology and Society*, 19: 67. www.ecologyandsociety.org/vol19/iss2/art67/ (last accessed October 8, 2018).

Office of the Regulator of Community Interest Companies (2013) *Community Interest Companies: Case Studies*, available at www.gov.uk/government/collections/community-interest-companies-case-studies (last accessed January 15, 2018).

Ostrom, E. (1990) *Governing the Commons: The Evolution of Institutions for Collective Action*, available at https://wtf.tw/ref/ostrom_1990.pdf (last accessed October 8, 2018).

Ostrom, E. (2005) "Unlocking public entrepreneurship and public economies," Working Paper DP2005/01, World Institute for Development Economic Research (UNU-WIDER), available at www.wider.unu.edu/publication/unlocking-public-entrepreneurship-and-public-economies (last accessed January 7, 2018).

Ostrom, E. (2010a) "Beyond markets and states: Polycentric governance of complex economic systems," *American Economic Review*, 100: 641–672.

Ostrom, E. (2010b) "Polycentric systems for coping with collective action and global environmental change," *Global Environmental Change*, 20: 550–557.

Ostrom, V., Tiebout, C.M. & Warren, R. (1961) "The organization of government in metropolitan areas: A theoretical inquiry," *American Political Science Review*, 55: 831–842.

Paris Mayor (2014) *Participatory Budget of The City of Paris*, available at: https://budgetparticipatif.paris.fr/bp/ (last accessed October 8, 2018).

Patti, D. & Polyak, L. (2017) "Funding the cooperative city," available at www.cooperativecity.org (last accessed January 21, 2018).

Pilastro (2016) Available at https://pilastro2016.wordpress.com/2016/03/14/lagenzia-ce-e-nata-sviluppo-pilastro-nord-est-onlus/ (last accessed October 15, 2017).

Portugali, J., Meyer, H., Stolk, E. & Tan, E. (2012) *Complexity Theories of Cities Have Come of Age*. Berlin, Heidelberg: Springer.

Rae, Douglas. 2003. *City: Urbanism and Its End*. New Haven, CT: Yale University Press.

Regies des quartier (2017) Available at www.regiedequartier.org/regies-de-quartier/carte/ (last accessed November 10, 2017).

REIC (2018) Available at www.nycreic.com (last accessed 6 January 2018).

Rose, C.M. (1988) "The several futures of property: of cyberspace and folk tales, emission trades and ecosystems," *Minnesota Law Review*, 83: 129–182.

Sabel, C. & Dorf, M. (1998) "A constitution of democratic experimentalism," *Columbia Law Review*, 98: 267–473.

Sassen, S. (2015) *Who Owns the City?* www.theguardian.com/cities/2015/nov/24/who-owns-our-cities-and-why-this-urban-takeover-should-concern-us-all (last accessed October 8, 2018).

Scruggs, G. (2017) "Can Swiss housing cooperatives introduce their approach in developing countries?" *CitiScope*, available at http://citiscope.org/story/2017/can-swiss-housing-cooperatives-introduce-their-approach-developing-countries (last accessed October 8, 2018).

Social Finance Group (2017a) "Peterborough," available at www.socialfinance.org.uk/projects/peterborough (last accessed November 15, 2017).

Social Finance Group (2017b) "Bright futures fund," available at www.socialfinance.org.uk/projects/bright-futures-fund (last accessed November 15, 2017).

Sylvain, O. (2012) "Broadband localism," *Ohio State Law Journal*, 73: 795–838.

Sylvain, O. (2016) "Network equality," *Hastings Law Journal*, 67: 443–498.

SynAthina Platform (2013) Available at: www.synathina.gr/en/ (last accessed October 8, 2018).

Tang, B. (2015) "'Not rural but not urban': Community governance in China's urban villages," *The China Quarterly*, 223: 724–744.

Thompson, M. (2015) "Between boundaries: From commoning and guerrilla gardening to community land trust development in Liverpool," *Antipode*, 47, 4: 1021–1042.

UK Government (2017) *Guide on Community Interest Companies*, available at: https://goo.gl/rAbXSE (last accessed October 9, 2018).

Unnikrishnan, H., Manjunatha, B. & Nagendra, H. (2016) "Contested urban commons: Mapping the transition of a lake to a sports stadium in Bangalore," *International Journal of the Commons*, 10, 1: 265–293.

Un Passo per San Luca (2014) Available at www.unpassopersanluca.it/ (last accessed October 20, 2017).

Zheng Lu, Peña-Mora, F., Xiaodong, R.W., Shen, C.Q. & Riaz, Z. (2015) "Social impact project finance: An innovative and sustainable infrastructure financing framework," *Procedia Engineering*, 123: 300–307.

20

INFRASTRUCTURE AND ITS GOVERNANCE

The British Broadcasting Corporation case study

Brett Frischmann

Introduction

The BBC is a publicly funded and regulated corporation, a complex media organization, a community, a public space, and even an iconic cultural and social brand. It is a massive £3.7bn annual investment in the knowledge economy (BBC 2014: 4), which according to some studies has an economic impact in excess of £8bn per year (BBC 2013a). The scale and scope of the BBC and its activities are incredibly expansive, spanning the globe and reaching almost all UK citizens on a regular basis. Ninety-seven percent of UK citizens use the BBC each week, and they use it extensively (approximately 18 hours per week, on average) (BBC 2015a: 2).

First and foremost, the BBC is a public service broadcaster incorporated under a Royal Charter and financed primarily by a license fee paid by UK citizens who require a TV license. It is a very unusual public entity, widely known and respected for its independence from commercial and political influence and its dedication to the public interest. The BBC exists to serve the public interest, and its core mission to inform, educate and entertain shapes the range of activities undertaken by the BBC. Specifically, the Charter states that "BBC's main activities should be the promotion of its Public Purposes through the provision of output which consists of information, education and entertainment, supplied by means of—(a) television, radio and online services; (b) similar or related services which make output generally available" (Royal Charter 2016). In other words, the BBC's main function is the generation and distribution of content that informs, educates and entertains the public and serves the following Public Purposes, which are enumerated in the Charter:

(a) sustaining citizenship and civil society
(b) promoting education and learning
(c) stimulating creativity and cultural excellence
(d) representing the UK, its nations, regions and communities
(e) bringing the UK to the world and the world to the UK
(f) in promoting its other purposes, helping to deliver to the public the benefit of emerging communications technologies and services and, in addition, taking a leading role in the switchover to digital television.

Thus, in stark contrast with competing broadcasters and content producers who also produce content that inevitably informs, educates or entertains in one way or another, the BBC must focus on allocating resources toward specific public ends, each of which entails the development of public and social goods.

A complex governance regime is articulated in the BBC's Framework Agreement and other regulatory instruments, some of which focus specifically on the BBC and others of which apply industry-wide. The BBC Trust, the BBC Executive Board, and OFCOM, among others, hold the BBC accountable to the public through evaluation of its strategic direction, operational management, sustained commitment to independence, and regulatory compliance. The Charter itself requires review and renewal after ten years. This timeframe gives the BBC a relatively long time horizon for planning and investment while subjecting the BBC to intense public scrutiny on a periodic basis.

This chapter aims to examine the infrastructural nature of this system, primarily by identifying and examining infrastructural resources within the BBC and made available by the BBC to the public. The following section provides a brief discussion of the economic theory of infrastructure and commons. It does not comprehensively examine or evaluate the BBC or advocate a specific position in the ongoing debates about the future of the BBC.

The reasons for undertaking this inquiry are to bring to the surface something that is latent and unappreciated yet incredibly important about the BBC. The BBC provides *public infrastructure*, and as is quite common with all sorts of infrastructure, it is all too easy to take it for granted. Bringing this aspect of what the BBC does to the surface is relevant to ongoing debates about the future of the BBC and public service broadcasting more generally.

The BBC began as a monopoly, regulated as a public utility and financially supported through a levy on all households that received broadcast signals. Like many regulated monopolies of the 20th century, the BBC was tasked with serving the public interest and providing universal service. These core commitments remain. Also like many regulated monopolies of the 20th century, the BBC has had to adapt to significant changes in the political, economic and technological environment. Deregulation, advances in communications technologies, and the transition to competitive markets have made public utility theory less relevant to the BBC. Still, it is worth noting that the BBC's history and experience is similar to that of many other infrastructure industries (Frischmann 2012).

In the past, the BBC has been justified through appeals to the economic theory of public goods. The textbook definition of a public good requires the good to be *nonrivalrous in consumption*, meaning one person's consumption does not reduce the consumption opportunities of others, and *nonexcludable*, meaning it is too costly to exclude people from consuming the good. Nonrivalrous consumption also means that the marginal costs of allowing another person to consume the good is zero, and thus it is cheaply, often freely, sharable (Frischmann 2012: ch. 3). Information, education, and entertainment ("content") are classic examples of a public good, regardless of medium.

Public goods theory is a useful place to start, but it is not enough. It fails to describe fully what role the BBC plays in British society. For example, some have suggested that since technology now makes exclusion relatively cheap (e.g., encrypted broadcast signals), content is no longer properly classified as a public good and the BBC's role as public good provider should shrink because markets should no longer fail. This argument is flawed, however. Low cost exclusion only means that a specific type of market failure can be ignored—specifically, the risk that free riding will prevent investors from recovering their costs in market transactions for the specific good (e.g., encrypted content). Accordingly, with low cost exclusion, markets will produce some types of commercial content more efficiently than otherwise. But it does not mean that markets will produce *all* types of content efficiently. For example, the market may under supply certain types of content (e.g., regional programming) because it is not sufficiently profitable. Simply put, even with low cost exclusion, a host of market failures remain. Society demands a wide range of content, the value of which is not fully reflected in price signals and market transactions (Mediatique report for Ofcom 2014). In this context, the BBC continues to play a critical role in ensuring society is provided with a wide range of public and social goods that are underprovided by markets.[1]

To appreciate how the BBC helps to resolve these wider market failures, as well as providing a bridge between markets, government, and society, it is helpful to move beyond simple public goods theory and look to a recently developed economic theory of infrastructure managed as commons. Commons management is a mode of governance for infrastructure that can be especially attractive from an economic and social perspective. The core principle of commons management is sustainable sharing on nondiscriminatory terms within a defined community (Frischmann et al. 2014). With respect to infrastructure,[2] this means nondiscrimination among users and uses, and the corresponding preservation of equality and general purposiveness.

An overview of infrastructure commons theory

A shift in how we think about infrastructure

"Infrastructure" generally conjures up the notion of a large-scale physical resource made by humans for public consumption. Familiar examples of "traditional infrastructure" include (1) transportation systems, such as highway systems, railway systems, airline systems, and ports; (2) communication systems, such as telephone networks and postal services; (3) governance systems, such as court systems; and (4) basic public facilities, such as school, sewers, and water systems. The list could be expanded considerably; daily, we rely on a wide range of traditional infrastructure resources. Recent law and economics research recognizes a more capacious understanding of infrastructure is warranted for functional economic analysis. In particular, recognizing that infrastructural resources are *shared means to many ends* expands the definitional scope beyond traditional infrastructure categories. A host of "nontraditional infrastructure" resources exist and also serve as the underlying frameworks or foundations of various economic and social systems. Examples include environmental infrastructure, such as oceans and the atmosphere, as well as intellectual infrastructure, such as basic research, ideas, and languages. As explored later, the BBC may be a decent example of a nontraditional infrastructure.[3]

Characteristics of infrastructural resources

Infrastructural resources are intermediate capital resources that serve as critical foundations for productive behavior within economic and social systems. Infrastructural resources effectively structure in-system behavior at the micro-level by providing and shaping the available opportunities of many actors. In some cases, infrastructural resources make possible what would otherwise be impossible, and in other cases, infrastructural resources reduce the costs and/or increase the scope of participation for actions that are otherwise possible. The difference between these two sets of cases may depend on the existence of other infrastructure resources. So, for example, some form of transportation infrastructure may be a necessary precursor to the regular movement of people and/or goods from Region A to Region B and thus to economic and cultural engagement between the regions. Once such an infrastructure is in place, additional forms of transportation infrastructure may be effective in lowering costs or increasing opportunities to participate in various activities enabled by the initial transportation infrastructure. Road and rail systems have a long history that illustrates these points. Both systems also depend on the advancement and availability of complementary technologies, institutions, and social practices.

Infrastructural resources often structure in-system behavior in a manner that leads to spillovers (Frischmann 2012; Frischmann and Hogendorn 2015). That is, infrastructural resources facilitate productive behaviors by users that affect third parties, including other users and even non-users of the infrastructure. The third party effects often are accidental, incidental, and not especially relevant to the infrastructure provider or user. To put it another way, the social returns on infrastructure investment and use may exceed the private returns because society realizes benefits above and beyond those realized by providers and users.

According to Frischmann (2012):

> Infrastructural resources satisfy the following criteria:
>
> (1) The resource may be consumed nonrivalrously for some appreciable range of demand.
> (2) Social demand for the resource is driven primarily by downstream productive activities that require the resource as an input.
> (3) The resource may be used as an input into a wide range of goods and services, which may include private goods, public goods, and social goods.

Traditional infrastructure, such as roadways, telephone networks, and electricity grids, satisfy this definition, as do a wide range of resources not traditionally considered as infrastructural resources, such as lakes, ideas, and the Internet. The first criterion captures the consumption attribute of public and impure public goods (Samuelson 1954; Cornes and Sandler 1996). In short, this characteristic describes the "sharable" nature of infrastructural resources. The second and third criteria focus on the manner in which infrastructural resources create social value. The second criterion emphasizes that infrastructural resources are capital goods that create social value when utilized productively downstream and that such use is the primary source of social benefits. Thus, societal demand for infrastructure is *derived demand*. The third criterion emphasizes both the variance of downstream outputs (the genericness of the input) and the nature of those outputs, particularly, public goods and social goods.

Demand-side market failures that infrastructural resources help resolve and why they arise

Infrastructural resources are not special-purpose resources optimized for a specific user or use to satisfy the demand derived from a specific downstream market. Instead, and importantly, they provide basic, multipurpose functionality. An electricity grid, for example, delivers power to the public, supporting an incredibly wide range of uses, users, markets, and technologies. It is not specially designed or optimized for any specific use, user, market, technology, or appliance; it provides nondiscriminatory service for a toaster and a computer, for staples and a pizzeria, and so on. The same could be said of the national highway system, the Internet, or even a language. Users determine what to do with the capabilities that infrastructure provides. Genericness implies a range of capabilities, options, opportunities, choices, freedoms. Subject to standardized compatibility requirements, users decide what to plug in, run, use, work with, play with. Users decide which roads to travel, where to go, what to do, who to visit. Users choose their activities; they can choose to experiment, to innovate, to roam freely. Users decide whether and what to build. Users decide how to use their time and other complementary resources. Infrastructure (providers) enable, support, and shape such opportunities (Figure 20.1).

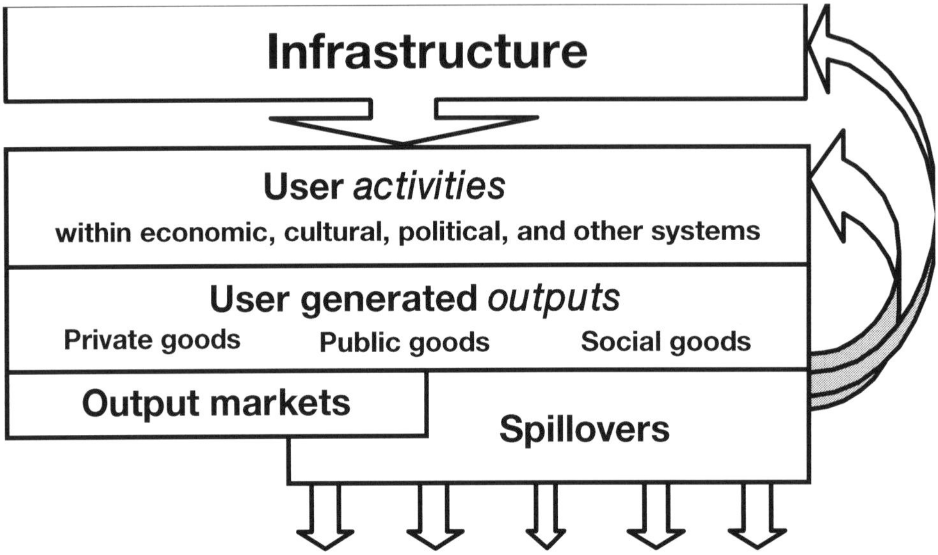

Figure 20.1 Infrastructure

Ultimately, the value of an infrastructural resource is realized by producers and consumers of these outputs. It is thus the demand for these outputs that determines demand for the infrastructure. For this reason, the third criterion emphasizes the nature of the downstream outputs and, more specifically, the potential production of public and social goods. When infrastructure supports these productive activities, markets will not work well in assessing demand and supplying infrastructural resources; a gap between private demand and social demand arises because the social value created by allowing users to access and use the resource to produce public and social goods is underrepresented in the prices people are willing to pay for infrastructure (Musgrave and Musgrave 1984). The social value may be substantial but extremely difficult to measure. This is the crucial market failure unaddressed by the conventional frame for looking at the BBC, namely public good theory. The BBC—as a provider of public infrastructure—addresses this wider market failure in a number of ways, and appreciating this is critical to understanding the BBC's role as a bridge between markets, government, and society. It is, therefore, worth considering why this demand-side market failure arises.

The reason for the gap between private demand and social demand is relatively straightforward: Infrastructure users' willingness to pay reflects private demand—the value that they expect to realize—and does not take into account value that others might realize as a result of their use. That is, it does not account for external effects associated with the production of public and social goods. This means that infrastructure users who produce public and social goods are not necessarily optimal purchasers of access and use rights, because they do not themselves capture the full social value of their use.

Not only is there a gap between private and social demand but, to make matters even more complex, the shape of the social demand curve can be quite different from that of the private demand curve due to the variance in producers and outputs. This added complexity gives rise to specific and substantial demand-side market failures that have static and dynamic consequences (see Frischmann 2012).

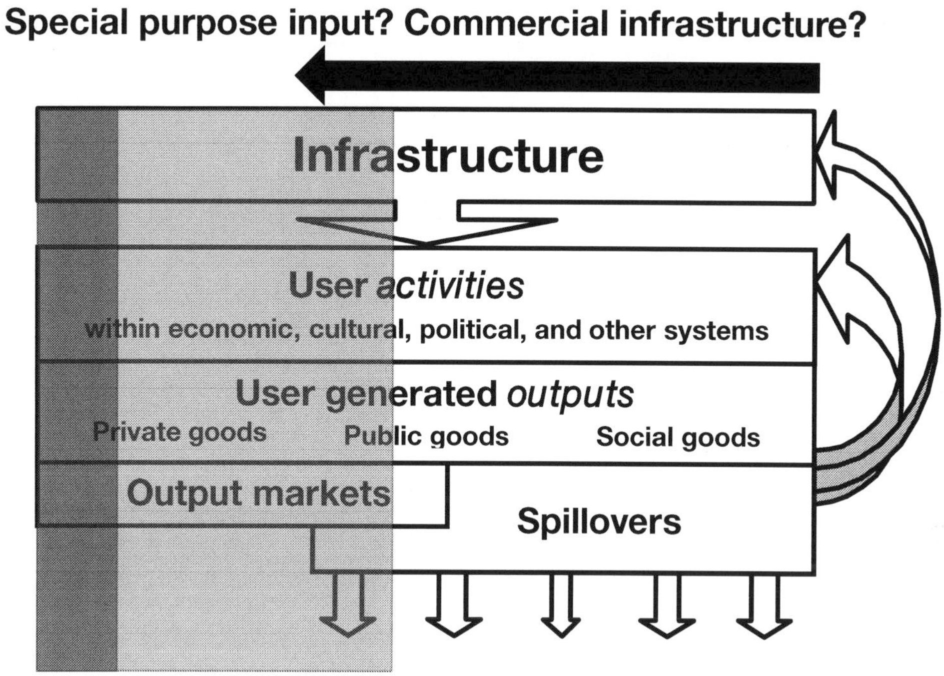

Figure 20.2 Infrastructure optimized for market demand

Difficulties in measuring and appropriating value generated in output markets translates into a problem for infrastructure suppliers and, consequently, the public. This "demand-manifestation" problem may affect infrastructure allocation, design, investment, and management, as well as other supply-side decisions. At least in market contexts, infrastructure suppliers base such decisions in large part on the prospect of foreseeable returns in downstream markets. Demand signals manifested in those markets, aggregated and, in a sense, communicated upstream by dynamic operation of the price mechanism, provide critical raw information for making assessments about prospective returns. To society's detriment, demand-manifestation problems can lead to the undersupply of infrastructure essential to various producers of public and social goods, and this undersupply can lead to an optimization of infrastructure design or prioritization of access and use of the infrastructure for a narrower range of uses than would be socially optimal (Figure 20.2).

As discussed later, what this means for the BBC is that private demand manifest in markets by consumers' willingness to pay is likely to understate the social value of the BBC infrastructure and public and social good outputs generated by the BBC and its audience. We should expect the private for-profit media infrastructure and outputs generated by users of such infrastructure to be systematically biased towards content that satiates, entertains, and otherwise satisfies consumptive demands while maximizing profits.

Why commons management of infrastructure resources can help avert market failures

Commons management is a mode of governance for infrastructure that can be especially attractive from an economic and social perspective. The core principle of commons management

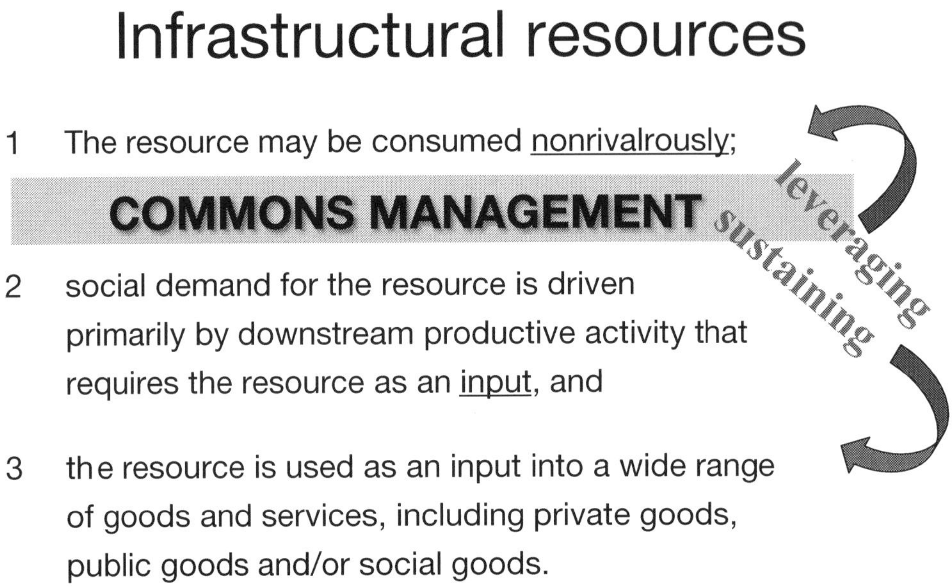

Figure 20.3 Infrastructure with commons management

is sustainable sharing on nondiscriminatory terms within a defined community (Frischmann et al. 2014). With respect to infrastructure,[4] this means nondiscrimination among users and uses, and the corresponding preservation of equality and general purposiveness. From an economic perspective, managing infrastructural resources as a commons often makes sense because doing so permits a wider range of downstream producers of private, public, and social goods to flourish. "The high variability in value of using both transportation and communications facilities from person to person and time to time have made a commons-based approach to providing the core facilities immensely valuable" (Benkler 2001). The point is not that all infrastructural resources should be managed as commons. Rather, for certain classes of resources, the economic arguments for managing the resources as commons vary in strength and substance (Figure 20.3). (For further development of those arguments, see Frischmann 2012; see also Benkler 2013.)

By precluding optimization and prioritization based on market demands alone, commons management sustains the social option value of the infrastructure.[5] But commons management is not only a buffer from market pressures. It also serves as a buffer from government pressures to optimize or prioritize infrastructure (use), which may originate from market actors lobbying for government interventions or from other pressures to regulate. Finally, managing public infrastructure as commons also reduces reliance on government to pick winners and direct subsidies to users who produce public and social goods. In other words, there is less reliance on government grants and subsidies. The government also faces demand manifestation problems because it does not know the shape of the demand curve and may struggle to identify which public or social goods to subsidize and which producers to fund. Government is often better off supporting open infrastructure as a means for supporting the capabilities of productive users.

In the end, the basic idea is that commons management leverages nonrivalry and the sharable nature of the infrastructure (first criterion) to sustain the general-purpose nature of the infrastructure and support the wide range of user activities that generate private, public and social goods (second and third criteria).

Applying infrastructure commons theory to the BBC

Conceiving of the BBC as producing and managing public infrastructure is helpful in two ways. First, it helps to explain what the BBC has done effectively and what distinguishes the BBC from other broadcasters, media companies, and potential competitors. Of course, the BBC and the public do understand what is special about the BBC, but as is common with much infrastructure, it is easy to lose sight of its value to society. The infrastructure framework can be a powerful reminder of—and a lens with which to identify and examine—hard-to-appreciate social value, for example, from the generation of spillovers. It also connects the resources—the *infrastructure*—to a specific management and governance regime—*commons*.

Second, thinking of the BBC as producing and managing infrastructure provides a useful and inspiring vision for the future. The BBC should continue to play and perhaps expand its infrastructural role in the global networked information environment. This section primarily addresses the first of these areas; in this section, I select a few examples among many that illustrate ways in which the BBC acts as a public infrastructure. I pick up the second area in the next section and set out a few tentative suggestions for the future of the BBC as well as for future research and discussion.

As discussed earlier, to society's detriment, demand-manifestation problems can lead to the undersupply of infrastructure essential to various producers of public and social goods, and this undersupply can lead to an optimization of infrastructure design or prioritization of access and use of the infrastructure for a narrower range of uses than would be socially optimal. What this means for the BBC is that private demand manifesting in markets by consumers' willingness to pay is likely to understate the social value of the BBC infrastructure and public and social good outputs generated by the BBC and its audience. We expect the private for-profit media infrastructure and outputs generated by users of such infrastructure to be systematically biased towards content that satiates, entertains, and otherwise satisfies consumptive demands while maximizing profits.

In the first subsection, below, I describe the core, internal capital resources shared within the BBC organization and not necessarily with the public. Like many organizations, the BBC produces and sustains an internal system of shared, complementary capital resources, and the system functions as infrastructure. After describing the resource system, I turn to the wide range of public and social goods the BBC produces and focus on four categories of outputs—content, R&D, human capital and social capital.

In the following subsection, I examine governance and evaluate the extent to which the BBC manages its infrastructural resources as commons. The BBC's special governance regime, including the funding mechanism, public interest orientation, and commitment to independence, achieve a similar outcome as the core principle of nondiscrimination at the heart of commons management. Both independence and nondiscrimination aim to achieve insulation from the systemic biases in market and political systems.

BBC infrastructure

One approach to analyzing the BBC infrastructure would be to black box the BBC and focus on the wide range of private, public and social goods that the BBC generates. This approach

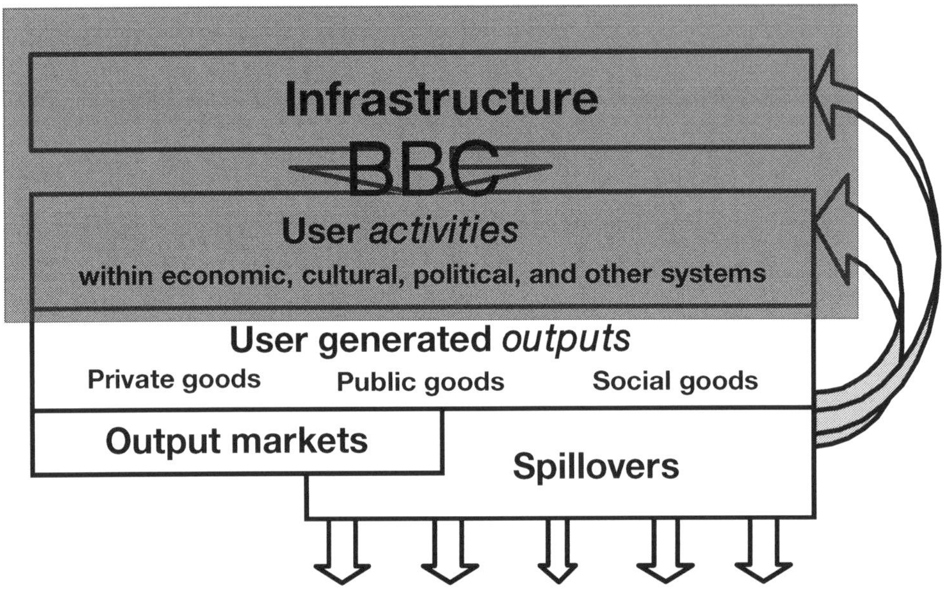

Figure 20.4 BBC infrastructure with commons management

treats the BBC itself as the infrastructural resource. The internal details may not matter so much to outsiders, and the black box would simplify the analysis and focus our attention on the third criterion—the wide variety of outputs. This is reflected in Figure 20.4.

The BBC is a means to many ends, and those ends, specified as public purposes in its 2007 Charter, correspond to an array of public and social goods. These ends are "high-level, serious, ambitious and comprehensive," and they also constitute criteria for evaluating and holding the BBC accountable (Hughes 2015: 11). Yet, as the House of Lords Communications Committee review of Public Service Broadcasting reported:

> Neither Ofcom's nor the BBC's set of objectives is sufficiently discriminating to distinguish between those programmes and services that merit public finance and special regulatory treatment and those that do not. Few programmes screened by any UK broadcaster would not satisfy at least one of the objectives.
>
> *(Hughes 2015, quoting House of Lords Select Committee on Communications (2009))*

Of course, that may be the point. The BBC governance regime insulates the micro-level decisions from discriminating judgments on a programme by programme basis and forces external evaluation at a more macro-level. This makes sense if the BBC is sustaining public infrastructure internally and managing such infrastructure as a commons requires leaving freedom for its internal users to determine what to do with the capabilities that infrastructure provides. This resonates with a point made by Lord Birt while giving evidence to the House of Lords Communications Committee. Public service broadcasting "is a programme tradition [with] the citizen rather than the consumer in mind," and "If you are in the system, you truly know the difference" (Hughes 2015: 11).

Lord Birt's reference to programme tradition and citizens rather consumers implies some normative constraints on how BBC employees use the internal infrastructure. We need to peek under the hood and get a better sense of the internal system. Unfortunately, the black box approach obscures too much. (To the extent that this concern seems overstated and the black box approach seems attractive, skip the discussion of internal infrastructure and focus on the outputs and impacts.) Among other things, the black box approach leaves ambiguous which resources are infrastructural, who are the users, and how resources are shared and used productively. Moreover, except at a very abstract level, it doesn't make sense to describe the BBC as a resource used as an input into a wide range of goods and services. It makes more sense to describe the BBC as a vertically integrated infrastructure provider and user that facilitates a wide range of goods and services.

A better approach, therefore, may be to conceptualize the BBC as a complex system with many different complementary capital resources used internally within the organization. I took a similar approach with respect to university science and technology systems (Frischmann 2009). In many ways, the BBC functions like a university science and technology system, except that the BBC aims to produce mass media content rather than research. The internal infrastructural system is comprised of at least five different sets of related, complementary resources, including:

1. Human capital, including complementary networks of people with knowledge, skills, and talent[6]
2. Governance capital, such as rules, norms, policies and other collective constraints that guide system participants' behavior
3. Physical capital, such as land, facilities, and equipment
4. Intellectual capital, such as knowledge, information, and ideas[7] and
5. Financial capital, such as license fee revenue.

Most organizations have some mix of these capital resources, and certainly, all of the BBC's competitors in broadcasting have internal infrastructural systems. The capital resources within the system are "sharable" in the sense that multiple users may access and use the system resources to engage in productive processes. Some components, such as intellectual and governance capital, are purely nonrival in consumption, while others, such as physical, financial, and human capital, are rival in consumption. The scarcity of these latter types of capital resources drives resource allocation decisions. To some extent, rivalrousness within the system is what puts pressure on the BBC's competitors to tailor or even optimize their systems for commercial outputs; the appropriable benefits (revenues) generated by such outputs provide the resources necessary to sustain the system. Of course, this is an attractive feedback loop for commercial entities producing goods and services to satisfy consumer demand. Market forces provide discipline and better ensure that broadcasters are allocating internal resources to the production and distribution of content for which people are willing to pay.

The BBC is not under the same pressure, however, and as a result, it does not optimize its internal infrastructure based on commercial demand. The BBC's special governance regime, including the funding mechanism, public interest orientation, and commitment to independence, distinguishes the BBC from its competitors. This means that the BBC does not rely on prospective revenue generation in commercial markets when deciding how to allocate its resources because it has a stable source of revenue from the license fee. So, for example, the BBC does not rely on expected financial returns from projects when deciding how to allocate an employee's time among those projects.

Critically, this does not mean that the BBC can avoid the pressures imposed by resource constraints, audience demand, and external governance. The BBC makes allocation decisions based on its mission, values, priorities and its governance regime, which includes and depends significantly upon internal governance capital—the rules, norms and decision-making hierarchies and processes that have developed and evolved within the BBC organization over decades. As Lord Birt explained, "If you are in the system, you truly know the difference" (Hughes 2015). You know the difference because you have learned and internalized it through training and experience. Just as the inclusion of human capital as a component of the internal infrastructural system integrates resource and resource user, the inclusion of governance capital as a component of the internal infrastructural system integrates resource and resource management. Such integration blurs lines that we might like to keep distinct for purposes of modeling, but it is a necessary reality and reminds us that we are evaluating a complex, dynamic system.

The BBC's bundle of capital resources has grown and evolved over decades. Its long history, stable funding, scale and scope have enabled the BBC to develop a considerably valuable mix of capital resources. It is worth noting that some of these capital resources depreciate with use and over time (e.g., physical capital), and others do not (e.g., governance capital). Some types of intellectual and human capital may even appreciate with use because of learning, network effects from standardization, and so on. I do not explore this point in this chapter, but the implication is important: There are considerably valuable capital assets internal to the BBC. Past public investments are not simply sunk costs; they are long-term investments in public infrastructure.[8] Aggregated together within the BBC, these resources are used and reused collectively and continuously as inputs into a variety of production processes and services to generate a wide variety of private, public and social goods.

The BBC provides a wide range of public and commercial services. The public services comprise nine domestic free-to-air television channels, some contributions to a Welsh language channel (S4C) and a Scottish Gaelic language channel (BBC Alba), over 50 radio stations, various online services through BBC.co.uk and the BBC iPlayer, support for orchestras, educational services, and many other smaller scale initiatives. The BBC provides commercial services through BBC Worldwide and BBC Studios. BBC Worldwide comprises various broadcasting channels, sales and distribution of content and merchandising, and other related activities. BBC Studios supplies production facilities and the BBC World News television channel.

For each service, the BBC manages a supply chain that typically involves at least three stages—the production of content, the aggregation of content into particular channels, and the distribution of content. Each stage involves BBC employees working within and across departments and with third parties (e.g., independent contractors), drawing on the resources noted above and using them productively to generate various outputs. Of course, distribution itself depends upon different communications infrastructures for various media channels (radio, television, and Internet). For brevity's sake, I do not discuss these infrastructures much in this chapter. Most of the BBC's spending is allocated toward the first two stages in the supply chain.

I discuss four categories of outputs: content, R&D, human capital, and social capital. For each, my aim is to classify the outputs in economic terms, provide a few examples, and suggest how the outputs relate to the BBC's mission and public purposes. I also consider how these outputs produce value in economic, cultural, and political systems. Some of this value is related to pure consumption by audiences who are entertained, and some of this value is related to the productive impact of the content on audience members, their participation in economic, cultural and political systems, and spillovers that would not be fully captured in market transactions. In many cases, the BBC-generated outputs also are infrastructural.

1. Content

All content producers and broadcasters compete for audience attention and thus seek to satisfy the demands of audience members. The BBC is different, however, in important ways. One differentiating factor is that the BBC aims for *universality*, meaning not only that everyone obtains free access to the content produced but also that BBC produces "something for everyone." This commitment pushes the BBC to produce a more diverse range of content than commercial competitors. There is plenty of evidence in both television and radio that demonstrates such diversity. Success in achieving the universality objective is reflected in the public's incredibly high usage of the BBC coupled with its strong reputation for quality and audience satisfaction. According to a 2015 report, 97 percent of UK citizens use the BBC each week, and they use it extensively, approximately 18 hours per week, on average (BBC 2015a: 2, 8–9).

Another differentiating factor is public trust in the impartiality of the BBC (BBC Trust 2008 (describing surveys); BBC 2015a). Such trust may explain why much larger audiences tune into the BBC for certain categories of content even when competitors offer much more of it. For example, the BBC shows only 22 percent of all TV News minutes broadcast in the UK but attracts 74 percent of the viewing. Similarly, in sports, the BBC shows only 3 percent of all the hours broadcast on UK TV but attracts a full 45 percent of all the viewing. By contrast, Sky Sports provides 53 percent of output but attracts only 24 percent of viewing (BARB 2014, cited in BBC 2015a). Of course, there may be other explanations.

Another differentiating factor is the *nature and quality* of the content produced by the BBC. BBC content generally is the final product of a rather complex process that involves a host of different creative, technical, and other professionals who make a wide range of different contributions. When produced in-house, the different contributions draw on all of the capital resources that comprise the system. When commissioned, only some of the internal capital resources are necessarily used; the indie producer supplies some as well. But how does the BBC allocate its resources? How does it choose which projects to support?

While necessarily attentive to the demands of audience members, the BBC also is attentive to social demands reflected in its mission, which is to produce high quality content that informs, educates and entertains the public and serves the public purposes enumerated in its Charter. Moving from mission to decision-making depends on the governance capital, including internal norms, strategies, hierarchies and decision-making processes, as well as external governance. The BBC Trust provides external oversight of the BBC strategic direction. In its 2010 submission to the BBC Trust, *Putting Quality First: The BBC and Public Space,*[9] the BBC articulated a strategy for:

> [h]igh-quality, distinctive content . . . built around four characteristics—the combination of which will define the BBC's distinctiveness across a broad range of genres and output:
>
> - *Excellence*: the highest production standards and the best talent to keep British content competitive with the best in the world
> - *Originality*: original production providing range and depth; fresh and new ideas; and creative and editorial ambition to meet public expectations
> - *Trustworthiness*: meeting the public's expectations in editorial standards and impartiality, and providing a benchmark for broadcasting in the UK
> - *Building on British talent*: focusing on content made for the UK, finding the next generation of British writers, directors and performers and helping support UK creative industries across the country.

This strategy provides a glimpse into the decision-making criteria, although these characteristics remain quite generic and leave considerable discretion to those making production decisions. Such discretion might be necessary given the range of different channels and audiences that the BBC seeks to reach.

In *Putting Quality First* (BBC 2010a), the BBC also articulated five clear content priorities for its investment decisions:

1. *The best journalism in the world*: Informing civic and democratic life at home and abroad—through independent, impartial and accurate news, current affairs and information.
2. *Inspiring knowledge, music and culture*: Enriching people's lives—bringing knowledge, music and culture to new minds, eyes and ears.
3. *Ambitious UK drama and comedy*: Stimulating and entertaining audiences—with stories about their lives and the world around them.
4. *Outstanding children's content*: Delighting and surprising young audiences—helping children explore their world in a safe public space.
5. *Events that bring communities and the nation together*: Being there for the whole UK—in moments of crisis, commemoration and celebration through landmark events, sport and entertainment.

Broadly, these priorities reflect commitments to particular categories of content that are under-provided by markets. A comprehensive analysis and breakdown of the wide variety of content produced and distributed by the BBC is beyond the scope of this chapter.

In terms of applying the infrastructure theory, the key points are: The BBC sustains an infrastructure comprised of various capital resources and uses those resources to produce a wide range of content, much of which generates spillovers and would be undersupplied by markets. How much is difficult to assess. I should note that I have not said much about the alternative to market supply—government supply. To the extent that markets predictably undersupply certain categories of content, for example educational content, there is a case to be made for a more direct government intervention—government subsidies for the specific content. In a sense, the BBC reflects a meso-level intervention in the form of an investment in an infrastructure for selecting and producing categories of such content rather than a series of micro-level interventions that select and fund project by project. There are several reasons why governments may fail in picking winners at the more micro-level.

Content produced by the BBC must be sufficiently entertaining to attract audiences, and its popularity and ratings strongly suggest that the BBC succeeds in generating substantial consumptive value for audience members. But the BBC does much more than entertain; it generates substantial, additional social value in the form of spillovers as a consequence of the nature and quality of the content produced and the productive impact of such content on audience members and society.

Consider, for example, the first priority—*the best journalism in the world*. The BBC is unparalleled in its production of "independent, impartial and accurate news, current affairs and information." It has a strong national and global reputation for independence and trustworthiness. Nationally, the BBC is the most trusted source of news—57 percent of the public picked the BBC when asked to name the one source they trust most, far ahead of the next nearest (ITV on 11 percent), and this holds across all demographic groups across the whole of the UK. Similarly, according to a study done for the BBC (Ipsos MORI 2015), the BBC is named first when people think of the source they turn to first for impartial news coverage. Its reputation for trust and impartiality translates into both audience attention and learning. The BBC shows only 22 percent of all TV News minutes broadcast in the UK but attracts 74 percent of the viewing. Eighty percent of people surveyed indicated that the BBC helps them to understand what is happening in the UK and the world today (BBC 2015a).

Journalism involves the production of a special type of public good because consumption of the good impacts consumers in a manner that itself produces social value. Consumers of "independent, impartial and accurate news, current affairs and information" gain knowledge and become more capable participants in civic, democratic, economic and other social systems, and this benefits not only the person who consumed the good but also other people as well. This is what economists call a spillover or an externality. Markets tend to underprovide goods that generate positive externalities because the external effects are not reflected in the consumers' willingness to pay in market transactions. For example, the amount of money Joe is willing to pay to watch a show tends to reflect the benefits he expects to realize from watching the show and none of the benefits that his neighbors might realize indirectly because Joe becomes better informed about some issue that affects the neighborhood and being better informed may affect Joe's behavior. There is a substantial literature on the systemic distortions or biases in the market for journalism (See, e.g., Baker 1994: 202).

Sometimes the incremental added social value from a specific bit of news may be small and seemingly insignificant, but when the news is distributed (and the value aggregated) across nearly the entire population and the analysis is extended to encompass not just a particular bit of news but instead a regular stream of "independent, impartial and accurate news, current affairs and information," the social value can be substantial,[10] although difficult to capture in conventional economic measures of value. For particularly important public events, such as political campaigns and elections, the incremental added social value of high-quality news and extensive public reach can be quite significant. While the net social value is impossible to measure, it is possible to evaluate quality, impact, and reach for such important public events: For example, surveys indicate that BBC election coverage rated ahead of Sky and ITV on all quality and most impact metrics, and cross-platform reach of BBC election coverage exceeded 80 percent each week of the campaign, peaking in the last week at 89 percent (BBC 2015a: 19).

Each of the categories focuses on content that generates social value not fully reflected in or captured by markets because of externalities. The mechanism by which externalities are produced varies across the categories (Frischmann 2008, 2012: ch. 3). Audience members *consume* the content, but they also *use* the content and produce intellectual, human, and social capital external to the BBC and within the audience or broader public. For example, with respect to the second and fourth priorities, the BBC emphasizes that building knowledge through factual and educational content is central to its mission. In addition to the news, the BBC produces a wide range of factual and educational content for all media (television, radio, and Internet) and the high quality and utility of such content is well-recognized by its audience and external experts and regulators. For reasons similar to those discussed above with respect to journalism, consumers of factual and educational content gain knowledge and human capital that benefit them as well as others within the broader community.

With respect to the third priority, the BBC emphasizes its role as a platform for supporting and cultivating UK writing and performing talent and developing content that connects audience members with each other and their shared UK experiences and culture. Thus, in addition to the consumptive value of the content, this involves the development of both human and social capital. I discuss each below. With respect to the fifth priority, the BBC emphasizes its role in promoting social capital through content that resonates nationally and brings the nation together "at times of celebration, commemoration and entertainment" (BBC 2010a, 30). I discuss this below as well.

In fact, many of the spillover-generating outputs are themselves infrastructural and are used productively by users. For example, audience members who consume BBC content build knowledge, human, and social capital, and these types of capital support public capabilities to participate in economic, cultural, political, and other social systems—to the benefit of society.

The aggregation and bundling of content into channels also constitutes an output to note. The various BBC channels perform a certain type of intermediation, acting as filters with recognizable brands for quality and subject matter. When coupled with the trust cultivated by the BBC with its audience, these channels can reduce search costs and make it easier for consumers to find content that they want. Yet it also permits the BBC to mix into the various channels new or educational material that the audience might not otherwise know about or select.

2. Research and Development (R&D)

The BBC engages in significant research and development. Pursuant to its 2006 Agreement with Her Majesty's Secretary of State for Culture, Media and Sport, the BBC has produced technology, innovations, ideas, industry standards, and other related outputs. The Agreement required the BBC to engage in research and development activities:

> [g]eared to the promotion of the BBC's Public Purposes and which aim to maintain the BBC's position as a centre of excellence for research and development in broadcasting and other means for the electronic distribution of audio, visual and audiovisual material, and in related technologies.

Further, the Agreement directed:

> [t]he Executive Board [to] pay particular attention to the desirability of supporting actively in national and international forums the development of "open standards" (that is to say, technologies where opportunities to participate in their creation and to use them are made widely available, free of charge or on terms that must be fair, reasonable and non-discriminatory).

The Agreement also suggested that the BBC cooperate with suitable partners. (These same commitments are reflected in the more recent December 2016 Agreement.)

As with content, these outputs are textbook examples of public goods. They can be shared and consumed nonrivalrously. Consumption often entails productive use. R&D outputs are generally means to other ends rather than ends in themselves, and this often means that consumption can generate spillovers. The manner in which the R&D outputs are managed often affects the degree to which the spillover potential is realized. The 2006 BBC Agreement provides some guidance on how intellectual property should be managed, specifically, directing the:

> Executive Board [to] keep the BBC's research and development activities under review, and [to] ensure that an appropriate balance is struck between—(a) the potential for generating revenue through commercial exploitation of its intellectual property, and (b) the value that might be delivered to licence fee payers and the UK economy by making new developments widely and openly available.

(Again, the 2016 Agreement has the same balanced commitment.)

The BBC has pursued a mixed set of strategies for managing its R&D outputs. For example, on one hand, it makes some R&D outputs openly accessible in its BBC R&D White Papers and through development of open standards, and on the other hand, it appears to protect some research outputs with patents and presumably other forms of intellectual property (e.g., trademarks and copyright) and license such intellectual property in various arrangements.

BBC R&D has improved the distribution stage of the supply chain in ways that directly benefit the BBC and its audience and also have broader societal and market impacts. Consider, for example, the BBC iPlayer, which is an online platform that allows users to watch television and listen to radio programmes on various electronic devices, including PCs, tablets, consoles and smart phones. In the mid-2000s, the BBC invested in the iPlayer platform despite considerable risks that hindered market investors. Risks included uncertain consumer demand as well as difficulties in developing and coordinating various services, complementary markets, and technologies. The BBC overcame these obstacles in part because of the core commitment, enumerated in its Charter, to deliver to the public the benefit of emerging communications technologies and services, and in part because of its stable funding, governance capital, and longer time horizon for planning and investment. Because of its leadership and risk taking, the BBC developed the technology, tested the market, and spurred demand, which paved the way for other entrants. As Reed Hastings (Netflix Chief Executive and co-founder) said "The iPlayer really blazed the trail. That was long before Netflix and really got people used to the idea of on-demand viewing" (*The Telegraph* 2014).

3. Training and human capital

The BBC's investment in training and human capital development is quite substantial, greatly exceeding comparable investments by its competitors. In a sense, the BBC performs training and related professional development services for the creative sector that can be seen as an extension of university training and education.

Human capital is incredibly important for human development and economic productivity and growth. There is a significant body of research on the economic and social value of human capital and development, yet human capital remains a difficult resource to define, measure and evaluate in economic terms. One reasonable but broad definition of human capital is the skills, talent, knowledge or other related capabilities associated with a "human's capacity to be meaningfully productive." Though not technically a public good, human capital does have public good features and the potential to generate spillovers.

Through the BBC Academy, the BBC provides extensive training programs for its employees and for many people outside the BBC, including partners and even competitors. The BBC expanded its training programs to freelancers, a group that has been the subject of chronic underinvestment. The BBC training programs cultivate a diverse set of skills and knowledge. People possess and retain the skills and knowledge—the human capital—and can use those resources productively throughout their lives, often, though not exclusively, to create public goods within the creative sector.

The BBC's substantial impact on the creative sector, discussed in more detail below, provides further evidence of the BBC's investment in training and human capital development and its broader societal impacts.

4. Social capital through trust, public space, brand, and cultural and national identity

The BBC produces a set of social goods that may be more amorphous and difficult to measure than content, but nonetheless are quite significant. The social goods include social capital, trust, a sense of identity connected to the BBC itself and to the UK more generally, and other such social and relational goods. For example, surveys indicate that the British public values the trust engendered by the BBC and the public space it creates by virtue of its independence from commercial and political biases (BBC Trust 2008). The BBC may create a sense of community among license fee payers who support the BBC and rely on it quite regularly (and perhaps also among those who do not). These outputs contribute to social capital—the networks, relationships, and social structures that play an integral role in shaping opportunities for and interactions among members of society. Social capital often is infrastructural as it enables people and groups to work together in many ways.

Some of the BBC content priorities discussed above reflect a commitment to national identity, UK talent and stories, and shared national experiences and conversations. These all reflect a social dimension to the content where audience members relate to the content, the content producers, and to each other. Bonds formed are an important type of social good that contributes to social cohesion, communication, and the creation of social capital. The relationships formed obviously vary, and it may be impossible to measure the value of such relationships. But I suspect there is considerable social value based on the wide scope and reach of the BBC coupled with its popularity and reputation for quality and trust.

Another related way in which the BBC produces social capital is through its brand. Brands are well-recognized means for building trust, goodwill, reputations, and social capital. The BBC's development of a strong global brand, known for high quality content and trustworthy journalism, has direct economic benefits in terms of the BBC's contribution to UK exports, but it also has indirect effects, such as positively influencing global attitudes toward the UK. One study found, for example, that nearly two-thirds of international business leaders surveyed found out about the UK through the BBC and that daily BBC users reported a significantly higher propensity to trade with the UK than those who use the BBC less often (BBC 2013b).

5. Case study: BBC impact on the creative sector

The BBC occupies a central position in the complex network of arts, design, music, technology and broadcast sectors. This reflects the infrastructural nature of the BBC, supporting, shaping, and contributing to the UK creative industries.

Since the late 2000s, various economic studies demonstrate the BBC's infrastructural role. For example, a report by Deloitte in 2010 estimated the Gross Value Added to the UK economy because of BBC activities was £7.739m, and the Net Value Added (based on a comparison of the license-funded BBC against the baseline counterfactual of an advertising-funded BBC) was £2.551m. Deloitte also estimated the economic impact that the BBC had on the independent production sector and found that £1.433m indie sector GVA would be foregone in the counterfactual scenario. (BBC 2011).

In 2011, the BBC reported that across "all of these activities, the BBC added over £8bn of value to the UK economy in 2009/10—generating over two pounds of economic value for every pound of the license fee" (BBC 2011). The report highlighted a range of BBC activities that contribute to the creative sector, including investments in content, training and developing creative talent, and technological innovation. The report also explained how the "the BBC acts as a quality leader in the UK broadcast market" and that "the stability of license fee funding of independent production companies arguably causes their cost of capital to be lower than it otherwise would be, helping them to compete more effectively at home and abroad" (BBC 2011).

In 2015, Frontier Economics prepared a report for the BBC that thoroughly evaluated its contributions to the UK creative industries. The report develops an economic framework to evaluate how BBC investments and activities "transmit" or flow "through the creative industries and beyond." The framework envisions three stages for the transmission mechanism: first round impacts of BBC investments in content, technology, and other supporting activities; second round impacts associated with further activities in the creative industries and other related industries; and finally, spillovers associated with wider benefits realized by other organizations in the creative industries. (Notably, the report takes a conservative approach to estimating these spillovers because it focuses only on benefits realized within the industries, for example, in terms of productivity gains.) The report strongly supports the claim that the BBC plays an infrastructural role in the creative sector. The first-round impacts include £2.2bn annual investments in the creative sector, £1.2bn of which is outside the BBC itself. The BBC invests heavily in small businesses in the creative sector as well as on and off screen talent, and the BBC spreads its investments widely across the creative industries. According to Frontier's analysis of BBC finance data, the BBC's external spending of £1.2bn includes £905m to 1,489 suppliers in film, TV, radio and photography; £144m to 233 suppliers in IT, software and computer services; £138m to 518 suppliers in music, performing and visual arts; £12m to 174 suppliers in publishing; and £43m to 292 suppliers in other creative industries. This reflects the breadth of the industry and the BBC's impact.

Frontier Economics evaluated the second-round impacts and spillovers in the context of two case studies, one focused on the music industry and BBC support for new British talent and the second focused on technological innovation and the BBC iPlayer. Here, I highlight and then briefly comment on a few findings from the music industry case study. The report found that the BBC supported and promoted new British music talent in four ways:

1. The BBC's talent schemes help to discover and champion new talent.
2. The BBC takes risks on new artists, providing exposure for the songs of new and emerging artists before release.
3. BBC radio promotes new artists and musicians by playing their songs on shows with the biggest audiences.
4. The BBC promotes new British music talent across its TV, radio and online services helping to drive record sales and build profile.

Each of these is supported with numerous examples. As discussed above, the BBC infrastructure supports the creative activities of many different actors both within and outside of the BBC itself. As a result, the BBC supports a wide range of UK talent and produces a wide range of content ignored or just unappreciated by its competitors. This is a critically important differentiating feature of the BBC as infrastructure. The BBC's resource allocation decisions—or what we might refer to as its selection process—are focused on quality, audience demands, and its mission and public purposes, but are decidedly insulated from market pressures to maximize revenues. This is what allows the BBC to take risks and invest in new talent and their knowledge and human capital.

Commons management of BBC infrastructure

Commons management refers to resource management/governance institutions that enable sustainable shared use of certain resources within a community. With respect to many infrastructure resources, especially traditional infrastructure, the relevant community that shares the resource consists of the public at large, such that there is no meaningful boundary between members and nonmembers of that community, and sustaining the resource for everyone on nondiscriminatory terms is the basic objective. Yet in some contexts, including the BBC context described in the previous section, the relevant community of users sharing the infrastructure is not the public at large and instead is a defined community, in the BBC context, BBC employees and contractors allowed access to the BBC's internal resource system. The basic objective of sustainable sharing of the infrastructure among members of the community remains, but additional dilemmas often arise and require governance. While I do not delve into these issues here, I note them to make clear that commons management does not always entail open access for the public or pure nondiscrimination rules.

The BBC's special governance regime, including the funding mechanism, public interest orientation, and commitment to independence, aims to achieve a similar outcome as the core principle of nondiscrimination at the heart of commons management. Both infrastructure governance institutions aim to achieve insulation from the systemic biases in market and political systems. Of course, in practice, there are constant pressures on the BBC from both markets and politicians, and independence is not perfectly maintained. Jacquie Hughes (2015: 30), for example, examines the BBC's vulnerability to political pressure and "governments repeatedly and unashamedly" seeking to exert pressure. For example, "as part of a bruising and hurried financial settlement between government and the BBC—the current fee was frozen until the Charter [was] renewed at the beginning of 2017."

Still, the BBC's complex governance regime sustains the social option value of the system of capital resources built and sustained within the BBC. Users of the BBC infrastructure retain considerable discretion when deciding what to do with the capabilities that the BBC infrastructure provides, but their discretion is cabined by the governance regime and the unavoidable need to allocate certain scarce resources (e.g., financial and physical capital as well as certain aspects of human capital, particularly time and attention). To some degree, the BBC must discriminate and optimize in the allocation of its resources, and the governance regime forces the BBC to do so in pursuit of its mission and enumerated public purposes. I do not want to overstate the observation that the BBC must discriminate because in many ways the BBC implements open nondiscriminatory systems for allocating resources and making selection decisions. The BBC's mission to inform, educate and entertain shapes the range of activities undertaken by the BBC, as does the express enumeration of public purposes. As described in the previous section, an important consequence of managing the BBC infrastructure in this fashion is the tremendous social value produced by the BBC.

BBC infrastructure for public use

This section briefly considers BBC's provisioning of infrastructure for public use. Many BBC-produced outputs are infrastructural and used by the public to generate a wide range of public and social goods. I briefly discuss two examples—the BBC Academy and the BBC Archive—and then make a few suggestions about how the BBC might think differently about its role in the future.

BBC Academy

The BBC Academy is a form of public infrastructure and represents a model of managing knowledge resources and developing human capital in ways that benefit the public. The BBC Academy creates innovative training programs that keep pace with rapid technological advances. These training programs are not created solely to instill skills in BBC employees; instead the BBC Academy offers opportunities for the larger public to participate in these training programs, thus providing the raw materials for building new programming within and outside the BBC.

The BBC Academy pays specific attention to providing knowledge resources to communities that are often deprived of resources for production. For instance, the BBC Academy targets cities and groups that are often underrepresented in media and has funneled training resources towards these communities. Providing these training resources ensures that British media will have a diverse set of individuals creating programming that reflects the diverse interests and demographics of the UK.

BBC Archive

The BBC Archive provides a glimpse of what it would mean for the BBC to provide infrastructure for public use. The BBC makes archived collections of radio and television broadcasts available to the public over the Internet. "The BBC believes that those sections of this library that have exceptional cultural, historic or educational value should, over time, be opened up within BBC Online as a publicly accessible 'permanent collection'" (BBC 2010b). The infrastructural resource consists primarily of content, which can be used for educational, historical, and other purposes. Though smaller in scale and scope, the BBC Archive is infrastructural in a manner comparable to Google Books, "a mixed infrastructure with substantial spillover potential associated with dramatically improved public access to millions of books and the ideas, knowledge, stories, and so on contained in them" (Frischmann 2012: 359–360).

BBC as provider of infrastructure for public use

This chapter comes at an interesting point during public debate over the future of the BBC, and the question therefore presents itself whether there is an opportunity to enhance ways in which BBC serves the public. The British public has invested in an incredibly valuable public infrastructure within the BBC, and the public could benefit from leveraging its existing resources, including the internal system described in the previous section as well as its governance regime and brand. Further, the BBC might consider expanding its infrastructural role in the global networked information environment. There are many possibilities, and it is beyond the scope of this chapter to explore them in detail. However, there are some general approaches to exploring options that make sense.

First, the low hanging fruit. The BBC could make more of the content it generates freely available to users on open terms in certain areas. This could include R&D as well as content made available on various platforms. As described in the previous section, much of the content the BBC produces is infrastructural, and managing such content as a commons could be as simple as utilizing creative commons, open source, and similar licenses whenever feasible. As the BBC recognizes with respect to the BBC Archive, the public already paid for the content and accordingly ought to be entitled to access and use the content without restriction, absent a strong and proven countervailing consideration. The BBC Archive could be expanded to make more content more easily available for public use, although such efforts may be constrained by existing licensing arrangements. Looking to the future, however, the BBC might reevaluate its copyright licenses and related contracts in general to facilitate future contributions to the BBC Archive. Moreover, there may be plenty of content not subject to restrictive licensing; for example, the BBC generates substantial data (e.g., weather data) that it could make publicly available on a real-time or close to real-time basis. The BBC Archive could evolve into a public repository for history and history in the making.

Second, the BBC could increase public access to existing infrastructure. This would depend on available resources and capacity and may be more difficult to do for some resources than others, but it is worth considering. Increasing public access to the BBC Academy, for example, might be difficult to do on a large scale, but there may be opportunities to explore, subject to concerns about potential negative market impact. The BBC also might be able to leverage its existing distribution infrastructure by taking advantage of its sharable capacity and enabling a wider range of content producers to reach an audience.[11] This could involve opening up part of the distribution infrastructure to a wider range of partners and non-commissioned independent producers.

Third, the BBC could develop new types of infrastructure for public use. This final category merits further attention.

The BBC claims to create and sustain a "public space," a shared and trusted environment for the public, and surveys indicate that the British public values the public space the BBC creates by virtue of its independence from commercial and political biases. Yet it is not clear what exactly is the *space* created by the BBC. It might be described metaphorically, as a space within the media landscape or as a space between markets and government. But these metaphors do not go far enough and fail to envision a space used by the public—like a public park or town square, or more relevant to the BBC, the Internet.

Within the global networked information environment, the BBC could leverage its resources and develop a shared and trusted space for an active and productive public. This public space would be built upon infrastructure—general-purpose, sharable platforms—for public users. Presumably, the BBC would leverage its existing resources, experience, reputation, and expertise and provide infrastructure that relates to its core mission and public purposes. Thus, we can envision the BBC providing infrastructure for the public generation and distribution of content that informs, educates and entertains and serves the Public Purposes enumerated in the Charter.

This could entail inviting the public into the content generation parts of the BBC business, in a sense, expanding the scope of the independent production sector. Or it could entail opening certain content production processes and networks. For example, citizen participation in journalism through social media and other channels has grown significantly since the late 2000s. Smartphones alone are incredibly powerful tools for producing and distributing content. Figuring out how to involve the public in journalism is a challenge that the BBC may be well-positioned to tackle.

The BBC infrastructure need not be limited to content production and distribution, however. The BBC might move beyond its role as a conduit for content. One can imagine the BBC developing more general-purpose infrastructural platforms for knowledge-production, communication, collaboration, and socialization. The Internet is perhaps the best example of such an infrastructure, and of course, there are many such infrastructural platforms available on the Internet, ranging from sites for sharing user-generated content like YouTube to social networks like Facebook (Frischmann 2012: ch. 13). Like the Internet, these existing infrastructural platforms provide users with basic capabilities to be productive in ways that benefit themselves and others because users produce and share public and social goods. But these infrastructural platforms also come with social costs, a full discussion of which is beyond the scope of this chapter. However, it is worth considering how the BBC might alleviate some of the social costs.

The BBC has something to offer that existing infrastructural platforms do not—independence, trust, and a commitment to core public values. There are very few entities in the media and knowledge economy that are, or even can be, unbiased, in the sense that they are not dependent on political or market funding mechanisms. The independence generated by the license fee arrangement, ten-year Charter cycle, and other governance institutions allows the BBC to articulate and pursue different conceptions of merit and public values in its activities. Thus, the BBC might offer many of the same basic capabilities—knowledge-production, communications, collaboration, networking, socialization—as existing platforms but without the attendant risks that flow from market and political pressures. Most prominent among these risks may be the loss of privacy that accompanies pervasive surveillance by private and public entities.[12] Alongside the loss of privacy is the risk of manipulation, bias, and discrimination by intermediaries. While most intermediaries hold themselves out as neutral and trustworthy, market and political pressures often introduce biases and incentives to shape experiences and optimize delivery of traffic (consumers to advertisers, content to consumers, and so on). The BBC has successfully cultivated public trust and internal governance capital and a culture that is responsive to that trust, and the BBC might leverage these assets. Trusted intermediation is a particularly important public service that the BBC could provide.

Consider a few examples. The BBC is fundamentally committed to "promoting education and learning." Historically, the BBC has satisfied this commitment by producing and distributing high quality educational content. In recent years, the BBC has done so both in the traditional media contexts of radio and television and in the digital environment. The BBC has successfully developed educational content freely used in schools and by families online. Leveraging its past successes, experience and expertise, its digital technology and distribution infrastructure, and its independence and public trust, the BBC could develop an educational infrastructure for public use. This could entail significantly expanding the scale and scope of educational materials (and content generally) made available for public use, especially for children, their families, and educators. It also could entail developing trusted digital environments within which children, their families, and educators could collaborate in pursuit of education and learning. The trusted digital environments could enable learning through play, social interaction, and independent creation and sharing of their own content, among other things. There is an incredible flurry of activity and investment in the education technology sector, and while much of the emerging technology is exciting, it is difficult to evaluate independent of the market and political biases driving the sector. In selecting learning apps and related digital learning tools, for example, the BBC might play a key role as trusted intermediary.

Consider one more far-fetched example (for purposes of illustration): Imagine that the BBC created an independent social networking platform to rival Facebook. The platform would provide the same basic capabilities—knowledge-production, communication, collaboration, and social networking. But it would not be subject to the same market pressures that drive Facebook. Simply put, the BBC social networking platform could provide its users with a trusted, privacy-protecting alternative for social networking. Users would know that the BBC would not sell user data, sell users' attention to advertisers, experiment on users to develop emotional contagion technology, or otherwise seek to profit from its users. The BBC social networking platform also could provide many other useful services, for example, a trusted login ID for various other digital services. It seems hard to imagine that Facebook users would switch to a BBC social networking platform, although the incredible reach of and trust in the BBC makes it somewhat easier to imagine. I do not want to belabor the point because it seems politically infeasible to imagine this happening in the near future, but the example illustrates how the BBC can think differently about its role in modern society and about its users.

Historically, the BBC has seen itself and been understood by the public as a producer and distributor of content, and its audience is usually described as license fee payers and consumers. To be a provider of infrastructure for public use, the BBC must re-conceptualize its audience as users, participants, co-owners, and community members. This is one of the critical lessons since the late 2000s, and it is one that the BBC and UK public should take to heart during the next round of public debates about the future of the BBC.

Conclusion

The BBC provides public infrastructure, and as is quite common with all sorts of infrastructure, it is all too easy to take it for granted. Bringing this aspect of what the BBC does to the surface is relevant to the ongoing debate about the future of the BBC. What is the social value of the BBC? This is a surprisingly difficult question to answer. The BBC is a publicly funded and regulated corporation, a complex media organization, a community, a public space, and even an iconic cultural and social brand. It is a massive £3.7bn annual investment in the knowledge economy, which according to some studies has an economic impact in excess of £8bn per year (BBC 2013a). The scale and scope of the BBC and its activities are incredibly expansive, spanning the globe and reaching almost all UK citizens on a regular basis. Ninety-seven percent of UK citizens use the BBC each week, and they use it extensively, approximately 18 hours per week, on average (BBC 2015a: 2). This information tells us something about the social value of the BBC, but still it does not fully answer the question, nor does it highlight adequately the role of commons management in enabling the production of social value. We need to grapple with the functional relationships between the infrastructural resources the BBC produces and manages as commons and the various outputs infrastructure users produce. This chapter briefly does so, focusing on four categories of outputs, content, R&D, human capital, and social capital. These outputs often generate spillovers because they are public and social goods, and sometimes the outputs themselves are infrastructural. Still, the infrastructure theory discussed and applied in this chapter does not fully answer the valuation question either. Instead, it provides a lens with which to see the different contributions the BBC makes and how those contributions generate substantial social value, often because of the productive activities of users enabled by commons. The BBC's special governance regime, including the funding mechanism, public interest orientation, and commitment to independence, achieves the paramount infrastructural objective of commons management—*insulation from the systemic biases in market and political systems*. In our current hyperpolarized political and economic climate, the BBC-style of commons governance reveals a counterintuitive but inspiring vision for the future (Frischmann and Selinger 2018: 279–83: suggesting that a "BBC-style social media platform" might be an attractive solution to fake news and reality jamming problems).

Notes

1 "Social goods generate value through their impact on social systems and interdependencies . . . Their value is inherently social because it depends on the existence and nature of interdependent social relations. . . . Social goods change environmental conditions and social interdependencies in ways that affect social welfare." Frischmann, *Infrastructure* (2012), ch. 3.

2 The defined community may be the public, in which case commons management converges with open access. The community might also be more limited in scope, in which case there may be members and nonmembers. See Frischmann 2012; Frischmann et al. 2014.

3 Both traditional and nontraditional infrastructures function as infrastructural capital. (Frischmann 2005, 2012; Frischmann and Lemley 2007).

4 The defined community may be the public, in which case commons management converges with open access. The community might also be more limited in scope, in which case there may be members and nonmembers. See Frischmann 2012; Frischmann et al. 2014.

5 Where future sources of social value are uncertain, there are good reasons to sustain infrastructure as commons. See Frischmann 2012, ch. 5.

6 One reasonable definition of human capital is the skills, talent, knowledge or other related capabilities associated with a "human's capacity to be meaningfully productive." See Frischmann and McKenna, Systems of Human and Intellectual Capital, *Texas Law Review* (2015).

7 Intellectual resources are pure public goods (meaning the resources are non-rivalrous in consumption for the entire range of demand), often a form of capital, often the source of various types of externalities, often integral parts of cultural, intellectual, and social progress, and generally constitutive of dynamic and complex resource systems that are not easily reducible to discrete elements or parcels. See Frischmann, *Infrastructure* (2012), pp.253–314.

8 This doesn't mean that scaling back on the license fee and scope of BBC activities necessarily would result in a loss of value or waste of assets. It might lead to increased efficiencies or some of the internal resources might be repurposed within or perhaps external to the BBC, for example as employees went elsewhere. But it might not. There are good reasons to think that the BBC is reasonably efficient in its use and re-use of its capital resources. The BBC appears to do much more (e.g., investment in original content or training) with less revenues than its competitors.

9 BBC, Putting Quality First: The BBC and Public Space (2010a), p.33. This document is admittedly dated, but it is not stale. Rather, the strategies and content priorities seem relevant to understanding the basic orientation of the BBC. More recently, in its 2015 charter review, British, Bold, Creative, the BBC announced its intention to "modernise the BBC to preserve what is best about public service broadcasting, and hand on a BBC that is bold, British and creative. BBC, *Bold, British, Creative. Charter Review* (2015b) p.99. The strategies to achieve these goals included creating "an open BBC," that will inform, educate and entertain. The concept of "an open BBC" comes from an understanding that the BBC must adapt to become more responsive and integrated with "content made by others, or to feedback from its audience." Ibid. at p.57.

10 I say, "can be" rather than "is" because citizens must still choose to exercise their capability to participate, and such decisions may be influenced by other factors beyond access to "independent, impartial and accurate news, current affairs and information."

11 "On-demand technology, far from separating people, has added to that shared experience: audiences can all catch up on the programmes their friends and family have enjoyed." BBC. Bold, British, Creative. Charter Review (2015b) p.18 "Platforms bring together content from many providers, not just one. Platforms thrive by encouraging everyone to create, share and participate." Ibid. at p.59.

12 My BBC may depend upon personalization and thus tracking of user behavior, and this might run counter to the suggestion I have made.

References

An Agreement Between Her Majesty's Secretary of State for Culture, Media and Sport and the British Broadcasting Corporation (2006).

An Agreement Between Her Majesty's Secretary of State for Culture, Media and Sport and the British Broadcasting Corporation (2016).

Baker, C. Edwin (1994) *Advertising and a Democratic Press*. Princeton, NJ: Princeton University Press.

Baker, C. Edwin (2002) *Media, Markets, and Democracy*. Cambridge, UK: Cambridge University Press.

BBC (2010a) *Putting Quality First: The BBC and Public Space*.

BBC (2010b) *BBC Strategy Review*.

BBC (2011) *Helping Drive Growth in the UK Creative Economy*.

BBC (2013a) *The Economic Value of the BBC*: 2011/12.

BBC (2013b) *The Economic Return to the UK of the BBC's Global Footprint*.

BBC (2014) *BBC Full Financial Statements* 2013/14.

BBC (2015a) *Public Service Content in a Connected Society*. Submission to OFCOM's Third Review of Public Service Broadcasting.

BBC (2015b) *Bold, British, Creative. Charter Review*.

BBC Trust (2008) *The BBC Trust Conclusions: The Economic Impact of the BBC on the UK Creative and Broadcasting Sector* (including an independent assessment and report by PricewaterhouseCoopers—PwC).

Benkler, Yochai (2001) "Property, commons, and the First Amendment: Towards a core common infrastructure," White Paper for the First Amendment Program, Brennan Center for Justice at NYU Law School, available at www.benkler.org/WhitePaper.pdf (last accessed October 8, 2018).

Benkler, Yochai (2013) "Commons and growth: The essential role of open commons in market economies," 80 *University of Chicago Law Review*.

Cornes, Richard & Sandler, Todd (1996) *The Theory of Externalities, Public Goods, and Club Goods* (2nd ed.). Cambridge, UK: Cambridge University Press.

Frischmann, Brett (2008) *Speech, Spillovers, and the First Amendment*. Chicago, IL: University of Chicago Legal Forum.
Frischmann, Brett (2009) "The Pull of Patents," 77 *Fordham Law Review*, 2143.
Frischmann, Brett (2012) *Infrastructure: The Social Value of Shared Resources*. New York: Oxford University Press.
Frischmann, Brett and Hogendorn, Christiaan (2015) "Retrospectives: The marginal cost controversy," *The Journal of Economic Perspectives*, Vol. 29, No. 1, pp. 193–205.
Frischmann, Brett M. and Lemley, Mark A. (2007) "Spillovers," *Columbia Law Review*, Vol. 100, No. 2, pp. 101–143.
Frischmann, Brett and McKenna, Mark (2015) "Systems of intellectual capital," *Texas Law Review*, 93, p. 231.
Frischmann, Brett, Madison, Michael and Strandburg, Katherine (2014) *Governing Knowledge Commons*. Oxford, UK: Oxford University Press.
Frischmann, Brett and Selinger, Evan (2018) *Re-Engineering Humanity*. Cambridge, UK: Cambridge University Press, pp. 279–283.
Frontier Economics (2015) *The Contribution of the BBC to the UK Creative Industries: A Report Prepared for the BBC.*
House of Lords Select Committee on Communications (2009) 2nd Report of Session 2008–09 "*Public Service Broadcasting: Short-Term Crisis, Long-Term Future?*" HL paper 61. 6.
Hughes, Jacquie (2015) *Broadcasting by Consent: The BBC, Public Service Broadcasting and Charter Renewal in 2017.*
Ipsos MORI for the BBC (2015) *UK Adults 15+ Who Follow the News* (1,906).
Mediatique report for Ofcom (2014) "*'PSB Review: Investment in TV genres' for a wider review of the gross profit margins for different genres invested in by commercial PSBs*, available at http://stakeholders.ofcom.org.uk/binaries/broadcast/reviews-investigations/psb-review/psb3/Investment_in_TV_Genres.pdf (last accessed October 8, 2018).
Musgrave, Richard and Musgrave, Peggy (1984) *Public Finance In Theory And Practice* (4th ed.). New York: McGraw Hill.
Royal Charter for the Continuance of the British Broadcasting Corporation (2016).
Samuelson, Paul A. (1954) "A pure theory of public expenditure," *Review of Economics and Statistics*, Vol. 36, No. 4, pp. 387–389.
The Telegraph (2014) "Netflix chief Reed Hastings takes on telcos, cinemas and global expansion." October 5.

21
MEDICAL INFORMATION COMMONS

Mary Anderlik Majumder, Peter D. Zuk and Amy L. McGuire

Introduction: the "commons" trend in biomedical research

In the U.S., the term "commons" is increasingly employed to describe a range of efforts to build large-scale resources containing human data for research, clinical decision-making, and other purposes. For example, in 2016, the National Cancer Institute launched a "Genomic Data Commons" to provide "the cancer research community with a unified data repository that enables data sharing across cancer genomic studies in support of precision medicine" (NCI 2017). The Genomic Data Commons includes a data submission portal, a data access portal, and a variety of tools to support researchers in working with data. Most recently, in 2017, under the umbrella of its Big Data to Knowledge (BD2K) initiative, the National Institutes of Health (NIH) funded a "Data Commons" pilot with a bioinformatics and data science focus (NIH Data Commons Pilot 2017). Indeed, the power of the term is such that in 2015 a public-private partnership initially known as the "Open Cloud Consortium" was rechristened the "Open Commons Consortium" (OCC 2017).

In the private sector, the non-profit technology company Sage Bionetworks has been a pioneer in commons creation and sponsorship of efforts to work out the implications of the commons concept for biomedical research. Sage has created the position of "Chief Commons Officer" and hosted a "Commons Congress," leading to development of a set of "Commons Principles." Those principles include promoting "collaborative discovery through the creation and support of a broadly accessible digital Commons consisting of curated data and methodological tools in which analytical results are shared in a transparent, open fashion" (Sage Bionetworks 2017).

The historical backdrop for such initiatives both within the U.S. and internationally includes the development, in the context of the Human Genome Project (HGP), of the principle that DNA sequence information should be "freely available and in the public domain" to encourage research and development and maximize societal benefit (Summary of Principles 1996; Contreras 2011; Cook-Deegan, Ankeny and Jones 2017). Lack of access to data, and to methods of analysis and other metadata, has been identified as a barrier to the realization of goals such as furthering the appropriate clinical integration of genomic sequencing technologies and supporting the development of precision medicine (Messner et al. 2016; Cook-Deegan and McGuire 2017).

The linkage to precision medicine is especially strong, owing to an influential report of the National Academy of Sciences (NAS) entitled *Toward Precision Medicine* (National Academy of Sciences 2011). The report identified creation of an "Information Commons" as a necessary foundation for a new taxonomy of disease and new approaches to medical treatment tailored to patients' individual characteristics.

In this chapter, we develop the concept of a medical information commons, laying out its central features and exploring connections to several strands of commons theory. We then home in on normative questions raised by efforts to create widely-shared informational resources in the biomedical research space. Addressing these questions involves examination of the claims of key stakeholders and excavation of the moral foundations of governance mechanisms. We believe this analysis yields several important lessons for leaders of current efforts. First, they should consider investing resources in empirical studies to inform evaluation of the trade-offs associated with alternative strategies for involving participants. Second, there is intrinsic as well as instrumental value to the engagement of all stakeholders on terms that approximate as closely as possible norms of mutual regard and equal respect, a standard that is demanding but also flexible. Third (and related), consistent with principles of adaptive governance, nesting, and institutional variety, differences of approach across efforts should be expected and are indeed desirable. Finally, while it is possible to combine insights from the public domain and common-pool resource strands of commons theory, there are tensions between the two approaches. The rhetoric of "broad sharing" and "openness" may obscure the need to make difficult decisions about the boundaries of a commons, allocation of resultant benefits and burdens, and enforcement of rules.

What is a medical information commons?

Building on the *Precision Medicine* report, we define a medical information commons or MIC as a networked environment in which diverse sources of health, medical, and genomic data on large populations become broadly available for research use and clinical applications.[1] While the NAS report said very little about the structure of information commons, our research has led us to a pluralistic understanding: the concept of an MIC can be used to describe an ecosystem of independent but interconnected data initiatives oriented toward making data more widely available. It can also be used to describe certain data-sharing initiatives within the ecosystem. In other words, the concept encompasses institutional arrangements for sharing health-related data that are "complex, redundant, and nested in many layers" (Dietz, Ostrom and Stern 2003: 1910).

The NAS report points to additional desirable features of an MIC. Most significantly, the report emphasized the value of linking together various types of data for individuals and following those individuals longitudinally. The report also envisioned connections to learning health care systems and to efforts to understand and improve population health. The Kaiser Permanente-UCSF Program on Genes, Environment, and Health, including the Genetic Epidemiology Research on Adult Health and Aging (GERA) cohort study, is cited in the report as an example of a research initiative already underway that exemplifies its vision of an information commons collecting data from point of care settings.[2]

What can the MIC learn from commons theory?

Two strands of commons analysis appear especially relevant to MICs. We turn first to a strand of commons scholarship associated with a group of legal scholars that includes Carol Rose, James Boyle, and Yochai Benkler. Rose, Boyle, and Benkler use the concept of a commons to capture certain kinds of non-market, open-access regimes that contribute to the public domain.

Their work provides insight into the rationale for and contours of openness in the context of an MIC. Second, as discussed in more depth in the next section, Elinor Ostrom and colleagues study the social dilemmas that confront groups seeking to manage shared resources, providing insight into governance challenges facing MIC designers. Their work and the work of another group of legal scholars, Katherine Strandburg, Brett Frischmann, and Michael Madison, suggest a number of features and strategies that can be adopted to increase the likelihood that MICs will be sustainable over time.[3] We also believe that the work of the political philosopher John Rawls illuminates several normative questions related to the creation and operation of MICs, questions we address in the final section of this chapter.

As noted earlier, many of the projects in the biomedical realm that have adopted the "commons" label appear to do so to underline their commitment to wide sharing of data. For Rose, Boyle, and Benkler, support for the commons as a public space is justified where there is an expectation of greater value with increased access to and use of a resource, or "the more the merrier" (Rose 1986: 768). Benkler suggests several criteria for selecting a commons-as-public domain approach (Benkler 2014: 89–90):

(a) Efficient allocation of the resource once created, across many uses, is not a paramount management concern. Data is generally considered to be a nonrivalrous good, meaning additional users add little or nothing to costs, and inefficient users should not crowd out efficient users (Frischmann 2012; OECD 2015). Hence, in line with this criterion, MIC designers' list of top priorities need not include identifying and giving exclusive access to the individuals or entities that might be predicted to use the data to maximum advantage.
(b) Wide accessibility can be anticipated to give rise to significant positive externalities. This should also be the case for an MIC, given the success of open science efforts like the HGP in fueling scientific productivity and innovation.[4]
(c) The resource is used as an input into many goods and services. In line with the findings in the NAS *Precision Medicine* report, an MIC has the potential to increase knowledge about health and disease, lead to improvements in health care through means including development of targeted therapies and better diagnostic tests, and expand the evidence base for public health and health policy.
(d) The range of potential users is diverse and includes non-market actors. Again, an MIC may be of value to a range of users, including non-market actors such as health-focused charities and public health officials.

Accordingly, MIC designers seeking to maximize positive externalities—and innovative uses that may be unforeseeable due to their very novelty (Boyle 2011)—should take truly wide sharing of data very seriously. This might encompass giving serious thought to developing processes that facilitate international data sharing and permitting access by citizen scientists, for example.

At the same time, Benkler (2014: 71) stresses that "open" is not equivalent to "anything goes." For some resources, openness may entail that access is available to all legitimate, credentialed users without discrimination, rather than no-constraints access for all comers. Indeed, what makes MICs distinctive are the special challenges encountered when the resource consists of *human* data. To begin with, the resource managed by an MIC is data about individual human persons. This feature gives rise to conceptual, technical, normative, and policy issues related to privacy, including developing and implementing standards and best practices for de-identification and evaluating and addressing the potential for re-identification. Further, in contrast to fisheries or other natural resources, this resource is the product of human practices and institutions. Medical data describe interactions between patients and clinicians, are generated by clinicians and other members of care teams, are supported by significant investments by institutions in technology, and speak to the quality of care received and not simply the biological make-up and characteristics of patients. Further, researchers and research institutions generate and invest in other types of data that ideally flow through an MIC.

We have found broad consensus that an MIC should be "participant-centric" (Deverka et al. 2017: 2). This consensus reflects a shared understanding that the data resource managed via an MIC is about (if not always directly extracted from) people (patients with a range of conditions as well as the currently healthy), who thereby become participants in the resource, and its circulation exposes these participants to risks (although in some cases with potential for benefit as well). However, we have found less consensus on what the principle of participant-centricism requires. Given the complexities alluded to earlier, this is perhaps unsurprising. In the next section we touch on the role of participants in governance. In the final section, we examine the ongoing debate about how best to balance respect for autonomy and protection of the privacy of participants with a commitment to openness.

What makes an MIC sustainable?

Recently, commentators have begun to elucidate the implications of Ostrom's work for MIC development (Cook-Deegan and McGuire 2017). Also, some scholars have applied a version of Ostrom's Institutional Analysis and Development (IAD) framework and related concepts to specific biomedical research initiatives or specific kinds of data.[5] In short, it is becoming clear that MIC designers have much to learn from this institutional design-focused strand of commons analysis.

In addition to highlighting a set of governance requirements, commentators have drawn attention to the advantages of decentralization.[6] In particular, decentralization facilitates adaptive governance of resources. Adaptive governance is the key to the resilience of any commons as a collaborative approach to resource management that is neither dictated entirely by the state nor determined by the assignment of private property rights and the operation of markets. It encompasses the creation of governance structures with features that are responsive to the needs and interests of stakeholders and anticipate the possibility of change. This approach runs counter to pressures to maximize efficiency in the short-term and tendencies toward bureaucratic rigidity. Thomas Dietz, Elinor Ostrom, and Paul Stern note that "systems that guard against the low probability, high consequence possibilities and allow for change may be suboptimal in the short run but prove wiser in the long run" (Dietz et al. 2003: 1909[7]). At the same time, the benefits of some degree of centralized coordination and standardization for MICs (and the MIC as an ecosystem) should be acknowledged (Lee 2017).

In this strand of commons analysis, three governance principles have been identified as particularly relevant to large-scale efforts, meaning they are particularly relevant to an MIC: (1) "Involve interested parties in informed discussion of rules" (*analytic deliberation*); (2) "Allocate authority to allow for adaptive governance at multiple levels from local to global" (*nesting*); and (3) "Employ mixtures of institutional types" (*institutional variety*) (Dietz et al. 2003: 1910). Regarding analytic deliberation, one of the major challenges associated with an MIC is involving participants in governance. This involves crafting a defensible process for selecting individuals to represent participants as a collectivity and persuading representatives of other stakeholder groups with established power and influence (e.g., funders, large academic medical centers) that participants should be engaged in a meaningful rather than pro forma manner. The commons funded by NIH are typically designed to be governed by and for professional scientists.[8] The work of Ostrom and colleagues suggests that this reflects an overly-restrictive vision of the rule formation process. Fortunately, developing tools and rules for meaningful engagement itself has been a project of the Patient-Centered Outcomes Research Institute (PCORI 2017). It is at the center of the mission of the newly created People-Centered Research Foundation. In a related development, Barbara Evans has been in the forefront of efforts to shift power through the formation of consumer-driven health data commons (Evans 2016a, 2016b, 2017).

The concept of nesting is valuable in illuminating what is occurring in the biomedical research space. It also suggests how certain efforts should be understood and guided. One of the most important recent developments in this space has been the creation of the Global Alliance for Genomics and Health (GA4GH). GA4GH has the backing of influential stakeholders from a variety of sectors, including researcher funders such as the NIH, the Canadian International Data Sharing Initiative, and the Wellcome Trust; representatives of industry such as Affymetrix, Allscripts Health Solutions, and Amazon Web Services (just to cite a few examples from the beginning of the alphabet); and professional societies, health systems, academic medical centers, research organizations, and non-governmental organizations. It serves as the umbrella for four multi-stakeholder working groups, including a Regulatory and Ethics Working Group. That working group has already published policies addressing accountability, consent, ethics review, and privacy and security. The nesting principle suggests that the challenge for GA4GH is to produce documents that are not so high-level and abstract as to be completely vacuous, and yet not so granular as to preclude adaptive governance at the regional and local levels. So far, it seems to be meeting that challenge, for example, including provisions in its Privacy and Security Policy that have some bite but preserve flexibility.[9] Finally, institutional variety is another principle that may seem inefficient in the short-term, as compared to institutional isomorphism or monoculture, but may be key to building trust by permitting tailoring to context and keeping alive the potential for innovation that contributes to long-term productivity and sustainability (Olson 2017).

Strandburg, Frischmann, and Madison provide an additional tool to MIC designers in the form of a series of questions about a commons linked to the IAD framework. In the area of governance, the questions include (Strandburg et al. 2017: 17):

- "What are the governance mechanisms (e.g., membership rules, resource contribution or extraction standards and requirements, conflict resolution mechanisms, sanctions for rule violation)?"
- "Who are the decision makers and how are they selected?"
- "What are the institutions and technological infrastructures that structure and govern decision making?"
- "What informal norms govern the commons?"

These questions were developed to structure case studies of commons. Posing them to designers as an MIC is being created presents opportunities for those involved to be more conscious of when a critical choice is being made and more thoughtful in evaluating options.

In the next section we will argue that many of the principles and considerations we have noted as relevant to MIC design have a normative dimension. They surely affect sustainability or other typical measures of effectiveness. When employed well, they also demonstrate respect for the multiple constituencies of an MIC, including participants, and increase the odds that MICs will be accountable to participants and worthy of their trust.

Normative questions

We now turn to normative aspects of commons creation in the biomedical research space. In particular, we consider four critical normative questions: (1) Under what circumstances is it acceptable for different types of human data to be included in an MIC? (2) What are the rights and interests of data generators and holders and how should they be accommodated, if at all? (3) How should specific governance features be understood and justified in light of their political character? (4) What are the implications of considerations of fairness for management of an MIC? Given space constraints, we cannot synthesize all of the material relevant to these questions. We do hope to provide enough information to give the reader a sense of the range of positions in current debates related to these questions and points to consider in assessing possible answers.

1. Under what circumstances is it acceptable for different types of human data to be included in an MIC?

The first normative question, regarding the acceptability of data inclusion, requires consideration of the rights and interests of individual participants in relation to the social benefits of widely sharing information about an entire population. Some have made the case for putting some health-related information in the public domain without individual authorization or consent. Typically, an important proviso is that the information be de-identified, meaning *not* that it would be impossible for a third party to tie the information to an individual, but that given the steps taken to protect privacy it is very unlikely that this would happen. In these circumstances, one commentator has argued that "de-identified (anonymized) data need not be considered as relating to the underlying data subject at all—unless and until their data has been re-identified" (Yakowitz 2011: 65). On this view, in the context of rigorously de-identified data, concerns about autonomy are attenuated, if they come into play at all. In addition, there are reasons for believing that a re-identification attack on this kind of data would indeed be very unlikely—if only because miscreants have easier routes to accessing data of value—and open access to such data can generate large social benefits (Yakowitz 2011).

However, an MIC is usually envisioned as incorporating robust data linked to individuals, precisely because support for sophisticated longitudinal analyses of complex health-related phenomena is considered a highly desirable feature. Accordingly, individuals should have some say related to access to and use of data about them by an MIC (Deverka 2017). Expansion of a public health model that dispenses with consent would be particularly problematic in countries with health care systems that are inegalitarian and exclusionary. In the U.S., for example, a claim that individuals can legitimately be conscripted into contributing to the public welfare via an MIC intended to fuel development of new drugs and devices that may be beyond their reach, while exposing them to the risk of losing any health benefits they currently enjoy and other forms of discrimination, is rightly met with skepticism (Rothstein 2017). There are also good reasons for rejecting an option at the other end of the autonomy spectrum: recognizing patients as the sole legal owners of all health-related data about them. Giving individuals that kind of exclusive control over health-related data would ignore the complex human practices and institutions involved in data generation and would be in tension with the concept of a commons as a set of institutional arrangements for managing shared resources.

Individuals are typically given a say about research involvement through an informed consent process, but the nature and specificity of that process has been the subject of much debate. Some proponents advocate that "broad consent" is sufficient to satisfy obligations to individuals, whereas others argue that to truly respect autonomy and the contributions of participants to an MIC we need "dynamic consent" (Vayenna and Blassimme 2017). Broad consent has been defined as "consent for future unspecified research studies," with ethicality ensured through independent oversight (Siu et al. 2016: 468). The terms broad consent and "blanket consent" are often used interchangeably, but some commentators reserve the term blanket consent for open-ended consent without any content or process limitations (Grady et al. 2015). Dynamic consent is an "approach for engaging individuals about the use of their personal information" that allows them to "alter their consent choices in real time" (Kaye et al. 2015: 142). A misconception about dynamic consent, which is also sometimes termed "dynamic, granular consent," is that it necessarily enmeshes individuals in micromanagement of data. In fact, approaches such as the Platform for Engaging Everyone Responsibly (PEER) allow users to set what might be called *meta-preferences*, including deferring to a trusted entity (although consent is dynamic, so those preferences can always be reset) (Genetic Alliance 2017).

We believe that respect for persons and for autonomy requires that MIC designers adopt an informed consent process that exceeds a certain threshold. In order for individuals to have some notion of what they are getting into, they must be provided with basic information about governance. They must also have an opportunity to choose whether or not to participate that meets core conditions such as voluntariness and a right to withdraw (at least prospectively). Choosing the ideal consent process involves trade-offs between three major ethical considerations: maximizing concordance with individual preferences, minimizing risks of harm (linked to re-identification concerns), and maximizing the promotion of the public welfare. At present, there is a dearth of empirical evidence to inform analysis of these trade-offs. For example, it would be helpful to have a study that examines the extent to which participation and data sharing are limited (and bias potentially introduced) with a dynamic consent approach compared to a broad consent approach.

Finally, the principle of nesting suggests that within the MIC ecosystem different data-sharing initiatives will hold different kinds of data, fulfill different purposes, and address the challenge of reconciling the principle of respect, concerns about privacy, and commitment to data sharing in different ways. For example, ClinVar, an NIH-funded database of variant-level information, serves clinical as well as research purposes (with cautions against direct diagnostic use and use in medical decision-making without involvement of a genetics professional). It has a policy of not accepting any personally identifiable information and restricting inclusion of potentially sensitive information combined with a policy of open access (ClinVar 2017). BRCA Exchange is a component of the GA4GH's flagship BRCA Challenge and is intended to be a comprehensive, global data repository cataloging variation within the BRCA genes and evidence related to classification. It has adopted a three-tiered structure as a way of minimizing risks of harm while promoting the public welfare through open sharing of some valuable data and controlled access to more sensitive data. Specifically, *any* Web user can access variant-level data from the original submitters. Researchers, clinicians, and patients can access curated, expert interpretations of *BRCA1/2* genetic variants, as well as supporting evidence, via a simple interface. The third tier is only available to registered users working on variant interpretation who need access to potentially identifiable case-level data (GA4GH 2016). In keeping with this kind of layered approach, the NIH recently issued a Request for Comments describing a plan to move summary statistics in the Database for Genotypes and Phenotypes (dbGaP) and other NIH-designated repositories out of controlled access into a "rapid access" tier (NIH Request for Comments 2017).

2. What are the rights and interests of data generators and holders and how should they be accommodated, if at all?

The entitlements and legitimate interests of data generators and holders, such as clinical trialists and other researchers, are also at the center of an intense debate that has played out in the pages of the

New England Journal of Medicine as well as other fora (Longo and Drazen 2016; Rosenbaum 2017). In general, data are considered non-rivalrous goods (Frischmann 2012; OECD 2015). However, data generators and data holders have interests and concerns that may lead them to withhold data. Evans has written extensively about the costs of data contribution (Evans 2016a, 2017). Data generators and holders, whether scientists or health care providers, may also have concerns about free-riding by peers, use of data to their disadvantage by competitors or regulators, and giving up the opportunity to exploit data as a proprietary asset.[10] Finally, these data generators and holders may also be concerned about liability or harm to reputation if a patient or research subject is harmed by a breach of confidentiality or a misuse of data. Thus, while some commentators have expressed skepticism regarding the legitimacy of data generators and holders' interests, and accordingly, resistance to recognition of any entitlements benefiting them, others have argued that data generators and holders should have some control over access to and use of "their" data.

MIC designers can address this issue by acknowledging interests that impact behavior and amount to more than egregious data-hoarding-coupled-with-free-riding and by modifying incentives where feasible. This may involve retaining periods of data exclusivity and using new technologies to track usage of contributed data sets and rewarding contributors of frequently used data sets, which also meets a need of funders to direct money to the highest-value resources. Jorge Contreras suggests that evolution in this direction by commons in the genomics research space delivers an important lesson about respecting perspectives and taking account of interests in order to maintain cooperation (Contreras 2011, 2014). Nonetheless, the weight given to these interests relative to the interests of other stakeholders (e.g., the interests of participants as well as funders in not having data tied up or siloed) must be continually renegotiated. Here a governance requirement, "dealing with conflict," merits attention. As Ostrom and Charlotte Hess note in their elaboration on this requirement, there may be a positive dimension to conflict, managed well:

> Conflicts among perspectives and views, if they do not escalate to the point of dysfunction, can spark new understandings and better ways of accomplishing outcomes. The core problem is designing conflict-resolution mechanisms that enable participants to air differences and to achieve resolutions that they consider legitimate, fair, and scientifically sound.
>
> *(Ostrom and Hess 2007: 67)*

3. How should specific governance features be understood and justified in light of their political character?

Our third normative question concerns the justification of various MIC governance mechanisms. Ostrom offers us principles for the *effective* governance of common-pool resources. This idea can be taken in two distinct ways. One might understand the idea of effectiveness purely in terms of creating the right sorts of incentives for sustainable resource development and use. To understand effectiveness in this way is to think of it purely descriptively. Alternatively, one might take the idea of effectiveness to also include conformity with certain normative standards. Whether or not normative considerations are built into the very idea of effectiveness, however, conformity with Ostrom's principles has normatively desirable results. This is due not only to the instrumental benefits that sustainability by definition brings, but also to the fact that adherence to Ostrom's principles yields forms of decision-making and cooperation that are valuable in their own right. Ostrom's principles not only work; they involve intrinsically worthy states of affairs in their workings.

This can be brought out by asking what justifies (in a normative sense) the various governance procedures that might be put in place by MIC designers. It will be helpful to distinguish between governance mechanisms that are designed to secure *prior, independently justified aims* and governance mechanisms that are themselves for *deciding upon aims*. These two kinds of mechanism can be differentiated along the lines of Rawls' (1971: 84–86) distinction between *imperfect* procedural justice and *pure* procedural justice.[11] Imperfect procedural justice involves following of a set of rules that reliably (though not infallibly) achieve an aim, as the rules of criminal trials are supposed to do in determinations of guilt. Pure procedural justice, by contrast, involves no such independent aim; rather, "there is a correct or fair procedure such that the outcome is likewise correct or fair, whatever it is, provided that the procedure has been properly followed" (Rawls 1971: 86). This is what Rawls' theory of justice as fairness is supposed to involve; he attempts to lay out a hypothetical contract situation possessing features such that, whatever outcome is agreed upon by the contracting parties, it is a just outcome in virtue of arising from the contract situation.

So, we can ask: which potential MIC governance mechanisms are akin to the general schema that the rules of criminal trials are supposed to embody, and which are akin to contractual agreements with initial parameters that themselves justify the resulting agreement (whatever its particular character)? It seems clear that MIC designers would do well to employ both sorts of mechanisms. One indispensable employment of imperfect procedural justice takes the form of mechanisms for compliance with pre-existing legal and ethical norms. Mechanisms for the resolution of conflicts arising between or otherwise involving MIC users and data generators are also likely to be instances of imperfect procedural justice.[12] So, too, are applications of sanctions for rules violations. Now, though the idea of imperfect procedural justice is most obviously applicable to governance mechanisms that are broadly judicial in character, it is by no means limited to them. Mechanisms embodying imperfect procedural justice are appropriate whenever their implementation will serve to protect clearly demarcated rights and interests.

An MIC must clearly involve many ongoing mechanisms of imperfect procedural justice in order to function. But one might be tempted to think of pure procedural justice as having its proper place at the initial stages alone, for the purpose of securing appropriate input and agreement from various MIC stakeholders in a way that realizes Ostrom's idea of analytic deliberation. This would be a kind of "wind-up" model for an MIC: wind it up and let it go. But we would do well to remember Ostrom's stress on the idea that common-property regimes *evolve* over time. As Ostrom reminds us, effective governance is adaptive governance, and one of the ways in which adaptation occurs is via stakeholder input. Such input in MIC governance is required not just for public oversight (a crucial accountability mechanism, especially given the nature of the resources involved in an MIC), but also to perpetually take account of the evolving interests of an evolving body of stakeholders.

Ongoing analytic deliberations of this sort, one might think, realize an important form of democratic engagement. They seem to realize what Anderson (1999) calls *democratic equality*, especially in virtue of involving an important form of mutual regard between the relevant interlocutors. In characterizing the idea of democratic equality, Anderson writes, "In seeking the construction of a community of equals, democratic equality integrates principles of distribution with the expressive demands of equal respect" (Anderson 1999: 289). Understanding MIC stakeholders as constituting a community of democratic equals not only fosters public trust of the kind necessary for sustainable governance, but also renders that public trust *justified* due to the governance's inclusive and responsive character. MIC designers will need to determine how best to realize these values in a way that is conducive to yielding the benefits that an MIC makes possible. Ostrom's attention to the benefits of institutional variety suggests that a combination of different approaches is called for here, especially given that different groups of stakeholders will be best able to voice their interests and concerns in different ways.

4. What are the implications of considerations of fairness for management of an MIC?

Our final normative issue concerns fairness in the allocation of benefits and burdens. Knowledge commons include a diverse array of stakeholders: users, providers, and policy-makers (Ostrom and Hess 2007). This diversity of stakeholders is particularly evident in an MIC, where stakeholders include participants, data generators and holders, researchers of various kinds, policy-makers at a variety of levels, and the public at large. This diversity leads to a correspondingly diverse array of potential benefits and burdens. MIC designers will need to identify the various benefits and burdens characteristic of an MIC and arrive at workable principles for their distribution.

A relevant body of research in political philosophy concerns what has come to be called *the principle of fairness*. Rawls (1971: 112) puts the principle particularly well: "We are not to gain from the cooperative efforts of others without doing our fair share" (Rawls 1971: 112). Receiving a benefit produced through collective action, the thought goes, obligates one to share in the burdens that make the benefit possible. The principle thus concerns the normative impermissibility of free-riding. Complicating matters is the fact that different thinkers give the principle different scopes. Klosko (2004), for example, applies the principle to a wide array of goods. This includes, especially controversially, cases in which there is an absence of prior agreement to participate in the cooperative schema that produces the goods, a form of application that Nozick (1974: 90–95) regards as a non-starter.

In this context, the distinction between common-pool resource and public domain conceptions of an MIC potentially has important normative consequences. Failure to meet fairness obligations related to a common-pool resource can be justifiably met with exclusion from further use, while public goods are characterized by their non-excludability. Furthermore, the diverse array of stakeholders (and so diverse array of benefits and burdens) that an MIC involves suggests the need for a more sophisticated application of the principle of fairness. This topic will be a particularly important one for future normative reflection on the part of MIC designers as they consider how to facilitate an arrangement that is both workable and equitable.

Acknowledgements

Work on this chapter was supported by a grant from the National Human Genome Research Institute ("Building the Medical Information Commons: Participant Engagement and Policy"; R01 HG008918), and we thank our collaborator Robert Cook-Deegan for reviewing the manuscript and providing insightful comments.

Notes

1 In the biomedical informatics and data science literature, definitions tend to be technology-focused. For example, Open Commons Consortium director Robert Grossman and colleagues define a "data commons" as "cyberinfrastructure that co-locates data, storage, and computing infrastructure with commonly used tools for analyzing and sharing data to create an interoperable resource for the research community" (Grossman et al. 2016: 11).

2 The other example mentioned in the report is the Electronic Medical Records and Genomics (eMERGE) Network.

3 From the HGP to the NIH Data Commons Pilot, expressions of commitment to open science are common. There are also expressions of support for community-generated rules and practices. However, connections to commons analysis of any kind are seldom made explicit, nor are the implications of analysis for practice typically discussed. For example, the NAS *Precision Medicine* report does not contain a single reference to Ostrom (NAS 2011). Sage Bionetwork's founder did reportedly read Ostrom's work (Timmerman 2010).

4 The human genes initially sequenced and kept as a proprietary resource by Celera Corporation were cited by 20–30% fewer research papers and resulted in fewer diagnostic tests than the genes mapped by the HGP and subject to its rapid release policies (Williams 2013). Concerning some of the complexities of assessing the size of the benefits of the HGP and public investment in biomedical research commons generally, see Wadman 2013 and Frischmann 2012.

5 This body of work includes many of the chapters in *Governing Medical Research Commons* and law review articles by Contreras and Evans. At least one legal scholar has invoked Garrett Hardin's analysis of the tragedy of the commons in work on data commons (Yakowitz 2011). We focus here on a series of governance challenges, but we recognize that a bedrock sustainability challenge for an MIC is financial—securing long-term funding or finding a way to monetize the resource.

6 Mark Pennington explores Ostrom's relationship to the political tradition of classical liberalism with its presumption against central planning. For example: "[C]entral regulatory authorities often lack knowledge about the specific character of the assets to be managed and the nature of the incentives facing resource users" (Pennington 2012: 31–32). Sometimes this knowledge can be acquired by regulators, but at considerable cost. Other forms of knowledge may be unavailable to such authorities, such as "knowledge of cultural norms and values that structure the way in which people perceive and respond to resource-management issues" within particular communities (Pennington 2012: 32; see also Hayek 1945).

7 This classic article focuses on the environment and natural resources, but Ostrom has carried the idea of adaptive governance and related concepts into the sphere of knowledge commons in her work with other collaborators. See, e.g., Ostrom and Hess 2007.

8 For example, in the list of activities for the NIH Data Commons Pilot, the first is "establishing community-endorsed unifying principles and guidelines to govern how the Data Commons operates and what it means for digital objects in the Commons to be FAIR" (NIH Data Commons Pilot 2017). From the context it is very clear that "community" refers to data scientists and other researchers, not participants or the public.

9 Two illustrations. The Privacy and Security Policy adopts a principle of "Ensuring Proportionate Safeguards," which states that "[d]ata privacy safeguards should be proportionate to the sensitivity, nature, and possible benefits, risks, and uses," and further provides that "assessments of privacy risks should involve not only disclosure issues, but also reasonably likely harms, which may include individual or group discrimination or stigmatization" (GA4GH 2015: 5). Thus, the Policy neither shrugs off privacy risks nor imposes an impossible standard, and it focuses attention on important points to consider. Concerning "Vulnerable Populations," the Policy asks initiatives to consider conducting a population-specific "Data Privacy Impact Assessment," which involves working with stakeholders including those affected to identify and reduce privacy risks, and developing a confidentiality agreement (GA4GH 2015: 8). The Policy also states that research findings that identify vulnerable persons or populations should not be disclosed without consent.

10 Frischmann et al. make the general point that "the nonrivalry of knowledge and information resources often rides on top of various rivalrous *inputs* (such as time or money) and may provide a foundation for various rivalrous *outputs* (such as money or fame)" (Frischmann et al. 2014: 17, emphasis added).

11 Rawls also identifies a third form: *perfect* procedure justice. Unfortunately, procedural justice of this kind "is rare, if not impossible, in cases of much practical interest" since it involves *guaranteed* achievement of an independent aim (Rawls 1971, 85).

12 This depends, of course, upon the nature of the conflict. If a conflict is such as to be resolvable by mutual agreement of a certain kind, the relevant mechanism may be an instance of pure procedural justice.

References

Anderson, E.S. (1999) "What is the point of equality?" *Ethics*, 109(2): 287–337.

Benkler Y. (2014) "Between Spanish huertas and the open road: a tale of two commons?" in Frischmann, B.M., Madison, M.J. and Strandburg, K.J. eds *Governing Knowledge Commons*. Oxford, UK: Oxford University Press, 69–98.

Boyle J. (2011) "Mertonianism unbound? Imagining free, decentralized access to most cultural and scientific material," in Ostrom, E. and Hess, C. eds *Understanding Knowledge as a Commons*. Cambridge, MA: MIT Press, 123–144.

ClinVar (2017) www.clinicalgenome.org/data-sharing/clinvar, accessed 15 October 2017.

Contreras, J.L. (2011) "Bermuda's legacy: policy, patents and the design of the genome commons," *Minnesota Journal of Law, Science & Technology*, 12: 61–125.

Contreras, J.L. (2014) "Constructing the genome commons," in Frischmann, B.M., Madison M.J. and Strandburg, K.J. eds *Governing Knowledge Commons*. Oxford, UK: Oxford University Press, 99–135.

Cook-Deegan, R., Ankeny, R.A. and Jones, K.M. (2017) "Sharing data to build a medical information commons: from Bermuda to the Global Alliance," *Annual Review of Genomics and Human Genetics*, 18: 6.1–6.27.

Cook-Deegan, R. and McGuire, A.L. (2017) "Moving beyond Bermuda: sharing data to build a medical information commons," *Genome Research*, 27: 897–901.

Deverka, P.A., Majumder, M.A., Villanueva, A.G., Anderson, M., Bakker, A.C., Bardill, J., Boerwinkle, E., & . . . McGuire, A.L. (2017) "Creating a data resource: what will it take to build a medical information commons?" *Genome Medicine*, 9: 84. https://genomemedicine.biomedcentral.com/articles/10.1186/s13073-017-0476-3.

Dietz, T., Ostrom, E. and Stern, P.C. (2003) "The struggle to govern the commons," *Science*, 3012: 1907–1912.

Evans, B.J. (2016a) "Barbarians at the gate: consumer-driven health data commons and the transformation of citizen science," *American Journal of Law & Medicine*, 42: 651–685.

Evans B.J. (2016b) "Power to the people: data citizens in the age of precision medicine," *Vanderbilt Journal of Entertainment & Technology Law*, 19(2): 243–265.

Evans B.J. (2017) "Genomic data commons," in Strandburg, K.J., Frischmann, B.M. and Madison, M.J. eds *Governing Medical Knowledge Commons*. Cambridge, UK: Cambridge University Press, 74–101.

Frischmann, B.M. (2012) *Infrastructure: The Social Value of Shared Resources*, Oxford, UK: Oxford University Press.

Frischmann, B.M., Madison, M.J. and Strandburg, K.J. (2014) "Governing knowledge commons," in Frischmann, B.M., Madison, M.J. and Strandburg, K.J. eds. *Governing Knowledge Commons*. Oxford, UK: Oxford University Press, 1–43.

Genetic Alliance, PEER Platform (2017) www.peerplatform.org, accessed 15 October 2017.

Global Alliance for Genomics and Health (GA4GH) (2015) "Privacy and security policy" www.ga4gh.org/wp-content/uploads/Privacy-and-Security-Policy.pdf (accessed 22 October 2018).

Global Alliance for Genomics and Health (2016) "A federated ecosystem for sharing genomic, clinical data," *Science*, 352: 1278–1280.

Grady, C. et al. (2015) "Broad consent for research with biological samples: workshop conclusions," *American Journal of Bioethics*, 15: 34–42.

Grossman, R.L, Heath, A., Murphy, M. and Patterson, M. (2016) "A case for data commons: toward data science as a service," *Computing in Science & Engineering*, 18(5): 10–20.

Hayek, F.A. (1945) "The use of knowledge in society," *American Economic Review*, 35(4): 519–530.

Hess, C. and Ostrom, E. eds (2007) *Understanding Knowledge as a Commons*. Cambridge, MA: MIT Press.

Kaye, J., Whitley, E.A., Lund, D., Morrison, M., Teare, H. and Melham K. (2015) "Dynamic consent: a patient interface for twenty-first century research networks," *European Journal of Human Genetics*, 23: 141–146.

Klosko, G. (2004) *The Principle of Fairness and Political Obligation*. Lanham, MD: Rowman & Littlefield.

Lee, P. (2017) "Centralization, fragmentation, and replication in the genomic data commons," in Strandburg, K.J., Frischmann, B.M. and Madison, M.J. eds *Governing Medical Knowledge Commons*. Cambridge, UK: Cambridge University Press, 46–73.

Longo, D.L. and Drazen, J.M. (2016) "Data sharing," *New England Journal of Medicine*, 374: 276–277.

Messner, D.A. et al. (2016) "Barriers to clinical adoption of next generation sequencing: perspectives of a policy Delphi panel," *Applied & Translational Genomics*, 10: 19–24.

National Academy of Sciences (NAS) (2011) *Toward Precision Medicine: Building a Knowledge Network for Biomedical Research and a New Taxonomy of Disease*. Washington, DC: National Academies Press.

National Cancer Institute (NCI) (2017) Genomic Data Commons https://gdc.cancer.gov, accessed 15 October 2017.

National Institutes of Health (NIH) (2017) "NIH Data Commons pilot phase explores using the cloud to access and share *FAIR biomedical big data," https://commonfund.nih.gov/bd2k/commons, accessed 15 October 2017.

National Institutes of Health (NIH) "Request for comments: proposal to update data management of genomic summary results under the NIH genomic data sharing policy," NOT-OD-17-110, 20 September 2017.

Nozick, R. (1974) *Anarchy, State, And Utopia*. New York: Basic Books.

Olson, M.V. (2017) "Precision medicine at the crossroads," *Human Genomics*, 11: 23. https://doi.org/10.1186/s40246-017-0119-1.

Open Commons Consortium (OCC) (2017) "From clouds to commons (letter from Robert Grossman, 31 October 2015)," http://occ-data.org/Commons_Clouds, accessed 15 October 2017.

Organisation for Economic Cooperation and Development (OECD) (2015) *Data-Driven Innovation: Big Data for Growth and Well-Being*. Paris: OECD Publishing.

Ostrom, E. and Hess, C. (2007) "A framework for analyzing the knowledge commons," in Hess, C. and Ostrom, E. eds *Understanding Knowledge as a Commons*. Cambridge, MA: MIT Press, 41–82.

Patient-Centered Outcomes Researcher Institute (PCORI) (2017) "What we mean by engagement," www.pcori.org/engagement/what-we-mean-engagement, accessed 15 October 2017.

Pennington, M. (2012) "Elinor Ostrom, common-pool resources and the classical liberal tradition," Ostrom, E., Chang, C., Pennington, M., and Tarko, V., *The Future of the Commons: Beyond Market Failure and Government Regulation*. London: Institute of Economic Affairs, 21–47.

Rawls, J. (1971) *A Theory of Justice*. Cambridge, MA: Harvard University Press.

Rose, C.M. (1986) "The comedy of the commons: custom, commerce, and inherently public property," *University of Chicago Law Review*, 53: 711–781.

Rosenbaum, L. (2017) "Bridging the data-sharing divide: seeing the devil in the details, not the other camp," *New England Journal of Medicine*, 376: 2201–2203.

Rothstein, M. (2017) "Structural challenges of precision medicine," *Journal of Law Medicine & Ethics*, 45: 274–279.

Sage Bionetworks (2017) "The commons," http://sagecongress.org/WP/2011congress/2011agenda/draftprinciples, accessed 22 October 2018.

Siu, L.L. et al. (2016) "Facilitating a culture of responsible and effective sharing of cancer genome data," *Nature Medicine*, 22: 464–471.

Strandburg, K.J., Frischmann, B.M. and Madison, M.J. (2017) "The knowledge commons framework," in Strandburg K.J., Frischmann, B.M. and Madison, M.J. eds *Governing Medical Knowledge Commons*. Cambridge, UK: Cambridge University Press, 9–18.

Summary of Principles Agreed at the First International Strategy Meeting on Human Genome Sequencing (Bermuda, 25–28 February 1996) as reported by HUGO, http://web.ornl.gov/sci/techresources/Human_Genome/research/bermuda.shtml#1, accessed 15 October 2017.

Timmerman, L. (2010) "Stanford, UCSD biologists take plunge into Arpanet-style project with Sage Bionetworks," www.xconomy.com/seattle/2010/10/06/stanford-ucsd-biologists-take-plunge-into-arpanet-style-project-with-sage-bionetworks, accessed 15 October 2017.

Vayenna, E. and Blassimme, A. (2017) "Biomedical big data: new models of control over access, use and governance," *Bioethical Inquiry*, doi:10.1007/s11673-017-9809-6.

Wadman, M. (2013) "Economic return from Human Genome Project grows," *Nature*, doi:10.1038/nature.2013.13187.

Williams, H. (2013) "Intellectual property rights and innovation: evidence from the Human Genome Project," *Journal of Political Economy*, 121: 1–27.

Yakowitz, J. (2011) "Tragedy of the data commons," *Harvard Journal of Law & Technology*, 25: 1–67.

22

ETHICAL STANDARDS FOR UNCONSENTED DATA ACCESS TO BUILD GENOMIC AND OTHER MEDICAL INFORMATION COMMONS

Barbara J. Evans

Introduction: the importance of large-scale medical information commons

Large-scale medical information commons provide infrastructure to advance the basic science of human genetics, clinical applications of genomic testing, and precision medicine (National Academy of Sciences 2011; Messner et al. 2016; Cook-Deegan and McGuire 2017; Majumder et al., Chapter 21, this volume). This chapter uses "genomic data commons" to refer to the subset of medical information commons containing at least some gene sequencing data along with other health-related information. Genomic data commons compile diverse sources of data about individuals—medical, genetic, environmental, lifestyle, and other data—and include data for large groups of people. They exemplify the big data resources on which 21st-century science increasingly relies.

Earlier works acknowledge that the genomic material human beings carry within their bodies is, in a very literal sense, a natural resource, so that the commons theory of Elinor Ostrom and her successors has considerable relevance to genomic data commons (Evans 2014, 2017). This chapter highlights a feature that distinguishes genomic data commons from the natural resource commons Ostrom wrote about. The human genome, unlike most other natural resources such as water and iron ore, contains deeply personal information that stirs concerns about individual privacy. Sheep grazing on a common meadow never put the adjacent villagers in terror that they might lose their insurance coverage or face social rejection for harboring an undesirable genetic mutation. To assemble genomic data commons, it is first necessary to resolve privacy and related civil rights concerns in a way that makes people comfortable with letting the commons exist.

Creating genomic data commons involves transparent and at times unconsented sharing of people's sensitive genetic and health-related information (Kulynych and Greely 2017). Genomic data commons usually are not open-access in the sense of being openly available to all, but the data resources they hold are accessible to a fairly broad group of qualified users (Majumder et al., Chapter 21, this volume). Potential users might include, for example, researchers studying the human genome (Kulynych and Greely 2017), regulators evaluating which tests and treatments are safe and effective for which population subgroups (Food and Drug Administration 2014), and healthcare providers seeking evidence to guide diagnostic and treatment decisions, including for the treatment of patients other than those whose information is being shared (Evans and Jarvik 2017; Rothstein 2018).

This level of data sharing with third parties offends many people's sense of data privacy. Surveys show that most people wish to be asked before their data are swept into data commons (Institute of Medicine 2009). This desire to be consulted is especially intense when the data are identifiable or re-identifiable, as deeply descriptive genomic and longitudinal health records often are (ibid.; National Committee for Vital and Health Statistics 2017). Yet honoring people's desire to control access to their data may inhibit the formation of data commons or diminish their utility for scientific research, in ways this chapter briefly explores.

There is a spectrum of bioethical views on whether informed consent should be required before people's identifiable data can be used for research, public health, and other socially beneficial purposes. Franklin Miller once described a view, advanced by some bioethicists and members of the public, that the individual's interest in data privacy outweighs the public's interest in access to identifiable personal information. According to this view, individuals should have complete sovereignty over potentially identifiable data that describe them. Even if research has high social value, even if consent is difficult or impossible to obtain, and even if a consent requirement may reduce the scientific validity of results, these facts "do not in themselves constitute valid ethical reasons for waiving a requirement of informed consent" (Miller 2008). At the opposite extreme, some bioethicists suggest that individuals have an ethical or civic duty to contribute their data for certain types of research that offer important benefits to society as a whole (Faden et al. 2013). A more centrist position views unconsented access to identifiable data as ethically justifiable under certain, carefully defined circumstances, provided various conditions are met (see, e.g., Privacy Protection Study Commission 1977; Department of Health, Education, and Welfare 1978).

In the United States, federal privacy law takes a centrist approach, mentioning informed consent but then allowing some unconsented access to people's data—including, at times, their identifiable data—for various socially beneficial purposes. This chapter briefly describes the range of uses that major U.S. federal regulations allow without individual consent. This chapter does not advocate for unconsented data use. It simply reports that unconsented data use, in many instances, is expressly authorized by law and is surprisingly common. In light of this reality, it is crucial to examine the ethical soundness of law's alternatives to explicit consent.

Unconsented data access raises a question that is the starting point for this chapter: What ethical protections are owed to people whose personal information is used, without their consent, to build genomic data commons for research and other socially beneficial purposes? Many people, if asked, might simply reply that their sensitive data should not be used without consent, but that is not the question. The question is how to make the process of unconsented data access as ethical and as worthy of public trust as it possibly can be, given that law already allows such access in a variety of circumstances. Designing alternative protections—a set of ethical standards for unconsented data use, as it were—feels like an oxymoron or a second-best solution, akin to debating the most ethical way to poison a baby. There is reluctance to deliberate ethical alternatives to consent, lest doing so suggest complicity in undermining the hallowed—if largely imaginary—informed consent norms that people love but law does not requite. This chapter draws on past attempts to identify ethical practices for unconsented data use and gleans a handful of insights relevant to the challenge of building genomic data commons.

The limits of informed consent for populating large data commons

Integrating genomic test results into clinical health care is still at an early stage. Typical individuals have about 3 to 3.5 million genetic variants (mutations) in their whole genomes

(Kohane et al. 2012; Food and Drug Administration 2014). Knowledge of how these variants affect human health is only starting to emerge and requires ongoing scientific study. Up to half a million of a person's variants often turn out to be novel or rare: that is, the variants were never seen before or only seen a few times in previous patients (Kohane et al. 2012). With rare and novel variants, there usually are too few past observations to support reliable conclusions about how the variants affect people's physical characteristics and health. A 2014 study found that fewer than 200 variants have enough evidence available for scientists to state, with confidence, that the variants are medically significant, such that people who have those variants should seek a follow-up clinical evaluation (Dewey et al. 2014). The clinical validity of the remaining—in other words, almost all—genetic variants is still uncertain or unknown.

Clinical validity

> [a]ddresses whether there is a strong and well validated association between having a particular gene variant and having a particular health condition and whether knowing that a person has the gene variant offers meaningful insight into the person's health or reproductive risks.
>
> *(Secretary's Advisory Committee on Genetic Testing 2000)*

If a well validated association of this sort exists between a variant and a health condition, then a test that detects the variant may be useful as a diagnostic tool (Fabsitz et al. 2010). Otherwise, it could be misleading or even dangerous to rely on the test as the basis for making medical decisions. Modern genomic tests are able to spot the variants that a person has, but determining what the variants mean—that is, assessing their clinical validity—requires the use of large-scale genomic data commons to search for possible associations between specific variants and the occurrence of various health conditions.

To study rare genetic variants, data for tens to hundreds of millions of individuals may be required (Shirts et al. 2014). For each individual, the data commons ideally should include the person's genomic and other diagnostic test results, extensive data from the person's clinical health records, and other available health-related information such as data from mobile and wearable health devices that shed light on the person's lifestyle and environmental factors that influence health (Evans 2011). The data need to be longitudinal in the sense of providing a fairly detailed chronological account of how each person's health has evolved over time and in response to various environmental stimuli and lifestyle choices.

For this purpose, all data are potentially health data. Parameters that are not usually regarded as health-related sometimes have important impacts on health. An example is that education and wealth confer advantages in terms of how long, and how healthily, people live. Including such data in medical information commons would allow their potential associations with health to be explored: Does wealth promote health merely because it enables better access to health care services, or are other factors at work? Could the stress of poverty trigger the expression of deleterious gene variants that, under better circumstances, might never have made the person sick? It will take far richer and more inclusive datasets than currently exist in order to discover and confirm such impacts, if they exist. Genomic data commons now on the drawing board will advance scientific understanding but may not advance it as far or as quickly as could be done with larger, richer data commons. Unfortunately, every new parameter added to a data commons presents incremental risks to people's privacy.

One way to create genomic data commons—and it is probably fair to say that most people view this as the ethically preferred way—is to obtain individuals' consent before their personal data are placed into a data commons. This is the model chosen by the National Institute of Health's All of Us Research Program, which will incorporate data for "1,000,000+ people" (National Institutes of Health 2018). A million-person cohort is a rich resource for scientific discovery and marks a significant advance over data resources available in the past. Yet it is small relative to the hundred-million-person and even multi-billion-person data resources that scientists ultimately may need in order to unlock the secrets of the human genome. An "all of us" research program ultimately should include data for all of us—not just for one million consenting Americans but for a truly representative sample of the 7.6 billion human beings who live on Earth today (Worldometers 2018). The sheer scale of the data commons needed for genomics, precision medicine, and other 21st-century science programs raises concern about the scalability of traditional informed consent models. Moreover, any data-acquisition plan that relies on individual informed consent raises questions about how representative the resulting sample will be.

Data commons for some applications—such as studying rare variants and detecting rare adverse events associated with the use of drugs and tests—need to be highly inclusive in the sense of incorporating data for most or even all persons, or at least a representative sample of them (Evans 2011; Shirts et al. 2014). Inclusive data commons capture rare events and provide results relevant to all population subgroups. Inclusive data commons also avoid consent bias (selection bias), which can arise if the people who volunteer their data are unlike the population at large (Casarett et al. 2005; Buckley et al. 2007; Institute of Medicine 2009). For example, patients suffering symptoms of disease might be more willing to volunteer their data for research, in the hope of finding answers, than are healthy, asymptomatic patients who never had a reason to wonder what causes the disease. If sicker people preferentially consent to research, then the research may produce biased results that overstate how often a given variant causes full-blown illness. Consent bias reportedly caused some early studies to overstate the lifetime risk that certain BRCA variants will lead to breast and ovarian cancer in women who have those variants (Bloom 2016). Consent bias can arise regardless of whether the consent framework employs "opt-in" consent (where people are excluded from the dataset unless they expressly agree to join), "opt-out" consent (where people are included unless they express a desire for their data to be removed) (Evans 2011), or broad consent (where people agree to allow a wide range of unspecified future uses of their data). The reliance on self-selection, rather than specific details of the consent process, is what causes the problem.

Recent amendments to a major U.S. federal research regulation, the Common Rule (45 C.F.R. pt. 46, subpt A), enable the use of broad consent (U.S. Department of Health and Human Services et al. 2017). These changes, scheduled to take effect in July 2018 (U.S. Department of Health and Human Services et al. 2018), will let researchers seek broad consent to store, maintain, and use people's identifiable information or biospecimens for secondary research (that is, research with data that originally were collected for a non-research purpose or for a prior, unrelated research project) (U.S. Department of Health and Human Services et al. 2017, at 7266–67, 45 C.F.R. § 46.116(d)). There is hope that broad consents may enhance access to data for research, but it is not clear that they will necessarily foster access to the large, representative datasets that genomic science needs. The rare breed of individual who would willingly allow broad, unspecified future uses of their data may differ from the rest of us. Broad consents have not been widely used in the past, so the potential biases they introduce are not well understood.

Apart from the possibility that informed consent requirements may reduce the availability or quality of data to populate genomic data commons, there are ongoing doubts about how effective consent really is as a tool to protect people's rights. The notion that informed consent is truly voluntary is a fiction in many contexts (Privacy Protection Study Commission 1977). Consent has never been an effective tool to ensure stringent data security, data destruction, and data use and transfer policies, which are what privacy protection actually requires (President's Council of Advisors on Science and Technology 2010). An additional problem in modern, interconnected data environments is that privacy is interdependent: your privacy depends not just on whether you consent to share your own data, but on whether other people consent to share theirs (Humbert 2015). This is not just a problem within families (for example, your sister's data reveal genetic information about you). Data scientists have demonstrated re-identification attacks that can guess—with surprisingly high probabilities—the person to whom a de-identified genome belongs. These attacks work by ruling out identified individuals (and relatives of those individuals, and their relatives, and so forth) to whom the data could not possibly belong (Evans 2016). If a stranger consents to share her variant data in identifiable format, her consent affects whether your variants can be linked to you. As big data environments reveal the extent of our interconnectedness, individual consent breaks down as a means of protecting individual rights. Individuals' privacy is "affected by the decisions of others and could be out of their own control" (Biczók & Chia 2013).

Populating the commons without informed consent

Imposing strong individual consent requirements has the potential to produce medical information commons that are too small, incomplete, or unrepresentative to meet the needs of 21st-century scientists, regulators, and other users. Major U.S. federal regulations such as the Common Rule and the Health Insurance Portability and Accountability Act of 1996 (HIPAA) Privacy Rule (Pub. L. No. 104–191, 110 Stat. 1936 (August 21, 1996); see 45 C.F.R. pts. 160, 164) foresaw these problems and offer legal pathways to disclose and use people's data, under some circumstances, without their permission. The HIPAA Privacy Rule, in particular, is often criticized as having broken its promise of privacy by facilitating unconsented data uses (see, e.g., Ohm 2010). In reality, however, the Privacy Rule never promised privacy. It expressly serves a mix of two competing values: privacy and data transparency (Evans 2018).

The 1996 HIPAA statute ordered the U.S. Department of Health & Human Services (HHS), by 1997, to make recommendations to Congress about the content of separate health privacy legislation that Congress envisioned enacting as a follow-up to HIPAA (see HIPAA statute at §§ 264(a)–(c)). These 1997 HHS recommendations gave great weight to data transparency, in the sense of sharing people's health data with third parties for socially beneficial purposes. HHS recommended that "[f]ederal health privacy law should permit limited disclosures of health information without patient consent for specifically identified national priority activities" (U.S. Department of Health and Human Services 1997). The priority activities included sharing people's data to support medical treatment and healthcare payments, regulatory oversight of the healthcare system, public health activities, research, and various judicial, law enforcement, forensic, and public safety uses of data. HHS confirmed that it was fully "aware of the concerns of privacy and consumer advocates about these uses" (ibid.) but nevertheless called for "a balancing of privacy and other social values" (ibid.).

The HIPAA statute gave Congress until August 21, 1999 to enact health privacy legislation in response to HHS's recommendations. If Congress failed to do so, the statute authorized HHS to take matters into its own hands and promulgate the HIPAA Privacy Rule (see HIPAA statute at §§ 264(a)–(c)). After reviewing HHS's recommendations, Congress chose to let HHS proceed with this rulemaking. This suggests that Congress concurred with the overall thrust of those recommendations: individuals, notwithstanding their strong desire to control access to their sensitive data, should not have a legal right to block the priority data uses identified in the 1997 HHS recommendations.

Sixteen years after HHS made these recommendations, Ruth Faden and co-authors elaborated an ethical basis for the view that people have a duty to contribute their data for socially beneficial purposes (Faden et al. 2013). According to their analysis, people have no ethical obligation to subject their bodies to the physical risks of clinical trials, but people may have an ethical duty to contribute their health records to studies that advance useful knowledge, provided these studies offer reasonable data security (ibid. at S23). Faden et al. suggest that this obligation is justified by a "norm of common purpose . . . a principle presiding over matters that affect the interests of everyone" (ibid.). "Securing these common interests is a shared social purpose that we cannot as individuals achieve" (ibid.). The authors acknowledge that these views "depart in significant respects from contemporary conceptions of clinical and research ethics" and may imply an obligation for patients to participate in activities that use their data (ibid. at S18).

The ethical principles Faden and her co-authors set out in 2013, even if controversial at the time, had already been embodied in the HIPAA Privacy Rule for many years. The Privacy Rule was first promulgated in December 2000 and, after minor adaptations in 2002, became effective in 2003–2004 (U.S. Department of Health and Human Services 2000, 2002). The Privacy Rule strikes a balance that lets individuals control many uses and disclosures of their data by granting or withholding consent, which HIPAA calls "authorization" (45 C.F.R. §164.508). People have no right, however, to block third parties' access to their data for a set of enumerated national priority data uses (45 C.F.R. §§164.502, 164.512). For these uses, a norm of common purpose trumps the individual's right to control data access. This is not simply an ethical norm, but a legal norm: U.S. federal law already allows this unconsented access.

The Privacy Rule's national priority activities span three types of data transparency that Frederick Schauer once identified (Schauer 2011). Schauer's concept of "transparency as regulation" empowers the recipient of information to regulate, monitor, or control the information provider (Schauer 2011, at 1347). The Privacy Rule serves this type of transparency through provisions allowing unconsented disclosure of people's protected health information for various FDA regulatory purposes, for health care system oversight, and for HHS's own privacy enforcement activities (45 C.F.R. §§164.512(b), (d); 164.502(b)(vi)). A second concept, "transparency as efficiency," regards data flows as instrumental to efficient markets and resource allocations (Schauer 2011, at 1350). This is exemplified, in the Privacy Rule, by provisions that allow unconsented data disclosures to support medical treatment, payments, healthcare operations, organ donations, and the processing of workers' compensation claims (45 C.F.R. §§ 164.502(a)(iii), 164.512(h), (l)). Schauer's third concept, "transparency as epistemology" (Schauer 2011, at 1350) sustains the creation of non-market and public goods such as scientific discovery, public health, well-functioning law enforcement and judicial systems, public safety, and national defense (45 C.F.R. §§ 164,512(i), (b)(1)(i), (c), (e)–(g), (j), (k)).

The Privacy Rule allows transparent sharing of data for these purposes without the individual's permission. The people whose protected health information is disclosed will sometimes, but not always, share in the broad social benefits, but they bear the full brunt of the risks that accompany data transparency. These risks might include, for example, a potential for stigmatization, discrimination, identity theft, or other harms if their data, once disclosed, are inappropriately used or re-disclosed. The Privacy Rule does not empower individuals to manage these risks by controlling access to their data. For the national priority data uses, the Privacy Rule enables a regime of unconsented data access by a fairly broad range of third parties. The challenge, then, is to find other ways to protect the interests of people whose data are shared without their consent.

Historical consent practices and the ethics of unconsented data use

Since the early 1970s, Congress has repeatedly commissioned studies to explore ways to protect people's rights in situations where they lack a right of informed consent. These studies, some of them forty years old, offer thoughtful commentary on a question that hangs over today's discussion of genomic data commons: What ethical protections are owed to people whose data are used for socially beneficial purposes without their consent?

The Fair Credit Reporting Act of 1970 allowed unconsented access to people's financial data to facilitate a well-functioning credit market; in return, it offered various protections for individuals, such as a right to inspect, correct, and receive copies of their data (15 U.S.C. § 1681). A 1973 Code of Fair Information Practices developed by HHS's precursor, the Department of Health, Education, and Welfare (HEW), recommended various protections—again including an individual access right along with other protections—for people whose data are stored, used, and shared (U.S. Department of Health, Educaiton and Welfare Services 1973). The Privacy Act of 1974, which protects data held in governmental databases, included an access right and other protections for individuals (5 U.S.C. § 552(a)(d)), and it established a Privacy Protection Study Commission (PPSC) to examine the use and protection of personal data more generally (Privacy Protection Study Commission 1977).

The PPSC construed its mandate broadly and included medical, financial, insurance, employment, and diverse other types of data that, together, span today's concept of big data. Concurrently with the PPSC's work, Congress created a National Commission for the Study of Ethical Problems in Medicine and Biomedicine ("National Commission") to recommend ethical principles for human-subject research (National Research Act of 1974, Pub. L. No. 93–348, §§ 201–02, 88 Stat. 342). The National Commission, while focusing mainly on clinical research, also touched on research uses of data and biospecimens (U.S. Department of Health, Education, and Welfare 1978). Another important study, mentioned earlier in this chapter, was the 1997 recommendations HHS prepared pursuant to the HIPAA statute (U.S. Department of Health and Human Services 1997). Those recommendations focused strictly on health data and informed the design of the HIPAA Privacy Rule. The Genetic Information Nondiscrimination Act of 2008 later expanded HIPAA's definition of health data to include genetic information—both clinically significant medical genetic data and genetic data unrelated to health (Evans 2018).

These past studies offer an important insight: the expectation that people should consent to uses of their data is of very recent origin. This is in sharp contrast to the modern bioethical requirement of informed consent for clinical research, which reflects centuries-old legal and cultural traditions against the unconsented touching of a person's body. At common law, unconsented touching of the body was both a battery and a tort, and the Supreme Court, in 1891, recognized "the right of every individual to the possession and control of his own person" (Union Pac. Ry. Co. v. Botsford, 141 U.S. 250, 251 (1891)). This right "is 'so rooted in the traditions and conscience of our people,' . . . as to be ranked as one of the fundamental liberties protected by the 'substantive' component of the Due Process Clause" (Newman v. Sathyavaglswaran, 287 F.3d 786, 789 (9th Cir. 2002) (citing Schmerber v. California, 384 U.S. 757, 772 (1966)). "The integrity of an individual's person is a cherished value of our society" (Rochin v. California, 342 U.S. 165, 174 (1952)).

The modern practice of seeking consent for the use of people's data does not, apparently, rest on longstanding legal or cultural traditions. The PPSC and National Commission both found that it was common for investigators to use people's medical records without informed consent in the late 1970s. The PPSC observed that "federal rules governing the funding of medical research require the informed consent of individuals who participate in it as research subjects, but do not require their consent when medical records are reviewed and abstracted for retrospective epidemiological research studies" (PPSC 1977, at 280). Investigators surveyed by the National Commission expressed the view that consent was unnecessary for studies "based exclusively upon existing records, data or materials gathered for other purposes" (U.S. Department of Health, Education, and Welfare 1978).

Indeed, if there are any longstanding norms relating to data, they seem to favor unconsented collection and use. One of the earliest examples of a large data commons was England's Domesday Book, a detailed statistical record of the inhabitants of England, compiled in the years 1085–86 (Anglo-Saxon Chronicle 1085). The data collection was extensive, intrusive, and apparently done with no consent whatsoever. Instead:

> [h]ad the king a large meeting, and very deep consultation with his council, about this land; how it was occupied, and by what sort of men. Then sent he his men over all England into each shire; commissioning them to find out 'How many hundreds of hides were in the shire . . . how much, each man had, who was an occupier of land in England, either in land or in stock, and how much money it were worth.' So very narrowly, indeed, did he commission them to trace it out, that there was not one single hide, nor a yard of land, nay, moreover (it is shameful to tell, though he thought it no shame to do it), not even an ox, nor a cow, nor a swine was there left, that was not set down in his writ.
>
> *(ibid.)*

Big data 900 years ago set the stage for big data today. The unconsented touching of a person's body and the unconsented touching of a person's data were viewed as two different things: one having profound ethical and legal consequence; the other, less so.

As part of its 1977 report, the PPSC took a fresh look at whether it is ethical to use people's data without consent. The PPSC concluded that, as a general rule, medical care providers—who in those days were the principal repositories of people's health information—should not "disclose, or be required to disclose, in individually identifiable form, any information" about an individual who has not consented to the disclosure (PPSC 1977, at 306). The modern concept that data uses require informed consent emerges very clearly in the PPSC's 1977 report. The PPSC did, however, state various exceptions to this general consent requirement (ibid.). In particular, the PPSC recognized that unconsented research uses of data, under certain circumstances, are ethically justified (ibid.). It then set out to identify ethical protections people should have when their data are used without their consent.

Four points to consider

The next discussion highlights just four of the PPSC's recommendations—a small subset of the recommendations PPSC offered. What makes these four recommendations interesting is that they were never implemented, or else were implemented but are now encountering challenges when applied in modern, big data environments. These four PPSC recommendations offer food for thought as scholars and policy-makers debate unconsented access to data to create genomic and other medical information commons.

The need for a public benefit standard

In order to be ethical, unconsented data disclosures should be held to a public benefit standard (PPSC 1977, at 306, rec. 10(c)(iii)). This is conceptually similar to the public use criterion seen in eminent domain jurisprudence, which requires takings of property to fulfill a socially beneficial, rather than a purely private, purpose (Evans and Meslin 2017). The PPSC called for the "medical-care provider maintaining the medical record [in more modern terms, the data-holder]" to ensure "that the importance of the research or statistical purpose for which any use or disclosure is to be made is such as to warrant the risk to the individual from additional exposure of the record or information contained therein" (PPSC 1977, at 306).

The PPSC's public benefit standard is distinct from the familiar criterion, at 45 C.F.R. § 46.111(a)(2) of the Common Rule, which requires IRBs not to approve research unless the risks are reasonable in relation to the anticipated benefits of the research—if any—to the individual and the importance of the knowledge that may reasonably be expected to result from the research. The Common Rule's section 46.111(a)(2) test asks whether research is so lacking in scientific merit that an IRB reviewing the research should disapprove it and not allow people to consent to it under any circumstance. The PPSC's public benefit criterion, in contrast, asks whether research has such great public importance that it can ethically be justified even in the absence of individual consent.

Over the years, many bioethicists have suggested that unconsented data uses should be held to a public benefit standard (See, e.g., Casarett et al. 2005). Even so, the Common Rule and HIPAA Privacy Rule include no such standard in their waiver provisions, which allow an IRB to approve unconsented access to data for research (see 45 C.F.R. §46.116(d) of the traditional Common Rule; 45 C.F.R. §46.116(f) of the post-2018 revised Common Rule; 45 C.F.R. §164.512(i) of the Privacy Rule). The National Commission that helped design the Common Rule cited the PPSC's recommendation but, without explanation, HHS did not require a finding of public benefit as part of the Common Rule's waiver conditions. This was an accident of history: the Common Rule's waiver provisions were not originally designed as a pathway for unconsented data access and only later came to be so used (Evans and Meslin 2017, at 114). The story with the HIPAA Privacy Rule is different: HHS closely heeded the PPSC's recommendations and included a public benefit standard in the Privacy Rule's proposed waiver provisions. This drew negative public comments expressing doubt that IRBs, with their narrow institutional perspective, would be able to apply such a standard, and the troublesome requirement was eventually dropped (ibid., at 115). Without a public benefit standard, it might be argued that every waiver that any IRB has ever approved under those two regulations was ethically suspect, at least according to the PPSC's criteria.

Unconsented "takings" of people's data will never merit public trust unless individuals can be sure that the burdens they bear for the sake of data transparency genuinely serve the broader public good. There are various ways to achieve this. One way is for legislatures to grant a Blackstonian consent on behalf of the people by passing statutes that enable unconsented access for enumerated purposes. This approach is common in state public health laws and in various federal statutes, such as the Fair Credit Reporting Act, that deem specific data uses worthy of unconsented access. Unfortunately, legislatures cannot enunciate general rules of data access that are sufficiently granular to distinguish, at a fine scale, which of the many conceivable uses of data offer public benefits and which do not. Local bodies, such as IRBs, lack the global perspective and the procedural transparency that this task requires. Scholars have proposed various alternatives (Evans & Meslin 2017, at 118–20). One would be to form a single, national data commons governance body to approve unconsented data uses of people's data under transparent rules that make the body accountable to the public whose data are involved. Another approach would be to limit unconsented access to uses meeting specific criteria, such as providing "reciprocity of advantage" so that those who bear the burdens have a fair chance of reaping benefits from the data commons. Still another approach would focus on identifying data uses that presumptively *do not* benefit the public—for example, proprietary data uses in which the user refuses to publish or otherwise share results—and would ban the use of consent waivers for those uses (ibid.). Non-public data uses would still be allowed, but only with consent.

Respecting individual transparency entitlements

Whenever people's data can be shared without their consent, the people are owed various transparency entitlements. This is a longstanding and well-established principle of U.S. federal privacy law, embraced by the PPSC and the 1997 HHS recommendations to Congress, and implemented in the Fair Credit Reporting Act, the Privacy Act, the HIPAA Privacy Rule, and GINA. It also appears in the law of other jurisdictions, such as the European Union's General Data Protection Regulation (Regulation (EU) 2016/679, Art. 15).

The specific transparency entitlements vary somewhat from one context to the next, but generally include granting people a right of access to their own data so they can understand what may be shared and take steps to mitigate risks they may face as a result; a right to correct errors in data that purport to describe them; and a right to an accounting for unconsented disclosures so that they know what types of data were shared and with whom. Finally, if people's data are used without their consent to make decisions that may affect them, the people may have a right to an explanation of those decisions and the data on which they were based (PPSC 1977; EU GDPR, Art. 20). In modern data environments that apply machine-learning algorithms, there is debate whether a right of explanation remains feasible, due to the inherent limits of algorithmic transparency (Goodman and Flaxman 2016).

For health and genetic information, the HIPAA Privacy Rule grants individuals a legally enforceable right to inspect and obtain copies of data about themselves that HIPAA-regulated entities maintain in storage (45 C.F.R. §164.524)—a right that, as it relates to genomic data, has encountered considerable resistance and is imperfectly enforced (Evans 2018). HIPAA also provides a right to request corrections (45 C.F.R. § 164.526) and a right to an accounting for unconsented disclosures (45 C.F.R. § 164.528). This latter right is imperfect because it does not require an accounting for disclosures of data that have been de-identified according to HIPAA's rather lax standards; it thus fails to notify people of the true extent of re-identification risk they may face. Moreover, data holders can account for disclosures by broad category (e.g., to researchers, to public health agencies). This may leave individuals unable to assess whether their data were disclosed to specific downstream users that may have good (or bad) reputations for protecting privacy. To merit public trust, HIPAA's existing transparency entitlements need strengthening.

Schauer's transparency framework recognizes a fourth, essential type of transparency: "transparency as democracy," the sharing of data with members of the public to empower the governed to monitor and manage their governance structures and to celebrate individual control as an end in itself (Schauer 2011, at 1349). In healthcare contexts, the term "transparency as respect for autonomy" seems more appropriate, because this fourth type of transparency promotes many of the same values served by the bioethical principles of respect for persons and respect for the autonomy of patients and research participants (U.S. Department of Health, Education, and Welfare 1979). Transparency with the people whose data are being shared is an essential element of public trust. To be ethical, data transparency entails more than simply sharing people's data with third parties; it requires transparency with the people themselves. This principle is well recognized in law but remains less honored in the field of bioethics. The Common Rule, at present, provides no individual transparency entitlements, and IRBs acting under the Common Rule at times take steps that limit people's access to their own data, ostensibly for the people's own good.

Modernizing the minimum necessary standard

Unconsented disclosures of data should be subject to a minimum necessary standard: no more information should be shared than is reasonably necessary to serve the purpose for which the data were requested. This principle is embodied in the HIPAA Privacy Rule and, as a long-standing principle of medical privacy, it is widely observed under the Common Rule. The main challenge, today, is to adapt this principal for the modern context of hypothesis-free data analytics. The minimum necessary standard arose in a past era when research tested specific hypotheses, for which specific data requirements could be identified in advance. For many machine-learning and hypothesis-free analytic applications, the minimum necessary data may be everything we know about everybody. It is time to update the minimum necessary standard to clarify when the disclosure of entire medical records and entire genomes is ethically justified, and what data security and other privacy protections users must implement as a condition of access to large data sets.

Restrictions on downstream re-disclosure and re-use of data

In its 1997 recommendations, HHS exhorted Congress to pass new privacy legislation, instead of having HHS proceed with rulemaking under the HIPAA statute (U.S. Department of Health and Human Services 1997). The problem was that HHS's jurisdiction under the HIPAA statute only extends to healthcare providers, insurers, and other entities in the healthcare payment chain. Socially beneficial data transparency requires them to share data with many entities that are not HIPAA-covered—for example, non-HIPAA research facilities and regulatory agencies. HHS has no authority to regulate downstream uses and re-disclosure of data once it moves beyond the reach of HIPAA regulation. This lack of authority to regulate downstream recipients of HIPAA-protected data is an enduring weakness of the federal health and genetic privacy framework. This is in addition to the other glaring problem: modern genomic data commons may include various types of data—such as data from people's fitness devices and other data bearing on their lifestyles—that were never HIPAA-regulated in the first place. Public trust in large-scale data commons ultimately may require comprehensive new privacy legislation to address the needs of modern big data environments.

Conclusion: building trust when control is absent

Trust becomes relevant precisely at the point when people are denied individual control. Trust matters greatly in a commercial airliner when you place your fate in the hands of a pilot you never met; it is less relevant when you drive your car and presumably have full information about your own sobriety and driving skills. The dialogue about public trust in medical commons all too often focuses on optimal consent procedures—opt-in versus opt-out, narrow versus broad. Yet the central problem in building public trust is to implement strong ethical and privacy standards to govern the many *unconsented* uses of our data. Past efforts, forty years ago, to envision a trustworthy *non-consent* framework have ongoing relevance today. When law allows widespread sharing of people's data without their consent—and law does allow this, whether or not we like it—people need assurance that the burdens they bear confer a public benefit. They need data transparency to extend not just to others but also to themselves. They need meaningful oversight of how much of their personal information is disclosed, and what can be done with it after it is disclosed. What else they may need, beyond these minimal ethical requirements, is a topic that scholars now need to discuss.

Acknowledgements

The author received support under a grant from the National Institutes of Health/National Human Genome Research Institute R01 HG008918, "Building the Medical Information Commons: Participant Engagement and Policy," with additional support from R01 HG008605, "LawSeq™: Building a Sound Legal Foundation for Translating Genomics into Clinical Application." The author thanks Mary Majumder and Peter Zuk for reviewing the manuscript and providing helpful comments. Views expressed are those of the author rather than the funding agencies.

References

Anglo-Saxon Chronicle: Eleventh Century (1085) available at: The Avalon Project, Lillian Goldman Library, Yale Law School, http://avalon.law.yale.edu/medieval/ang11.asp (accessed June 1, 2018).

Biczók, G. and Chia, P.H. (2013) "Interdependent Privacy: Let Me Share Your Data," available at http://fc13.ifca.ai/proc/10-1.pdf (accessed June 1, 2018), at 1.

Bloom, T. (2016) "Next Generation Sequencing and Bioinformatics: A Researcher's Perspective" (statement at the American Association of Law Schools Annual Conference, BioLaw Section, January 8, 2016).

Buckley, B., Murphy, A.W., Byrne, M. and Glynn, L. (2007) "Selection Bias Resulting from the Requirement for Prior Consent in Observational Research: A Community Cohort of People with Ischaemic Heart Disease," *Heart* 93(9): 1116–1120.

Casarett, D., Karlawish, J., Andrews, E. and Caplan, A. (2005) "Bioethical Issues in Pharmacoepidemiologic Research," in Strom, B.L. ed. *Pharmacoepidemiology*, 4th ed. Hoboken, NJ: Wiley, 587–598.

Cook-Deegan, R. and McGuire, A.L. (2017) "Moving Beyond Bermuda: Sharing Data to Build a Medical Information Commons," *Genome Research* 27: 897–901.

Dewey, F.E., Grove, M.E., Pan, C. et al. (2014) "Clinical Interpretation and Implications of Whole-Genome Sequencing," *The Journal of the American Medical Association* 311: 1035–1045.

Evans, B.J. (2011) "Much Ado About Data Ownership," *Harvard Journal of Law & Technology* 25: 69–130.

Evans, B.J. (2014) "Mining the Human Genome after Association for Molecular Pathology v. Myriad Genetics," *Genetics in Medicine* 16: 504–509.

Evans, B.J. (2016) "Power to the People: Data Citizens in the Age of Precision Medicine," *Vanderbilt Journal of Entertainment and Technology Law* 19: 243–265.

Evans, B.J. (2017) "Genomic Data Commons," in Strandburg, K., Frischmann, B. and Madison M. eds. *Governing Medical Knowledge Commons*. Cambridge, UK: Cambridge University Press, 74–101.

Evans, B.J. (2018) "The Genetic Information Nondiscrimination Act at Age 10: GINA's Controversial Assertion that Data Transparency Promotes Privacy and Civil Rights," *William & Mary Law Review* 60 (forthcoming), available at: https://papers.ssrn.com/abstract=3154009.

Evans, B.J. and Jarvik, G.P. (2017) "Impact of HIPAA's Minimum Necessary Standard on Genomic Data Sharing," *Genetics in Medicine* e-pub ahead of print September 14, 2017, doi:10.1038/gim.2017.141.

Evans, B.J. and Meslin, E.M. (2017) "Biospecimens, Commercial Research, and the Elusive Public Benefit Standard," in Lynch, H.F., Bierer, B.E., Cohen, I.G. and Rivera, S.M. eds. *Specimen Science*. Cambridge, MA: MIT Press, 107–124.

Fabsitz, R.R., McGuire, A.L., Sharp, R.R. et al. (2010) "Ethical and Practical Guidelines for Reporting Genetic Research Results to Study Participants: Updated Guidelines from a National Heart, Lung, and Blood Institute Working Group," *Circulation Cardiovascular Genetics* 3: 574–580.

Faden, R.R., Kass, N.E., Goodman, S.N., Pronovost, P., Tunis, S. and Beauchamp, T.L. (2013) "An Ethics Framework for a Learning Health Care System: a Departure from Traditional Research Ethics and Clinical Ethics," *The Hastings Center Report* Special Issue S16–27, at 16.

Food and Drug Administration (2014) Optimizing FDA's Regulatory Oversight of Next Generation Sequencing Diagnostic Tests: Preliminary Discussion Paper, available at www.fda.gov/downloads/medicaldevices/newsevents/workshopsconferences/ucm427869.pdf (accessed June 1, 2018).

Goodman, B. and Flaxman, S. (2016) "European Union Regulations on Algorithmic Decision-Making and a 'Right To Explanation,'" *ArXiv* (June 28, 2016), https://arxiv.org/abs/1606.08813 (accessed June 1, 2018).

Humbert, M. (2015) "When Others Impinge upon Your Privacy: Interdependent Risks and Protection in a Connected World," available at https://infoscience.epfl.ch/record/205089/files/EPFL_TH6515.pdf (accessed June 1, 2018), at 66.

Institute of Medicine (2009) Nass, S.J., Levit, L.A., Gostin, L.O. eds. *Beyond the HIPAA Privacy Rule: Enhancing Privacy, Improving Health Through Research*, at 209–214.

Kohane, I.S., Hsing, M. and Kong, S.W. (2012) "Taxonomizing, Sizing, and Overcoming the Incidentalome," *Genetics in Medicine* 14: 399–404.

Kulynych, J. and Greely, H.T. (2017) "Clinical Genomics, Big Data, and Electronic Medical Records: Reconciling Patient Rights with Research When Privacy and Science Collide," *Journal of Law and the Biosciences* 4: 94–132.

Messner, D.A. et al (2016) "Barriers to Clinical Adoption of Next Generation Sequencing: Perspectives of a Policy Delphi Panel," *Applied & Translational Genomics* 10: 19–24.

Miller, F.G. (2008) "Research on Medical Records Without Informed Consent," *Journal of Law, Medicine & Ethics* 36: 560–566.

National Academy of Sciences (2011) Toward Precision Medicine: Building a Knowledge Network for Biomedical Research and a New Taxonomy of Disease Washington, DC: National Academies Press.

National Committee for Vital and Health Statistics (2017) *Recommendations on De-identification of Protected Health Information under HIPAA*, available at www.ncvhs.hhs.gov/wp-content/uploads/2013/12/2017-Ltr-Privacy-DeIdentification-Feb-23-Final-w-sig.pdf (accessed June 1, 2018).

National Institutes of Health, All of Us Research Program (2018) *The Future of Health Begins with You*, available at https://allofus.nih.gov/ (accessed June 1, 2018).

Ohm, P. (2010) "Broken Promises of Privacy: Responding to the Surprising Failure of Anonymization," *UCLA Law Review* 57: 1701–1777.

President's Council of Advisors on Science and Technology (2010) Report to the President: Realizing the Full Potential of Health Information Technology to Improve Healthcare for Americans: The Path Forward. Washington, DC: U.S. Government Printing Office.

Privacy Protection Study Commission (PPSC) (1977) *Personal Privacy in an Information Society*, available at www.ncjrs.gov/pdffiles1/Digitization/49602NCJRS.pdf (accessed June 1, 2018).

Rothstein, M.A. (2018) "Reconsidering the Duty to Warn Genetically At-Risk Relatives," *Genetics in Medicine*, e-pub ahead of print February 1, 2018, doi:10.1038/gim.2017.257.

Schauer F. (2011) "Transparency in Three Dimensions," *University of Illinois Law Review* 2011: 1339–1358 at 1347–50.

Secretary's Advisory Committee on Genetic Testing (2000) *Enhancing the Oversight of Genetic Tests: Recommendations of the SACGT*. Washington, DC: U.S. Government Printing Office, at 15, note 11.

Shirts, B.H., Jacobson, A., Jarvik, G.P. and Browning, B.L. (2014) "Large Numbers of Individuals are Required to Classify and Define Risk for Rare Variants in Known Cancer Risk Genes," *Genetics in Medicine* 16: 529–534.

U.S. Department of Health, Education, and Welfare, Secretary's Advisory Committee on Automated Personal Data Systems, Records, Computers, and the Rights of Citizens (1973) *Records, Computers, and the Rights of Citizens* 41, available at www.justice.gov/opcl/docs/rec-com-rights.pdf (accessed June 1, 2018).

U.S. Department of Health, Education, and Welfare (1978) "Protection of Human Subjects: Institutional Review Board; Report and Recommendations of National Commission for the Protection of Human Subjects of Biomedical and Behavioral Research," *Federal Register* 43: 56174–56198 (November 30, 1978), at 56188.

U.S. Department of Health, Education, and Welfare, National Commission for the Protection of Human Subjects of Biomedical and Behavioral Research (1979) *Ethical Principles and Guidelines for the Protection of Human Subjects of Research, pt. B*, available at www.hhs.gov/ohrp/regulations-and-policy/belmont-report/index.html (accessed June 1, 2018).

U.S. Department of Health and Human Services (1997) Confidentiality of Individually-Identifiable Health Information: Recommendation of the Secretary of HHS Pursuant to Sec. 264 of the Health Insurance Portability and Accountability Act of 1996 § I.I, available at https://aspe.hhs.gov/report/confidentiality-individually-identifiable-health-information (accessed June 1, 2018).

U.S. Department of Health and Human Services (2000) "Standards for Privacy of Individually Identifiable Health Information," *Federal Register* 65: 82461–82829 (to be codified at 45 C.F.R. pts. 160, 164).

U.S. Department of Health and Human Services (2002) "Standards for Privacy of Individually Identifiable Health Information," *Federal Register* 67: 53182–53273 (to be codified at 45 C.F.R. pts. 160, 164).

U.S. Department of Health and Human Services et al. (2017) "Federal Policy for the Protection of Human Subjects: Final Rule," *Federal Register* 82: 7149–7274 (to be codified at 45 C.F.R. pt. 46 and in various other regulations of implementing agencies).

U.S. Department of Health and Human Services et al. (2018) "Federal Policy for the Protection of Human Subjects: Delay of the Revisions to the Federal Policy for the Protection of Human Subjects," *Federal Register* 83: 2885.

Worldometers (2018) *Current World Population*, available at www.worldometers.info/world-population/ (accessed June 1, 2018).

23
TECHNOLOGY DEPENDENT COMMONS

Nina Wormbs

Many commons are central to our existence. The air, the water and the forest, to name but a few, are commons necessary for survival, and we value them as crucial parts of life on earth. When they are compromised, polluted or deteriorated in some way, they cannot serve us and our normal way of life is threatened. Often this is due to technological change and the expansion of humanity, two things tightly coupled. The shrinking of forests in the Roman Empire was connected to the enlargement of the territory and high demand for wood for building and cooking. The London smog, a result of industrialization and the growth of coal burning factories, from time to time made it most difficult to breathe and caused deaths. The acidification of present-day seas, that poses unprecedented threats to the salt waters of the world, is part of climate change and the increased amount of carbon dioxide in the atmosphere. Climate change in turn is anthropogenic and the so-called Great Acceleration of our impact after the 1950s has been made possible through science and technology (Steffen et al. 2007).

Science and technology also change the commons themselves and humankind's potential to appropriate them. There are even commons that would not exist without science and technology. One such example is radio, that today is used for a large number of wireless services necessary for the information society. Furthermore, these commons are often wide ranging and spanning several nations or entire continents. Sometimes they are even planetary. The changing scales of these resources from local to global poses challenges to sustainable use. Since the impact of science and technology changes over time, a historic dimension is also essential. To understand the interdependence between technologies and resource use, we need longitudinal studies that can highlight how new practices and systems are superimposed on old ones.

This chapter will discuss transnational technology dependent commons within a historical context, with the aim of furthering our theoretical understanding of commons in general. I will start with a very short example regarding wireless technologies in the early 20th century. Then I will move to a discussion of radio waves in general, moving to a more detailed case of broadcasting in Europe in the interbellum and after World War II. A theoretical overview of the sophisticated understanding of technology and social change then functions as the foundation that allows me to finally discuss how this can serve commons studies more generally.

Early use of the radio spectrum for communication

Scientific work that preceded the use of radio waves for practical communication dates back to the 19th century and the discoveries of physicists like Faraday, Maxwell and Hertz. Often the Italian engineer Marconi is credited for having invented radio as a communication means in 1895. He then spent the following years trying to transmit over longer distances like the Atlantic and succeeded in the early years of the new century. He was awarded the Nobel Prize in Physics in 1909. Very soon the practical applications became apparent and work to improve the technology was also undertaken by others. A common way of naming the technology was wireless, based on the idea that the foreseen use was telegraphy but without the wire.

A truly revolutionary use of radio telegraphy was in places where having a wire was impossible, like at sea. To communicate between ships meant not only that the captain became tied to the company he worked for, being never beyond reach even at sea, but more importantly that ships could communicate with other ships out of sight. In fact, this was such an important application of the new technology that the Marconi Company came close to establishing a de facto monopoly on radio communication at sea through extensive contracting of equipment, including a clause stating non-communication with systems other than the Marconi system. Deliberate disturbance of the communication of others was also reported (Tomlinson 1979, 18). Unconditional communication was an important reason to call the first international radio conference in Berlin in 1903 and a follow up in 1906 leading to a Convention signed by 27 countries on three continents (Tomlinson 1979, 27).

One historic instance when new technology was brought to public attention was the sinking of the *Titanic* in 1912. The ship had been equipped with the latest technology and the radio telegraphists on board were employed by the Marconi company. The distress signal picked up by the *Carpathia* arguably saved the lives of 703 people. However, there were smaller ships in the vicinity of the *Titanic* not equipped with radio, and on the *Californian*, a large ship that was somewhat closer, the telegraphist had gone to bed (Tomlinson 1979). The *Titanic* accident illustrated both the usefulness of radio transmissions as well as the need to regulate the practice. In the international radio conference in London 1912, only two months after the *Titanic* accident, several proposals were put forward to increase the usefulness of the technology and avoid future disasters. The Marconi Company also declared that it would communicate regardless of system. The London conference was an important step towards international agreement on the use of radio, even though it almost exclusively dealt with maritime services (Codding 1952).

The first decades of wireless telegraphy clearly illustrated how the radio spectrum and its use could be viewed as a commons. It had the two central characteristics of a low threshold to access, and that the use affected others' use. To be able to transmit radio signals one needed transmission equipment. However, the transmission was very likely to cause interference and thus affect the transmission of others and subsequently the possibility of reception, a key feature of communication. Below I will discuss a few central features of this resource that affect how it can be used.

The radio spectrum: some basics

There are many types of electromagnetic waves, from ionizing gamma rays, to light visible to the human eye, to radio waves. Gamma rays have a very short wavelength and high frequencies whereas the longest radio waves have extremely low frequencies. The radio spectrum is part of the electromagnetic spectrum and has been discovered and appropriated for communication over time. A telling example is when radio amateurs were referred to wavelengths below 200 meters because those wavelengths were believed to be of less practical use. Experiments on these shorter wavelengths with higher frequencies, however, led to the discovery and subsequent use of short wave for communication over very long distances.

Further research and experimentation on radio waves have subsequently enlarged our understanding of the radio spectrum and its possible use. Before I return to the technology broadly understood, a few things should be said about the basic nature of the radio spectrum.

The radio spectrum is the part of the electromagnetic spectrum from 3 kHz to 3000 GHz.[1] A common division of this space is in logarithmic steps with very low frequencies 3–30 kHz, low frequencies 30–300 kHz, medium frequencies 300 kHz–3 MHz, etc. up to the extremely high frequencies. Lower frequencies can penetrate buildings but are more sensitive to disturbances. Higher frequencies, of which there are also more, need line-of-sight and suffer absorption. Today, radio services in the Gigahertz area are typically satellite communication whereas those in the kilohertz area are maritime communication. The fact that some services are concentrated to specific areas in the spectrum is partly due to the features of the frequencies themselves, having different propagation properties. However, there are also practical reasons not to mix very different technologies. For example, you do not want your remote to the garage to change channels on your TV-screen. With growing globalization, telecom operators and device producers are also interested in selling in every possible market, which is simplified if radio regulation is harmonized between countries. Other divisions of the radio spectrum into bands are based on their use rather than frequencies, like bands for air-traffic communication or marine telecom. More common and better known for general users are probably a further subdivision sometimes based on technology, like the AM-band or the FM-band, on which broadcasting takes place with different types of modulation.

Central to the appropriation of the radio spectrum is the fact that the appropriation operates in two different dimensions, or spaces. One is the frequency on which the transmission occurs, discussed above. The other is the geographical position of the transmission. When there are very few transmitters at work, neither matters much for the reception of the signal, unless they are close in both dimensions. But as transmitters multiply, frequency and geography become increasingly important as radio waves know of no national borders and sometimes reach well beyond the intended audience. The technology of transmission, including issues pertaining to frequency stability, bandwidth and power, interplays with organizational principles of location. A peculiarity of the spectrum, which arguably, however, does not affect its use, is also that the very instant we stop appropriating it, it becomes pristine again.

Radio technology and international organization

When Marconi first started his experiments in the back yard of his home in Bologna, the transmission was made with a spark gap which was a technology that gave a very broad and "dirty" signal. It was sufficient for sending Morse signals but littered the spectrum and made the communication of others hard. Sometimes it meant that signals could be intermingled with each other, or that the resulting interference made any listening impossible. However, soon technology was developed that allowed for the transmission of "cleaner" signals on designated frequencies (Aitken 1976, 1985). With the implementation of a continuous wave, the possibilities of utilizing the spectrum increased. Another important development was modulating the signal for transmission of speech or music on a carrier wave rather than just Morse code.

In the early 1920s, the technology to transmit music on a carrier wave of a specific and stable frequency existed. To go from there to broadcasting was a change of mind rather than a change of matter (Douglas 1987). Broadcasting is a radio service that mirrors newspapers and books communicating from one to many, in contrast to telegraphy that is a point-to-point medium, which mirrors a letter. The success of broadcasting relies on an audience, meaning people with receivers that can pick up an audible signal. This public service, which also became the name of a particular and dominant way of organizing broadcasting in Europe, brought the commons features of the radio spectrum to general attention as will be seen below.[2]

The innovation to broadcast music or speech quickly spread in both Europe and the United States in the early 1920s. There was talk of a radio fever and the interest grew fast as stations were established and transmissions commenced in many countries. This was helped by the fact that the wavelengths used had long reach and broadcasting in England could be heard in Sweden, for example. In a polylingual area with many nations like Europe, the reach of the radio waves also had appreciable side effects for autonomous nation states. Soon the issue of how to organize this activity nationally was raised and at about the same time efforts to coordinate internationally were also made (Lommers 2012).

In 1924, the first meeting of what was to become the International Broadcasting Union (IBU) took place in Geneva. The convener was a private broadcaster himself, the Belgian Maurice Rambert, but some countries sent national representatives from their post and telecom authorities. The IBU was constituted in 1925 with its main office in Geneva, aiming for international recognition in the city that housed the headquarters of the newly established League of Nations. The first and central task was to discuss a frequency plan for Europe. This was necessary since the airwaves were already crowded and reception difficult due to interference.

In 1925, the first European engineering conference of the IBU took place with the aim of crafting a frequency plan that would benefit broadcasting in Europe. The participants agreed that the area of the country, the size of the population and the advancement in telecommunications, based on international economic statistics, were all to be decisive parameters when available spectrum was divided between the appropriators. This meant that those who had begun broadcasting early should be allowed more space.

The boundaries of the resource were set to the medium wave band, 200–600 meters. The definition of Europe and thus the definition of appropriators was not as clear cut, however, and would eventually come to challenge the plan. This is a peculiarity of the spectrum as a commons since the geography of the appropriators can change without the resource increasing. It was soon clear that given the technology at hand with a bandwidth of 10 kHz it was not possible to give every transmitting station an exclusive frequency. Instead they had to work with both exclusive and shared frequencies in allocating the resource. The plan was ratified and put into operation in 1926, improving broadcasting in Europe. Also important was the establishment of a Technical Committee of the IBU and a Technical Centre in Brussels which monitored adherence to the plan and installed equipment that improved frequency stability in transmissions (Wormbs 2011; Fickers and Griset forthcoming).

However, interest grew rapidly and soon the process had to be repeated. In fact, the interbellum saw a great number of frequency conferences trying to improve the planning for the benefit of broadcasting in Europe. At the same time, the IBU was challenged by efforts on the supra-international level. In 1927, the telegraph authorities gathered for a plenipotentiary Radio Conference in Washington, the first since the London conference in 1912 mentioned above. The results were important for the regulation of radio and many of the principles established had long term effects, among them the establishment of the International Radio Consultative Committee and the decision to dedicate specific services to specific bands (Tomlinson 1979; Codding 1952).

The Washington conference resulted in the shrinking of the medium wave band for broadcasting in Europe to 200–545 meters, compared to the Geneva plan. One way of handling the decrease was to make the bandwidth of each channel smaller and thereby fitting more channels. Instead of giving each 10 kHz, the width was set to 9 kHz. In a similar manner, the incorporation of stations from the Western Soviet Union into the plans was achieved by placing them between existing stations, 4.5 kHz from any other transmitter. This was arguably possible since the Soviet stations were geographically located far away from the stations closest in frequency. In general, close attention was paid to the two spatial dimensions of the appropriation, frequency and geography (Wormbs 2011).

During World War II the broadcasting service was under severe pressure, breaking down in parts of Europe. Deliberate interference such as jamming was part of war strategy. When peace was reached, Europe saw a different order, which also affected broadcasting. The first efforts to clean up the air waves were made in Copenhagen in 1947 at a conference that opened the day after the start of the Berlin Blockade and with recent propaganda stations erected by the west and jammed by the east. Germany was not allowed to defend its interests in this conference, which instead were guarded by the occupying powers. It was soon apparent that ideas of how to best use the radio spectrum for broadcasting in Europe differed between France and Britain on the one hand and the US on the other. In fact, it appears Soviet interests were not necessarily at odds with the rest of Europe on certain levels, which could be explained by the fact that, contrary to the US, the Soviet Union was not just an occupying power but also a nation which was dependent on functioning broadcasting for its own sake. The same was true for France and Britain. Even though the Soviet Union voiced severe criticism, it signed the final treaty whereas the US stated it would not adhere to the plan. True to this statement the US broke the plan and started broadcasting both on frequencies that were allotted to others and those that were not allotted at all. The resulting disorder was partly due to efforts to increase power in the affected stations, thereby breaking the plan even more. A few years into the 1950s, it was declared a failure, but efforts to renegotiate were not made until the 1970s (Craig 1990; Wormbs 2013).

Broadcasting did not break down totally during the Cold War. Bi-lateral negotiations were important in solving some of the interference, but the introduction of transistor technology improving reception was also central. That the Copenhagen plan was maintained was, however, most likely due to an enlargement of the resource through migration to new frequencies. After the war, higher bands were used for broadcasting and the modulation method also changed. A great expansion of broadcasting was possible with this move to FM on VHF allowing for more channels and eventually new actors. Moving from AM to FM meant implementing a modulation method less susceptible to interference. Also of great importance was that these frequencies did not have as long a range and thus did not disturb surrounding countries as much.

As illustrated by the frequency chart from the Swedish Post and Telecom Authority (Figure 23.1), newer technologies find themselves on higher and higher frequencies. It is striking how stable this organization of the resource has been over time and one might think that the non-materiality of radio waves would result in a more volatile distribution of services along frequencies. However, both their respective properties, discussed above, and the fact that enormous amounts of resources are tied to transmitting and receiving equipment, makes this a technological system with great inertia. Let me elaborate briefly on the different ways of understanding features like inertia in relation to technology and society.

Technology and social change

Even though there is a grand narrative in present day main stream discourse which holds that technology is a main driver of social change, historical study has not engaged with technological change to a very high degree. The same is true for natural science which is often found under specific headings in history books. Certain strands of history have indeed addressed technological change, such as agrarian history, economic history, business history or military history. However, the lack of incorporation of science and technology in the sophisticated understandings of change has been and is notable, and it was probably one reason for the establishment of history of technology in the mid-1950s as yet another sub-discipline of history.

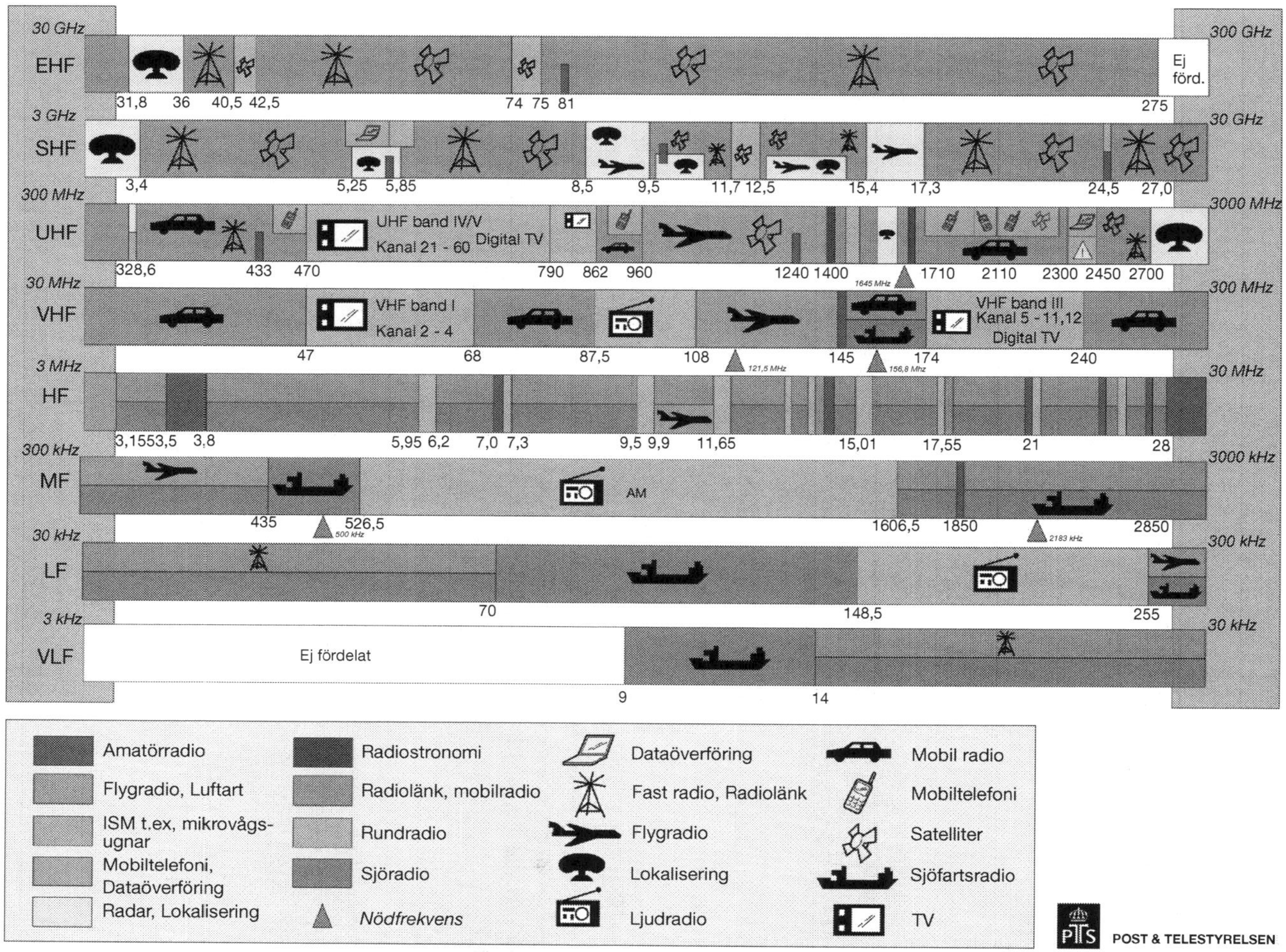

Figure 23.1 Spectrum chart from the Swedish Post and Telecom Authority (PTS)

Credits: Swedish Post and Telecom Authority.

The central argument for much of the scholarship in the history of technology is that technology and society are co-produced and that technology never should be regarded as exogenous to change (Jasanoff 2004). Precisely how the processes of change occur is hard to generalize and there are shifting modes of analyses and explanations just like in most disciplines. Many of these discussions now take place within the transdisciplinary field of science and technology studies, growing out of, and partly incorporating, history and sociology of science and technology. Readers and overviews abound (see, for example, Jasanoff 1995; Biagioli 1999; Hackett 2008).

Co-production is only one way of describing the interplay between technology and society. Of central importance has also been a systems approach to, for example, large societal infrastructures like transport, communication or energy. Large technological systems grow in phases which have different kinds of growing problems. Systematic studies have showed that these phases and problems can be generalized across national borders. However, it is also possible to detect technological styles, pointing to the contextual importance. By terming these systems socio-technical, the dependence on the context of economy, law and culture is highlighted (Hughes 1983; Kaijser 1994; Kaijser and Hedin 1995). Discussions on the momentum and inertia of these systems and their importance for societal change also stirred theoretical discussions regarding to what extent the understanding is technologically determinist (Bijker and Law 1992).

Discussions on technological determinism has occupied a central role in historically and sociologically informed scholarship on science and technology for a long time, sometimes framed as an issue of whether technology drives change (Smith and Marx 1994; Wyatt 2008). The ideas of the social shaping of technology (MacKenzie and Wajcman 1985) and the social construction of technology (Bijker et al. 1987) are arguably reactions to hard determinism which stipulates that technology is the most important agent of societal change.

The social construction of technology (SCOT), was not only inspired by classic constructivist theory (Berger and Luckman 1967), but perhaps more importantly by the sociology of scientific knowledge and its strong programme (Bloor 1976). Central tenets are to take interest in conditions for knowledge production, to be impartial in respect of what is true or false, to be symmetrical and use the same explanatory means for both successes and failures, and finally to be reflexive and allow for the methodology and theory of the scholarly effort to also be used on sociology itself. These principles were slightly altered and carried on to what became known as actor network theory, or ANT (for example Callon et al. 1986; Law 1986, Latour 1987). Just like the sociology of scientific knowledge, ANT had some challenging properties – the postulate not to differentiate between human and non-human actors was regarded provocative by many – but quickly became enormously influential in science and technology studies. Agnosticism or impartiality, generalized symmetry and free association have been applied by numerous scholars who have put their analytical focus on relations and networks. Scholars have carefully tried not to essentialize and imbue meaning and value a priori but rather rest on empirical work (Law and Hassard 1999). Moreover, the conclusion that science and technology are not always clear-cut and easily separable activities, but rather interdependent and messy, have led to the often-used compound "techno-science".

This theoretical relation to technological change over time also affects the understanding of resources and commons to which I will now again turn.

Technology as intrinsic to sustainable and unsustainable commons

That the radio spectrum indeed has a history is evident from Figure 23.1 where old technologies occupy low frequencies, and as we move up in the spectrum we have technologies closer to our own time. This also illustrates the central point of the spectrum being created as a resource through technology itself. If it was not for transmission and reception equipment we would not be able to use it at all.

A few commons resemble the radio spectrum in that the need for rather advanced technology for appropriation is absolute, like air traffic and space flight. The air as an infrastructure for transport is created through the different means of flight. When balloons were used, their speed and number were limited, and in general the commons of the air was non-subtractable, not affecting the use of others in a negative way. With the advent of airplanes, however, numbers increased rapidly and the need to organize was realized early on. Planes could fly faster and on higher altitudes which increased the infrastructural resource but were at the same time dependent on new communication and navigation technology. As with radio, international coordination soon became key. The intricate ways in which this coordination relied on large and heterogeneous networks that worked with and around changing technology has been detailed by Eda Kranakis in the case of early European commercial flight (2013).

Space flight is similar in that we cannot go into space without advanced and costly technology, for many nations impossible to afford on a national level and thus demanding international cooperation. Once we managed to go into space, on 4 October 1957, a new commons emerged, which is still under development. Space is indeed vast, but there are also limitations to our use, some of a physical nature. The geostationary orbit is such an example. It is the orbit around the earth where satellites are stationary in relation to a specific longitude. Or, differently put, the satellite uses 24 hours to circle the earth and thus appear stationary in relation to earth. This orbit is particularly attractive for telecommunications that serve the same geographic area, such as broadcasting satellites. The transmitters and receivers on earth can be stationary and the task for the satellite is to relay the signal. Even though the orbit is large, located at 36,000 km above the equator, crowding can occur and collisions must be avoided. Thus, international coordination of the orbit is imperative and has been on the agenda for the International Telecommunications Union since the 1970s.

It can be argued that any resource appropriation is dependent on some kind of technology, broadly understood. Even if you fish with your bare hands, the practice of catching the fish could be regarded as a technology.[3] As fishing develops with tools for the catch, vessels for getting out on deep waters and scientific and technological means to localize, sense and calculate the fish, so does the ability to catch larger and larger amounts. However, the creation and appropriation of commons by technological means only partly captures the importance of technology and science in this respect. The institutions involved are also profoundly affected, including how to value resources and how to monitor the appropriation.

A central effort to bring sophisticated history of technology analyses to commons studies is the volume *Cosmopolitan Commons: Sharing Resources and Risks across Borders* edited by Nil Disco and Eda Kranakis (2013). The book contains several case studies, all transnational and all technology dependent in one way or another. In their introduction, Disco and Kranakis carefully detail the different ways in which technology is part and parcel of resource use and control. At the same time, they pay attention to how technology changes over time and the fact that larger, technology dependent commons are very often transnational and thus demand institutional cooperation that national or local commons might not.

Disco and Kranakis point out that just as access to commons often is technologically dependent, so are the means of limiting access. Examples can be standards, de facto or de jure. Marconi's original attempt to monopolize maritime radio communication is an example. Had he succeeded, access to maritime communication would have only been possible through the purchase of particular patented equipment.

On the other hand, certain standardization is also necessary if the resource shall function as a commons in the first place; frequency plans illustrate that point clearly but it is valid for many areas. In fact, most standards are so familiar to us that we do not think of them anymore as standards and thus the results of meticulous and often tedious work. How internet addresses are written or how information is coded through internet protocols, allowing for computers and browsers to connect, is just one example.

In the case of space, access to outer space is not only limited by the fact that building satellites is very expensive, but the places and means to get them into orbit are few. During the 1980s the capacity to launch satellites into orbit was severely hampered by the failures of the US shuttle due to the explosion of *Challenger*, and on the European side the Ariane project was facing major problems. Thus, whoever controls launching sites also controls use.

Another example that Disco and Kranakis bring forward is the ability to surveil and monitor use. In the case of radio this was put in place early on, with a technical center belonging to the International Broadcasting Union. It kept track of breaches of the plan and engaged in public shaming by making those brakes public through the community newsletter (Fickers and Griset forthcoming). In the case of acid rain in Europe and efforts to regulate and abate the consequences, advanced technology and information handling was crucial to establishing the basics of the phenomena in the first place, as some pollution can travel far and the origins are difficult to track (Kaijser 2013).

Finally Disco and Kranakis bring out the complexities of many commons in that they house several resources and thus appropriation activities at once and that these are dependent on historic change. Rivers, for example, are places not only where people and goods are moved and fish are caught but also where drinking water is collected and waste is displaced. The simultaneous management and development of the Rhine over centuries can also be regarded as an "avenue of modernity" according to Disco (2013). Over time the possibilities to exercise these activities change as the different commons change in the face of earlier use.

The North Sea is another case in point where, just as with many rivers, several nations have stakes. Håkon With Andersen argues that the North Sea is actually three distinct and superimposed commons for transport, fishing and fossil-fuel extraction, highly dependent on the institutional and technological means available (2013). Moreover, the three types do not just involve different actors and technologies; they each have specific physical properties making the resulting commons different from each other. The sea surface is for example easily accessible and non-depletable, whereas the water itself is vulnerable both to pollution and depletion. The sea-floor, finally, is in many respects unexplored as it is extremely difficult to access – yet, appropriation occurs. Moreover, these distinct commons, even though they are distinct and have different properties, are also connected and mutually dependent. Andersen's (2013) analysis of the North Sea case beautifully illustrates the intricate ways in which nature is ordered, compartmentalized and appropriated through large techno-scientific installations and advanced systems allowing for new understanding of the sea as a commons.

Conclusion

The extraction of a resource is tightly coupled to technology. In fact, sometimes the resource itself is created through technology, like airspace or radio. A longer perspective on resources, their use and the functionality and spread of the technologies sustaining appropriation, raises questions which can be further elaborated by viewing the resource as a commons. Thus, commons studies and its tools and practices have entered the toolbox of historians of technology.

However, that toolbox already contains a great number of rather sophisticated instruments to analyze technological change, having been the objective of the field for many decades. Central normative assumptions are, for example, that technology and society are co-produced and that technology is not an exogenous force but a result of negotiations and struggles between many actor categories with various agendas working over shorter or longer periods of time. Arguing that technology is socially constructed and should be understood contextually and historically means to engage with these struggles and their material conditions and outcomes. This is also true for commons. Thus, a commons approach to technology should also engage seriously with how the conditions for sustainable appropriation changes with technology, which in turn is partly an outcome of the negotiations of that appropriation.

Incorporating tools and theories from sophisticated transnational history of technology into commons' studies means complicating things further. However, I would argue that the result will be worthwhile. Technologies – highly advanced and difficult to grasp or straightforward and easy to understand – are part and parcel of resource appropriation and monitoring. Sustainable commons will be easier to achieve if we understand the intricacies of those relationships.

Notes

1 In the early days of radio, it was more common to talk about wavelengths than frequencies, a practice which is still visible in terms like long wave or short wave. The frequency for any given wavelength can be calculated though dividing the speed of light by the wavelength.
2 The way in which broadcasting is organized as commercial or public service might arguably affect appropriation; however, I do not explore that in this chapter.
3 I do not distinguish between technology and technique in this chapter but take them as synonyms.

References

Aitken, H. G. J. (1976) *Syntony and spark: The origins of radio*. New York: Wiley.

Aitken, H. G. J. (1985) *The continuous wave: Technology and American radio, 1900–1932*. Princeton, NJ: Princeton University Press.

Berger, P. L. and Luckmann, T. (1967) *The social construction of reality: A treatise in the sociology of knowledge*. London: Allen Lane Penguin.

Biagioli, M. (1999) *The science studies reader*. New York: Routledge.

Bijker, W. E. Hughes T. P. and Pinch T. J. eds. (1987) *The Social Construction of Technological Systems : New Directions in the Sociology and History of Technology*, Cambridge, MA: MIT Press.

Bijker, W. E. and Law, J. eds. (1992) *Shaping technology/building society: Studies in sociotechnical change*. Cambridge, MA: MIT Press.

Bloor, D. (1976) *Knowledge and social imagery*. London: Routledge.

Callon, M., Law, J. and Rip, A. (1986) *Mapping the dynamics of science and technology: Sociology of science in the real world*. Basingstoke, UK: Macmillan.

Codding, G. A. (1952) *The International Telecommunication Union: an experiment in international cooperation*. Leiden, the Netherlands: Brill.

Craig, R. S. (1990) "Medium-wave frequency allocations in postwar Europe: US foreign policy and the Copenhagen conference of 1948," *Journal of Broadcasting and Electronic Media* 43: 119–135.

Disco, N. (2013) "'One touch of nature makes the whole world kin': Ships, fish, phenol, and the Rhine, 1815–2000" in Disco, N. and Kranakis, E. eds. *Cosmopolitan commons: Sharing resources and risks across borders*. Cambridge, MA: MIT Press, 271–316.

Disco, N. and Kranakis, E. eds. (2013) *Cosmopolitan commons: Sharing resources and risks across borders*. Cambridge, MA: MIT Press.

Douglas, S. J. (1987) *Inventing American broadcasting, 1899–1922*. Baltimore, MD: Johns Hopkins University Press.

Fickers, A. and Griset, P. (Forthcoming) *Communicating Europe*. Basingstoke, UK: Palgrave Macmillan.

Hackett, E. J. ed. (2008) *The handbook of science and technology studies*. Cambridge, MA: MIT Press.

Hughes, T. P. (1983) *Networks of power: Electrification in Western society, 1880–1930*. Baltimore, MD: Johns Hopkins University Press.

Jasanoff, S. ed. (1995) *Handbook of science and technology studies*. Thousand Oaks, CA: Sage.
Jasanoff S. (2004) *States of knowledge: The co-production of science and the social order*. London: Routledge.
Kaijser, A. (1994) *I fädrens spår: Den svenska infrastrukturens historiska utveckling och framtida utmaningar*. Stockholm: Carlsson.
Kaijser, A. (2013) "Under a common acid sky: negotiating transboundary air pollution in Europe" in Disco, N. and Kranakis, E. eds. *Cosmopolitan commons: Sharing resources and risks across borders*. Cambridge, MA: MIT Press, 213–242.
Kaijser, A. and Hedin, M. eds. (1995) *Nordic energy systems: Historical perspectives and current issues*. Canton, MA: Science History Publications.
Kranakis, E. (2013) "The 'good miracle': building a European airspace commons, 1919–1939" in Disco, N. and Kranakis, E. eds. *Cosmopolitan commons: Sharing resources and risks across borders*. Cambridge, MA: MIT Press, 57–96.
Latour, B. (1987) *Science in action: How to follow scientists and engineers through society*. Cambridge, MA: Harvard University Press.
Law, J. ed. (1986) *Power, action, and belief: A new sociology of knowledge*. London: Routledge.
Law, J. and Hassard, J. eds. (1999) *Actor network theory and after*. Oxford, UK: Blackwell.
Lommers, S. (2012) *Europe: On air – Interwar projects for radio broadcasting*. Amsterdam: Amsterdam University Press.
MacKenzie, D. A. and Wajcman, J. eds. (1985) *The social shaping of technology: How the refrigerator got its hum*. Milton Keynes, UK: Open University Press.
Smith, M. R. and Marx, L. (1994) *Does technology drive history? The dilemma of technological determinism*. Cambridge, MA: MIT Press.
Steffen, W., Crutzen, P. J. and McNeill, J. R. (2007) "The Anthropocene: are humans now overwhelming the great forces of nature" *AMBIO: A Journal of the Human Environment* 36: 614–621.
Tomlinson J. D. (1979) *International control of radiocommunications*. New York: Arno Press.
With Andersen, H. (2013) "Changing technology, changing commons: freight, fish, and oil in the North Sea" in Disco, N. and Kranakis, E. eds. *Cosmopolitan commons: sharing resources and risks across borders*. Cambridge, MA: MIT Press, 245–270.
Wormbs, N. (2011) "Technology-dependent commons: the example of frequency spectrum for broadcasting in Europe in the 1920s" *International Journal of the Commons* 5: 92–109.
Wormbs, N. (2013) "Negotiating the radio spectrum: the incessant labor of maintaining space for European broadcasting" in Disco, N. and Kranakis, E. eds. *Cosmopolitan commons: Sharing resources and risks across borders*. Cambridge, MA: MIT Press, 97–122.
Wyatt, S. (2008) "Technological determinism is dead: long live technological determinism" in Hackett, E. J., Amsterdamska, O., Lynch, M. and Wajcman, J. eds. *The handbook of science and technology studies*. Cambridge, MA: MIT Press, 165–180.

24
FROM HISTORICAL INSTITUTION TO *PARS PRO TOTO*

The commons and their revival in historical perspective

Tine De Moor

The commons are currently witnessing a true revival in many parts of the world. The term "commons" is hereby used in a context where citizens solve personal and/or societal problems as a group, and thereby also create a collective product or service, which they try to govern collectively. The historical background of the term commons, which forms the starting point of this chapter, will show that in the current context the term is used as a *pars pro toto*, whereby many more varieties of institutions for collective action are considered in a local context, on the basis of their common features such as self-governance, collective action and high member-participation. In this chapter we will generally use the term institutions for collective action (or ICAs) when considering institutional arrangements that are formed by groups of people in order to overcome certain common problems over an extended period of time by setting certain collective rules regarding access to the group (membership), use of the resources and services the group owns collectively, and management of these resources and services. Perceiving the commons as an institution for collective action is by now widely accepted. However, the usefulness of widening the time perspective, to include hundreds of years in our search for the holy grail of what makes institutions resilient, is often still underestimated. Commons-scholars such as Elinor Ostrom herself found it unlikely that the commons in a historical setting, taking us back several centuries, could be studied with the same depth as present-day commons, However, as we will demonstrate in this article, with the right methods and resources, the incredibly rich account of centuries-long successful resource management may deepen and possibly alter our understanding of sustainable cooperation fundamentally.

Paradoxically, whilst citizens and local governments in the Western world increasingly put the commons forward as an alternative governance regime for both public and private good provision, examples of commons in the developing world have been and are still under pressure to dissolve, in order to be replaced by a market-driven individualized use of land and resources. The latter process can be considered as an example of history repeating itself, though on another continent: most western commons have gone through a very similar process of top-down instigated dissolution of commons and varieties thereof in the 19th century, at the latest. Commons in England and Wales were already going through a dissolution process—also referred to as

"Enclosures"—from the 15th century onwards. The current "return of the commons" is however often seen as an entirely new movement, not taking into account the lessons to be learned from history. If history is taken into account in the present debate on commons, it is often used to provide an overly romantic picture of commons and their commoners, usually not going beyond anecdotal references. Current "application" of historical knowledge often lacks the historical nuance about, for example, the way in which commons dealt with inequality, and the dynamics in the importance of commons as a governance regime over time are taken too lightly. If taken seriously, the richness of commons history may offer a goldmine for those studying the commons, in particular for those willing and apt to use the commons in a nuanced and critical way, regardless their original disciplinary training.

Instead of providing a narrative history of the commons in general—an exercise that has been done elsewhere already extensively (though not exhaustively; e.g. De Moor et al. 2002)—or of some commons in particular, this chapter will provide a broader historiographical perspective by explaining why a historical lens is important to use when studying commons. Furthermore, it will provide insights into how we should perceive the dynamics of commons taking a longitudinal perspective and give some examples of the application of historical research methods that focus on that dynamic character of institutions for collective action.

Gaining deeper insights into the factors explaining institutional resilience of commons over the very long run and satisfactorily understanding the feedback mechanisms, synergies, and learning and adaptation processes underlying institutional resilience unavoidably requires a long-term perspective and, in turn, such a long-term approach inevitably implies the work of historians and historically-trained social scientists (Ostrom 2014; Laborda Pemán & De Moor 2016). As such this chapter should also be read as a call to historians to engage in the research field of commons in an interdisciplinary way, and also as a call to commons-researchers without a historical training to dig deeper and to use history not as a mere eclectic spicing up of their studies on present-day commons. Despite the many voices stressing the relevance of empirically well-grounded analyses on how collective action institutions evolve over time, stress on static approaches, attention to relatively short time spans, limited empirical evidence often derived from secondary sources, and methodologies not explicitly designed for long-run analyses have prevented scholars from embarking upon such an endeavour. It is precisely here where the work by historians and other historically-trained social scientists is highly relevant. Their use of primary sources and their focus on the long run allows them to fill many of the gaps observed in the previous generation of studies on institutional resilience. Despite three decades of intensive commons-research, our knowledge about how institutions develop and change over time still remains limited (Ostrom 2014).

Until the end of the 20th century contributions to the analysis of institutional change oscillated between sophisticated theoretical exercises and the in-depth analysis of a small number of case studies, mostly ignoring their long-term development. For a long time the work of historians also did not manage to get integrated into work on commons done in other disciplines. It mainly revolved around the issue of how the rich heritage of commons exhibited by Western Europe before 1800 impacted on its social and economic performance. Traditional views stressing the constraints that common land put on the efficient functioning of land markets only started to be contested in the 1970s and 1980s. A number of works not only began to question the supposedly beneficial impact of enclosures but also suggested that the prevalence of historical commons was fully compatible with economic rationality and social welfare (Allen 1982; McCloskey 1991). Only after this re-evaluation of the commons' heritage of Europe began, did historians start paying attention to aspects that until then had been neglected, such as the origins of these collective governance mechanisms or their functioning and evolution over time (De Moor et al. 2002; De Moor 2009). Since the late 2000s historical evidence has continued to grow and historians are now much more aware of the richness and diversity of commons across Europe (De Moor 2015). Notwithstanding the quantitative progress—in terms of the number of methodological studies—in multidisciplinary approaches that incorporate longer time-frames, there is still a considerable gap to fill between disciplines using different time-frames.

Though the temporal perspective in this chapter is extremely broad—more or less the past millennium—the geographical scope is limited to mainly Western Europe. Western Europe forms not just the cradle of both commons-practice and of the historical political-philosophical discussion of commons' relevance for economy and society,[1] but also has been the most extensively studied. But even more so, understanding the European history of the commons is important since the geographical origins of the negative implications of commons that culminated in Hardin's work—though himself an American biologist—were found in Europe. Even before Hardin's publication, collectivities had been negatively associated with inefficiency and backwardness, most prominently in the work of 18th-century Physiocratic philosophers in France, and later on also in treatises written by the economist William Forster Lloyd, among others, in the 19th century. But as will be explained further, European commons' history goes back much further in time, all the way to the late middle ages, in times when commons were considered a solution, not a problem.

The long-term history of commons' evaluation gives us another reason why the historical approach must go beyond providing a nice narrative to current examples of so-called "new commons." With many examples of commons that demonstrate a life-time of several centuries, this type of institution for collective action is an excellent example of a "resilient" institution, whereby a constant adaptation of the institutional design was needed to deal with the sometimes harsh conditions, in terms of climate, weather, political turmoil and demographic pressure. Notwithstanding the fact that we need to take into account that our current perception of the historical common might be biased by a lack of sources of commons with a shorter lifespan—due to a lack of archival preservation—the vast richness of historical material left behind by many commons gives us an excellent view of what it took of the commoners to deal with external pressures. The commons of historical Europe, and in particular their archival remains, give us a chance to understand how commoners managed to cooperate, as a collectivity, for the benefit of the collectivity. If we consider the vast challenges that lie ahead for the world population right now, commons' history may provide inspiration to find solutions for those problems that are based on the similar social dilemmas that also project long-term effects of their current behaviour on future generations. Understanding the dynamics of commons-governance over long periods of time may give us a better understanding of achieving resilience in the long run, which demands more than a regular check of a list of design principles. After more than 25 years of doing research on commons without much interaction between different disciplines, we need to start building more on multi-method comprehensive analyses based on multi-faceted theoretical explanations, that include long timespans (Poteete & Ostrom 2008). The historical examples we are about to discuss in this chapter demonstrate that, contrary to what late 18th century contemporaries and many after that believed, commoners were constantly adapting their institutions to the changing ecological, economic and political environment.

A bird's eye view on historical and "new" commons: "Waves of collective action" in the past 1000 years

Understanding the dynamics of small-scale cooperation within a changing environment represents a fundamental challenge for 21st-century scholars—with findings likely to have immediate impact on the design of more resilient communities around the globe. As a consequence of the serious limitations exhibited over the past decades by market- and state-based alternatives and encouraged by growing opportunities for communication, the ability of individuals to self-organize and arrange the production and distribution of goods and services collectively has become one of the emerging features of society. A "wave" of new, collective initiatives taken by "ordinary" citizens has been emerging since 2005 across Europe, although countries' activity may vary across sectors and through time.[2] Citizens are uniting to set up energy cooperatives across Europe, providing in some countries, like Germany, nearly half of the renewable energy (Morris & Pehnt 2012: 29); Alternative forms of insurance and care provisioning are developing rapidly.[3] In many places where public goods are endangered, citizens collectively try to provide an alternative, sometimes in cooperation with local governments, but often without.[4] The attempts of citizens' collectivities or institutions for collective action (ICAs) currently emerging across Europe to provide public and private goods in various sectors are often considered as new and revolutionary. However, when we dive into long-term European history there seem to be more periods when organizations that thrive on collective action, self-governance and high member-participation were set up with rather similar institutional design as those we see emerging today.[5] For the late Middle Ages this has been highlighted by various historians (Greif, Reynolds), and for some countries this has also been tested on the basis of a quantitative analysis. For the Netherlands, a "wave" of collective action has been described to emerge around 1200 (De Moor 2008, 2013: 12–14), when commons and waterboards in the countryside and guilds and beguinages in the cities were set up to regulate access to goods and services collectively among stakeholders. Rapid development of merchant and craft guilds, commons, collective irrigation systems, and other forms of ICAs has also been described for countries as Italy (Mocarelli 2008), Spain, Belgium (Soly 2008), and Germany (Ehmer 2008). However, if we consider, for example, the Dutch case again for the period after the top-down dissolution of guilds, commons and other types of institutions for collective action (18th and early 19th centuries),[6] another period of accelerated development of organizations based on very similar principles surfaces. In the decades between 1880 and 1920 cooperatives were set up in farming, banking, and many other sectors to provide access to working material and capital and other benefits such as social security.[7] Case studies and country-wide studies on similar historical forms of collective action elsewhere in Europe suggest that similar "waves" may have indeed existed throughout Europe in more or less the same two periods, but with a variety in appearance, depending on the local problems collectivities of citizens tried to address. In some countries, cooperatives appeared somewhat earlier. In 1844 the Rochdale Society of Equitable Pioneers in Rochdale, England, put down the so-called Rochdale principles[8] (see e.g. Fairbairn 1994)—which have until today been perceived as the ideals that cooperatives should pursue in their operation, although many coops with their origins going back to the second wave of ICAs seem to struggle to honour these principles. Building on the same ideas, the first Swedish consumer cooperative was established in 1850 (Fjørtoft & Gjems-Onstad 2013: 583); the German Raiffeisen Bank developed in the 1860s, with many other countries soon to follow its example.

The current surge in ICAs has been attributed to market imperfections that arise after periods of accelerated development of the free market, in which privatization plays an important role (De Moor 2013: 16–18; Verhofstadt & Maertens 2014; Dedeurwaerdere et al. 2015). The first wave of ICA-formation was preceded by an accelerated development of the agricultural and urban labour markets during the Middle Ages (van Bavel 2007, 2011). Market imperfections also seem to have been the main stimulus of rural cooperatives in the 19th century, following a strong wave of liberalization and privatization since the middle of the

19th century. In historical literature, the rise of the cooperatives in agriculture in the 19th and early 20th centuries has been explained by the advantage cooperatives provide over private factories in reducing transaction costs with suppliers (O'Grada 1977; van Zanden 1994; Henriksen 1999; O'Rourke 2006).[9] Many advantages such as scale, risk avoidance, and risk sharing are put forward, as explanation, though the primary cause should rather be considered the lack of solutions provided by the state and the market for the daily needs of citizens. An often referenced example of this is the development of the Raiffeissen across Europe in the second half of the 19th century, which offered to provide loans and savings facilities to farmers and other citizens living in the countryside who could not gain the trust of commercial city banks. The cooperative bank became a popular solution to this market failure, with many other types of cooperatives developing in other sectors (insurance, dairy produce, energy, etc.) as well as mutuals, which function along the same principles as cooperatives. Similar market failures, following the privatization of public providers in the 1990s and 21st century, as a consequence of the EU-policy in this domain, also seem to be the main drivers behind the current surge in the formation of new collectivities.

If we consider the meso-level, that of the organizations, we do need to acknowledge however that not all groups of users were in fact as successful in keeping their organization going in the long run, although long-term plans may have been made when the organization was set up. To a large extent the variation over time in longevity of the ICAs is related to the chosen strategy of dealing with the increasing scale of the individual institutions. If we consider again the Dutch case, there is substantial variation in terms of longevity of individual organizations both within each "wave," as well as when both waves are compared. Previous in-depth analysis of cases showed, for example, how some commons survived only a century, whereas others easily surpassed 600 years (De Moor & Tukker 2015). ICAs developing in the second "wave" from the late 19th century onwards however were on the whole shorter-lived: of all cooperatives that emerged between 1880 and 1920 in the Netherlands, only one-fifth are still active today, while more than 90 per cent of the eastern Netherlands' agricultural commons survived much longer, easily passing the 150 year mark. About 62 per cent of their urban counterparts (guilds) survived over 150 years, also considerably more than the cooperatives of the second wave (De Moor 2013: 16–18[10]). Although some may argue this could be a sign of inertia or a consequence of monopolization (Richardson 2001), longevity of an institution can be an indicator of the resilience of the governance regime in place, or of the ability to recover from setbacks, to adapt well to change, and to keep on going in the face of adversity. Longevity in itself is recognized by scholars in business as an advantage since an older institution on the whole tends to be more efficient and has been implicitly considered by commons-scholars—including Elinor Ostrom—as a valid indicator for a robust/resilient institution, though often not in relation to the governance regime, but to the ecological situation of the resources held in common.[11]

Longevity as an indicator of resilience has been approached so far mainly from the institutional perspective. In particular the Bloomington school of thought influenced by Elinor and Vincent Ostrom stressed that a good institutional design, including a specific combination of rules (the famous eight design principles of Ostrom 1990: 90) could have a positive effect on the cooperative behaviour of members and consequently also on the resilience of ICAs.[12] Much of the Bloomington school literature reacted against the assumptions about human cooperation as laid down by Garret Hardin in his 1968 *Science* article, the "Tragedy of the Commons." However, although scholars from the Ostrom school have indeed been able to identify what can make an ICA long-enduring, most of the (predominantly non-historical) studies on the resilience of ICAs have considered institutions with a relatively short life-span (e.g. of less than a century), and as such these cases miss the variation needed to understand why some live (considerably) longer than others. As explained above, a longer time perspective can contribute considerably to our understanding of the dynamics of institutional change (e.g. Lana 2013, 2014a, 2014b; Lana & Iriarte-Goñi 2015) but also demonstrates that a focus on institutional development alone is insufficient to understand why institutions which all have a very similar institutional design still tend to differ in "age at dissolution."

Building a resilient organization is a time-dependent process, of which the impact can only become visible if considerable time has passed. Setting resilience as a goal for institutional developments today will, given this time-dependency, demand a longitudinal perspective. Digging deeper over time delivers more insights into the evolution of design principles, under stress, and also into the balance between various interactions between design principles. A third reason why the history of commons is important is that it brings to the fore forgotten tactics and strategies developed by commoners in the past. The need to think in a collectivity has, in many western countries, been set aside as unproductive, or has been—partially—taken over by the providers of the welfare state. Reciprocal thinking, at the basis of any form of institution for collective action, is not an easy or straightforward "mind-set": it demands intensive interaction, thinking ahead, and being able to build on a memory of previous actions, of which some may have positive outcomes and others not. Creating rules that take into account "best practices" of the past, but in a new context, and take into account new members of the collectivity, demands creativity. This "historical creativity" will be addressed in the third part of this chapter.

Besides these fundamental reasons why we should be taking longitudinal change into account, change and time-lags between a commoner's contribution to and his appropriation from the commons are other reasons why historical development may give us a better understanding of commons. One of the major reasons why we find understanding the collectivity as a functional unit so hard is the difficulty in uniting the permanent care for the collectivity versus the variety in time-lag between investment and individual benefit for the members. Economic theory tells us that we should not expect individuals to invest in collectivities. Individuals should not be expected to accept collectively made rules that may affect their individual (short-term) benefits, let alone adhere to them. This is based on the expectation that individuals are not capable of rationally arguing for the collectivity rather than their individual benefits. And yet, in reality, both historical and present-day, we see that individuals are in fact capable of doing so.

Framing the commons in a dynamic way, allowing for longitudinal studies

As mentioned earlier, resilience has been a major issue in the study of commons, and the long-term dynamics of commons in the past may be very helpful in contributing to achieving a resilient organization. However, for the analysis of a complex governance system whereby instead of one party deciding on resource use and management (the state or a private owner) multiple users are deciding on a collective resource involved, we need a framework that allows us to identify the factors affecting the willingness of members to *continue* contributing to the commons over substantial periods of time, preferably covering several generations. This is however a complex task, especially if we take into account the changing contextual factors that may have influenced commoners' willingness and capacity to cooperate, including factors such as demographic change, economic hardship and political turmoil. In the model below, we split the commons as a governance regime into three components: resources, users and institutions.

In literature the resources commoners deal with are also referred to as common pool resources. In some cases the resources are also used as a *pars pro toto*: it is assumed that the resources as such are the commons. However, as already stated earlier, the commons are a concept that refers to a governance regime, whereby a group of people holds a collective property and needs to decide collectively over the common pool resources. In order to make sure that members of the group act reciprocally, the members have to make sure everyone that is part of the group gets sufficient benefits out of the collective resources. If done on a continuous basis, this leads to the formation of an institution, a Common Pool Institution, necessary to make sure that rules are developed and adapted over time, to the changing environment. These multiple interrelated factors make the study of the internal functioning of commons complex: whereas in private or public property arrangements decisions are made by individuals or the government, in an ICA a collectivity of entitled users may be involved in both the use of the benefits and the design of the institution to regulate use, access, management, and governance of these benefits. With ICAs relying heavily on group norms and reciprocal behaviour between their members, the interaction between users, resources and institutions is all the more important to understand periods of crisis and dissolution of the commons (Bruynis et al. 2001; Birchall & Simmons 2004; Jussila et al. 2012). In Figure 24.1, all three dimensions that together constitute the governance regime are shown. Bringing together these dimensions, and visualising their interaction, is important, as the current debate over commons often tends to focus on one aspect in particular. Figure 24.1 depicts the three dimensions (institutions, users, and resources) of commons as governance regimes established by a group of resource users themselves to regulate and provide order in the use and management of a common-pool resource, but also to prevent its over-exploitation and guarantee its sustainable provision over time. As it becomes evident from the self-governance and collective features, such a governance model differs markedly from a state-driven organization as well as from market-based appropriation. Among the challenges usually faced by this governance regime, two are particularly important. On the one hand, the common-pool nature of the resource requires from the commoners constant assessment and re-design of the rules in place in order to balance individual incentives and group objectives. On the other, the institutional arrangement must be permanently adapted to the changing environmental, economic, and political environment in which it is embedded. As such we need to especially focus on the interplay between internal (i.e. the incentives embedded in rules, the commoners' motivation, and the behaviour resulting from the interaction between rules and motivation) and external (e.g. environmental, political, demographic, economic, and technological forces operating outside the domain of the institutional arrangement) factors, and on how this interaction explains the resilience of the commons over the long term.

The risks to which commoners were exposed historically were manifold, though the actual effect they had in practice remains understudied. But we can hypothetically describe a number of potential effects external shocks may have had on the commons and their users. Some of the most straightforward external threats to commons in the past are related to climate and weather-related hazards, ranging from snow and hail to storms and floods. Such events may have an effect on the accessibility of grazing facilities on the commons, though we would not expect an effect on harvesting of subsoil resources, unless, for example, in the case of a flood which makes the whole area inaccessible. In terms of the use of the commons, such events may result in changes in the rules that prescribe a reduced number of cattle each member can let graze on the commons, or a belated opening date of the common grazing facilities. The effect on agricultural production in general may be that the number of cattle that can be fed during winter and the resulting number of cattle allowed on the commons (referred to as the practice of "levanchy and couchancy") is altered. In terms of access rules, commoners may decide that if the conditions/consequences of bad weather are severe, a complete restriction on access to the commons could be ordained, for all commoners.

In case a cattle disease or plague takes place this may have other effects on the commons. If the commoners themselves suffer severe cattle loss, this may affect access rules: in case of potential underuse, the commons may be opened to non-entitled users. It may also affect management rules: in order to limit the spread of the plague, cattle may be concentrated in a particular part of the commons and isolated from other cattle populations. Demographic shocks may also lead to substantial changes in the management. The demand for use of the commons' resources may alter substantially, demanding a change of the access rules. If the population grows rapidly, supplementary restrictions on access might be introduced, in particular rules related to access of heads of household (and thus possibly related to marriage-based restrictions to access). Similarly, access rules might be loosened, in case of population decrease/shock due to disease. Similar restricting or loosening of the rules may take place in relation to use, in terms of maximum number of resource units allowed per user.

Although one may probably assume that warfare and political turmoil may have similar effects on the rules as a population decrease, there are in fact more effects in other domains to be expected, assuming that warfare would involve primarily heads of households. Warfare could substantially affect the regularity of meetings and as such demand an adjustment of the governance rules. Heads of household would be absent, which would make meetings of the governors of the commons less easy to organize, thus leading to an effect on the governance and management as a whole. This in turn may warrant more flexibility in rotation schemes and taking up managerial roles. Through all of these proposed potential causal relationships between external shocks and crises and rule changes, which may not take place for each and every commons, they clearly present the multiple effects various external factors may have on the commons, the commoners and the resources they govern. It also demonstrates that multiple dynamics are at hand, potentially involving changes in various rules designed to govern the commons.

The challenge for the members of a self-organizing institution like a commons is to keep the system in balance, regardless of external factors, for example, to keep the utility sufficiently high even when the supply of resources is shrinking or when the group of users is growing. A change in one of the domains may have a clear influence on the other domains; if, for example, a group of users grows this may lead to lower equity and put pressure on the group to change the institutional design. Also, a decreasing supply of resources may affect the rules that constitute the commons as an institution. Understanding long-term dynamics thus means not just taking into account the changing context but also changes in both the resources and in groups of users over time, as ethnicity, religion, wealth, occupation, dependency on resources, and other factors may affect the individual members' willingness to contribute to the collectivity. In the case of a pasture commons, impoverishment of the commoners may lead to a situation whereby some of the resources commoners are entitled to use are no longer useful; if some of the commoners no longer have the cattle to graze on the commons, the utility for at least part of the members will decrease substantially, which in turn may affect their willingness to act reciprocally. A lack of utility may encourage them to freeride or at least refrain from contributing to the collective good.[13] In organizations relying on self-governance, where members can influence the governance of their collective resources, this may in the end also lead to a change in governance: commoners may collectively decide to dissolve their organization as they do no longer find it useful to invest in it.[14] Similar developments may take place if commoners (feel they) are insufficiently involved in the decision-making process; they may perceive—justly or not—the decisions taken as less fair to them individually. In social science literature it has been amply demonstrated that a governance regime that is perceived as less equitable also may lead to defective behaviour,[15] though such studies often are only snapshots and limited in time-perspective. The historical commons were constantly challenged to involve their members in the decision-making process, but also saw their groups change in size and composition. An increasing group of members may have a positive influence on the total amount of capital available within an ICA but may have a negative effect on the social control as large groups make it harder to recognize members

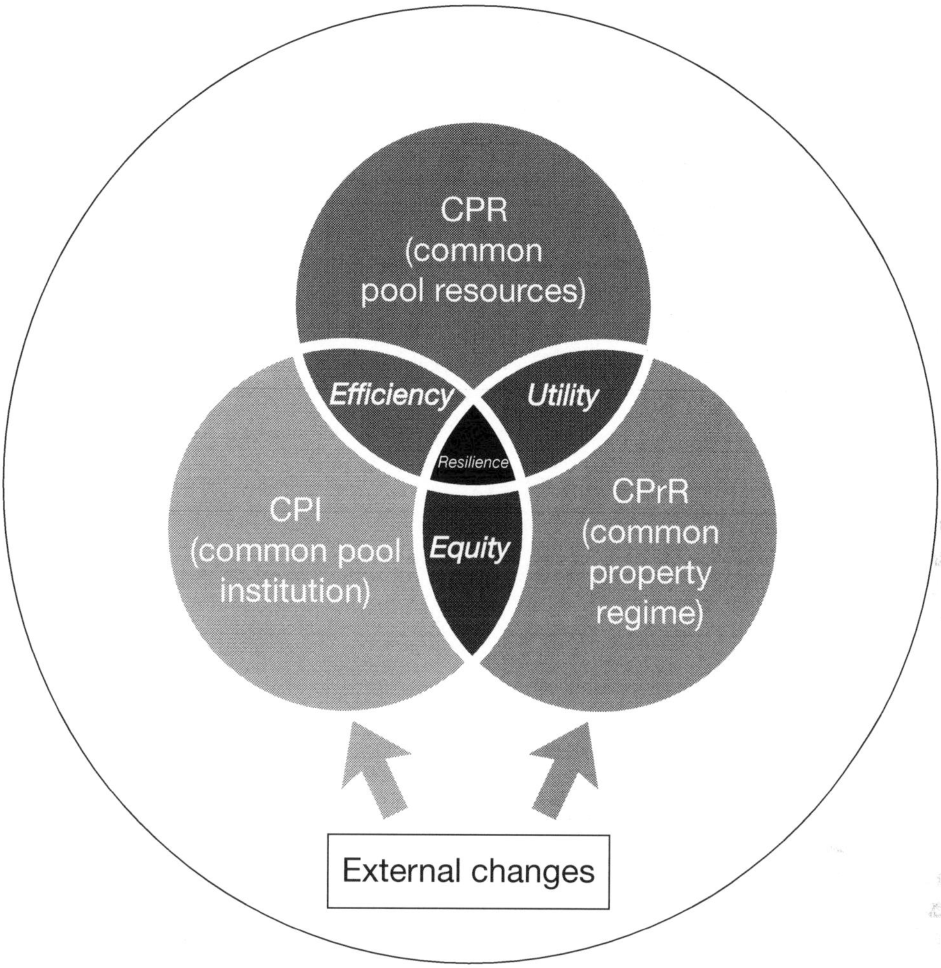

Figure 24.1 Micro-perspective: how to achieve resilience on the commons?

of the group. In social science literature it has been described that cooperative behaviour is promoted if the other people can observe one's personal choice behaviour (Jorgenson & Papciak 1981), and that this "social control" mechanism may be responsible for the fact that people are more willing to work hard under conditions of high visibility than in more anonymous settings (Latané et al. 1979). All these elements demonstrate that since the members of the group may change, as well as the needs and expectations of the individual members towards the ICA, the resilience of an institution cannot be understood if the interaction between the group dynamics, the institution and the resources an ICA is not studied over substantial periods of time.

Better understanding the mechanisms behind ICAs—and commons in particular—and how not only institutional design but also group dynamics may affect the longevity of an institution and how these interact, is of great importance to the current societal change we are undergoing. The answers we may find in the historical analysis, in combination with the efforts put into knowledge dissemination and dialogue with stakeholder-citizens, should contribute to building better ICAs that can—as in early modern history—survive over many centuries.

Only recently, attempts to empirically understand the dynamics of institutional change within a commons setting in the longer run have started (De Moor 2015). An international team of historians recently presented unprecedented data recording historical regulations for 25 European commons over 7 centuries (De Moor et al. 2016). In total, the database contains over 5000 detailed rules defining these institutions (e.g. regulating resource use, management, and sanctioning) and records the changes made during these centuries. Studying the regulation on the micro-level is not new in itself, and in fact forms the basis of the work by Ostrom and her colleagues, but the historical approach allowed an entirely different spectrum of research possibilities, though without the opportunity to work with living interviewees.

The richness of historical data offers far more insights into cooperation within the commons than what can be comprised in one single chapter, so the focus will be here on some specific aspects of commons' functioning, related to each of the three parameters that we introduced earlier that affect resilience.

Utility: avoiding resource use beyond what is needed

Hardin's classic story was based on a number of premises, one of them being that use of a collective resource would invariably lead to overuse by the *homo economicus*. Humans are tempted to commercialize, and hence lead to overexploitation of their resources. Whereas humans might feel temptations to do so, history shows us that collectivities of users find ways to institutionally limit the commercialization of their resources. What is most interesting about the rules formulated to achieve this is that commercialization of goods is often not directly addressed, but indirectly, by addressing the drivers that may lead to commercialization.

Equity: involvement of users is prime to achieve cooperative behaviour

Self-governance is usually presented as one of the most important features of institutions for collective action. But who is involved in that governance? And how democratic is the formulation of the regulation? Equitable governance is hard to track historically, but its importance surfaces when looking, for example, at sanctioning systems. Many experimental studies have tried to establish what an effective and efficient sanctioning system should look like by trying to determine the "right" level of a fine, to what extent people are willing to punish each other, what the structure of a fine should be (pool punishment or not; the choice between "carrots" or "sticks") and to what extent external enforcement can be efficiently used to implement sanctions in an effective way. Elinor Ostrom and others argued that specific types of sanctioning—such as graduated sanctioning—can have a preventive effect as they also serve as a warning to current defectors of "worse to come" (Ostrom 1990). Nevertheless, the option to use tools other than sanctioning to prevent freeriding is hardly ever seriously considered in the literature, although it is generally acknowledged that sanctioning can be a costly affair. However, when we explored the use of sanctioning in a number of very long-lived commons in the Netherlands (De Moor & Tukker 2015), with longevity varying from 236 to 695 years of documented existence, the conclusions were rather surprising. Archival sources allow us to retrieve the regulation devised to control who the commoners were (access rules), how the use of their common resources was regulated (use rules), how access and use were managed (management rules) and how the governance of the institution as a whole was arranged (governance rules). We started from the premise that both rule-making and sanctioning are costly affairs, in particular because the rules needed to be discussed and agreed upon by the whole group of members. Therefore, we approached the complete body of rules found per commons as the "total effort" of a group of commoners. We used this concept as a way to link the rules and accompanying sanctions to the longevity of each of the cases.

All commons in our case study analysis made frequent changes to their regulation system, although some more often than others. There was little difference between the cases in terms of the effort they spent on adjusting the rules—on average, 61 per cent of the rules were adjustments to previously mentioned rules. This and the continuous effort to adjust their rules show that all cases were essentially "dynamic." But still, their longevity varied substantially. This can be explained by several factors. First of all, it appears that sanctioning was not a decisive factor in the longevity of these commons. Rather the contrary seemed to be true: the longer-lasting cases seem to have had a lesser need to come up with sanctions for their rules. Second, it seems that most of their rule-making effort went into creating rules to ensure a well-functioning management. There is also a clear difference in terms of the repetition of the rules required: longer-lasting institutions needed to repeat less, and instead concentrated on adjusting the rules. Taking these results together, one might claim that the secret of a long-lived institution seems to be ensuring that people meet frequently so that they "internalize" new rules and adjustments easily, rather than threatening people with sanctions. Whenever commoners wanted to change the rules, they had to convene and approve of the changes. If this is done frequently, commoners will also be more frequently confronted with their moral duty to behave well towards others than if they meet only once in a while. High levels of participation and effective involvement of the users in the governance of the institution, consequently, may have been more important for the longevity of the institution than sanctioning.

Future research on the basis of commons' regulation may help us to disclose the mechanisms behind institutional change in the long run more clearly. Using historical data may help us to unravel why and how institutions have been modified in response to changes in their social and natural environment. It may also tell us whether there are typical developmental paths that long-lasting institutions take, using a perspective similar to the evolutionary process leading to adaptations (Richerson & Boyd 2005; Ostrom 2014), focusing on the relationship between the different systems of rules historically recorded and analysing their development over time.

Conclusions

The commons of today differ from their historical counterparts in two major ways. First, the context in which they develop is fundamentally different, with new technologies and ways of communication influencing interaction among users/members and with other needs and desires of the latter. Second, the resources that are governed in institutions referred to as "commons" are substantially different from the agricultural context we associate with the historical commons. However, when it comes to the behavioural aspects of those setting up, using and managing resources collectively, not that much seems to have changed. Throughout history we see that through self-governance and collective action citizens try to fulfill their needs and desires. If we compare the waves of collective action described earlier, we do see some differences that deserve further research. Taking the bird's perspective, looking at the long-term development of institutions for collective action in the long run, we may say that institutions for collective action have increasingly become mono-purpose organizations. Whereas citizens today choose to join an energy coop, a care coop and many other types of coops in order to fulfill their daily needs in a collective way, the ICAs of the first, medieval-early modern, wave combined many different purposes from those related to their occupation to care and even charity. Though indeed this is partly related to changes in production and consumption patterns, combining multiple purposes does allow ICAs to balance more easily different types of demands from their members. Offering multiple services to members may also help to prevent freeriding as members become dependent on various needs through the same collectivity. Having to participate in several different collectivities reduces the opportunities for using reciprocal behaviour as a complementary incentive.

To conclude, there are some thoughts on the future of the commons as a governance regime, based on the historical analysis of commons in the European past, which we need to keep in the back of our minds when studying the commons in the present day. Seen from the historical perspective, the use of the term commons is basically a *pars pro toto*, as it refers to many different types of collective resource management, on various levels. However, the constant "stretching" of the term that has taken place the past few decades leads to a number of paradoxes, which keep coming back in the debate without being resolved, or even properly acknowledged. First of all, current day proponents of developing commons as an alternative governance model often see opportunities in the domain of public good provision, now that many welfare states are increasingly trusting the market to take over their role. However, the original forms of institutions for collective action were not set up to provide public goods. In many cases there were indeed activities that came to the benefit of the whole local community (mainly in the form of poor relief activities supported by commoners), but it was not its prime intention to do so. Second, increasingly commons are also expected to solve resource management problems on a much larger, even global scale. The term commons itself is on the one hand increasingly applied to resource management that goes far beyond the local level, with references even to global resources, but on the other hand the features of the resources managed as commons have been strictly local. It is rather logic that out of the research done on the history of commons lessons can be drawn that are restricted to local resources, but the potential of scaling these lessons up to deal with global resource issues is far less obvious. Third, in the current political philosophical writings, the commons are often seen as a constant factor throughout history without much fluctuation in their presence or importance. "Commons have always been part of our history" is, however, a false claim. Throughout history we have seen periods in which new commons emerged more rapidly than before, but we also see periods when they have been dissolved, mostly under external pressure orchestrated by national governments. History does show that regardless of the attempts of governments to eliminate forms of self-organization among citizens, the latter will keep doing so. The degree to which forms of cooperation are institutionalized is, however, variable throughout time.

This chapter has shown that the new wave of institutions for collective action builds upon an extremely rich heritage of collective resource management in the past centuries. The ability of such small-scale, self-governed governance mechanisms to deliver satisfactory results in the environmental and social realms has attracted the attention of both academic and policy-making circles—which have gradually shifted from dismal views on collective action (e.g. Hardin's famous "Tragedy of the Commons" (1968) or Olson's (1965) analysis on the logic of collective action) towards much more optimistic assessments in which sustainable cooperation is not only possible but also socially beneficial (e.g. Ostrom 1990). The speed of technological, economic, political, and environmental changes in our current societies and the need of anticipating reactions to those changes makes it indispensable to gain deeper knowledge about the mechanisms behind the processes of institutional change, resilience and failure of the commons—all of them processes unfolding over the (very) long term. Given the growing importance of collective governance regimes in our 21st-century societies, both across developed and developing regions, and the rapid pace of change in the surrounding environment, a better understanding of these issues is likely to provide scholars, policy-makers, and citizens alike with relevant insights about how to design and adjust institutions that are able to simultaneously deliver economically efficient, socially inclusive, and environmentally sustainable outcomes.

Notes

1 See amongst others the discussion on the commons in in particular the philosophical tractates of the Physiocrates in the 18th century, which created the foundation for the top-down dissolution of the commons in in particular continental Europe.
2 The analysis by De Moor (2013: 10–11) shows a particular leap forward since 2005, with dozens of new cooperatives formed every year in particular in professional services, industry, energy, transport, and care.
3 See www.accesstoland.eu/, accessed 20 October 2018.
4 See for example the Bologna "Regulation on Collaboration between citizens and the city for the care and regeneration of urban commons," adopted by the city of Bologna in 2014 (Collaborare è Bologna 2014).
5 Some current initiatives also like to promote historical parallels between their own activities and organization today and those organizations they consider their predecessors, as is visible in the name they choose. E.g. the *Energiemarke* ("Energy common") in Haarlose Veld (ROM3D 2016) combines references to the historical use with generating renewable energy through landscape.
6 The ICA-concept has in the meanwhile also been applied to other historical phenomena, e.g. communities of Dutch maritime transporters (Scheltjens 2015).
7 In the same period, we also see the development of other forms of collective action such as the labour unions and mutual insurance funds (De Moor 2013: 16–17).
8 Today the International Cooperative Alliance still considers—an adjusted version of—the Rochdale principles as their guiding principles (International Cooperative Alliance 2018).
9 For a recent confirmation of this theory see the article by Koen Frenken (2014) who demonstrated this with a study using data on 1130 dairy factories in the Netherlands that cooperative factories indeed performed significantly better than private factories.
10 See De Moor (2013: 16–18); analysis based on data from project "Data Infrastructures for the Study of Guilds and Other Forms of Collective Action" (Institutions for Collective Action 2012).
11 See her work on long-enduring Common-Pool Resources, as e.g. *Governing the Commons* (Ostrom 1990).
12 In particular since the Nobel Prize in Economics was in 2009 awarded to Elinor Ostrom, who wrote the seminal *Governing the Commons* (see also previous notes).
13 As demonstrated for a Flemish case study at the end of the 18th century (De Moor 2010).
14 On the same Flemish case studied by De Moor, the increasing dominance of the passive users over the active ones lead to a substantial change in governance, and nearly to the dissolution of the common (De Moor 2015: 139 ff).
15 For a literature review, see Diekmann (2004).

References

Allen, R. (1982) "The Efficiency and Distributional Consequences of Eighteenth Century Enclosures" *The Economic Journal* 92: 937–53.

Birchall, J. & Simmons, R. (2004) "What Motivates Members to Participate in Co-operative and Mutual Businesses?" *Annals of Public and Cooperative Economics* 75: 465–95.

Bruynis, C.L. *et al.* (2001) "Key Success Factors for Emerging Agricultural Marketing Cooperatives" *Journal of Cooperatives* 16: 14–24.

Collaborare è Bologna (2014) *Regulation on the Collaboration Among Citizens and Administration for the Care and Regeneration of Urban Commons*. Bologna, Italy: Comune di Bologna.

Dedeurwaerdere, T., Polard, A. & Melindi-Ghidi, P. (2015) "The Role of Network Bridging Organisations in Compensation Payments for Agri-Environmental Services under the EU Common Agricultural Policy" *Ecological Economics* 119: 24–38.

De Moor, T. (2008) "The Silent Revolution: A New Perspective on the Emergence of Commons, Guilds, and Other Forms of Corporate Collective Action in Western Europe" *International Review of Social History* 53: 179–212.

De Moor, T. (2009) "Avoiding Tragedies: A Flemish Common and its Commoners under the Pressure of Social and Economic Change during the Eighteenth Century" *Economic History Review* 62: 1–22.

De Moor, T. (2010) "Participating Is More Important Than Winning: The Impact of Socio-Economic Change on Commoners' Participation in Eighteenth- and Nineteenth-Century Flanders" *Continuity and Change* 25: 405–33.

De Moor, T. (2013) *Homo Cooperans: Institutions for Collective Action and the Compassionate Society*. Utrecht, the Netherlands: Utrecht University.

De Moor, T. (2015) *The Dilemma of the Commoners. Understanding the Use of Common-Pool Resources in Long-Term Perspective*. Cambridge, UK: Cambridge University Press.
De Moor, T., Laborda-Pemán, M., Lana-Berasain, J.M., van Weeren, R. & Winchester, A. (2016) "Ruling the Commons. Introducing a New Methodology for the Analysis of Historical Commons" *International Journal of the Commons* 10: 529–88.
De Moor, T., Shaw-Taylor, L. & Warde, P. (2002) *The Management of Common Land in North West Europe, c. 1500–1850* (CORN publication series nr. 8). Turnhout, Belgium: Brepols.
De Moor, T. & Tukker, A. (2015) "Survival Without Sanctioning: The Relationship Between Institutional Resilience and Methods to Deal With Freeriding on Early Modern Dutch Commons" in: Grüne, N. (ed.), *Rural History Yearbook*, vol. 12. St. Pölten, Austria: Institute for Rural History, 175–204.
Diekmann, A. (2004) "The Power of Reciprocity: Fairness, Reciprocity, and Stakes in Variants of the Dictator Game" *Journal of Conflict Resolution* 48: 487–505.
Ehmer, J. (2008) "Rural Guilds and Urban-Rural Guild Relations in Early Modern Central Europe" *International Review of Social History* 53: 143–58.
Fairbairn, B. (1994) *The Meaning of Rochdale: The Rochdale Pioneers and the Co-Operative Principles* (Occasional Papers series), Centre for the Study of Co-operatives. Saskatoon, SK: University of Saskatchewan.
Fjørtoft, T. & Gjems-Onstad, O. (2013) "Norway and Scandinavian Countries" in: Cracogna, D. Fici, A. & Henrÿ, H. (eds), *International Handbook of Coöperative Law*. New York/London: Springer, 563–83.
Frenken, K. (2014) "The Evolution of the Dutch Dairy Industry and the Rise of Cooperatives: A Research Note" *Journal of Institutional Economics* 10: 163–74.
Hardin, G. (1968) "Tragedy of the Commons" *Science* 162: 1243–48.
Henriksen, G. (1999) "Avoiding Lock-in: Cooperative Creameries in Denmark, 1882–1903" *European Review of Economic History* 3: 57–78.
Institutions for Collective Action (2012) *Projects: NWO Project (Summary)*, www.collective-action.info/projects_NWOProject, accessed 29 May 2018.
International Cooperative Alliance (2018) *Co-operative Identity, Values & Principles*, https://ica.coop/en/whats-co-op/co-operative-identity-values-principles, accessed 29 May 2018.
Jorgenson, D. and Papciak, A. (1981) "The Effects of Communication, Resource Feedback and Identifiability on Behavior in a Simulated Commons" *Journal of Experimental Social Psychology* 17: 373–85.
Jussila, J.J., Kärkkäinnen, H. & Leino, M. (2012) "Learning from and with Customers with Social Media: A Model for Social Customer Learning" *International Journal of Management, Knowledge and Learning* 1: 5–25.
Laborda Pemán, M. and De Moor, T. (2016) "History and the Commons: A Necessary Conversation" *International Journal of the Commons* 10: 517–28.
Lana, J.M. (2013) "Agricultural Transformation, Land Ownership, and Markets in Inland Spain: The Case of Southern Navarre, 1600–1935" in: Hillbom, E. & Svensson, P. (eds), *Agricultural Transformation in a Global History Perspective*. London/New York: Routledge, 26–50.
Lana, J.M. (2014a) "Dormant Agriculture? Some Evidence of Agricultural Change in Nineteenth-Century Navarre" in: Béaur, G. and Chevet, J.-M. (eds), *Measuring Agricultural Growth. Land and Labour Productivity in Western Europe from the Middle Ages to the Twentieth Century (England, France and Spain)*. Turnhout, Belgium: Brepols, 87–106.
Lana, J.M. (2014b) ""Esta Especie de Socialismo Campesino. . . Manso y Tranquilo" Un Estado de la Cuestión desde la Historia Rural Española" *Revista Mexicana de Sociología* 76: 167–97.
Lana, J.M. & Iriarte-Goñi, I. (2015) "Commons and the Legacy of the Past. Regulation and Uses of Common Lands in Twentieth Century Spain" *International Journal of the Commons* 9: 510–32.
Latané, B., Williams, K. & Harkins, S. (1979) "Many Hands Make Light the Work: The Causes and Consequences of Social Loafing" *Journal of Personality and Social Psychology* 37: 822–32.
McCloskey, D. (1991) "The Prudent Peasant: New Findings on Open Fields" *Journal of Economic History* 51: 343–55.
Mocarelli, L. (2008) "Guilds Reappraised: Italy in the Early Modern Period" *International Review of Social History* 53: 159–78.
Morris, C. & Pehnt, M. (2012) *Energie Transition: The German Energiewende*. Berlin: Heinrich Boell Foundation.
O'Grada, C. (1977) "The Beginnings of the Irish Creamery System, 1880–1914" *Economic History Review* 30: 284–305.
Olson, M. (1965) *The Logic of Collective Action. Public Goods and the Theory of Groups*. Cambridge, MA: Harvard University Press.

O'Rourke, K.H. (2006) "Late 19th Century Denmark in an Irish Mirror: Land Tenure, Homogeneity and the Roots of Danish Success" in: Campbell, J.L., Hall, J.A. & Pedersen, O.K. (eds), *National Identity and the Varieties of Capitalism: The Danish Experience.* Montreal, QC: McGill-Queen's University Press, 159–96.
Ostrom, E. (1990) *Governing the Commons, The Evolution of Institutions for Collective Action* Cambridge, UK: Cambridge University Press.
Ostrom, E. (2014) "Do Institutions for Collective Action Evolve?" *Journal of Bioeconomics* 16: 3–30.
Poteete, A.R. & Ostrom, E. (2008) "Fifteen Years of Empirical Research on Collective Action in Natural Resource Management: Struggling to Build Large-N Databases Based on Qualitative Research" *World Development* 36: 176–95.
Richardson, G. (2001) "A Tale of Two Theories: Monopolies and Craft Guilds in Medieval England and Modern Imagination" *Journal of the History of Economic Thought* 23: 217–42.
Richerson, P.J. & Boyd, R. (2005) *Not By Genes Alone: How Culture Transformed Human Evolution.* Chicago, IL: University of Chicago Press.
ROM3D (2016) Energie Marke Haarlose Veld, www.rom3d.nl/portfolio/project1/, accessed 29 May 2018.
Scheltjens, W. (2015) *Dutch Deltas: Emergence, Functions and Structure of the Low Countries' Maritime Transport System, ca. 1300–1850.* Leiden, the Netherlands: Brill.
Soly, H. (2008) "The Political Economy of European Craft Guilds: Power Relations and Economic Strategies of Merchants and Master Artisans in the Medieval and Early Modern Textile Industries" *International Review of Social History* 53: 45–71.
Van Bavel, B.J.P. (2007) "The Transition in the Low Countries. Wage Labour as an Indicator of the Rise of Capitalism in the Countryside, 1300–1700" *Past & Present* 195(S2): 286–303.
Van Bavel, B.J.P. (2011) "Markets for Land, Labor, and Capital in Northern Italy and the Low Countries, Twelfth to Seventeenth Centuries" *Journal of Interdisciplinary History* 41: 503–31.
Van Zanden, J.L. (1994) *The Transformation of European Agriculture in the 19th Century: The Case of the Netherlands.* Amsterdam: VU University Press.
Verhofstadt, E. and Maertens, M. (2014) "Smallholder Cooperatives and Agricultural Performance in Rwanda: Do Organizational Differences Matter?" *Agricultural Economics* 45(S1): 39–52.

25

CUSTOMARY AUTHORITY AND COMMONS GOVERNANCE

Frank Matose, Phil René Oyono and James Murombedzi

Introduction

Debates are raging about whether we are witnessing the re-emergence, resilience or merely recasting of customary authority, and the emasculation of customary authority in Africa (Lund 2008; Nuesiri 2014; Baldwin 2016). Customary authority is used with reference to some locally or ethnic-based rule systems, that may have layered jurisdictional boundaries, and whose governance is presided over by some 'traditional' leadership, including chiefs. This is particularly relevant in relation to customary authorities' engagement with state and non-state actors in re-inventing their identities and roles in a globalizing world and due to democratic transitions and wide-ranging tenure reforms (Baldwin 2016), on the one hand, and, on the other, their performance in community representation in commons governance (Ntsebeza 2006; Cousins 2008; Taha 2016). As the above-mentioned three processes have been taking hold, what are the implications and impacts for commons governance in Africa? How have these three processes shaped or been shaped by commons governance? Who has gained and lost as these processes have evolved?

Thus, this chapter questions the role of customary authority in commons governance in Africa. As shown in the next section, first, customary institutions themselves have had many twists and turns over many centuries of colonial and post-colonial crafting. Second, commons have largely endured as a result of the intricate authority over their governance as imbued by ambivalent expression of customary authority (Perrot and Fauvelle-Aymar 2003; Ntsebeza 2006; Nuesiri 2014).

From a critical sociology perspective, following Baldwin's ambivalent results in the assessment of answerability outcomes espoused by its exercise, we argue that customary authority is central to the governance of commons in Africa—insofar as it, among other things, guarantees collective action in the defence of the commons and equity (see Kouassigan 1966; Tobin 2011; Nuesiri 2014). Also, the historical analysis of its expression and exercise puts forward limits in terms of democratic representation and downward accountability, thus giving rise to decentralized despots and *free riders* (Mamdani 1996; Ntsebeza 2006; Cousins 2008; Baldwin 2016).

Before problematizing the issue addressed by this chapter, it is pertinent to provide a brief synopsis of the frameworks that have been prevalent in analysing customary authorities in the governance of commons in Africa. These include: legal pluralism (Larcom et al. 2016), decentralized despotism (Mamdani, 1996); (for lack of a more encompassing term), 're-emergence' (Nuesiri 2014);

the *retour des rois* (in French, meaning the 'return of kings') (Perrot and Fauvelle-Aymar 2003); unrepresentative representation (Ntsebeza 2006; Oyono and Deng-Athoi 2015; Koechlin et al. 2016); and the ambivalence of the exercise of customary authority (Baldwin 2016).

The chapter is organized as follows. The next section provides a historical account of customary authorities in Africa, placing emphasis on their interaction with colonial authorities. Such a background is important in locating the fraught relationship between the state and customary authorities which in turn provides a context for the alienation of resource governance from the authority by the state. Following that, we situate African commons in the new millennium, in highlighting a set of key challenges. The subsequent section then delves into good practices and pitfalls of the expression of customary authority in commons governance. The penultimate section discusses the vexed issue of democratizing customary authority before some conclusions are provided at the end of the chapter.

A brief history of customary authorities in Africa

Customary authority has always played a pivotal role in resource governance in African history. This historical perspective makes it possible to understand the present of the question and to envisage the future of resource governance. The pre-colonial context already offered an interesting scene for commons governance. This section traces the evolution of customary authority in resource governance in Africa, albeit in a rather synoptic manner to situate the contemporary practices that are discussed in subsequent sections.

Precolonial practices

Pre-colonial resource management practices have tended to be romanticized by Africanists and other scholars, arguing that commons governance through customary authority was definitely better, that it is more equitable and grounded in societies (Maquet 1965; Belaunde 2010; Eldredge 2015; Bandyopadhyay and Green 2016). Overall, this school of thought argues that pre-colonial resource governance practices through customary authority ensured equitable access and collective interests (Kouassigan 1966; Reader 1998; Alden Wily 2011).

There is a dearth of studies about these practices, although available evidence does indicate that as precolonial society became first regimented then stratified, access to and use of natural resources also came to be stratified, and conservation practices began to reflect the attempts to balance competing interests (Copet-Rougier 1998; Oyono 2002; Perrot and Fauvelle-Aymar 2003). For example, it is pointed out that colonization led to the fragmentation of governance of commons by colonial authority, insofar as the latter had been weakened and dislocated by colonial administrations (Maquet 1965; Crowder 1968; Iliffe 1995; Alden Wily 2011).

Nevertheless, recorded pre-colonial conservation practices, such as the demarcation of sacred areas, the allocation of totems, the expropriation of labour for conservation etc., did not necessarily reflect egalitarian and consensual conservation, but rather the exercise of power over people and resources by dominant clans or classes (Baldwin 2016; Igboin 2016). In his analytical exploration, Igboin (2016) shows how traditional leaders in pre-colonial West Africa were corrupted and offered customary land to slave merchants and traders. Some cases of contemporary personal rule are seen as the historical prolongation of the personal accumulation and control of natural wealth and land by pre-colonial kinglets and kings (Jackson and Rosberg 1982; Hall 1987; Oyono 2013; Igboin 2016).

In very early times, evidence exists to suggest that when resources came under pressure from, say, increased human populations, or economic activity, a typical response was for whole populations to move to new uncolonized and resource abundant areas in Equatorial Africa, as indicated by Vansina (1984, 1990). These political responses to ecological phenomena resulted in the several waves of migrations from central into southern Africa and back, from east to west Africa, including the Luba-Lunda dispersions, the Mfecane of the late 18th century and the Fulani across West Africa, respectively.

As this response became restricted by widespread settlements, new political, religious and technological innovations were developed to deal with ecological concerns. These included such innovations as pastoralism, slash and burn agriculture, water harvesting, and the development of institutional regulation of resource use. In relation to big fauna and flora, for instance, numerous evidence exists to demonstrate that because of technological limitations, indigenous hunter gatherers were not equipped to hunt big game in large numbers nor cut large trees over large areas (Matowanyika 1989). As such, during various times, there were settlements, large or small, under the leadership of some authority that would exist for a time before moving on elsewhere.

Colonial practices

With colonization, new regimes of commons governance were put in place (Hall 1987; Copet-Rougier 1998; Alden Wily 2011). When European settlers arrived in Africa, large herds of wildlife and indigenous forests within the landscape were seen as a bounty to be exploited as food and timber for the enrichment of individuals through trade (Crowder 1968). In those days, there were no strong central governing authorities to effect any controls, and several species were hunted to the brink of extinction. The blue buck, the quagga and the Cape lion were wiped out in the Cape region with other species facing similar fate across many parts of Africa. Wildlife and forest resources were placed under the purview of the state—a form of privatization—to stem the overexploitation of resources by settlers who sought to maximize profits from this wild bounty newly found in 'virgin' Africa (Matowanyika 1989; Colchester 1994).

As a result, protected areas were created to stem this pillaging of nature. When protected areas were carved out of communal or tribal lands, the surrounding populations were treated as unimportant in the process (Oyono 2005). These parks were to be rigorously administered along technical lines, and managed within rigid boundaries, often separated from the surrounding populations (Rangarajan 2003). The colonization of nature and adjacent people was thus begun and relied on a range of conceptual strategies that were employed to support supremacism of nation, gender (the white male) and race (Plumwood 2003).

These human/nature dualist anthropocentric conceptual strategies, which exaggerate differences while at the same time denying commonalities between humans and nature, include (Plumwood 2003: 54–9, *en passim*) the following. First, radical exclusion, which functions to mark out the 'Other' for separate and inferior treatment. Nature is treated as Other, and humans are separated from nature and animals (Rangarajan 2003). Nature is a separate lower order, lacking any real continuity with the human. At the same time, the colonizing groups associate themselves with mastery of nature. This was very central to colonial governance of commons (see also Anderson and Grove 1987).

Second, homogenization/stereotyping in which the Other is not an individual, but a member of a stereotyped class, thus making it interchangeable, replaceable and homogenous. Nature is treated as interchangeable units, as resources (Colchester 1994). Radical exclusion and homogenization work together to produce a polarized understanding in which human and non-human spheres correspond to two quite different substances or orders of being in the world. This was important in the colonial perception of nature as a resource to be exploited (Rodney 1983; Oyono 2005; Eldredge 2015).

Third, polarization, where radical exclusion and homogenization work together to produce a polarized understanding in which overlap between the human and non-human spheres are denied and discouraged. Nature is only nature if it is 'pure', untainted by human influence, as untouched 'wilderness', while human identity is separate from and outside of nature—in other words an Edenic view of African commons (Anderson and Grove 1987). Fourth, denial, backgrounding in the sense that once the Other is marked as separated and inferior, there is strong motivation to represent them as unessential. In ecology, the colonized are first denied as uncivilized. Their prior ownership of the land is denied, and their land is viewed as *terra nullius*—with no pre-existing regimes of rights to it. Nature is thus considered as a basic non-essential constituent of the universe.

Plumwood (2003) also raises a fifth issue around 'assimilation' in which the colonized are devalued as lacking the colonizer's essential quality—reason. Differences are judged as deficiencies, and therefore as grounds of inferiority. The order of the colonized is represented as disorder. Thus, the colonized and their disorderly space are available for assimilation and use by the colonizer. Similarly, the intricate order of nature is presented as disorder, to be replaced by human order in governance of commons. The sixth and last point concerns 'Instrumentalism'.

The colonized Other is reduced to being a means of the colonizer's ends. The extent to which indigenous peoples were ecological agents who actively managed the commons is denied, and they were presented as largely passive in the face of nature. In ecology, nature's agency and independence are denied, subsumed in, or remade to coincide with human interests. Since the non-human sphere is empty of purpose and devoid of agency of its own, it is appropriate that the human colonizer impose their own goals. As such, it is significant that the definition of places as wild played an equally important part in colonial and post-colonial commons governance practices.

Although by the late 19th century this ultimate title over land had become symbolic in Europe, the paradigm was conveniently revitalized in the European capture of Africa (Crowder 1968; Hall 1987; Iliffe 1987; Vansina 1990). Setting aside exceptions (such as Ghana's forest kingdoms), whether they knew it or not, most Africans became legal tenants of the state overnight on establishment of colonies (Murombedzi 2003; Alden Wily 2011; Oyono 2013). The trend towards suppressing customary rights was dominant, undermining their legitimacy as property and in turn increasing central government control over disposition of land and other commons, thus implementing a new politics of the commons (Murombedzi 2003; Wilder 2005; Eldredge 2015).

Post-colonial practices

Two trajectories can be discerned of what African post-colonial states have been doing in relation to customary authorities. First, disbandment of customary authorities with movement towards democratically elected ones. The first few years of post-colonial state control witnessed attempts to do away with customary authorities on the basis of their collaboration or assimilation by colonial governments. Attempts were made across the continent to replace customary authority with some elected representatives with mandates that included resource governance. The extent of success in this regard varied across the continent with most notable success being in Tanzania where village assemblies eventually took hold.

When post-colonial governments' attempts at disbanding customary authorities in commons governance failed, especially in land administration (Okoth-Ogendo 1989, 2008; Mafeje 2003), many governments re-instituted customary authority in rural areas. Through this second trajectory, the resuscitation or re-emergence of customary authority has been witnessed in which they have become central to the governance of resources and civil matters in rural areas (Nuesiri 2014). Post-colonial Africa is thereby meaningful of a *retour des rois*, for—mainly—political reasons: control of local communities by central authorities, electoral strategies, clientship, redistribution, etc. (see Koechlin *et al.* 2016).

This section has provided the background to understanding the relationship between customary authority and commons governance by addressing three salient points. First, that there is a history of undermining customary authorities as states struggled with who to place in charge of governing commons. Second, in many countries and across many commons, the resilience of customary authorities is witnessed, despite the undermining by both colonial and post-colonial governments. The final point is the lack of consistency amongst post-colonial states in relation to the status they accord to customary authorities in resource governance.

There has been pre-varication between doing away with customary governance in some countries and yet in others attempts have been made to revive their former roles in overall governance which has to be located in the overall state project. The contention is that African states struggle to locate themselves as Western democratic institutions or some bricolages of westernized modern polities with features of their African heritage or some constellations of their own African identities with their own heritage. Customary authorities and their place in commons governance becomes a reflection of this lack of direction or failed resolve. Continued exclusion of customary practices of resource management in preference for privatized control of resources through the state to facilitate capital accumulation is unabated within post-colonial states.

African commons in the new millennium

- The first thesis locates customary governance of commons within a plural context in which customary authority is only a part amongst many other forms of governance. In which case, this may manifest complex and overlapping jurisdictions over resource governance leading to vacuums or circumstances that may not favour access for poorer people. Resource commons users in this case rely on a multiplicity of authorities for gaining access to and control over resources. Challenges often then arise over which authority has primacy or preponderance over another in such plural contexts (Baldwin 2016).
- The second thesis is often associated with a critique of the efficacy of customary authority in general governance given their dubious democratic credentials (Mamdani 1996; Ntsebeza 2006). The main argument from this framing is that customary authorities tend to be 'despotic' and their governance is not benevolent to their 'subjects' (Perrot and Fauvelle-Aymar 2003). On many occasions, traditional rulers, attracted by gifts and other financial offers, have betrayed clanic communities by negotiating behind their backs the access of investors to customary lands.
- The third thesis postulates the re-emergence of customary authority in the governance of commons in Africa due to a variety of processes and developments. In other words, customary authority is argued to be regaining its significance in the governance of commons where previously it had declined or been undermined (Nuesiri 2014).

Commons on the African continent have faced many challenges, sometimes even threats to their survival, due to many processes and drivers, each of which have had different impacts across the continent. Perhaps the most influential challenge to customary authority is the accelerated democratic decentralization impetus, with funding support from outside partners, for African governments towards the end of the last millennium (DFID). Attempts were made to shift the locus of resource governance to elected representatives and away from customary authority. This created seismic shifts in the way resources were governed and managed at different locales across Africa. The undermining of a key institution for local level governance weakens commons governance. In this section, these challenges and processes that have been undermining customary authority in the new millennium are discussed in turn.

Resource extraction

Since the new millennium, Africa's preferential access to the United States through the African Growth and Opportunity Act (AGOA 2000), and China and India's industrial expansion, have all combined and contrived to place enormous demands on African minerals and other natural resources. The new millennium has witnessed increased demand for commodities from Africa which has led to increased resource extraction across many countries. This, in turn, has often been accompanied with greater infrastructure development, ultimately placing pressure on commons, such as the loss of rangelands for Maasai in Kenya, the loss of land in Oromia, Ethiopia, and the loss of land for agriculture in the Great Lakes Region. In Central Africa, economic development policies, built around a series of growth and employment strategies, are oriented towards the setting-up of huge and land-demanding infrastructures, such as extractive industries, dams and seaports (see Oyono 2015).

Land and green grabbing[1]

Increasing demand for large tracts of land in Africa for agricultural expansion and intensification, coupled with global financial crises, have opened investment opportunities in Africa through land deals that have adverse impact on land and resources held in common. Biofuels too have also played a part in this expansion. Food security, biofuels, rates of return from agricultural investment and general policy reforms have all led to increased land deals across the continent (Cotula *et al.* 2009). Many of these studies point out how many countries lack legal or procedural mechanisms to protect local rights to ensure livelihoods, interests and welfare. In other words, these arrangements adversely affect local capacities for managing commons (Tobin 2011). Forests and large tracts of land are also being taken from the remit of local governance through climate change mitigation mechanisms such as Reduction of Forest Degradation and Deforestation (REDD+) and Payment for Ecosystem Services (PES) (see, for example, Oyono 2015; Leach and Scoones 2015), which point to problems associated with informed consent in such deals.

Conflict and migration

Many regions of the continent have been plagued by conflicts that have endured for long periods resulting in displacement and forced movement of large populations within and across countries. Movement of large populations at relatively the same time puts pressure on resources in areas of settlement without the accompanying governance to cope with such influxes. Migration, forced movement and conflict disrupt existing institutional arrangements for resource governance which take time to build and become effective given the lack of cultural and social practices that are disturbed.

This section has provided an outline of where commons are placed in relation to broader political economic processes taking place and how they influence their governance. In the next section, we turn to the effect of these processes on customary authority

Customary authorities and the management of commons

Customary authorities have a chequered history of fraught relationships, not only with the state but with the populations they govern and those populations' access to commons. The authorities have either been altruistic or self-interested in terms of their governance over resources. Nonetheless, Alden Wily (2011) estimated that as much as 69.25 per cent of all land in Africa is under customary domain in some form of commons, whether forests, rangelands, land, rivers or wetlands. In the few places where their authority has been successfully replaced, the state has had to craft robust institutions in their place immediately following colonial administrations, as in the case of Tanzania. In this case elected village assemblies replaced traditional authority in land and natural resource administration (see Alden Wily 2011 for details).

With the introduction of the indirect rule system by the British in the early 1800s (Lugard 1922), the legitimacy of a customary authority was moved from one of personal association and recognition by community members to one of a technical nature revolving around land and power to administer it. Recognition then came through assignment and approval from the central state, which raises three issues surrounding customary governance. The first pertains to ordinary rural people's capacity and freedom to choose the cultural and legal authority under which they will live. The second pertains to people's ability to participate in their own governance and the common property resources they share with others around them (after Mnisi-Weeks 2013). The third issue surrounds the location of rule-making, dispute-resolution and general governance power whether it is concentrated in an individual customary leader or distributed amongst the people who assign it upward and at various levels of social organization.

Customary administration and dispute resolution over common resources are prominently recognized to take place at multiple levels of nested and layered authority (that align with people's social identities) (Cousins 2008: 129). Many scholars also recognize that such property rights and authority systems are 'relatively stable, [but] also flexible and negotiable' (Okoth-Ogendo 2008: 98). The negotiability aspect means that they can therefore be responsively amended to meet the circumstances and needs of the community as they arise, at the same time this may also give rise to self-interested distortion and abuse by those who have the economic and socio-political means to achieve it (Cousins 2008: 130).

The overall African picture requires that we cannot disregard the impact of distorting colonial policies on community governance of resources held under customary tenure (Mnisi-Weeks 2011). Such distortion mostly interfered with customary practices of authority and power, pertaining to resources held in common and capacities to resolve disputes. In turn, this interference resulted in several responses 'and adaptations from the communities they affected but have largely resulted in an imbalance of power between ordinary rural people and those forming part of traditional authorities' (Mnisi-Weeks 2011: 2).

However, there are a lot of cases of pitfalls. African customary authority is more and more characterized by the capture of the commons—a kind of elite capture. Capitalist land accumulation in Coastal West Africa is very often benefiting from the support of traditional chiefs, extremely active in the current explosion of land transactions for agro-industries and mining (Boamah 2014; Oyono 2015). In South Africa, the Native Land Act and the Communal Land Rights Act gave extended land powers to traditional chiefs, therefore generating the tribalization of land and authoritarian practices (Vircoulon 2003; Cousins 2008).

In the Congo Basin countries, community forestry revenue intended for community projects is mismanaged by traditional chiefs (Bigombé Logo 2003; Matata and Oyono 2015). In the Democratic Republic of Congo, traditional chiefs—called here *chefs terriens* (masters of the land)—are giving land away to urban elites and foreigners (Matata and Oyono 2015).

In customary logics, customary authorities are considered as the guardians of commons, but this is not usually the case in nowadays Africa (Baldwin 2016).

Recognition of customary authority in Africa

The most contentious issue amongst scholars is the democratization of customary authority in order to make it more accountable, transparent and inclusive and not largely a hereditary institution that serves the interests of a few people. However, the counter argument and observation are that customary authority tends to be embedded in cultural practices of the people they govern, which places emphasis on reciprocity and downward accountability (Okoth-Ogendo 2008). Alden Wily (2011) develops a categorization system to situate countries where there is significant recognition of commons under customary authority relative to countries with very little, with the year of recognition by the state in parenthesis.

Tanzania (1999), Uganda (1998), Ghana (1986, 1994), Mozambique (1997), Southern Sudan and South Africa fall into the category of countries with the highest level of recognition of customary authority. They are followed by Benin (2007), Cote d'Ivoire (1998), Burkina Faso (2009), Niger (1993, 2000), Angola (2004), Senegal (1964, 1996) and Lesotho (2010) in having significant recognition of property rights. All these states 'recognize customary tenure as a legal source of property and do not require formalization in registered entitlements for this to be upheld administratively or in the courts' (Alden Wily 2011: 746). These countries ensure customary tenure has equal standing with rights acquired through freehold and leasehold tenure, for example.

In contrast to colonial dispensation, individuals, families and communities are treated as legal persons able to lawfully own property. As such these countries institutionally provide for customary and/or community-based property administration, as a way of recognizing customary rights to commons. Institutions that include customary authority are recognized to administer customary rights at both the individual and collective levels. Such formalized and democratized legal recognition of the lowest level government provides a framework for not only protecting but enshrining customary property rights (Alden Wily 2011).

There are countries that are ambivalent toward customary rights and the recognition of customary rights. Botswana (1968), Namibia (2002) and Madagascar (2005), as well as Malawi (2002) fall into a category where the recognition is not extended to communal property where large tracts of community commons are vulnerable to private leases through state support rather than customary authority. Sudan (1984), Cameroon (1974), Mali (1993, 1996, 1997), Chad (1967, 2002), DR Congo (1967, 1973, 1980), Gabon (1967) and Somalia (1975) are the countries with the least recognition of customary property rights. Vast tracts of resources, forests and rangelands are essentially private property of the state such that evictions for large-scale investments or conservation purposes are common and lead to dispossessions of communities with customary rights (Alden Wily 2011).

As presented earlier in the section tracing the history of customary authority, beginning in the 1990s with the democratic decentralization impetus, changes to the form and practice of this authority have been made to align with governance practices that are current (Nuesiri 2014). This is particularly the case in West Africa, where chiefs and other customary authorities are elected rather than being hereditary along kinship lineages.

Conclusion

The overall and enduring challenge for customary authorities is determining the role they play in the governance of commons. Are they still relevant and what is their relationship with the state, in other words are they recognized? Alden Wily (2008: 213) documents over 30 or more new African constitutions since 1990 that are embracing community and customary rights over land, forests and other resources, increasing formal recognition of communal property rights that had been appropriated by colonial and post-colonial states and governments—for a long time these were considered or had been unnamed '*terres sans maîtres*' (in French, meaning 'lands without masters'); see Alden Wily (2008: 213). The result is that millions of hectares of forests, rangelands and wildlife habitat are formally recognized and have an upgraded status of customary rights across many states that previously had 'constrained the wilful appropriation of lands' (Alden Wily 2008: 214) and other resources by the state. Changing tenure policies, in particular customary property rights, as statutorily protected, improves the capacity of communities to hold on to their commonly held resources. Such assurance, through recognition, stops undue appropriation by governments and elites (Alden Wily 2008).

South Kivu in the Great Lakes region of central Africa and many other examples point to their status being enhanced by official recognition in which they are part and parcel of the governance systems through formal decentralized state arrangements (Ansoms and Marysse 2011). But whether there is recognition or not, their authority is prevalent across many states at the sub-national level and continues to play a significant role in the governance of local populations and resources (Ribot and Larson 2008; Romano and Reeb 2008; Larcom *et al.* 2016).

Whereas transformation of land and resources that were under community control through customary authorities is changing to centralized state control, largely for purposes of capitalist penetration, resilience of customary authority is still witnessed across many countries. Processes begun with European settlement on the continent in the 19th century have seen new formations through large scale acquisitions of land and other resources by local and foreign investors for food, oil palm and carbon trading (Alden Wily 2011; Leach and Scoones 2015). The state, using formal law, continues to retain 'their acquired position as majority landlords and controllers of most of the region's estate' (Alden Wily 2011: 752). These broader processes and drivers undermine the status of resources that are held in common under customary tenure. Using a commons lens, in which resources are held in common by groups of users, to analyse the use of resources in Africa has provided broader benefits to users, managers and decision makers especially given highly misunderstood and misrepresented common property tenure. Customary authorities in their various formations are an integral part of the governance of commons in Africa, given their embeddedness in social and cultural norms and practices of many people. It is very important for Africa to retain this form of resource governance.

Note

1 The phenomena of 'grabbing' have caught scholarly attention since their introduction by Borras *et al.* (2011, available at www.tandfonline.com/doi/pdf/10.1080/03066150.2011.559005) and Fairhead *et al.* (2012, available at www.tandfonline.com/doi/abs/10.1080/03066150.2012.671770), respectively, for land and natural resources.

References

Alden Wily, L. (2008) 'Are sacred groves in sub-Saharan Africa safe? The legal status of forests' in Sheridan, M.J. and Nyamweru, C. eds. *African Sacred Groves: Ecological Dynamics and Social Change*. Oxford, UK: James Currey, 207–20.

Alden Wily, L. (2011) '"The law is to blame": the vulnerable status of common property rights in sub-Saharan Africa' *Development and Change*, 42(3): 733–57.

Anderson, D. and Grove R. (eds.) (1987) *Conservation in Africa: People, Policies and Practice*. Cambridge, UK: Cambridge University Press.

Ansoms, A. and Marysse, S. (eds.) (2011) *Natural Resources and Local Livelihoods in the Great Lakes Region of Africa: A Political Economy Perspective*. London: Palgrave Macmillan.

Baldwin, K. (2016) *The Paradox of Traditional Chiefs in Democratic Africa*. New York: Cambridge University Press.

Bandyopadhyay, S. and Green, E. (2016) 'Precolonial political centralization and contemporary development in Uganda' *Economic Development and Cultural Change*, 64(3): 471–508.

Barnes, G. and Child, B. (eds.) (2014) *Adaptive Cross-Scalar Governance of Natural Resources*. Oxford, UK: Routledge.

Belaunde, S. (2010) *Land, Legitimacy and Governance in Cameroon, Study Report*, Institute for Research and Debate on Governance and Columbia University School of International and Public Affairs, New York.

Bigombé Logo, P. (2003) 'The decentralized forestry taxation system in Cameroon. Local management and state logic' Working Paper 10, Environmental Governance in Africa Series, World Resources Institute, Washington, DC.

Boamah, F. (2014) 'How and why chiefs formalise land use in recent times: the politics of land dispossession through biofuels investments in Ghana' *Review of African Political Economy*, 41(141): 406–23.

Colchester, M. (1994) *Slave and Enclave. The Political Ecology of Equatorial Africa*. Penang, Malaysia: Rainforest Movement.

Copet-Rougier, E. (1998) 'Histoire politico-economique de la Haute-Shanga' in Eves, H.E., Hardin, R. and Rupp, S. (eds.). *Utilisation des ressources naturelles dans la région tri-nationale de la Shanga en Afrique equatoriale: Histoires, savoirs et institutions*. Bulletin Series 102, New Haven, CT: Yale School of Forestry and Environmental Studies, Yale University, 56–77.

Cotula, L., Vermeulen, S., Leonard, R. and Keeley, J. (2009) *Land Grab or Development Opportunity? Agricultural Investment and International Land Deals in Africa*. London: IIED.

Cousins, B. (2008) 'Characterising "communal" tenure: nested systems and flexible boundaries' in Claassens, A. and Cousins, B. eds. *Land, Power and Custom: Controversies Generated by South Africa's Communal Land Rights Act*. Cape Town, South Africa: UCT Press, 109–37.

Crowder, M. (1968) *West Africa Under Colonization*. London: Hutchin & Co.

Eldredge, E.A. (2015) *Kingdoms and Chiefdoms of Southeastern Africa: Oral Traditions and History, 1400–1830*. Rochester, NY: University of Rochester Press, Boydell & Brewer.

Hall, M. (1987) *The Changing Past: Farmers, Kings and Traders in Southern Africa*. Cape Town, South Africa: David Philip.

Igboin, B. (2016) 'Traditional leadership and corruption in pre-colonial Africa: how the past affects the present' *Studia Historiae Ecclesiasticae*, 42(3): 142–60.

Iliffe, J. (1995) *Africans, the History of a Continent*. Cambridge, UK: Cambridge University Press.

Jackson, R.H. and Rosberg, C.G. (1982) *Personal Rule in Black Africa. Prince, Autocrat, Prophet and Tyrant*. Berkeley, CA: University of California Press.

Kouassigan, G.A. (1966) *L'homme et la terre. Droit foncier coutumier et droit de propriété en Afrique occidentale*. Paris: ORSTOM.

Koechlin, L., Quan, J. and Mulukutla, H. (2016) 'Tackling corruption in land governance'. Analytical Paper, London: DFID.

Larcom, S., van Gevelt, T. and Zabala, A. (2016) 'Precolonial institutions and deforestation in Africa' *Land Use Policy*, 51: 150–61.

Leach, M. and Scoones I. (eds.) (2015) *Carbon Conflicts and Forest Landscapes in Africa*. Oxford, UK: Routledge.

Lugard, F.J.D. (1922) *The Dual Mandate in British Tropical Africa*. Edinburgh, UK: William Blackwood and Sons.

Lund, C. (2008) *Local Politics and the Dynamics of Property in Africa*. Cambridge, UK: Cambridge University Press.

Mafeje A. (2003) 'The agrarian question, access to land, and peasant responses in sub-Saharan Africa' Civil Society and Social Movements Programme, Paper No.6, Geneva, Switzerland: UNSRID.

Mamdani, M. (1996) *Citizen and Subject: Contemporary Africa and the Legacy of Late Colonialism*. Johannesburg: Wits University Press.

Maquet, J. (1965) 'Droit coutumier traditionnel et colonial en Afrique Centrale' *Journal des Africanistes*, 35(2): 411–28.

Matata, P. and Oyono, P.R. (2015) 'Choix institutionnel, gestion autoritaire et privatisation de la rente communautaire en Province Orientale (RD Congo)' Document de Travail No.21, Dakar: CODESRIA.

Matowanyika, J.Z.Z. (1989) 'Cast out of Eden: peasants versus wildlife policy in savanna Africa' *Alternatives Journal: Perspectives on Society, Technology and Environment*, 16(1): 30–9.

Mnisi-Weeks, S. (2011) 'Layers of authority, boundaries of decision-making: controversies around the Traditional Courts Bill' Paper presented at the 13th IASC Biennial Conference, India: Hyderabad.
Mnisi-Weeks, S. (2013) 'Historical introduction, legal pluralism, traditional leadership, and traditional courts' in Himonga, C. and Nhlapo, T.R. eds. *African Customary Law in South Africa*. Cape Town, South Africa: Oxford University Press.
Murombedzi, J.C. (2003) Pre-Colonial and Colonial Conservation Practices in Southern Africa and Their Legacy Today. Harare: IUCN Regional Office of Southern Africa.
Ntsebeza, L. (2006) *Democracy Compromised: Chiefs and the Politics of Land in South Africa*. Leiden, Netherlands: Brill Academic Publishers, and Cape Town, South Africa: HSRC Press.
Nuesiri, E.O. (2014) 'The re-emergence of customary authority and its relation with local democratic government' Working Paper No.6, Dakar: CODESRIA.
Okoth-Ogendo, H.W.O. (1989) 'Some issues of theory in the study of tenure relations in African agriculture' *Africa*, 59: 6–17.
Okoth-Ogendo, H.W.O. (2008) 'The nature of land rights under indigenous law in Africa' in Claassens, A. and Cousins, B. eds. *Land, Power and Custom: Controversies Generated by South Africa's Communal Land Rights Act*. Cape Town, South Africa: UCT Press, 95–108.
Oyono, P.R. (2002) 'Les usages culturels de la forêt au Sud-Cameroun: rudiments d'ecologie sociale et materiau pour la gestion du pluralisme' *Africa*, 57(3): 334–55.
Oyono, P.R. (2005) 'The foundations of the *Conflit de langage* over land and forests in Southern Cameroon' *African Study Monographs*, 26(3): 115–44.
Oyono, P.R. (2013) 'The narratives of capitalist land accumulation and recognition in coastal Cameroon' LDPI Working Paper No.29, Cape Town, South Africa: Institute for Poverty, Land and Agrarian Studies, University of the Western Cape.
Oyono, P.R. (2015) Gouvernance climatique dans le Bassin du Congo: reconnaissance des institutions et redistribution' Document de Travail No.9, Dakar: CODESRIA.
Oyono, P.R. and Deng-Athoi, G. (2015) 'Land governance, local authorities and unrepresentative representation in rural South Sudan. *A Preliminary Exploration'* Working Paper No.27, Dakar: CODESRIA.
Perrot, C.-H. and Fauvelle-Aymar, F.X. (2003) *Le retour des rois. Les autorités traditionnelles et l'etat en Afrique*. Paris: Karthala.
Plumwood, V. (2003) 'Decolonizing relationships with nature' in Adams, W. and Mulligan, M. eds. *Decolonizing Nature: Strategies for Conservation in a Post-Colonial Era*. London: Earthscan.
Rangarajan, M. (2003) 'Parks, politics and history: conservation dilemmas in Africa' *Conservation and Society*, 1: 77–98.
Reader, J. (1998) *Africa. A Biography of a Continent*. London: Penguin Books.
Ribot, C.J. and Larson, A.M. eds. (2008) *Democratic Decentralization Through Natural Resources Lens*. Abingdon, UK: Routledge.
Rodney, W. (1983) *How Europe Underdeveloped Africa*. London: Bogle L'Ouverture Publications.
Romano, F. and Reeb, D. (eds.) (2008) 'Understanding forest tenure in Africa: opportunities and challenges for forest tenure diversification forestry policy and institutions' Working Paper No.19, Rome: FAO.
Taha, M.E. (2016) 'Land use, ownership and allocation in Sudan. The challenge of corruption and lack of transparency' Study Report, Khartoum: Sudan Democracy First Group.
Tobin, B. (2011) 'Why customary law matters? The role of customary law in the protection of indigenous people's rights' Unpublished PhD Thesis, Galway, Ireland: National University of Ireland.
Vansina, J. (1984) 'Western Bantu expansion' *Journal of African History*, 25: 129–45.
Vansina, J. (1990) *Paths in the Rainforests: Toward a History of Political Tradition in Equatorial Africa*. Madison, WI: University of Wisconsin Press.
Vircoulon, T. (2003) 'Les questions de la terre dans la nouvelle Afrique du Sud' *Hérodote*, 111: 99–115.
Wilder, G. (2005) *The French Imperial Nation-State: Negritude & Colonial Humanism Between the Two World Wars*. Chicago, IL: Chicago University Press.

26

THE ROLE OF PSEUDO-COMMONS IN POST-SOCIALIST COUNTRIES

Insa Theesfeld

Introduction

If Hardin's (1968) widely cited case of a pasture accessible to everyone were the standard for public natural resources, those resources would have the tendency to be depleted, degraded, destroyed or overexploited. Hardin explains that each herdsman found it more profitable to graze more animals than the pasture could support, since each took all the profit from an extra animal but bore only a fraction of the cost of overgrazing. Yet, what Hardin calls a "Tragedy of the Commons" is rather a "Tragedy of Open Access" (Feeny et al., 1990), since Hardin confuses the characteristics of a resource, such as low excludability and high rivalry in consumption, with its property rights regime which can take diverse forms. Hardin describes an open access system, meaning no rules and no property rights regime in place, that leads to overuse and degradation, instead of a resource that is held in common. Later, Hardin (1994) refers to the same case as an "unmanaged commons". With his pasture example, he aimed to contradict the laissez-faire attitude shared by economists in the late 1800s, which was that if each man pursued his own interest then the interests of all would be best served in the long run.

To motivate this chapter, I will recall four possible methods of natural resource management that can resolve this tragedy. I will also highlight the fourth, often overlooked approach, i.e. the public manager option. In fact, there is a continuum between the two polar ends of the spectrum, from governance enacted by a single central authority to a fully decentralized system of individual decision making. Between these two extremes lies a range of governance regimes that might involve higher levels of government along with local systems (Theesfeld, 2008a; Frey et al., 2016).

Hardin concluded that there are two methods to avert the tragedy: either state regulation, when a government would regulate the use of the pasture by establishing limits on the number of cattle each herdsman was permitted to graze; or privatization, when each herdsman would be assigned a section of the pasture for his exclusive use. The latter follows from the idea that each resident would then have a personal interest in conserving his part of the pasture because he would bear the entire cost of overgrazing it. Ostrom (1990, 2005) gives ample empirical cases for a successful third method of managing a natural resource in a sustainable, long-enduring way; namely, the self-organization of actors. With this, analysis of the commons expanded into the sphere of agri-environmental policy, since it offered a new perspective on policy measures that go beyond regulations and market mechanisms; namely, advisory and education measures.

Villagers would encourage one another to act responsibly in the interests of the community: in the pasture example by not adding cattle beyond the number the pasture could sustain. Evidence has accumulated over the last two decades that private, state, and common-property regimes are all potentially viable resource management options.

There is a fourth method to avoid the tragedy of open access which has received considerably less attention in the academic literature (Hardin, 1994; Mirovitskaya and Soroos, 1995). A resource manager, i.e. a civil servant directed by the community, could be hired to look after the herd and make decisions on its size. Any product of profits generated from the herd would be divided among the residents. If the community could instruct to maximize production under the condition that the pasture should be sustained and not degraded, this would prevent overgrazing. This is, according to Hardin in essence, a socialist approach (Hardin, 1994; Mirovitskaya and Soroos, 1995). Particularly this fourth method, which is slightly different from state management, additionally drives why we see so many pseudo-commons in post-socialist countries today. In contrast to the Ostrom notion of common-property regime, where private people with individualized bundles of property rights (Ostrom and Schlager, 1992) argue and agree on a set of rules for managing the natural resource, in a socialist approach, the publicly-owned resource is managed by a civil servant employed by the public. This public manager option is different from a state property regime in the respect that the natural resource itself stays in community – or "working people" ownership, not state ownership. The working people own it and employ a manager, whereas in a state-property regime, the state has ownership rights and regulates. Hardin, in a short article from 1994, also follows this idea and describes socialism as a system whereby resources are held as common property yet, instead of being managed by a group (the property-owners, i.e. the people), a manager is appointed to avoid exploitation. Hardin (1994) conceptualizes two general risks: that in practice a self-enriching incompetent manager cannot be fired; and bureaucrats can easily hide their mistakes when the group of people they represent is too big. This fourth method establishes the background for the later, often visible, pseudo-commons where resource users do not feel responsible and do not set their own rules, described in the remainder of this chapter, nevertheless it is widely cited as a commons regime.

Commons is defined in this chapter as a management system that allows the resource to be held in common and does not stand for the common-pool resource per se. Based on that, pseudo-commons are initiatives that use the notion of commons and a blueprint of a common-property regime as an artificial nutshell, not functioning on the ground. This could be due to the purposeful set-up of such pro-forma schemes to fulfill other individual benefit-strategies combined with a prevailing Soviet mentality and persisting legacies, hampering voluntary collective action approaches, as explained in the remainder. This chapter discusses the hypothesis of whether post-socialist countries are particularly prone to such misuses of the commons management system. This is widely referred to as commons management, but, as I will highlight, it only exists as a pseudo-format, posing severe risks for a proper implementation of such management regimes in the future.

It is questionable whether, based on Marx and Engels' theory, socialism is less destructive of the environment because it avoids the self-serving, profit-maximizing behavior of capitalist systems and facilitates a societal decision-making process that can effectively balance economic and environmental values (Mirovitskaya and Soroos, 1995, 78). Mirovitskaya and Soroos (1995) focus on the way in which the principles of Marx and Engels were implemented and distorted in the Soviet state by Lenin, Stalin, and more recent leaders. The fact is that first, by the way Marxist theory has been implemented, we rather observed environmental devastation (Cole, 1998, 149) and second, already the economic and anthropocentric attitude toward nature expressed in the theory itself contributed to these outcomes. Cole in his book (1998) particularly outlines how the labor theory of value by Marx is incompatible with environmentalism, seen in Soviet-style communism. In that sense, human history is interpreted as an effort by man to gain dominance over nature. For instance Engels (1940, 306) outlined that it is the alteration of nature by men, not solely nature as such, which is the most essential and immediate basis of human thought. Based on that legacy, common-pool resource protection is difficult. Marx's labor theory of value has implications on the economic concept of scarcity. For economists, natural resources have value because they are limited, one reason why excludability rules become central in common-property regimes. Yet, based on Marx, only human labor creates economic value, and thus natural resource scarcity could not make any difference in economic valuation (Cole, 1998, 158). This implicitly means, natural resources as such cannot be scarce, thus making many common-property regimes – or commons – rule settings obsolete. This theoretical underpinning, which influenced daily working habits towards natural resources, is another point to keep in mind, when analyzing why successful commons turn out to be so difficult in post-socialist countries.

Another point is that, beginning with Lenin's Decree on Land of November 8, 1917, resources, such as land, minerals, water, and forests, were made public resources and private ownership was dismantled. The natural resources were declared the common property of all working people, yet were managed nationally (Mirovitskaya and Soroos, 1995, 81), slowly driving to a de-facto system which could be provocatively named "bureaucratic private property". The difference for state property regimes is that the ownership of the natural resource stayed with the working people. In theory social ownership meant that everyone owns and is responsible for preserving resources. In practice, either the socialist state took direct control of the natural resources, leaving responsibility for managing them to political leaders, who put their perception of the necessary priorities of national development above the needs of the (local) people, or it turned out that no one was responsible for anything (Cole, 1998, 155). The socialist central planners acted rather economically shortsighted in their mission towards economic progress under socialism (Cole, 1998, 155) Instead of the self-organization of local groups – in the Ostrom sense of local common-property – the state controlled in the name of the national interest. It turns out that the particular problem with the Soviet system was that responsibility for acquiring and managing public property was too often scattered among relevant groups with a strong economic stake in projects involving large state investments in production that were given high priority. Mirovitskaya and Soroos (1995, 87) speak of a "de-facto bureaucratic private property". Indeed, during Soviet times a "system [emerged] of pseudo-public resource property in which environmental and resource management was in the hands of groups with economic interests" (Mirovitskaya and Soroos, 1995, 88).

It has been shown that the above mentioned fourth possible method of solving the tragedy of open access failed. The expanding expropriation of public property is given as an argument why, in the Russian Federation since 1994, Decrees on Sustainable Development – environmental reforms – have not reached their goals. Even after 1990 the socialist economic elite was able to reproduce its social position, for instance through its control of the implementation of privatization processes (Grancelli, 2011). Thus, the question of this chapter arises whether the particular legacy of state management with some economic groups pursuing profit is responsible for the likelihood of the post-socialist countries ending up in pseudo-commons management today.

Furthermore, Babajanian (2005, 449) introduces the notion of the "Soviet mentality" as a serious obstacle to developing active self-organizing communities. He refers to the public sector domination during the Soviet period that enforced citizen passivity and expectations that the authorities would be responsible for community welfare.

Thus, I want to explore here what the particularities of natural resource management in the history of socialist countries are that paved the way for pseudo-implementation or pseudo-collective action property regimes in contemporary post-socialist countries. Moreover, what are the risks emerging from that for future commons? I concentrate on the management of two natural resources, i.e. forestry and irrigation water, analyzed in various geographical regions of post-socialist countries. I show the appearance of pseudo-commons and outline reasons for such failed implementation.

The understanding of pseudo and commons

First of all, *pseudo* means not genuine, and by pseudo-implementation I understand the pro-forma implementation of formal rules, which have little influence on the de-facto behavior of the humans involved.

If we follow North's (1990) view of institutional change, he distinguishes between formal and informal institutions. North (1990, 36) discusses why so many aspects of a society persist despite a total change in the [formal] rules. Defining institutions as constraints, he notes that "informal constraints that are culturally derived will not change immediately in reaction to changes in the formal rules". This leads to a tension between altered formal rules and persisting informal constraints (North, 1990, 45). This pluralism of formal rights and customary claims may contradict itself, but likewise the two may influence each other (Meinzen-Dick, 2014). We often find property rights functioning on the ground that are made up of combinations of both formal secured rules and informal rules.

Understanding the implications of pseudo-implementation requires understanding the role of informal relationships and formal rules. These have been stressed in numerous analyses, including those whose particular focus is the socialist countries. For instance, Zdravomyslova and Voronkov (2002) state that informal and daily public life in Soviet society was regulated by a diversity of unwritten rules, in parallel with the official public realm which was regulated by official rules and controlled by the party-state bureaucracy.

Following Chavance (2008), I link the changes in formal rules, their enforcement mechanisms, and informal rules with four consecutive periods of reform. This is exemplified in the

Table 26.1 Four periods of change in Central Europe

Periods	*Formal rules*	*Enforcement of formal rules*	*Informal rules*
Pre-revolutionary	Significant but partial changes (in Poland and Hungary already in the 1980s)	Uneven, often decreasing	Fast and opportunistic changes
Revolutionary (after 1990)	Radical, wide-ranging and extended systemic changes	Generally weak	Fast, opportunistic and protective changes and innovations (high level of uncertainty)
Post-revolutionary	Continuing frequent changes (e.g. privatization)	Consolidation, uneven depending on domains considered	Opportunistic change, learning processes in stabilizing formal framework
Evolutionary (after EU accession)	Incremental changes	Consolidation, routinization	Changes principally through learning processes, gradual innovations

Source: adapted from Chavance (2008, 2).

following schematically for those Central and Eastern European countries which joined the European Union (EU) in 2004: the last decade of socialism as pre-revolutionary; the first three years after 1990 as revolutionary; the twelve following years as post-revolutionary (the latter two are often referred to as a transition phase); and the period following the 2004 EU accession as evolutionary.

Table 26.1 shows that, particularly during the revolutionary phase, there was a generally weak enforcement of formal rules combined with opportunistic and protective changes in the informal sectors. In addition, common to all post-socialist countries is the fact that in the post-revolutionary periods legislation changed frequently, often following external pressures, such as the preconditions for EU accession or for international development aid. Regarding reforms toward more local participation and self-governance, this had two consequences: 1) a huge discrepancy between formal and informal rules, resulting in pseudo-implementation of reforms; and 2) assuming the implementation was successful, it did not turn out to function under the Ostrom principles.[1] In particular, the opportunistic changes observed in the informal rules point to powerful groups or individuals that strive for personal benefits (in line with Mirovitskaya and Soroos (1995) and Theesfeld (2004)).

This leads me to the second aspect of defining *pseudo* in this context. Pseudo-commons or pseudo-collective action also means that formal rules have been implemented and an active organization has been built, yet there is no collective action or bottom-up self-organization. There is either top-down management or it rather resembles an open access situation with some powerful actor to strive for personal benefits (Theesfeld, 2004). In extreme cases, as described by Theesfeld (2004), for those water user associations, users officially listed as members were often unaware of the existence of those associations.

In this contribution, *commons* are understood as the management system that allows the resource to be held in common and not the common-pool resource per se. Following this understanding, a synonym is a common-property regime. Hardin (1994) refers to resources that are not held by anyone as "unmanaged commons". Unmanaged commons is linguistically very close to pseudo-commons, particularly if referring to socialism where the resource owners are the people. For those cases at risk, described by Hardin, the term commons should not, however, be used as this may lead to confusion. In his pasture example, what he refers to is open-access and not a common-property regime, which would be understood by his connotation of unmanaged commons.

Another term that looks at first like a synonym for pseudo-commons is "quasi-commons". This notion emerged in rural China, where we find many different forms of de-facto ownership and a high fragmentation of property rights. Quasi-commons describes a management system with often too much involvement by local governments (Xu, 2010). Xu (2010) shows how local governments have de-facto control over collectively-owned rural land. However, in contrast to the compulsory land acquisition in rural China which deprives farmers of their land, the quasi-commons describes a system where farmers can still keep their community land-use rights and use these rights in a profitable way (Xu, 2010, 568). The corresponding lack of self-governance by farmers and the lack of communal governance of resources still hinders the formation of real commons management in rural China.

Pseudo-commons in the agricultural water and forestry sectors

There are many studies that deal with farm restructuring in post-socialist countries and the evolution of new cooperatives (Rozelle and Swinnen, 2004; Gardner and Lerman, 2006) which often have little in common with cooperative principles (Grancelli, 2011).[2] Cooperatives are even said to be a misnomer for farms in transition countries (Gardner and Lerman, 2006), thus pseudo-commons would make a better fit. Their restructuring is often judged to be superficial (Rozelle and Swinnen, 2004). Generally, the notion of cooperation has been overstretched in the socialist legacy (Theesfeld, 2004).[3]

In contrast is the notion of "lacking a sense of ownership" described, for instance, in the Czech Republic (Prager et al., 2012). During the restructuring process, a decade of weak property rights prevented farmers from developing a sense of ownership and responsibility, especially in areas where a large proportion of land was still leased out. New owners of the often-scattered plots faced high transaction costs if they attempted to withdraw their plots in order to create a new private farm. Yet, without first feeling responsibility for the land, possible engagement in collective action is limited.

In the following section, the appearance of pseudo-commons is exemplified with the management of the natural resources: water and forest. First, I look at the breadth and diversity of problems related to water user associations (WUAs), across various geographic areas. Second, I present the depth and complexity of problems related to another particular case of natural resource management, forestry, in a single geographic area. Initiatives by donors and non-governmental organizations (NGOs) to establish WUAs or joint forest management from the late 1990s to the early 2000s were driven by the normative assumption that commons management would be the best regime in the current post-socialist situation. Thus, the chapter's structure provides a holistic overview of some typical governance problems leading to the pseudo-implementation of natural resource management policies.

Corresponding legal reforms in both sectors followed the first wave of privatization with a time delay of roughly a decade as a by-product of the land restitution and privatization process (Theesfeld, 2008b).

Pseudo-water user associations

First, in the agricultural irrigation sector 80% of the world's freshwater resources are used. Given population growth and climate change predictions, this demand will increase since agricultural production will not only have to intensify to feed the world's population, it will also have to expand to areas with less favored soil and precipitation conditions. For Bromley (1992, 14), irrigation systems represent the essence of a common-property regime:

> There is a well-defined group whose membership is restricted, there is an asset to be managed (the physical distribution system), there is an annual stream of benefits (the water which constitutes a valuable agricultural input), and there is a need for group management of both the capital stock and the annual flow (necessary maintenance of the system and a process for allocating the water among members of the group of irrigators) to make sure that the system continues to yield benefits to the group.

Approximately 76% of the world's irrigated area (227 million hectares, 40% of the world's food production) is managed by local WUAs (Frey et al., 2016). Considering irrigation systems in post-socialist countries, most share certain characteristics. First, during socialism irrigation water was regarded as abundant and irrigated agriculture was strongly promoted (Theesfeld, 2008a), and a sense of entitlement to free water still prevails (World Bank, 2009). This is for instance visible in the extreme destruction of the Aral Sea due to cotton irrigation in the past and prevailing non-adapted water withdrawal at present. According to the socialist production strategy, the more water a production cooperative used, the better it could prove its productivity to administrative bodies.

Second, the irrigation systems were built to serve large production units during socialism and could not meet the needs of the huge number of small-scale landowners who emerged following the land restitution process (Theesfeld, 2004, 2008b; Sehring, 2009). Thus, irrigation in transition countries has been drastically affected by the political and economic changes that happened after the collapse of the socialist system. During the first decade of transition the facilities largely deteriorated and the amount of water used for irrigation declined sharply. For instance, in Bulgaria only 5% of the fields equipped with irrigation devices were actually irrigated in this period (Theesfeld, 2004).

Third, collective action, i.e. WUAs, was seen, and strongly promoted by international donors (mainly the World Bank), as a way for post-socialist societies to secure sustainable agricultural water management.

Surface irrigation management in former EU accession and EU partnership countries

In most Central and Eastern European countries the agricultural sector buffered the national economic decline after transition. However, to a large extent, the agricultural sector was also struggling and at the same time highly dependent on irrigation for production. Following the World Bank's push for collective action management solutions in the agricultural water sector, many countries reformed their legislative system accordingly. Most of the associations arising from this, however, were only created formally. In practice, they were pseudo-WUAs, neither functioning nor familiar to the village farmers. Theesfeld (2004, 2008b) argues that there is no common ground where collective action can grow. Features specific to post-socialist countries, such as: 1) the discrepancy between formal political intentions and informal effective institutional change at the local level; 2) the high information and knowledge asymmetries; and 3) deteriorated social capital, lead to behavioral attributes that hamper collective action solutions (Theesfeld, 2004). In particular, the evolution of social capital is constrained by distrust among community members and trust is, in turn, a prerequisite for collective action to emerge (Theesfeld, 2004).

Bulgaria and Romania both joined the EU in 2007 when they were in transition and still reliant to a large extent on their agricultural sector. Despite legislative reforms and rehabilitation programs for the irrigation sector, I categorize their irrigation sector during the second decade of transition as a common-pool resource with the formal institutional arrangement of a common-property regime. But the effective local rule seemed to be open access; that is, no property regime supported by efforts of a formal institution to exert some authority. Theesfeld (2008a) even talks about pseudo-devolution in Bulgaria due to the country's actual implementation process of legislative reforms (the Bulgarian Water Law of 2000 followed by the Water User Association Act of 2001), which resulted in a concentration of property rights by the state authorities. Evidence shows that what gave the impression of facilitating collective action was, in fact, a further concentration of power in the irrigation sector. State actors perceived the WUAs as a threat to their power and reacted by undermining efforts to support their serious development (Theesfeld, 2008a).

Since 2004, the South Caucasus states have been part of the European Neighbourhood Policy, encouraging closer ties between them and the EU. For instance, in February 2017, a revised Comprehensive and Enhanced Partnership Agreement was announced between Armenia and the EU on deepening their political and economic ties in line with Armenia's wish to cooperate and work with both the EU and the Eurasian Union.

Also in Armenia the World Bank supported collective action by investment projects to promote the participation of local communities in their own economic and social development,[4] including the rehabilitation of irrigation systems. Yet, the existing patterns of local social organization turned out to be resistant to change. Babajanian (2005) concludes that changing patterns of interpersonal social relationships are not effective without changes in the nature of the country's governance. In Armenia, Babajanian (2005) finds well-established trust and reciprocity among the rural population which would be a strong basis for collective action yet, at the same time, limited community participation. The existing formal WUAs were established by the government and they have experienced limited participation in decision making with regard to the formulation and design of local policies and resource allocation (Babajanian, 2005). For the irrigation projects, community members were only involved in the identification meeting and for mobilizing community contributions for instrumental reasons, i.e. to satisfy formal requirements. As the reason for this implementation failure, Babajanian (2005) mentions the survival of the Soviet networks for allocating goods and services and redistributing resources. Moreover, norms of personalized relationships, unwritten rules, favoritism, misuse of public positions, and rent-seeking continue. As in the forestry sector described below, the inability of the state to control and enforce regulations effectively creates an environment where bureaucrats and local elites can take advantage of their positions for personal gain and is another reason for this pseudo-implementation (Babajanian, 2005).

Surface irrigation management in Central Asian states

Following its independence, the economic crisis in the Central Asian states resulted in a reduction of finance for the water sector to a fraction of that of the late 1980s (Soliev et al., 2015). This made the sector very dependent on donor money. Thus, the states decided on similar water governance reforms including the transfer of local irrigation management to WUAs. But, in Kyrgyzstan and Tajikistan, for instance, proper implementation is lacking – the corresponding land reform and decentralization reforms are even regarded as mainly cosmetic (Sehring, 2009). The primary incentives for conducting reforms seem to be donor pressure.

For example, across Tajikistan WUAs have been created supported by a law of 2006 (Sehring, 2009; World Bank, 2009). Set up by a top-down program, Sehring (2009) estimates that, in 2005, roughly 100 WUAs managed around one-fifth of the total irrigated land in Tajikistan. Yet, as in Bulgaria, a survey revealed that many rural people were not even aware of WUAs, and that those established were not functioning (Sehring, 2009). Since 2011, WUAs have been further supported in the region by the United States Agency for International Development (USAID), but water users still have little knowledge of the exact tasks of the WUA or awareness of the rights attached to membership. Also, there appears to be a lack of clarity about the method used for levying irrigation services fees (Balasubramanya et al., 2016).[5] In all, the result is a situation that would be inconceivable with effective local self-organization under a common-property regime, because there is no knowledge and experience in such governance forms to draw from.

For WUAs in Tajikistan it has been shown that if the organization is not functioning well in terms of a reliable water supply this further erodes user confidence in the regime (World Bank, 2009). WUAs could not meet their obligations, such as canal maintenance, due to the low rate of fee collection. The World Bank study further shows that much work is needed for the successful implementation of community-based water initiatives without the placing of favorite, but largely ineffectual, elites in key positions in the associations. A sign of a pseudo-WUA is that after donors disengage, a WUA fails due to a lack of sense of ownership by local people (World Bank, 2009). It seems that reforms were conducted only on paper in order to receive urgently-needed financial resources. Similar outcomes are reported for Kyrgyzstan where a law and a new water code approved in 2005 allowed for greater participation by stakeholders, but lacked ownership among experts, basically because the international donor had written the law (Sehring, 2009). In 2004, 59% of the irrigated land in Kyrgyzstan was managed officially by 353 WUAs (Sehring, 2009), but their development took place mainly within the framework of donor projects. For Uzbekistan, official numbers show that up to 70% of inter-farm and on-farm irrigation canals are served by WUAs, mostly established between 2003 and 2006. However, these WUAs represent a wide heterogeneity and largely do not fulfill their promised tasks (Abdullaev and Rakhmatullaev, 2015).

In Turkmenistan, agriculture is almost totally dependent on irrigation. After independence, due to a program of self-sufficiency in grain production, little of the limited government funds could be invested in the eroded infrastructure system. However, the Turkmen government has started to introduce measures aimed at improving water use and reducing demand (O'Hara, 1999). Over-watering represents a major problem here and the expectation of having abundant water available is an obstacle to the implementation of self-organized appropriation rules. Furthermore O'Hara (1999) doubts that farmers can change quickly from being purely laborers to decision makers. Thus the precondition of empowered private farmers needed for successful WUAs is in danger. Still, WUAs as part of farmers' associations are suggested for solving problems with collecting fees and water scheduling in the irrigation sector (O'Hara, 1999). There is hardly any literature following up on the implementation of WUAs in Turkmenistan.

For the Central Asian states in particular, the strong dominance of the president and his circle can be shown (Sehring, 2009). The unchallenged position of local patrons reflects a reliance on authorities as a main characteristic of the political culture. Thus, similarly to the Soviet mentality, people expect these local patrons to take care of them and are not used to being proactive themselves (Sehring, 2009). Sehring (2009) shows that clientelism, corruption, and personalistic leadership patterns undermine formal democratic structures. She further outlines that such an administration actively resists reforms that threaten the status quo. Again, WUAs have been established for the most part by international donor agencies.

Another reason for the pseudo-WUAs in Central Asia is the fact that effective local self-organization demands, first of all, independent farmers as the decision makers for their crop choice and water use. If farmers are not empowered, because, for instance, land reform results in quasi-privatization and has no impact in practice, then WUAs cannot be effective.

Groundwater irrigation associations in China

Similarly, in China, WUAs have been advocated as an organizational form for self-management. While most research (see our examples above) focuses on surface-water irrigation, this case presents the formation of WUAs in charge of groundwater irrigation in China. In areas with arid alluvial aquifers in South Asia, farmers adapt to less available irrigation water by both drilling deeper and shifting to salt-resistant cash crops. They expect to continue to make use of groundwater despite the deteriorating water quality and falling water table. The establishment of groundwater user associations in Minqin county from 2007 onwards to resolve the problem did not, however, lead to participatory solutions. The function of such pseudo-WUAs was instead to better implement direct regulatory measures from water authorities since such associations also reduce the number of actors with whom the administration needs to interact (Aarnoudse et al., 2012). WUAs have been identified as an important tool used by the county government to enable effective direct regulation measures. Improved control of water users facilitated the closure of wells and the per capita water use restriction (Aarnoudse et al., 2012).

Pseudo joint forestry management

In post-socialist countries the first goal after transition was to overcome poverty. In Russia in particular, since its economy is based on the extraction of natural resources, there is a specific danger from the myth that raw-material economies generate widespread wealth and reduce poverty (Murota and Glazyrina, 2010).[6] But the real impact of the raw-material economy (including the forestry sector) on human welfare is not very significant in Russia.

I focus on Russia as a post-socialist country because, since the early 1990s, Russia has experimented with the reallocation of authority between the center and the regions particularly in the forestry sector (Libman and Obydenkova, 2014). In addition, Russia's forest sector constitutes the largest forest area in the world, accounting for approximately 20% of global forest area (FAO, 2010). At the same time, Russian forests have a long history of mismanagement, resulting in both low productivity and the systematic overuse of resources (Carlsson, 2000; Eikeland et al., 2004) likewise culminating in extensive forest fires in 2010 (Libman and Obydenkova, 2014).

Unfavorable conditions for self-organization in forestry

Libman and Obydenkova (2014) study the allocation of jurisdiction between the center and the regions in Russia's forestry sector, and they call for the forest to be managed in a polycentric way instead of mere decentralization. They treat forests as a common-pool resource and therefore consider any management of that resource as commons management, which is not necessarily a common-property regime in the way I define commons management in this chapter. Libman and Obydenkova (2014) describe how the state capture alliances between regional governments and a number of privileged firms remained resilient after the 1990s and how the set-up makes local bureaucrats – even after reforms in 2006 –myopic and not interested in improving the quality of the forest. Ulybina (2015) also raises the fact of the historical predominance of top-down forest management and abundant regulations is not a favorable condition for the evolution of self-organized resource management. Therefore, there is no basis for an effective common-property regime and there never was an aim to develop one. Even effective decentralization turned out to be "pseudo-decentralization", due to the history of local alliances, the prevailing non-democratic system and the large country with inefficient control mechanisms (Libman and Obydenkova, 2014, 304). Andersson and Ostrom (2008) point out that decentralization may fail to deliver the expected outcomes when dominant groups of political and economic elites are further strengthened by the process.

As in the agricultural water sector, in Russia we find no cultural basis for the self-organization of appropriation rules due to prevailing local norms of unlimited resource use (Libman and Obydenkova, 2014). This, in fact, has its roots also in Marx's labor theory of value, which says that only labor gives economic value to a natural resource. Komarov (1980, 69–70) illustrated this for Soviet times in Siberia, where giant piles of trees could be seen rotting next to railroad tracks, awaiting trains that never came to take them to market. When asked to explain why loggers nevertheless continued harvesting, the local manager replied that they had a constitutionally guaranteed right to work. Another obstacle to more sustainable forms of natural resource management in Russia is the lack of a tradition of local ownership. If there is private ownership, owners often operate out of Moscow and have strong connections with the federal authorities (Eikeland et al., 2004). Eikeland et al. (2004, 285) refer to the combination of hierarchical structures and personal networks causing corrupt management practices and high transaction costs in the forestry sector as "Soviet heritage". This implies for the post-socialist countries that there might be an inherited behavior promoting a pseudo-implementation of rules and organizations for the management of common pool resources.

Furthermore, there is no central back-up for a real common-property regime. Official recognition is otherwise referred to as Ostrom's (1990) seventh design principle for stable long-enduring local common property management, i.e. self-determination of the community, recognized by higher level authorities. The new Russian Forest Code, passed by the Federal Parliament in 2006, does not require regional governments to include regional and local interest groups in the decision-making process (except for certain relatively broad conditions, that Libman and Obydenkova (2014) regard as easy to circumvent). Thus, local users' involvement – already at an initial stage in common-property management – is limited.

The new Forest Code is aimed at a decentralization process that to a great extent shifted the former comprehensive federal government's responsibilities to regional bureaucrats and politicians. These regional governments thereby gained power over the allocation of forest access rights and the setting of standards for forest use. Not only did the federal government lose its de jure right to allocate access rights, but also the de facto decision-making right became the duty of regional bureaucrats who were responsible for monitoring the process (Murota and Glazyrina, 2010). This means that the federal government may still establish basic standards for forest use but cannot ensure compliance with these rules at the regional level (Libman and Obydenkova, 2014). Consequently, it is not capable of ensuring adequate protection against illegal logging (Eikeland et al., 2004), forest fires or pollution. At the same time, large areas that once belonged to the state remain with de-facto, unclear management responsibilities. Murota and Glazyrina (2010, 38) name this "unmanaged commons", although they do not represent commons in the first place.

In many territories in Russia it became an informal practice that timber auctions were highly influenced by the strong ties between certain logging companies and the local administration, which allowed them to win the tender (Murota and Glazyrina, 2010). The problem is that informal "rent-seeking" alliances led to unsustainable and destructive forest use. This disregarded the replenishment of natural resources and kept the Russian forestry sector inefficient (Murota and Glazyrina, 2010). Only if the new forest owners recognize that it would be an advantage to collaborate can commons as a management system emerge.

Role of leskhoz, *models farms and national park zoning*

In the study of commons and forestry in Russia, the Soviet *leskhoz* is often referred to. *Leskhozy* (plural of *leskhoz*) are the territorial organizations responsible for both forest use and forest development (Libman and Obydenkova, 2014).[7] They began to appear after 1945 when the forest mining approach was diminished and there was a need for a permanent skilled labor force to develop local management organizations in remote areas. In the early 1990s, there were 1,831 *leskhozy* in Russia (Eikeland et al., 2004). These decentralized agencies are partly referred to as "community-based" Soviet-type institutions (Eikeland et al., 2004). Therefore, I want to analyze if they qualify as a common-property regime and what their current situation is. This is triggered by the fact that, for instance, Eikeland et al. (2004) name them as one of the very few community-based institutions that survived the post-socialist transition phase. Yet, when looking at their organizational set-up and responsibilities, they do not represent a commons management from a common-property regime perspective and rather resemble pseudo-commons. Carlsson (2000) argues that instead of community management which might contribute to social capital that could enhance more sustainable management, the *leskhozy* were rooted in a robust state forestry organization that included Communist Party involvement. The *leskhozy* supervised the planning of the state logging enterprises and oversaw their activities. They also directed the implementation of the new conservation principles introduced by the state, and thus did not set their own rules or resource management aims. A local incentive to fulfill ever-rising national production targets evolved during socialism. NGOs, such as the Russian branch of the World Wildlife Fund and Greenpeace, accused the *leskhozy* in the 1990s of illegal cutting and disregarding sustainable forestry practices (Eikeland et al., 2004), thus they represented a pseudo-commons – if it can be called a common-property regime at all. Industrial cutting was often hidden behind ostensible sanitary and thinning activities (Eikeland et al., 2004). After reforms, most *leskhozy* have been privatized in a nontransparent way, resulting in a highly fragmented logging industry with owners representing former Soviet managers and local administrators (Libman and Obydenkova, 2014).

Another rather weak attempt to establish common-property regimes in Russia's contemporary forestry sector can be observed in a small fraction of the existing model forests.[8] One of the four model forests studied by Ulybina (2015) aimed at facilitating local participation by forest users. The objective of engaging the local population in decision-making processes can be seen as an initial step towards commons management system. Yet, the idea of participatory environmental governance is still not widely accepted. Model forests in Russia were proposed as a new form of governance, which could potentially help harmonize the interests of different stakeholders, including forest users, regional and federal authorities, businesses, indigenous people and academia, and increase the role of the public in forest resource management. As such, model forests could be regarded as fulfilling the role of strengthening democracy and the development of local communities (Ulybina, 2015). An actual pseudo-organization in that respect is one model forest described in the study which was set up with the help of an international pulp and paper manufacturer, which also controls logging operations in Russia. Timber companies consider model forests to be a tool that may help them to avoid fines. According to interviews, the real intention of engaging in the local initiative has been to circumvent the law, since model forests grant the opportunity of complying with slightly different, less stringent laws (Ulybina, 2015).

Despite the study on the model forest showing that they failed to provide sufficient incentives for more active public involvement in forestry matters, it also shows underlying obstacles for management solutions based on common-property regimes in general. Following her study, Ulybina (2015) is very pessimistic about the possibility of local people self-organizing in Russia's forestry sector. The Soviet mentality (Babajanian, 2005) arises again when people are regarded as happy to delegate total authority to a few individuals. Ulybina (2015) even talks about a feeling of individuals' impotence to change practices and instead belief in the state's power to reform and improve forest use. The idea of public participation is even regarded as "exotic" (Ulybina, 2015, 481).

Although by law and in principle common-property regimes and local self-governance of forestry would be possible in Russia, these management solutions are also difficult due to the extremely bureaucratic formation process of such organizations.

Finally, Murota and Glazyrina (2010) discuss an exceptional case of common-property regime in the forestry sector, which has been facilitated by the establishment of a national park. The Alkhanay National Park in Eastern Siberia was created with strong support from Siberian scientists of the Russian Academy of Sciences. The park is arranged according to functional zones, one of which represents forest management run by local people. Previously the state held the property rights to the forest in the name of the people, but over-extracted. Yet, self-management of the natural resource can only happen according to scientifically-defined specifications that, for instance, delineate commercial timber cutting, and only allow local people to fell timber for home needs. This predefined common-property regime limits the freedom to develop other rules. The situation prior to, but also after, the establishment of the national park represents a pseudo-commons, either representing the state's mismanagement in the name of the people, or a top-down implementation of local management rules.

Conclusion

The practices of natural resource use under "real socialism" in the Soviet states often demonstrated that there was no working people resource property as in the Marxist vision. The old Soviet-style hierarchical networks at the local level, in combination with a lack of control and implementation capacity for environmental protection rules, partly due to features of socialist political-economy that inherently favored production over environmental protection (including the fact that the party-state, as owner of the means of production, participated directly in the economic risks), provide an environment in which bureaucrats and influential persons can take advantage of their position. This can even last into the transition phase. In fact, for both natural resources under consideration, I could show a pro-forma set up of commons management with some economic groups making profits or political actors strengthening their powerful positions. I conclude that it is the described Soviet legacy in particular that makes the post-socialist countries so likely to end up with pseudo-commons again, meaning structures that are named commons or common-property but show very few key-features of that governance regime.

In the post-socialist countries, the large collective farms were administered almost everywhere under the label "cooperatives", even though they had not evolved out of voluntary associations and had little in common with cooperative principles (Grancelli, 2011). Thus, the rural population started the transition phase with a strong psychological resistance to cooperation, originating from years of abuse of the whole concept by socialist regimes. Authors state that the use of the word cooperative[9] in Central and Eastern Europe not only creates the wrong impression but can even create barriers to progress (Gardner and Lerman, 2006).

Starting from this initial situation, any new pseudo-cooperation that pretends to use cooperation, but in fact either allows a small group to increase personal benefit or a government measure to be better implemented, destroys the trust in real cooperation. What is the starting point for such mistrust? Is cooperation among small producers quite rare as specified for Poland, for instance, because there is a lack of reciprocal trust outside the interfamily network (Grancelli, 2011)? Then distrust would hamper collaboration. Or, is the pseudo-implementation of organizations and the opportunistic behavior responsible for the decline in reciprocal trust among actors? Then distrust would be an outcome of the pseudo cooperation agreement. This vicious cycle of pseudo-collective action leads to further lack of trust in such forms of local self-organization (Theesfeld 2004).

Either way, if societies are not ripe for commons as a management system, their top-down implementation, whether pushed by donor organizations with good intentions directly or by governments in order to fulfill donor requirements, makes it worse. In the long term, it will destroy the opportunity to experiment with this property regime called in the beginning of this chapter a third method to solve natural resource overuse and degradation. The negative role model of pseudo-collective action organizations destroys the trust in such property rights regimes even further. In addition, the general sense of powerlessness within the local communities, together with the inherited norm of "Soviet mentality" of expecting the state to be responsible for solving problems, reinforces the current failure of commons management.

Notes

1 These are: 1) well-defined boundaries, 2) congruence between appropriation and provision rules and local conditions, 3) collective-choice arrangements, 4) monitoring, 5) graduated sanctions, 6) conflict-resolution mechanisms, 7) minimum recognition of rights, 8) nested enterprises (Ostrom, 1990).

2 The seven cooperative principles are: voluntary and open membership; democratic control; common property of all or part of the enterprise assets; autonomy from the public or private organizations with

which the cooperative interacts; training of the members and spread of information; cooperation among cooperatives; and finally, concern for the community (International Cooperative Alliance: www.ica.coop/en/cooperatives/what-is-a-cooperative ,accessed 08/2017).

3 Theesfeld (2004, 268), depicts actors' attitudes towards collective action with extracts from statements, among them: "I am sick of cooperatives and collective working; I only care about how my own things are doing".

4 In such community-driven social funds, beneficiaries are expected to play a significant role in the initiation, design and implementation, operation and maintenance, monitoring and evaluation of micro-projects.

5 In a study by USAID and IWMI, 52% of WUAs reported that irrigation fees are defined by crop type and area, and only 21% reported that these fees are defined by area. In contrast, when the farms were surveyed, 80% claimed that they paid their fees by area alone (Balasubramanya et al., 2016).

6 Whereas the share of financial flow from raw-material sectors in the Russian state budget is 75% (mostly from oil and gas extraction), the share of the raw-material industry in gross domestic product was only 10% in 2006 (Murota and Glazrina, 2010).

7 Libman and Obydenkova (2014) refer to it as lespromkhozy and focus on the industrial and logistics responsibilities in contemporary Russia as an outcome of transforming this organization.

8 Model forests test and demonstrate innovative national forest programs to promote sustainable forestry worldwide.

9 There is also a linguistic issue: in Russian the written word "Kooperativ" is not associated in the same way with cooperation or collaboration as in the English language.

References

Aarnoudse, E., Bluemling, B., Wester, P. & Qu, W. (2012) "The role of collective groundwater institutions in the implementation of direct groundwater regulation measures in Minqin County, China". *Hydrogeology Journal*, 20, 1213–1221.

Abdullaev, I. & Rakhmatullaev, S. (2015) "Transformation of water management in Central Asia: from State-centric, hydraulic mission to socio-political control". *Environmental Earth Sciences*, 73, 849–861.

Andersson, K. P. & Ostrom, E. (2008) "Analyzing decentralized resource regimes from a polycentric perspective". *Policy Sciences*, 41, 71–93.

Babajanian, B. V. (2005) "Bottom up and top down? Community development in post-Soviet Armenia: the social fund model". *Social Policy & Administration*, 39, 448–462.

Balasubramanya, S., Buisson, M.-C., Saikia, P., Macdonald, K., Aslamy, S., Horbulyk, T., Nannah, C., Yakubov, M. & Platonov, A. (2016) "Impact of Water Users Associations in water and land productivity; equity and food security in Tajikistan". *Baseline Technical Report*. Colombo, Sri Lanka: International Water Management Institute.

Bromley, D. W. (1992) "The commons, common property, and environmental policy". *Environmental and Resource Economics*, 2, 1–17.

Carlsson, L. (2000) "Towards a sustainable Russian forest sector". *Natural Resources Forum*, 24, 31–37.

Chavance, B. (2008) "Formal and informal institutional change: the experience of postsocialist transformation". *The European Journal of Comparative Economics*, 5, 57–71.

Cole, D. H. (1998) *Instituting Environmental Protection: From Red to Green in Poland*. Basingstoke, UK: MacMillan Press.

Eikeland, S., Eythorsson, E. & Ivanova, L. (2004) "From management to mediation: local forestry management and the forestry crisis in post-socialist Russia". *Environmental Management*, 33, 285–293.

Engels, F. (1940) "Dialectics of nature". In: C. Dutt (ed. and trans.) New York: International Press. www.worldcat.org/title/dialectics-of-nature/oclc/544969.

FAO (2010) *Global Forest Resource Assessment*. Rome: Food and Agriculture Organization of the United Nations.

Feeny, D., Berkes, F., McCay, B. J. & Acheson, J. M. (1990) "The tragedy of the commons: twenty-two years later". *Human Ecology*, 18, 1–19.

Frey, U. J., Villamayor-Tomas, S. & Theesfeld, I. (2016) "A continuum of governance regimes: a new perspective on co-management in irrigation systems". *Environmental Science & Policy*, 66, 73–81.

Gardner, B. & Lerman, Z. (2006) "Agricultural cooperative enterprise in the transition from socialist collective farming". *Journal of Rural Cooperation*, 34, 1–18.

Grancelli, B. (2011) "Local development in the rural regions of Eastern Europe: post-socialist paradoxes of economic and social entrepreneurship". *Journal of East European Management Studies*, 16, 31–53.

Hardin, G. (1968) "The tragedy of the commons". *Science*, 162, 1243–1248.
Hardin, G. (1994) "The tragedy of the unmanaged commons". *Trends in Ecology & Evolution*, 9, 199.
Komarov, B. (pseud.) (1980) *The Destruction of Nature in the Soviet Union*. White Plains, NY: M. E. Sharpe.
Libman, A. & Obydenkova, A. (2014) "Governance of commons in a large nondemocratic country: the case of forestry in the Russian Federation". *Publius: The Journal of Federalism*, 44, 298–323.
Meinzen-Dick, R. (2014) "Property rights and sustainable irrigation: a developing country perspective". *Agricultural Water Management*, 145, 23–31.
Mirovitskaya, N. & Soroos, M. (1995) "Socialism and the tragedy of the commons: reflections on environmental practice in the Soviet Union". *Journal of Environment & Development*, 4, 77–110.
Murota, T. & Glazyrina, I. (2010) "Common-pool resources in East Russia: a case study on the creation of a new national park as a form of community-based natural resource governance". *Environmental Economics and Policy Studies*, 11, 37–52.
North, D. C. (1990) *Institutions, Institutional Change and Economic Performance*. Cambridge, UK: Cambridge University Press.
O'Hara, S. L. (1999) "Irrigation and water management in Turkmenistan: past systems, present problems and future scenarios". *Europe-Asia Studies*, 51, 21–41.
Ostrom, E. (1990) *Governing the Commons. The Evolution of Institutions for Collective Action*. Cambridge, UK: Cambridge University Press.
Ostrom, E. (2005) *Understanding Institutional Diversity*. Princeton, NJ: Princeton University Press.
Ostrom, E. & Schlager, E. (1992) "Property-rights regimes and natural resources: a conceptual analysis". *Land Economics*, 68, 249–262.
Prager, K., Prazan, J. & Penov, I. (2012) "Soil conservation in transition countries: the role of institutions". *Environmental Policy and Governance*, 22, 55–73.
Rozelle, S. & Swinnen, J. (2004) "Success and failure of reform: insights from the transition of agriculture". *Journal of Economic Literature*, 42, 404–456.
Sehring, J. (2009) "Path dependencies and institutional bricolage in post-Soviet rural water governance". *Water Alternatives*, 2, 61–81.
Soliev, I., Wegerich, K. & Kazbekov, J. (2015) "The costs of benefit sharing: historical and institutional analysis of shared water development in the Ferghana Valley, the Syr Darya Basin". *Water*, 7, 2728–2752.
Theesfeld, I. (2004) "Constraints on collective action in a transitional economy: the case of Bulgaria's irrigation sector". *World Development*, 32, 251–271.
Theesfeld, I. (2008a) "Devolution in Bulgaria's irrigation system: contesting the public. In: T. Sikor (ed.) *Public and Private in Natural Resource Governance: A False Dichotomy?* London: Earthscan Ltd.
Theesfeld, I. (2008b) "Irrigation sector in Bulgaria: impact of post-socialist policy reforms". *Water Policy*, 2008, 375–389.
Ulybina, O. (2015) "Model forests in the Russian Federation: local perspectives, challenges and outcomes". *Environmental Policy and Governance*, 25, 474–485.
World Bank (2009) "Agricultural activities, water, and gender in Tajikistan's rural sector: a social assessment of Konibodom, Bobojon Ghafurov and Yovon". *Europe and Central Asian Studies*. Tajikistan: Social Development Department of the World Bank, Government of the Republic of Tajikistan.
Xu, T. (2010) "The end of the urban-rural divide? Emerging quasi-commons in rural China". *Archives for Philosophy of Law and Social Philosophy*, 96, 557–573.
Zdravomyslova, E. & Voronkov, V. (2002) "The informal public in Soviet society: double morality at work". *Social Research*, 69, 50–69.

27

FACILITATED SELF-GOVERNANCE OF THE COMMONS

On the roles of civil society organizations in the governance of shared resource systems[1]

Frank van Laerhoven and Clare Barnes

Questioning the "self" in self-governance

We observe that the common pool resource (CPR), or commons,[2] literature in general seems to focus on a particular form of governance – i.e. pure self-governance – that in reality seldom exists. There is a tension between theorizing in the CPR literature, which often implicitly assumes communities make decisions about their resources in isolation from other actors, and the presence of outsiders, or external actors, who present themselves to communities with the aim of supporting their self-governance efforts. In this chapter we frame this tension as a shortcoming in the CPR literature, which leaves unanswered important questions regarding the presence of outsiders, such as "How can outsiders, and civil society organizations (CSOs) in particular, support local communities in developing forms of self-governance that lead to sustainable outcomes for both resource systems and local livelihoods?" or "Can self-governance be facilitated, and if so, how?"

Commons scholarship seems predominated by an emphasis on *self-governing the commons*. Traditionally, its focus has been on showing that Hardin was wrong when pleading for either state or market solutions as the only way out of the social trap destined to result in a *tragedy of the commons*. An important and acknowledged result of this branch of study has been to show that commoners do not always need outsider-assistance in order to act collectively to avoid resource system collapse. Indeed, institutional arrangements based on self-governance principles are shown to be able to outperform its rival alternatives that do involve important roles for outsiders – i.e. institutions crafted and enforced by local resource system users themselves can under certain conditions pose a viable alternative to externally imposed rules or privatization (Ostrom 1990; Dietz et al. 2003).

At the same time, we notice that in practice local communities across the Global South hardly ever govern their shared resource systems in isolation. The heterogeneous group consisting of civil society organizations (CSOs) is one of the main groups of actors intervening in community level CPR institutions. Contrary to the implicit claim often found in commons scholarship that under certain conditions commoners can do very well without outsider-assistance, for CSOs, offering outsider-assistance in the form of organizing communities of commoners around the governance of CPRs is often at the heart of what they do (e.g. Shukla and Sinclair 2010).

On the one hand, commons scholarship has taught us a great deal about when communities are more likely to successfully self-govern the commons. It has attempted to explain variation in CPR governance outcomes, by following the constituent components of the Institutional Analysis and Development (IAD) framework (Ostrom 2005). Consequently, commons scholars posit that when certain resource and community attributes and institutional arrangements are in place, communities are better able to act collectively to govern their commons. However, only a small set of what Agrawal calls "*critical enabling conditions for sustainability on the commons*" (2001a: 1659) appear to be manipulable through the kind of interventions that outsider-assistance could implement (Van Laerhoven and Barnes 2014). For example, it would be difficult (and raise ethical questions) for CSOs to attempt to alter the size of the community. As a consequence, the insights gained through commons scholarship do not always translate easily into input for intervention strategies for the many CSOs worldwide who see supporting communities of commoners as their core business.

CSOs on the other hand, have a lot of experience in supporting grass root organization, but unfortunately their track record in starting collective action in a CPR context where it did not previously exist, or supporting its further development where it did, seems sparsely documented at best, or rather poor, at worst. As a consequence, the lessons learned by CSO practitioners in this realm, as well as the questions they might have about how to better support community collective action, seem to seldom make it onto the research agendas of academia in general, and of students of the commons in particular.

In view of the above, in this chapter we embark upon two questions: "How can durable forms of collective action in a context of CPR governance be stimulated by CSOs?", and "What potential impacts could CSO interventions have on resource systems and livelihoods?" To this purpose we distinguish situations in which CSOs are present, from the pure self-governance at the heart of CPR literature, by introducing the term *facilitated self-governance*. By this we mean a mode of governance in a CPR context in which external actors, such as CSOs, engage with communities composed of CPR users, with the broad aim of supporting the community's governance of their CPR and/or improving their livelihoods. This influence could be indirect, through involvement in raising issues up the political agenda (see Barnes et al. 2016) or influencing policy design and implementation at different levels, or direct, through working with communities. In this chapter we focus on the latter where engagement with CPR literature is most usefully applied.

In the subsequent sections we will propose a typology of CSOs that differentiates between the approaches CSOs take in their interventions in community CPR institutions. Then, we will present an analytical framework that accounts for the disaggregated impact of CSO intervention strategies on resource systems and livelihoods, respectively. First however, we will (i) share our conceptualization of communities and CSOs, (ii) briefly outline our understanding of tendencies in policies towards community involvement in resource system governance, and (iii) mobilize the theoretical building blocks we intend to include in our agenda for further developing the concept of facilitated self-governance.

Conceptualizing communities and CSOs

Advocates of communities being at the center of natural resource system conservation sometimes skirt over clearly defining what is meant with the term *community*. Many authors have found that multiple interests can be found within communities and that some individuals will be able to influence the decision-making process more than others (Agarwal 2001; Shyamsundar and Ghate 2011; Arnall et al. 2013). Ignoring these divisions and "romancing the commons"

(Imperial and Yandle 2005: 459) may undermine the arguments for community based commons management (Agrawal and Gibson 1999; Ostrom 2005). In this chapter, communities are seen as being composed of individuals attempting to act collectively to create and maintain the institutions required to provide for shared resource system and to govern the individual and cumulative appropriation of resource units from such systems.

External actors are defined broadly here as any actor beyond the community itself involved in shared resource system governance. By employing the descriptor *external*, a distinction is drawn between the individuals living in or near the resource system and those who come from a different locality to work directly or indirectly with these individuals in a context of shared resource system governance. This descriptor recognizes the heterogeneity of *external* actors and the often blurry and dynamic line between state and non-state actors in many developing countries, due to the often indistinct functions (Brass 2016), the complex interactions between groups (Agrawal 2001b; Lemos and Agrawal 2006), and the movement of individuals between sectors. However, a broad distinction between state actors and civil society is usefully made in identifying the actors.

State actors across levels and departments are often formally tasked with roles in natural resources related policy formation and implementation through decentralization processes. The types of processes, the particular decision-making authority and the outcomes of such processes have been found to vary greatly (Cook et al. 2017). The administrative departments, level and individuals' ascribed roles will depend on the country's historical and current administrative structures, culture and politics.

Civil society, understood here as associational life (see Ehrenberg 1999; Edwards 2009), can include non-governmental organizations (NGOs) (though see Bebbington 2005; Banks et al. 2015; Brass 2016), community based organizations (CBOs) and social movements. NGOs are understood here as professional organizations, whereas CBOs are smaller voluntary organizations operating in one locality, and social movements are networks of activists and/or organizations (Krinsky and Crossley 2014) where the defining characteristic is the degree of social ties between individuals (Tilly 2005). Through employing a broad definition of civil society organizations (CSOs) (following Edwards 2009, and Lund 2006) the aim is to avoid missing any relevant organizations. The focus of this chapter is on CSOs, although some of our propositions may be relevant for explaining and analyzing the relation between communities and state actors.

Community involvement in the governance of shared resource systems

How do current (natural) resource system governance arrangements perceive of the roles of local communities? Agrawal (2001a) found that over 60 countries had decentralized some aspect of their natural resource system management. Five main reasons for this shift can be identified. Firstly, many authors argue that state governments have not been effective governors of (natural) resource systems (Agrawal and Gibson 1999; Arts 2014; Hall et al. 2014), whereas local users are seen as better guardians of such systems due to the direct stake they have in its maintenance and their local knowledge (Andersson et al. 2014). Secondly, it is argued that increasing the importance of the role of local communities in resource system governance is cheaper than governance arrangements that place the locus of control with the state (Agrawal 2007; Somanathan et al. 2009). Thirdly, forms of resource system governance that afford important roles to resource system users fits within the wider discourse of decentralization, democracy, pluralism and rights (Agrawal and Ribot 1999; Larson and Ribot 2004; Agrawal 2007). Fourthly, scholars argue that political considerations and power struggles involving multiple actors such as donors and civil society movements, prompt and shape such reforms (Ribot et al. 2006; Blaikie and Springate-Baginski 2007). Finally, the growing

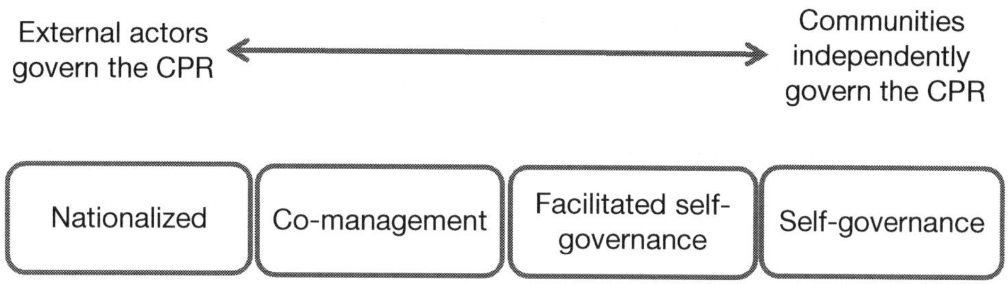

Figure 27.1 Facilitated self-governance in relation to other modes of CPR governance

influence of the CPR literature on policy making should be considered (Agrawal 2001a; Blaikie 2006; Mosse 2006; Van Laerhoven and Ostrom 2007).

To what extent does this shift towards community involvement in the governance of shared resource systems allow for the kind of pure self-governance that is often highlighted in the CPR literature? As the policies resulting from this shift towards community involvement in resource system governance have gained footing across many developing countries, they have often formally assigned facilitating roles to actors beyond the community – state or non-state – in their implementation (Blomley and Ramadhani 2006). Even where no formal roles are found in policy documents, *in practice* external actors have often taken it upon themselves to support their implementation through working on the ground with communities and/or as part of policy implementation processes (Blaikie and Springate-Baginski 2007; Wright and Andersson 2013). Examples of induced or facilitated community governance arrangements can be found across many developing countries, including India (Ghate 2003) Nepal (Agrawal and Ostrom 2001), the Philippines (Donoghue et al. 2003; Duthy and Bolo-Duthy 2003), Mexico (Bray et al. 2006), Cameroon (Lescuyer 2013), Tanzania (Babili and Wiersum 2013) and across the Amazon region (de Koning 2011). It is these situations which are not currently captured under CPR scholars' conceptualization of *self-governance*. The governance of shared resource systems can range from full regulation by external actors (i.e. authorities) to pure self-governance, with the middle-part probably being more reflective of what is found in practice (Figure 27.1). Facilitated self-governance is distinguished from comanagement (Carlsson and Berkes 2005) as in the latter governance mode, partnerships between external actors and communities are established which set out clear divisions of responsibilities. In facilitated self-governance, this is not the case.

Integrating perspectives on facilitated self-governance

Although our approach is rooted in the Ostrom traditions of engaging with the commons, we feel that we must complement this perspective with concepts and insights from *International Development Studies* and *Critical Institutionalism* in order to be able to move towards a practice of facilitated self-governance. These two bodies of literature have engaged explicitly with debates on the roles of CSOs in rural contexts of the Global South and therefore drawing on these insights can help us to more adequately conceptualize and analyze situations of facilitated self-governance.

The CPR literature

Until the 1970s it was generally understood that only clear public or (the public regulation of) private property rights over natural resource systems could ensure their sustainability. The work of Hardin (1968) was influential to this regard as he argued that (i) the benefits of appropriation from a CPR are private, whilst the costs are shared (appropriation dilemma), and (ii) the cost of provision of a functional commons are private, whilst the benefits are shared (provision dilemma). According to Hardin, this makes overharvesting of and underinvestment in the resource stock – and ultimately a resource collapse – unavoidable, and led him to influentially argue for public or private management of shared resource systems. The presumption was that shared resource systems in which appropriation and provision rules were not imposed by either the state or the market were devoid of any rules, and were used as open access (Wade 1988; Ostrom 1990; Ostrom and Cox 2010).

Since the 1980s a growing body of CPR literature has shown that what Hardin called *open access* may actually be being managed as common property, with communities crafting their own rules to deal with appropriation and provision dilemmas (Wade 1988; Ostrom 1990; Baland and Platteau 1996; Dietz et al. 2003). In order to develop and maintain such institutions, Ostrom (1990) poses that users must overcome three types of collective action dilemmas. They need to, firstly, agree on access and withdrawal rules (also see Schlager and Ostrom 1992), secondly, organize the monitoring of compliance to the rules, and thirdly, decide on the sanctioning of rule infractions (Andersson et al. 2014).

A rich body of literature has developed analyzing the variables seen to influence the likelihood of communities overcoming these collective action dilemmas to build robust institutions (Westermann et al. 2005; Agrawal 2007; Van Laerhoven and Berge 2011). Ostrom (1990) famously posed eight Design Principles to characterize robust institutions for managing CPRs, drawn from successful cases of community self-governance of the commons – though she stressed that they were not intended to serve as a blueprint for designing CFM (Ostrom 1990). These principles have been refined, supplemented and amended since (Agrawal 2001a; Agrawal and Chhatre 2006; Cox et al. 2010; Andersson et al. 2014).

In an extensive review of the burgeoning commons literature, Agrawal (2001a) states that the likely number of factors affecting the successful management of the commons will be between 30 and 40 with significant overlap between those facilitating the emergence of institutions and their continuation. He categorizes these factors into four groups: resource system characteristics, group characteristics, institutional arrangements and the external environment (the second category is further subdivided into economic and socio-political in Agrawal and Chhatre (2006) and Agrawal (2007)). Similarly, Cox et al. (2010) reports that scholars who found the institutional focus of the original design principles to be incomplete suggested adding factors covering the social values of the group, the resource characteristics and external socioeconomic factors. In Andersson et al.'s (2014) review of literature on institutional arrangements, the authors found that the institution rule-making process, characteristics of the rules, monitoring and enforcement of rule compliance and imposition of sanctions on rule violators, were important characteristics for success.

Commons scholarship seems wary of externally imposed rules, presenting evidence supporting design principle 3 that a high level of autonomy and discretion in decision-making is important for durable institutions (see Baland and Platteau 1996; Vedeld 2000; Sekher 2001). Aside from concerns about the local appropriateness of such rules, Agrawal and Chhatre (2007) also argue that they are more likely to be ignored or may cause internal conflicts. It is therefore maybe understandable that CPR literature has historically paid far less attention to the role external actors could play in CPR governance than the conditions for successful commons governance reviewed above (Agrawal and Gibson 1999; Agrawal 2001a; Wright and Andersson 2013). The extent to which these conditions are manipulable by external actors has not received serious attention. Indeed in their review of CPR literature, Cox et al. (2010) found that scholars often suggested greater attention for the "external environment" was required.

The limited number of empirical CPR studies of external actor involvement at the community level (Pretty and Ward 2001; Wright and Andersson 2013) mean empirical evidence of intervention effects is thin and sometimes contradictory. These mixed and inconclusive empirical results could be due to the different types of external actors, the approaches they take, and what the researchers themselves determine to constitute external actor involvement in the case studies under investigation. The studies rarely give the level of detail required to analyze interventions to this depth, or to draw out the causal pathways between external actor involvement and the different outcome variables used by CPR scholars. We therefore propose to adopt insights and concepts from International Development Studies and Critical Institutionalism, respectively.

International Development Studies (IDS)

IDS scholars have analyzed CPR governance as both one of the strategies communities could employ in order to improve their livelihoods and the object of external actor interventions (Chambers and Conway 1992; Bebbington et al. 2002; Thin and Van Gardingen 2004; Scoones 2009; Ingram et al. 2015). A useful distinction in intervention activities is often made between *service provision* e.g. delivering technology or knowledge, and those *facilitating institutional change*, which mediates access to livelihoods (Bebbington and Perreault 1999; Edwards 1999; Bebbington et al. 2002; Berkes 2007; Banks et al. 2015). Scholars generally call for a focus on the latter type of intervention by arguing that attention for institutions will mean greater participation of marginalized people (Ingram et al. 2015) and sustained and broader effects (Scoones 1998; Edwards 1999; Thin and van Gardingen 2004).

The IDS research tradition seems to pay greater attention to the role of shared (natural) resource systems in communities' wider livelihood portfolios. This is often conceptualized through employing the Sustainable Livelihoods Approach (SLA), which has been influential in rural development thinking since the late 1980s (Chambers and Conway 1992; Scoones 2009). The SLA presents an alternative to the single sector focus on production, employment and income as the sole concerns of livelihood development (Scoones 2009). Livelihoods are seen as being comprised of two elements: the different capitals communities can access, and the strategies communities can employ to improve their livelihoods (Scoones 2009).

This body of literature has also paid greater attention than most CPR literature to critiquing state and civil society organizations, and their policies and interventions in the interrelated fields of rural development and natural resource management, across multiple levels (but see Gibson et al. 2005a). Though IDS scholars do not seem to directly engage with the specific CPR dynamics and institutions outlined above, they provide insights into (i) different types of external actors, (ii) their possible approaches to working with community institutions at a local level, and (iii) effects of interventions on livelihoods.

Critical Institutionalism (CI)

CI questions the assumed insularity of communities, which was common in earlier CPR studies (Lund 2006; Mosse 2006; Cleaver and de Koning 2015). Whilst both CPR and CI scholars argue that institutions are important for understanding the behavior of individuals (Johnson 2004; Jones 2015), CI scholars question whether such institutions can be designed for a particular purpose, as they argue the CPR literature implies (Cleaver 2002; Cleaver and de Koning 2015; Jones 2015). Instead they favor *institutional bricolage* in which institutions are borrowed and adapted from elsewhere (Hall et al. 2014; Cleaver and de Koning 2015). They also question whether institutions determine behavior in the way envisaged by CPR literature (Cleaver and de Koning 2015), which necessarily results in win-win situations for both (shared) resource systems and livelihoods (Johnson 2004; Jones 2015). Additionally, they argue that institutions related with CPR governance cannot be viewed separately from the social, political and historical context specific realities into which they are entwined (Mosse 2006; Gutu et al. 2014; Cleaver and de Koning 2015; Haapala et al. 2016).

CI is therefore a helpful addition to our analytical framework as it (i) draws attention to the actions of external actors, (ii) discusses the social embeddedness of CPR institutions, which may be part of external actor interventions at the local level, (iii) and questions whether CPR institutions result in win-win situations as envisaged in CPR literature.

Towards a practice of facilitated self-governance

In what follows, we will elaborate on an approach that in our view can help bridge the divide between theory and practice – *theory* that so far has not really managed to provide CSO practitioners with hands-on tips and tricks to support local users of shared resources in terms of organizing collective action and self-governance, and *practice* that so far has not optimally managed to have its lessons learned converted into contributions to theory building. To that purpose, we draw on the CPR, IDS and CI literature outlined above and will proceed as follows:

1. We propose *a typology of CSO approaches to institutional change*: The sustainable self-governance of shared resources requires resource users to solve both provision and appropriation dilemmas, which in turn require certain institutions for collective action. The rationale for any type of CSO intervention is that such institutions are not yet optimally functioning (which subsequently leads to non-optimal resource and livelihood outcomes). As CSO interventions are targeting institutional change, we first need to understand the logic behind the CSO views on how local institutions for collective action can or should be adapted.
2. We present *a framework to study the interaction between CSOs and CPR users*: As noted earlier, on the one hand, commons scholarship has not provided CSO practitioners with a clear and sturdy handle on how to successfully approach the facilitation of self-governance. CSO practice, on the other hand, has not found a way to systematically track what they are doing in this respect and assess what works and what does not (and why). Our framework aims at facilitating the systematic analysis of *inputs* (i.e. CSO approaches to institutional change), *outputs* (i.e. CSO activities) and *outcomes* (i.e. impacts on resource systems and livelihoods) in a context where CSOs and communities meet.

A typology of CSO approaches to institutional change[3]

We propose to move away from static typologies of CSOs, which categorize them according to characteristics such as size, location, funding body, or stated objectives (see Yaziji and Doh 2009 for an overview of such categorizations). These static typologies have received criticism from scholars such as Chhotray (2007) who refers to the fallacy of the binary distinction between NGOs as "political entrepreneurs" and "development actors" and Thomas et al. (2010: 368) who pose that CSOs can show "multiple identities" encompassing selective collaboration with the state, gap-filling and posing alternatives. In contrast to these typologies, we argue that the institutional change literature allows us to create a typology of CSO approaches, which can capture time and context dependent diversity in approaches within individual CSOs in a meaningful way.

Firstly, we draw on the debate in the institutional change literature on the extent to which the potential for institutional change is predominantly determined by either structure or agency. Those advocating for the importance of structures in society in determining behavior argue that actors act strategically and through calculus based on and determined by structural features, such as pre-existing, nested institutional arrangements (Saravanan 2015). They emphasize that institutions need to be purposefully *designed* in order to steer actor behavior in a particular direction. CSOs holding this view would direct their efforts at institutional design through a focus on introducing rule structures. Alternatively, those that highlight the role of agency emphasize that efforts to bring institutional change need to be directed at enhancing the capability of CPR users to engage in do-it-yourself institutional bricolage (Cleaver 2002). This institutional *crafting* perspective holds that institutions which respect time-and-place particularities of, and interactions between, both the social and the biophysical system can be crafted, proactively. CSOs holding this view of institutional change would direct their efforts at enhancing the capabilities (agency) of communities to craft their own institutions.

Secondly, we draw on the notion that institutions can be either exogenous or endogenous to the change actor (Alexander 2005). When they are exogenous "the object of the undertaking – the institutional structures and/or practices that are to be changed – is outside the institutional change actors' own institutional context" (Alexander 2005: 211). When they are endogenous it is assumed that the intended institutional change would become effective only through reshaping the actors' perceptions and cognition. Applied here, a CSO holding an exogenous perspective would view themselves as not being part of the local level institution and therefore the ultimate change actor is the community. A CSO taking the endogenous perspective would see themselves as being part of the institution and therefore they themselves are a change actor.

The resulting typology leads to four archetypical perspectives on whether and how CSOs can change community institutions in such a way that a tragedy of the commons can be avoided (Figure 27.2).

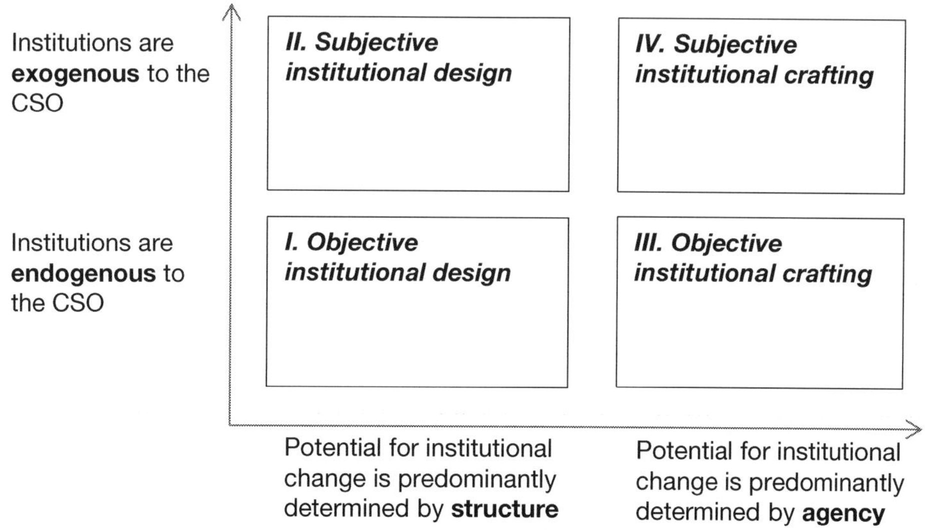

Figure 27.2 A typology of CSO approaches to institutional change

1. *Objective institutional design*: The NGO itself is the primary change actor. Activities are focused on creating incentives through designing institutions. It applies a generic approach, imposing institutional arrangements that have been proven to work elsewhere.
2. *Subjective institutional design*: Ultimately, the target community is the primary change actor. The NGO applies an approach that facilitates a reflective-dialogic process among resource users in order to design locally appropriate institutions.
3. *Objective institutional crafting*: The NGO itself is the primary change actor. It applies a generic approach using input from local analyses (e.g. participatory appraisal techniques) aimed at crafting customized training modules to empower local communities.
4. *Subjective institutional crafting*: Ultimately, the target community is the primary change actor. The NGO applies an approach that facilitates a reflective-dialogic process among resource users aimed at the empowerment of communities (e.g. through action research techniques).

In sum, in order to exert influence on institutions, external actors could focus on the rules determining structure (institutional design), or on the agency of individuals to effect change (institutional crafting). Approaches could also differ according to whether the institutional change is led by the external actor (objective) or the community itself (subjective).

This typology allows for analyzing the underlying approaches ingrained in CSO interventions.In the framework presented next we will refer to this as "input". This input, we argue, will determine the kind of activities (or, "output") that CSO will employ, which in turn will have an effect on "outcomes" related with (i) the communities' ability to deal with provision and appropriation dilemmas (and subsequently, their ability to avoid a resource collapse caused by under investment in the resource system, and over exploitation of resource *units* from that system), and (ii) the communities' livelihood portfolio (i.e. access to capitals and strategies).

A framework to study the influence of CSO interventions on CPR conservation and community livelihoods[4]

CSO interventions in a CPR context are essentially interventions into complex social-ecological systems with the aim of conserving shared (natural) resource systems and providing livelihood benefits. Such interventions have proven difficult to design (Gibson et al. 2005b; Bauch et al. 2014) and in this chapter, we argue that this is at least partly because social-ecological systems defy simple implementation pathways between CSO activities and positive outcomes in terms of CPR conservation and livelihoods.

CPR scholarship has stressed the central role of CPR institutions for successfully conserving resources and enhancing livelihoods. In the framework we present here, we align with the CI literature outlined above, thus avoiding the presumption that there is such a simple linear causal pathway between the self-governance facilitated by external actors and the outcome of interventions. We suggest broadening the commons literature's focus on CPR institutions by drawing on the IDS literature to develop an analytical framework encompassing multiple pathways between external actor interventions and the *outcome* variable: "sustainable livelihoods in a CPR context". The intervention activities are seen as the *output* variables that could lead to changes in the *outcome* variables. The choice of activities (*output*) is determined by the external actor's motivation and approach to institutional change – here the *input* variable. Below, each variable in our framework – input, output, and outcome – is discussed in turn.

Inputs

The typology developed earlier differentiates between four archetypical CSO approaches to institutional change in the communities with whom they work. This forms the first part of our input variable. The second part of our input variable comprises the motivations CSOs hold for directing efforts at certain activities, or working in particular ways. Such motivations can be influenced by real or perceived demand from donors for demonstrating impact (referred to as "production-oriented interventions" in Bebbington (2005: 945)). This creates incentives for actors to focus on narrow target groups where impact can be demonstrated in shorter timeframes, rather than investing in slower institution-building, and activities with tangible, measurable and easily-communicated outputs (Wright and Andersson 2013). Organizational expertise, culture and perceived division of responsibilities between an external actor and community (Barnes and Van Laerhoven 2015) also influence what is seen by the CSO as the "normal" way of working.

Outputs

We identify three different types of intervention activities: following the CPR literature, (i) *intervention activities directed at CPR institutions*, and following the SLA literature (ii) *activities aimed at directly affecting capital stocks and strategy choices (service provision)*, and (iii) *activities focused on strengthening or altering community institutions* which mediate access to livelihoods (Bebbington et al. 2002).

From the CPR literature, in order to analyze interventions directed at CPR institutions (activity type 1 in Figure 27.3), we adopt the useful distinction between involvement in *functioning* collective action (characterized by Poteete and Ostrom (2004) as (i) regular meetings; (ii) the presence of rules on entry, harvesting and monitoring; and; (iii) the presence of a system to enforce the rules) and activities directed at stimulating *durable* collective action (knowledgeable actors with management and communication skills, plus sufficient material and financial resources).

From the SLA literature we adopt the notion that interventions should help reconfigure institutions mediating access to resources (Edwards 1999: 372; Allison and Ellis 2001) (activity type 3 in Figure 27.3) as this will allow service provision activities (activity type 2 in Figure 27.3 – such as technology and resource transfers) to have wider, longer lasting effects (Scoones 1998; Bebbington and Perreault 1999; Mansuri and Rao 2013; Ingram et al. 2015). Combining activity types 2 and 3 is seen as favorable in the literature (Hulme 2000; Thin and van Gardener 2004; Westermann et al. 2005; Berkes 2007).

Outcome

Our outcome variable is comprised of the interrelated components: the community's ability to deal with appropriation and provision dilemmas (A&P dilemmas) (drawn from CPR literature) and their livelihoods portfolio (drawn from SLA literature). From a CPR lens, dealing with A&P dilemmas i.e. avoiding overharvesting and underinvestment of the CPR, requires collective action of the CPR users (Gardner et al. 1990; Dietz et al. 2003). From an SLA lens, the outcome of interventions should be improved capital stock (human, social, financial, natural, physical capitals) and livelihood strategies (seen as instrumental, hermeutic and emancipatory – see Bebbington 1999). Scoones (2009) draws on Sen's (1981) entitlements approach to emphasize the mediating role of institutions in defining access to these capitals and strategies. The types of institutions being referred to by SLA literature are socio-cultural and political processes, such as gender or caste. We refer to the institutions discussed in the SLA literature as *community* institutions to distinguish them from the *CPR* institutions discussed in the CPR literature.

The interwoven components exert influence in both directions i.e. the livelihoods portfolio influences the community's ability to deal with A&P dilemmas, and vice versa. A community which is better able to deal with A&P dilemmas may experience improved CPR conditions and livelihoods (Ostrom 1990; Agrawal 2001a; Pretty and Ward 2001) however meta-studies also point towards potential trade-offs between CPR conditions and livelihoods (Pagdee et al. 2006; Persha et al. 2011). Elements of the livelihoods portfolio can independently or collectively affect a community's ability to deal with A&P dilemmas as they alter the ratio of private benefits to shared costs (appropriation dilemma) or shared benefits to private costs (provision dilemma) and therefore the incentives to overharvest or underinvest.

The complexities of the interrelationships between the outcome components lead us to argue for a framework, which allows multiple causal pathways between interventions and outcomes (see Figure 27.3).

Future research

Lange et al. (2013) propose a meta-framework that sets out three dimensions along which the key features of modes of governance – such as facilitated self-governance – can be mapped. This initial step in sketching facilitated self-governance serves as a departure point for future researching the concept of facilitated self-governance within and across the dimensions.

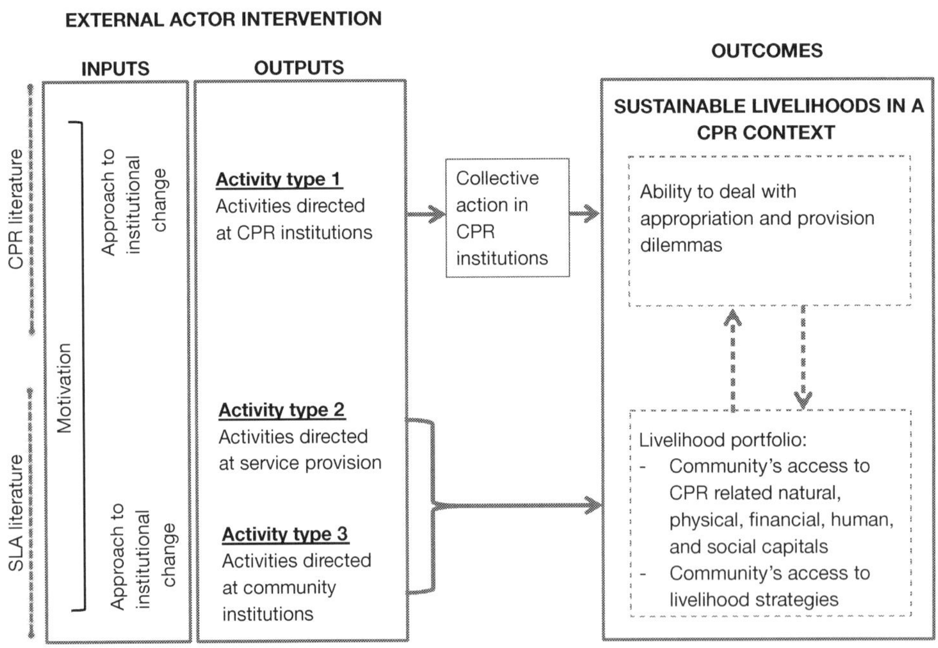

Figure 27.3 A framework to study the influence of CSO interventions on CPR conservation and community livelihoods

- The *politics dimension* draws attention to the actors and their interactions, preferences and power relations. In facilitated self-governance this would be the presence and actions of external actors who have the objective of working directly or indirectly (through influencing CPR related policies) with communities using shared resource systems to improve their livelihoods and/or the forest condition.
- The *polity dimension* refers to the institutional "rules of the game" that shape the interactions of actors. In facilitated self-governance, two types of institutions across levels are relevant. Firstly, community level CPR institutions retain a large degree of autonomy in the day-to-day functioning of collective action but are shaped by external actors. Secondly, CPR institutions affect and are affected by wider community institutions, which in turn may be the subject of external actor interventions.
- The *policy dimension* encompasses policy formulation and implementation including objectives and instruments of political steering towards outputs. In facilitated self-governance at the policy level, multiple non-state and state actors negotiate and influence CPR related policy formulation and implementation. At the community level intervention activities are varied, change over time and create effects beyond the initial objective.

Diverse theoretical perspectives could be taken to lead research in facilitated self-governance as no assumptions of relations between key variables (such as CPR institutions and resource system condition outcomes) have been made in laying out its key features. For example, scholars taking an institutionalism approach inspired by the IAD framework could explore the interactions between institutions across levels in the polity dimension, whilst those taking a critical institutionalism perspective could explore the power relationships between actors (politics dimension) and the extent to which communities amend the institutions designed or crafted by external actors over time (polity dimension). The ontological differences between these perspectives therefore serve as an interesting basis for rival research questions.

Further research both within and between the politics, polity and policy dimensions of facilitated self-governance could focus on: differentiating between external actors and communities, and within communities themselves (politics); the possible selection bias of external actors in apparently choosing to work in locations in which functioning collective action is present (polity); and in a more detailed understanding of how external actors draw on, align or reconcile their presence across policy and local levels (policy). Such lines of enquiry would help in fleshing out the concept of facilitated self-governance with the goal of more accurately conceptualizing the myriad of influences external actors can have on the self-governance of shared resource systems.

Notes

1 Disclaimer: This chapter contains ideas and excerpts from our earlier work on this topic, notably from Barnes and Van Laerhoven (2015), and Barnes et al. (2017).

2 Note: in this chapter we use the terms "CPR", "commons", and "shared resource systems" interchangeably.

3 This section leans on earlier work in which we both develop and apply this typology to community forest management (CFM) in a context of Joint Forest Management (JFM) in the states of Maharashtra, Andhra Pradesh and Odisha in India (Barnes and Van Laerhoven 2015).

4 This section leans on earlier work in which we both develop and apply this framework to community forest management (CFM) in a context of the Forest Rights Act (FRA) in the state of Andhra Pradesh in India (Barnes et al. 2017).

References

Agarwal, B. (2001) "Participatory exclusions, community forestry, and gender: an analysis for South Asia and a conceptual framework" *World Development* 29(10), 1623–1648.

Agrawal, A. (2001a) "Common property institutions and sustainable governance of resources" *World Development* 29(10), 1649–1672.

Agrawal, A. (2001b) "State formation in community spaces? Decentralization of control over forests in the Kumaon Himalaya, India" *The Journal of Asian Studies* 60(1), 9–40.

Agrawal, A. (2007) "Forests, governance, and sustainability: common property theory and its contributions" *International Journal of the Commons* 1(1), 111–136.

Agrawal, A. and Chhatre, A. (2006) "Explaining success on the commons: community forest governance in the Indian Himalaya" *World Development* 34(1), 149–166.

Agrawal, A. and Chhatre, A. (2007) "Institutions, co-governance, and forests in the Indian Himalaya" *Studies in Comparative International Development* 42, 67–86.

Agrawal, A. and Gibson, C.C. (1999) "Enchantment and disenchantment: the role of community in natural resource conservation" *World Development* 27(4), 629–649.

Agrawal, A. and Ostrom, E. (2001) "Collective action, property rights, and decentralization in resource use in India and Nepal" *Politics & Society* 29(4), 485–514.

Agrawal, A. and Ribot, J. (1999) "Accountability in decentralization: a framework with South Asian and West African cases" *Journal of Developing Areas* 33, 473–502.

Alexander, E.R., (2005) "Institutional transformation and planning: from institutionalisation theory to institutional design" *Planning Theory* 4(3), 209–223.

Allison, E.H. and Ellis, F. (2001) "The livelihoods approach and management of small-scale fisheries" *Marine Policy* 25, 377–388.

Andersson, K., Benavides, J.P. and León, R. (2014) "Institutional diversity and local forest governance" *Environmental Science and Policy* 36, 61–72.

Arnall, A., Thomas, D.S.G., Twyman, C. and Liverman, D. (2013) "NGOs, elite capture and community-driven development: perspectives in rural Mozambique" *Journal of Modern African Studies* 51(2), 305–330.

Arts, B. (2014) "Assessing forest governance from a 'Triple G' perspective: government, governance, governmentality" *Forest Policy and Economics* 49, 17–22.

Babili, I.H. and Wiersum, F. (2013) "Evolution and diversification of community forestry regimes in Babati District, Tanzania" *Small-Scale Forestry* 12(4), 539–557.

Baland, J. and Platteau, J. (1996) *Halting Degradation Of Natural Resources. Is There a Role for Rural Communities?* Oxford, UK: Clarendon Press.

Banks, N., Hulme, D. and Edwards, M. (2015) "NGOs, states and donors revisited: still too close for comfort? " *World Development* 66, 707–718.

Barnes, C. and Van Laerhoven, F. (2015) "Making it last? Analysing the role of NGO interventions in the development of institutions for durable collective action in Indian community" *Environmental Science and Policy* 53, 192–205.

Barnes, C., Claus, R., Driessen, P., Dos Santos, M.F., George, M.A. and Van Laerhoven, F. (2017) "Uniting forest and livelihood outcomes? Analyzing external actor interventions in sustainable livelihoods in a community forest management context" *International Journal of the Commons*, 11(1). ISSN 1875-0281.

Barnes, C., van Laerhoven, F. and Driessen, P.P. (2016) "Advocating for change? How a civil society-led coalition influences the implementation of the Forest Rights Act in India" *World Development* 84, 162–175.

Bauch, S.C., Sills, E.O. and Pattanayak, S.K. (2014) "Have we managed to integrate conservation and development? ICDP impacts in the Brazilian Amazon" *World Development* 64(S1), S135–S148.

Bebbington, A. (1999) "Capitals and capabilities: a framework for analysing peasant viability, rural livelihoods and poverty" *World Development* 27(12), 2012–2044.

Bebbington, A. (2005) "Donor–NGO relations and representations of livelihood in nongovernmental aid chains" *World Development* 33(6), 937–950.

Bebbington, A and Perreault, T. (1999) "Social capital, development, and access to resources in highland Ecuador" *Economic Geography* 75(4), 395–418.

Bebbington, A., Rojas, R., Hinojosa, L., Muñoz, D., Guerrero, E. and Martinez, E. (2002) *Contributions of the Dutch Co-Financing Program to Rural Development and Rural Livelihoods in the Highlands of Perú and Bolivia: Synthesis Report. Steering Committee for the Evaluation of the Netherlands Co-financing Programme.*

Boulder, CO: Ministry of Foreign Affairs, The Netherlands. www.research.manchester.ac.uk/portal/en/publications/contributions-of-the-dutch-cofinancing-program-to-rural-development-and-rural-livelihoods-in-the-highlands-of-peru-and-bolivia-synthesis-report(b25ad5c6-5ad9-4121-a45c-6dacb4384c21)/export.html, accessed 20 October 2018.
Berkes, F. (2007) "Community-based conservation in a globalized world" *Proceedings of the National Academy of Sciences* 104(39), 15188–15193.
Blaikie, P. (2006) "Is Small really beautiful? Community-based natural resource management in Malawi and Botswana" *World Development* 34(11), 1942–1957
Blaikie, P. and Springate-Baginski, O. (2007) "Understanding the policy process in participatory forest management". In: Springate-Baginski, O. and Blaikie, P. (eds.), *Forests, People And Power: The Political Ecology of Reform in South Asia.* London: Earthscan, 61–91.
Blomley, T. and Ramadhani, H. (2006) "Going to scale with Participatory Forest Management: early lessons from Tanzania" *International Forestry Review* 8(1), 93–100.
Brass, J. (2016) *Allies or Adversaries: NGOs and the State in Africa.* New York: Cambridge University Press.
Bray, D.B., Antonori, C. and Torres-Rojo, J.M. (2006) "The Mexican model of community forest management: the role of agrarian policy, forest policy and entrepreneurial organization" *Forest Policy and Economics* 8, 470–484.
Carlsson, L. and Berkes, F. (2005) "Co-management: concepts and methodological implications" *Journal of environmental management* 75(1), 65–76.
Chambers, R. and Conway, G. (1992) *Sustainable Rural Livelihoods: Practical Concepts for the 21st Century.* Brighton, UK: Institute of Development Studies. IDS Discussion Paper 296.
Chhotray, V. (2007) "Political entrepreneurs or development actors? An NGO's story of resistance and acquiescence". In: Bebbington, A., Hickey, S., Mitlin, D. (eds.), *Can NGOs Make a Difference? The Challenge of Development Alternatives.* London: Zed Books.
Cleaver, F. (2002) "Reinventing institutions: bricolage and the social embeddedness of natural resource management" *European Journal of Development Research* 14(2), 11–30.
Cleaver, F.D. and de Koning, J. (2015) "Furthering critical institutionalism" *International Journal of the Commons* 9(1), 1–18.
Cook, N.J., Wright, G.D. and Andersson, K.P. (2017) "Local politics of forest governance: why NGO support can reduce local government responsiveness" *World Development* 92, 203–214.
Cox, M., Arnold, G. and Villamayor Tomás, S. (2010) "A review of design principles for community-based natural resource management" *Ecology and Society* 15(4), article 38.
De Koning, J. (2011) "Reshaping institutions: bricolage processes in smallholder forestry in the Amazon", PhD Thesis. Wageningen, the Netherlands: Wageningen University.
Dietz, T., Ostrom, E. and Stern, P. (2003) "The struggle to govern the commons" *Science* 302, 1907–1912.
Donoghue, E.M., Cubbage, F.W. and Mercer, D.E. (2003) "Contract NGOs in community-based forest management in Philippines" *Journal of Sustainable Forestry* 17(4), 47–73.
Duthy, S. and Bolo-Duthy, B. (2003) "Empowering people's organizations in community based forest management in the Philippines: the community organizing role of NGOs" *Annals of Tropical Research* 25(2), 13–27.
Edwards, M. (1999) "NGO performance – what breeds success? New evidence from South Asia" *World Development* 27(2), 361–374.
Edwards, M. (2009) *Civil Society.* Cambridge, UK: Polity Press.
Ehrenberg, J.R. (1999) *Civil Society: The Critical History of an Idea.* New York: NYU Press.
Gardner, R., Ostrom, E. and Walker, J.M. (1990) "The nature of common-pool resource problems" *Rationality and Society* 2(3), 335–358.
Ghate, R. (2003) *Ensuring 'Collective Action' in 'Participatory' Forest Management* (Working Paper No. 3-03), Kathmandu, Nepal: South Asian Network for Development and Environmental Economics.
Gibson, C.C., Andersson, K., Ostrom, E. and Shivakumar, S. (2005a) *The Samaritan's Dilemma. The Political Economy of Development Aid.* Oxford, UK: Oxford University Press.
Gibson, C.C., Williams, J.T. and Ostrom, E. (2005b) "Local enforcement and better forest" *World Development* 33(2), 273–284.
Gutu, T., Wong, S. and Kinati, W. (2014) "Institutional bricolage as a new perspective to analyse institutions of communal irrigation: implications towards meeting the water needs of the poor communities in rural Ethiopia" *Pinnacle Agricultural Research & Management* 2(1), 1–14.
Haapala, J., Rautanen, S., White, P., Keskinen, M. and Varis, O. (2016) "Facilitating bricolage through more organic institutional designs? The case of water users' associations in rural Nepal" *International Journal of the Commons* 10(2), 1172–1201.

Hall, K., Cleaver, F., Franks, T. and Maganga, F. (2014) "Capturing critical institutionalism: a synthesis of key themes and debates" *European Journal of Development Research* 26, 71–86.
Hardin, G. (1968) "The tragedy of the commons" *Science* 162(3859), 1243–1248.
Hulme, D. (2000) *Protecting and Strengthening Social Capital in Order to Produce Desirable Development Outcomes. Social Development Systems for Coordinated Poverty Eradication Paper No. 4*. Manchester, UK: University of Manchester.
Imperial, M.T. and Yandle, T. (2005) "Taking institutions seriously: using the IAD framework to analyze fisheries policy" *Society and Natural Resources* 18(6), 493–509.
Ingram, V., Ros-Tonen, M. and Dietz, T. (2015) "A fine mess: bricolaged forest governance in Cameroon" *International Journal of the Commons* 9(1), 41–64.
Johnson, C. (2004) "Uncommon ground: the 'poverty of history' in common property discourse" *Development and Change* 35(3), 407–433.
Jones, S. (2015) "Bridging political economy analysis and critical institutionalism: an approach to help analyse institutional change for rural water services" *International Journal of the Commons* 9(1), 65–86.
Krinsky, J. and Crossley, N. (2014) "Social movements and social networks: introduction" *Social Movement Studies: Journal of Social, Cultural and Political Protest* 13(1), 1–21.
Lange, P., Driessen, P.P.J., Sauer, A., Bornemann, B. and Burger, P. (2013) "Governing towards sustainability – conceptualizing modes of governance" *Journal of Environmental Policy & Planning* 15(3), 403–425.
Larson, A.M. and Ribot, J.C. (2004) "Democratic decentralization through a natural resource lens: an introduction" *European Journal of Development Research* 16(1), 1–25.
Lemos, M.C. and Agrawal, A. (2006) "Environmental governance" *Annual Review of Environment and Resources* 31, 297–325.
Lescuyer, G. (2013) "Sustainable forest management at local scale: a comparative analysis of community forests and domestic forests in Cameroon" *Small Scale Forestry* 12(1), 51–66.
Lund, C. (2006) "Twilight institutions: public authority and local politics in Africa" *Development and Change* 37(4), 685–705.
Mansuri, G. and Rao, V. (2013) "Can participation be induced? Some evidence from developing countries" *Critical Review of International Social and Political Philosophy* 16(2), 284–304.
Mosse, D. (2006) "Collective action, common property, and social capital in South India: an anthropological commentary" *Economic Development and Cultural Change* 54(3), 695–724.
Ostrom, E. (1990) *Governing the Commons: The Evolution of Institutions for Collective Action*. Cambridge, UK: Cambridge University Press.
Ostrom, E. (2005) *Understanding Institutional Diversity*. Princeton, NJ: Princeton University Press.
Ostrom, E. and Cox, M. (2010) "Moving beyond panaceas: a multi-tiered diagnostic approach for social-ecological analysis" *Environmental Conservation* 37(4), 451–463.
Pagdee, A., Kin, Y. and Daugherty, P.J. (2006) "What makes community forest management successful: a meta-study from community forests throughout the world" *Society & Natural Resources* 19(1), 33–52.
Persha, L., Agrawal, A. and Chhatre, A. (2011) "Social and ecological synergy: local rulemaking, forest livelihoods, and biodiversity conservation" *Science* 331(6024), 1606–1608.
Poteete, A. and Ostrom, E. (2004) "Heterogeneity, group size and collective action: the role of institutions in forest management" *Development and Change* 35(3), 435–461.
Pretty, J. and Ward, H. (2001) "Social capital and the environment" *World Development* 29(2), 209–227.
Ribot, J., Agrawal, A. and Larson, A.M. (2006) "Recentralising while decentralizing: how national governments reappropriate forest resources" *World Development* 34(11), 1864–1886.
Saravanan, V.S. (2015) "Actors of institutional change: the contribution of new institutionalism in understanding water governance in India" *Environmental Science & Policy* 53, 225–235.
Schlager, E. and Ostrom, E. (1992) "Property rights regimes and natural resources: a conceptual analysis" *Land Economics* 68(3), 249–262.
Scoones, I. (1998) *Sustainable Rural Livelihoods: A Framework for Analysis* (Working Paper 72). Brighton, UK: Institute for Development Studies.
Scoones, I. (2009) "Livelihoods perspectives and rural development" *The Journal of Peasant Studies* 36(1), 171–196.
Sekher, M. (2001) "Organized participatory resource management: insights from community forestry practices in India" *Forest Policy and Economics* 3(3–4), 137–154.
Sen, A. (1981). *Poverty and Famines: An Essay on Entitlement and Deprivation*. Oxford, UK: Oxford University Press.

Shukla, A. and Sinclair, A. (2010) "Strategies for self-organization: learning from a village-level community-based conservation initiative in India" *Human Ecology* 38(2), 205–215.
Shyamsundar, P. and Ghate, R. (2011) *Rights, Responsibilities and Resources: Examining Community Forestry in South Asia* (Working Paper No. 59–11). Kathmandu, Nepal: South Asian Network for Development and Environmental Economics.
Somanathan, E., Prahbakar, R. and Mehta, B.S. (2009) "Decentralization for cost-effective conservation" *Proceedings of the National Academy of Sciences of the United States of America* 106(11), 4143–4147.
Thin, N. and Van Gardingen, P. (2004) "Participatory forestry: sharable lessons for better management of commons" *Forests, Trees and Livelihoods* 14(2–4), 229–242.
Thomas, B.K., Muradian, R., De Groot, G. and De Ruijter, A. (2010). "Confronting or complementing? A case study on NGO-state relations from Kerala, India" *Voluntas: International Journal of Voluntary and Nonprofit Organizations* 21(3), 358–370.
Tilly, C. (2005) *Identities, Boundaries, and Social Ties*. Boulder, CO: Paradigm Publishers.
Van Laerhoven, F. and Barnes, C. (2014) "Communities and the commons: the role of community development support in sustaining the commons" *Community Development Journal* 49(S1), i118–i132.
Van Laerhoven, F. and Berge, E. (2011) "The 20th anniversary of Elinor Ostrom's Governing the Commons" *International Journal of the Commons* 5(1), 1–8.
Van Laerhoven, F. and Ostrom, E., (2007) "Traditions and trends in the study of the commons" *International Journal of the Commons* 1(1), 3–28.
Vedeld, T. (2000) "Village politics: heterogeneity, leadership and collective action" *The Journal of Development Studies* 36(5), 105–134.
Wade, R. (1988) *Village Republics: Economic Conditions for Collective Action in South India*. Oakland, CA: ICS Press.
Westermann, O., Ashby, J. and Pretty, J. (2005) "Gender and social capital: the importance of gender differences for the maturity and effectiveness of natural resource management groups" *World Development* 33(11), 1783–1799.
Wright, G. and Andersson, K. (2013) "Non-Governmental Organizations, rural communities and forests: a comparative analysis of community-NGO Interactions" *Small-Scale Forestry* 12, 33–50.
Yaziji, M. and Doh, J. (2009) *NGOs and Corporations: Conflict and Collaboration*. New York: Cambridge University Press.

28

COMMONS, INDIGENOUS RIGHTS, AND GOVERNANCE

Iliana Monterroso, Peter Cronkleton and Anne M. Larson

Introduction

Scholarly work on the commons has illustrated the important role that common pool resource systems play in sustaining the livelihoods of forest dependent communities; in particular, vulnerable groups such as indigenous peoples and poor women (Beck and Ghosh, 2000; Beck and Nesmith, 2001; Chhatre and Agrawal, 2008, 2009). A recent assessment estimated that globally over 8.5 billion hectares of land can be categorized as common pool resources, either formally recognized or informally held (Wily, 2011: 11). The commons concept has been widely used in reference to common pool resources under collective property where exclusion is difficult and competition is prevalent (Ostrom and Hess, 2007).

While traditional peoples around the world have made claim to commons, this chapter focuses on Latin American cases because the extent of formal recognition in this region has been particularly significant and diverse. Globally, communities and indigenous peoples are estimated to hold as much as 65% of the available land under customary tenure but only 18% of this land has been formally recognized either as owned or designated for the use for indigenous peoples (RRI, 2015[1]). In sub-Saharan Africa, as much as 70% of the land can be categorized as customary common property;[2] nonetheless, only 3% of this land is formally recognized in law (Wily, 2011). In contrast, Latin America has witnessed widespread policy reform for many years now (Roldan, 2004), which has changed regulations that redefined tenure rights to increase access to indigenous and traditional commons that had previously been claimed by governments. Although the modes of transfer and the types of rights granted vary between and within countries, this devolution process has taken place mostly in large forested regions, drawing from a variety of institutional types: from collective land and territorial titling to co-management schemes and concession contracts that recognize lesser or temporary rights (Sunderlin et al., 2008; RRI and ITTO 2009; Larson et al., 2010; RRI, 2012, 2014).

Recent data analyzing changes in statutory regimes around the world show that Latin America is the developing region where the most significant changes appear to have occurred, at least on paper, in the transfer or recognition of rights designed for the use and management of communities. In Latin America, 59% of forest lands are owned and administered by the state, compared to approximately 92% in Africa and 62% in Asia (RRI, 2016). In total, as of 2013, nearly 33% of Latin American forests were under some type of collective tenure regime owned by communities, most of which are of indigenous peoples, and another 8% of the area had been designated for their use (RRI, 2016). This amounts to 232 million ha of forests in the hands of these groups – more than 85 million of which were recognized since 2002 – representing a major shift in tenure rights for local communities (RRI, 2015). Research indicates that common areas in the hands of indigenous groups and other forest dependent communities in this region represent over 60% of the global increase of forest under collective ownership or control during the period 2002–2013 (White and Martin, 2002; RRI, 2016). Considering available data, while the scope of indigenous rights struggles is much larger, this chapter focuses on the claims of Latin America's indigenous and traditional people to land and forests, from the perspective of the commons.

Our analysis of indigenous claims refers to territorial demands, which means that the "territory" includes the resource system that covers both the flow of resource units and the benefits produced at the broader scale. It refers to timber and non-timber forest products but also other ecosystem goods and services generated at the territorial level, including intangible ones such as carbon sequestration and traditional knowledge. The local, tangible resources alone are key for sustaining the livelihoods of over 350 million people across the globe, belonging to 5000 different ethno-linguistic groups across 90 countries that are internationally recognized as indigenous (Feiring, 2013). Commons have played a key role in sustaining indigenous knowledge, culture and traditional practices (Arifin, 1995; Berkes et al., 2000; Toledo et al., 2003). They are a central part of coping and adaptive strategies contributing to livelihoods and playing an important role in poverty reduction (Beck and Nesmith, 2001).

This chapter analyzes the evolution of indigenous claims to common pool resources and the recognition of indigenous rights to resource commons. The chapter is organized in two sections. The first section uses commons theory to analyze the evolution of indigenous claims to commons, drawing on cases from Latin American to illustrate the process of formalization of indigenous rights to land and forest resources. The following demonstrates and analyzes the challenges of establishing effective governance of indigenous commons. While governance challenges faced by indigenous commons are manifold, we focus on five that are directly linked to formalization processes as mechanisms to recognize indigenous rights: (1) reconciling the extent of rights granted with existing claims over resources; (2) addressing legal overlaps; (3) the difficulty of realizing benefits from acquired rights; (4) the need to build equitable governance institutions; and (5) the assumption that management and governance institutions in the indigenous commons should be based on collective action.

Indigenous peoples and commons

In Latin America, recognition of indigenous commons is the result of social movements in which indigenous peoples struggle for land and resources and pressure governments to recognize their rights, a process that has had significant influence on environmental governance (Roldan, 2004). In fact, in several Latin American countries, a large proportion of national territory has been formally recognized as either designated for or owned by indigenous groups and local communities, including Mexico (52%), Peru (35%), Colombia (34%), Brazil (23%) and Nicaragua (28%) (RRI, 2015; IBC, 2016; Larson et al. 2016; Herrera, 2017). Some of these reforms began many years ago. The recognition of indigenous rights to land in Mexico dates back to 1910 as part of the agrarian reform process (Bray and Merino, 2002; Bray, 2013). Panama established the first *comarca* in 1938 and six more were recognized following significant mobilization from indigenous groups during the 1950s (Velasquez Runk et al., 2011; Velasquez Runk, 2012).

In Peru, the recognition of native communities started in the 1970s as part of the reforms promoted by a revolutionary government and after significant mobilization following the formation of the Interethnic Association for the Development of the Peruvian Amazon (AIDESEP) (AIDESEP 2013; Monterroso et al., 2017). In the case of Colombia, after endorsing International Labor Organization Convention on Indigenous and Tribal Peoples 169 (ILO Convention 169) in 1991, the country enacted a new Constitution that recognized rights benefiting both indigenous and Afro-descendant communities, allowing a series of reforms that further advanced and consolidated the recognition of their rights (van de Sandt, 2003; Velasquez Ruíz, 2017; Guerrero et al., 2017). In Bolivia, indigenous marches in the 1990s were instrumental in pressuring the government to formally recognize millions of hectares of indigenous land claims as TCOs (*Tierras Comunitarias de Origen*) and to create mechanisms for recognizing additional indigenous claims (Lehm, 1999). Brazil accounts for most of this change in forest tenure distribution through lands designated for and owned by communities and indigenous peoples (RRI and ISA, 2015).

What are indigenous commons?

Since Hardin's publication of "The tragedy of the commons" in the 1970s, increased scholarly work on the commons, common pool resources and common property resources dramatically shifted debate, particularly in the 1990s (Van Laerhoven and Ostrom, 2007). Analysis of more than 2000 papers archived between 1990 and 2011 from global conferences of the International Study of the Commons found that more than 50% of the material corresponds to papers focusing on Asia (35%) and Africa (22%) (Robson and Lichtenstein, 2013). In contrast, in the case of Latin America, archived papers account for less than 13%. Nonetheless, over 60% of these papers focus on land tenure and property rights analyzing forests and territorial-based commons (ibid.). In the region, academic work using a commons theory perspective to examine indigenous peoples' commons has increased over the last twenty years. Existing works that analyze indigenous rights to lands and forests (Bray, 2013; Lauriola, 2013; Monterroso and Larson, 2013; Dell Angelo et al., 2017) and territorial claims (Bray et al., 2012; Larson and Soto, 2012; Larson et al., 2015) are useful to frame this work.

At the conceptual level, to analyze the evolution of indigenous claims over commons, consideration of recent work around social movements and "politic commons"[3] is of particular interest. Social movements across multiple regions have embraced the concept of commons to reclaim access to and control over resources, as well as to demand the protection and allocation of common goods against enclosures (Mattei, 2007, 2016; Bailey and Mattei, 2012; Bollier and Helfrich, 2012). Scholars in the commoning debates and the entitlement school are central in the analysis of the historical struggles that determine resource access and entitlement to indigenous peoples in the Americas (Johnson, 2004; Linebaugh, 2008). The commons concept provides an alternative paradigm to the mainstream property school, which supports the formalization of individual rights (De Soto, 2006), with the argument that common property promotes arrangements that allow for more equitable access and distribution of resources. The constitution of commons is possible amidst struggles of particular groups to defend, use and have access to resources (Bailey and Mattei, 2012). Therefore, commons cannot be considered public or private domains, their nature is rather determined by collective use. The use of commons here follows an "ecological-qualitative category based on inclusion and access" (Mattei, 2011: 11), where resources "belong to the people as a matter of necessity, claim free access, and radically oppose both the state and private property as shaped by market forces" (2011: 3).

We argue that the term indigenous commons refers to two different things. First, those resources, including the flow and benefits derived from them, claimed or held in possession by indigenous peoples, usually claimed at the level of territory. Second, the set of rules, including customary or traditional institutional arrangements that promote collective action and define resource access, use and control.[4] Customary systems contain the norms, customs, traditions and practices of a given community inherited from ancestors or developed locally and accepted, reinterpreted and enforced by indigenous groups to govern local resource access and use (Fitzpatrick, 2005; Wily, 2011). The state may or may not be recognized by these systems. Key in this analysis is the differentiation between the terms "rights" and "rules." When referring to the use of resources, "rights are the product of rules and thus not equivalent of rules" (Schlager and Ostrom, 1992: 250). Rights here are then defined as institutional arrangements that determine who is the rights holder, the scope of rights and the types of responsibilities and benefits one may obtain from them (Agrawal and Ostrom, 2001). Indigenous rights to land and forests are then the result of rules, norms and conventions that authorize the exercise of these rights (Schlager and Ostrom, 1992). In this sense, institutions defining – and legitimizing – indigenous rights to commons exist at different levels. For instance, the nation-state can establish legal regulations while communities have local rules concerning who gets to use which resources and how, which may or may not be written into bylaws of community or territory governments.

These two levels are sometimes in conflict and sometimes in agreement. The provisions defined in national law and international regulations are the basis for the formalization of indigenous rights (RRI, 2012; Larson et al., 2016). Customary rules existing at the territorial, communal and local level are the basis of territorial management regimes that govern resources at the local level, but they may or may not be recognized in formal regulations. When rights to indigenous commons are formalized as collective lands, these are referred to as communal lands systems or regimes (Wily, 2011: 4). Ostrom and Hess argued that common pool resources may be owned and managed by governments at the national, sub-national and local level, or by a community, private individuals or companies (2008: 9)[5]. We argue that ownership should not be a pre-condition to claim common resources as indigenous commons. In the following sections, we analyze the status of indigenous claims to land and forests irrespective of the actual ownership status.

The evolution of indigenous claims to commons

Historically, local communities and indigenous peoples under customary tenure regimes governed significant portions of land in rural areas. However, over time, with colonization and nationalization of natural resources, these areas were declared state lands under statutory law (Scott, 1998).[6] International legal frameworks for indigenous rights brought a shift in this process, as they established the obligation of nation-states to demarcate indigenous areas, and distinct regulations were enacted across different countries to formalize preexisting indigenous rights to respond to claims over lands and forests.

The current understanding of who is considered indigenous in international law evolved from the discussions around decolonization following the creation of the United Nations (Clech, 2004; Kwokwo Barume, 2014). This process was formalized by the adoption of ILO Convention 169 in 1989[7] and the United Nations Declaration of the Rights of Indigenous Peoples (UNDRIP) in 2007. The definition of indigenous peoples in these international conventions incorporates the anthropological, ethnographic and cultural concept defining indigeneity as a human rights construct. According to ILO Convention 169, the status of indigenous peoples is regulated wholly or partially by custom and/or tradition or by special laws or regulations that seek to protect cultural identities and cultural existence as community (ILO Convention, 1989).

The global indigenous movement has evolved rapidly since the 1960s, emerging initially from mobilization around civil rights (Kwokwo Barume, 2014). In the 1980s, the alliance with other civil society social movements, such as the alliance between rubber-tappers (*siringueiros*) and indigenous peoples in Brazil (Schmink and Wood 1992; Lauriola, 2013), the abandonment of assimilation policies and the rise of multicultural and identity policies promoted the adoption of ILO Convention 169 (Stavenhagen, 1992; Hale, 1997; Yashar, 2005; Charter and Stavenhagen, 2009). Mobilization centered on the recognition of ethnic identity and political autonomy and pushed for the recognition of land and resource rights claims, enabling the institutionalization of local customary regimes and indigenous community practices.

The 2007 UNDRIP is the outcome of many years of advocacy by indigenous leaders and civil society movements (Charter and Stavenhagen, 2009). It addresses a full range of civil, political, economic, social, cultural and environmental rights. Existing international legal frameworks include provisions that refer to the collective or group nature of the right: "indigenous peoples possess collective rights which are indispensable for their existence, well-being and integral development as peoples" (UN, 2007: 4). In this context, according to Feiring, indigenous rights include a "well-defined set of individual and collective rights, including to lands, territories, and resources" (2013: 14). In fact, the protection of these rights is recognized for indigenous both as groups as well as individuals (Clinton, 1990).

There are two significant elements for understanding the recognition process for indigenous rights to commons. First, the collective right to land forms an essential part of indigenous peoples' identity and ensures their cultural reproduction and political autonomy (ILO Convention 169, Part II, Arts. 13–19). The concept of territory, rather than land, is often used in claims as well as in some regulations, requiring a closer discussion of the concept of territory as a socially constructed space and how cultural and social elements are embedded in collective practices around land and forests (Davis and Wali, 1994; Hvalkof, 2002; Larson and Soto, 2012). In Latin America, formal recognition of indigenous territories in Bolivia, Nicaragua and Honduras are examples of how these international conventions have been mainstreamed in national regulations (Larson and Lewis-Mendoza, 2012; Larson et al., 2015, 2016). The use of territory as the legal form to recognize collective rights to land was adopted first in Amazonian countries like Brazil (1988), while formalization of indigenous territories in Bolivia and Ecuador was influenced by the adoption of the UNDRIP (A/HRC/EMRIP/2017/CRP.2). Other countries have chosen alternative legal mechanisms that allow the recognition of collective rights to land – either as collective lands in Panama, indigenous communal lands in Peru or indigenous reserves (*resguardos*) in Colombia (Monterroso et al., 2017; Velasquez Ruíz, 2017).

The second important element for understanding indigenous recognition of rights to commons is the coexistence of expansive forest lands and the areas that indigenous claim as their territories (Chapin et al., 2005), indicating existing social arrangements at the local level around the use and management of these natural resources.

Several provisions of the UNDRIP are relevant for understanding the evolution of territorial claims. The first is the right to self-determination (Art. 3), which is directly linked to autonomy and self-government (Art. 4). In this regard, the UNDRIP recognizes that indigenous peoples have the right "to own, develop, control and use the lands and territories. [In addition,] States shall give legal recognition and protection to these lands, territories and resources" (Arts. 26 and 27). UNDRIP also establishes that such recognition should be conducted respecting customs and existing land tenure systems. In ILO Convention 169, territory is defined as lands that indigenous people "traditionally occupy" and over which they should be granted "rights of property and possession" (Part II, Arts. 13 and 14).

While formal recognition processes are recent, these international instruments recognized that historically, indigenous peoples have been disenfranchised and have suffered loss of control over their lands and resources. It is recognized that the survival of indigenous peoples is tied to their identity, livelihood practices and their knowledge of and dependence on their natural environments

(Berkes and Folke, 1998). Therefore, the protection of ancestral territories is considered the basis of their ethnic identity, and their continued existence is linked to the protection of their right to continue to occupy and control their traditional territories and natural resources. This represents the basis for their "claim of territoriality" (Clech, 2004: 131), directly linked to their right to exercise self-determination, a fundamental criterion recognized in ILO Convention 169.

Remaining challenges to secure and support effective governance of indigenous commons

Although Latin American countries have made significant advances in recognizing indigenous tenure rights, implementation of reforms in the region have revealed gaps in progress, and many indigenous groups are still awaiting legal recognition. For example, in Brazil, it is estimated that more than 100 million hectares reserved as indigenous lands are still pending formalization (ISA, 2018). In Peru, the national indigenous organization AIDESEP argues that 20 million hectares claimed by indigenous people remain unrecognized by the state (AIDESEP, 2013). In Colombia, a recent analysis has found that 271 Afro-Colombian councils are pending recognition and formalization of their collective lands (Guerrero et al., 2017). Other important drivers of conflict endangering the livelihoods and cultures of indigenous peoples are the increasing pressures on contested lands targeted for development and/or extractive investment by external interests (Velasquez Runk, 2012). These ongoing threats to customary lands also affect the commons worldwide (Wily, 2011; Dell Angelo et al., 2017). Nonetheless, indigenous movements and their allies continue advocating for the formalization of indigenous rights to land and resources as the most important mechanism to secure indigenous rights and livelihoods, arguing that "when local communities and Indigenous Peoples lack formal, legal recognition of their land rights, they are vulnerable to dispossession and loss of their identities, livelihoods, and cultures" (RRI, 2015). Recognition of collective tenure rights is thus linked to tenure security and livelihood outcomes (Seymour et al., 2014).

Examining the ongoing recognition of indigenous commons also demonstrates the governance challenges emerging from the formalization of these rights. While these challenges are manifold, we focus on those that are directly linked to formalization processes as mechanisms to recognize indigenous rights. The first three challenges are related to the extent to which formalization allows local communities to take advantage of acquired rights. This highlights the risk of promoting formalization as a generalized normative prescription that assumes that tenure security will result. The fourth challenge is related to the need to operationalize indigenous claims as territory with clear boundaries and authorities. These characteristics are defined in contested landscapes, where ongoing resource conflicts frame the configuration of territorial practices and institutions, requiring improved understanding of factors that promote tenure security. A final challenge is that many legal frameworks defining rights to these indigenous commons assume that management is based on collective action when, in fact, the domestic production system that characterizes the livelihoods of residents using the commons is frequently individualized (Ankersen and Barnes, 2004; Cronkleton and Larson, 2015).

Reconciling the scope of rights granted with existing claims over resources

A significant challenge emerges during the design of reforms that shape tenure regimes and define the extent of formalized rights. A useful way to conceptualize the set of rights recognized during the formalization processes is the bundle of rights approach developed by Schlager and Ostrom (1992). The bundle of rights determines "who is allowed to use which resources, in what way, for how long, and under what conditions, as well as who is entitled to transfer rights to others and how" (Schlager and Ostrom, 1992; Agrawal and Ostrom, 2001). These bundles consist of user rights (access, use) and decision-making rights (management, exclusion and alienation). Use rights may also include the right to earn an income from resources (Andersson, 2012). Additionally, management can be broken down further into the power to define rules around how resources may be used, the ability to define and allocate control rights, rule compliance and monitoring, and the power to settle conflicts (Agrawal and Ostrom 2001; Sikor et al., 2017). Using this approach, RRI (2012) showed that, while reforms that recognize collective tenure rights are advancing across developing regions, regulatory frameworks often limit the scope of the rights granted during formalization, and may lack clarity or specific provisions to be able to exercise them in practice.

In comparison to other developing regions, Latin America has the broadest and most complex set of tenure systems that recognize rights of indigenous peoples to land and forests (RRI, 2012). Nonetheless, although the variety of tenure regimes across the region recognize withdrawal and management rights, restrictions on the use of forest resources often include additional cumbersome requirements to prove eligibility for these rights, which prevents communities from asserting them in practice (Larson and Pulhin, 2012; Notess et al., 2018). These often require compliance with management regulation to obtain harvest permits and/or approve plans. Additionally, indigenous communities may be required to provide evidence of traditional resource use to be able to exercise rights or may be required to comply with additional onerous requirements to benefit economically from these resources.

Moreover, beyond the scope of the rights granted during formalization, another problem is the extent to which the formalized rights pertain to different resources in the area claimed as indigenous territories. Frequently, regulations recognize rights to overlapping claims to different resources in the same area. For example, national regulations formalize rights to lands but are not clear regarding other resources in the territories, such as water, forest, petroleum, gas or the benefits these generate. Analysis of existing legal frameworks developed to recognize indigenous rights to lands and forests indicates that national regulations often fail to clearly define the rights over and use of natural resources in indigenous territories or provide specific provisions to ensure decision making over these resources (Roldan, 2004; RRI, 2012).

A recent study of emerging challenges from formalization found that while reforms recognized rights to land, often the ability to use or derive benefits from natural resources like water, forests, or practices such as carbon sequestration and biodiversity conservation, are governed under different legal regimes (Notess et al., 2018). This has happened in Peru: although regulations allow for the collective titling of lands classified as apt for agriculture and pasture, those lands classified as forests are still considered within the public domain. Only usufruct rights are recognized for indigenous peoples, which requires additional administrative procedures to manage timber resources for commercial benefit (Baldovino, 2016; Larson et al., 2016). A similar situation happened in Mexican *ejidos*, where until the 1990s *ejidos* had recognized rights to land but not to manage timber products (Bray, 2013; Hodgon et al., 2013). In Nicaragua, the landmark case of Awas Tingni illustrated the need to reform existing regulations to formalize indigenous territories in the Caribbean Coast, as government decisions to grant timber concessions to companies were undermining the ability of indigenous groups to govern their territories (Larson et al., 2016).

Addressing legal overlaps

As regulatory frameworks recognize rights to the same land for different stakeholders, additional types of legal overlaps affect indigenous territories. One type of overlap emerges in areas that have been categorized as both conservation areas and indigenous territories. For instance, according to recent data, this type of overlap exists in as much as 45% of the Amazon (RAISG, 2016). This includes over 36 million hectares in areas that have already been formalized and areas that have not been formalized (19 million hectares following data by RAISG, 2016: 28–29). Another type emerges when indigenous territories overlap with rights granted to third parties, mining interests, petroleum companies or other major infrastructure initiatives. A recent map produced to analyze ongoing pressures in the Amazon shows that there are 33 million hectares with overlaps between areas with mining potential or active mining concessions and indigenous territories; the map also found overlaps with potential or active petroleum concessions that affected 20 million hectares (RAISG, 2015). The lack of attention to these types of problems makes it difficult to protect borders from new claims. This includes ongoing government actions to recognize rights of third parties, such as concessions for subsoil resources or permits to extract commercial timber granted in areas that have been formalized as indigenous territories. While free, prior and informed consent mechanisms have been promoted to resolve these conflicts, they are often viewed as voluntary, are not legally binding, or lack proper grievance mechanisms (Notess et al., 2018).

Acknowledging these overlaps is essential for addressing territorial claims, as demands for land ownership are linked to the guarantee not only of subsistence needs but also of cultural continuity (Berkes and Folke, 1998; Roldan, 2004). Moreover, the compartmentalization of rights is contradictory to the notion of territorial rights inherent to the claims of indigenous peoples. It differentiates the set of rights, such as extraction of subsoil rights from investment, or extraction initiatives from the ecological functions and benefits these resource systems provide. Although governments claim ownership over water bodies and subterranean oils and mineral resources, indigenous peoples by custom consider them an integral part of their territories. This places the commons under dual, contradictory, and overlapping tenure resulting in what Bollier and Helfrich defined as the "tragedy of the anti-commons dynamic in which excessive, fragmented property rights impede innovation and cooperation," hindering the ability of indigenous groups to be able to exercise the rights gained (2012: 1). Moreover, overlaps may bring out differing views around resource management, sometimes exacerbating conflict.

Gaining tangible benefits from formalization

There are different factors that affect the ability of indigenous groups to gain tangible benefits from the recognized rights; two of those are related to how formalization processes are designed and implemented. First, even though countries may enact regulations that recognize rights to commons, these do not guarantee actual implementation. Problems associated with implementation include failure to develop necessary provisions to operationalize the rights guaranteed; time consuming, overly complex or poorly conceived procedures; and low budgets (Roldan, 2004: 15–16; Notess et al., 2018). Other factors include the failure to provide technical assistance or support to communities, the lack of specialized skills by government officials, such as the capacity to address inter-cultural tensions or conflict management, the lack of clear mandates among the government institutions involved, and the failure to prioritize formalization of indigenous rights in political agendas (Zamora and Monterroso, 2017). Some of these factors also explain the slow pace of formalization.

Second, while legal reforms provide the basis for accessing the commons and should offer an essential foundation for livelihoods development, too often regulatory frameworks do not provide the necessary conditions for indigenous people to exercise these newly acquired rights. Rights granted may be conditional and non-permanent, imposing further obligations and limitations on indigenous rights to the commons (RRI, 2012). This is particularly the case with forests, where rights are granted conditionally based on demonstrated compliance with forest management regulations. The resulting co-management regimes could potentially allow government oversight and support to local forest managers, but when poorly designed or implemented they impose costs and other burdens (Cronkleton et al., 2012). Even in cases where regulations exist, indigenous communities may lack the technical capacities or financial resources to invest in commons (Zamora and Monterroso, 2017). In some cases, it appears that governments feel "their work is done" once they provide the title, failing to provide technical assistance or support of any kind to the titled communities (Larson, 2017).

Sustained investment in commons is also needed to ensure that formalization processes deliver socio-economic and environmental returns (Gnych et al., 2018). Community forest enterprises in Mexico are a good example. Common property arrangements allowed the emergence of organizational structures that *ejidos* established to commercialize timber production, developing new sources of livelihoods around timber and non-timber forest products, with important results for income generation (Bray and Merino, 2002; Antinori and Bray, 2004). The government promoted reforms to legally recognize the rights of indigenous communities to manage and commercialize the forest products from their communal forests for their own benefit, financed forest management plans and provided forestry extension services (Antinori 2000).

Finally, even when governments grant formal rights to commons, national legal frameworks can leave indigenous rights contested, which relates to the first challenge discussed. Indigenous groups may lack the right to exclude some outsiders from their lands or have no recourse to compensation mechanisms for lost rights or resources. Granting unclear rights may undermine incentives to invest in long-term improvements, which also affects livelihoods options (Peña et al., 2017).

Operationalizing territories, boundaries and authority during formalization processes

While formalization of indigenous lands in Latin America has been promoted by national governments with the support of international donors, the analysis of formalization processes in Peru, Ecuador, Nicaragua and Colombia suggest that formalization is not a silver bullet for establishing functioning commons (Velez, 2011; Robinson et al., 2014; Cronkleton and Larson, 2015; Larson et al., 2015; Holland et al., 2017). The process towards constituting and strengthening collective territories marked the beginning of a new phase in the struggle for the political configuration of indigenous territories, the negotiation of boundaries and the constitution of authorities at different governance levels. Formalization does not resolve conflicts and internal disputes that put pressures on resource control and extraction of resources (Finley-Brook, 2016). Nor does it stop external pressures, such as land invasions or pressure from the private sector and the government for resource access, including the rollback of rights granted (Monterroso and Larson, 2013). Institutional arrangements are the result of constant negotiation between the state and indigenous groups, promoted by mobilization efforts not only to gain access to resources but also to increase local decision-making around forests (Lemos and Agrawal 2006; Larson, 2010). Once rights have been formalized, establishing the institutions and authority necessary to govern indigenous commons require clear, specific consensus on territorial boundaries, the nature of rights within them and agreement on authority (Larson et al., 2015).

Nonetheless, in some cases the establishment of fixed boundaries restricts fluid use patterns that nomadic indigenous groups have had for years. The intention to recognize only traditional use and occupation of land that is visible, permanent, and sedentary has overlooked the rights of many hunter-gatherers in the Amazon, who still remain isolated from society (IWGIA, 2015). With the exception of some attempts to create larger multi-community territories, which still do not fully address the problem, the essentialist intention also fails to recognize shared spaces, where more than one "community" accesses resources. In this process, the formalization of fixed rigid boundaries has also affected the adaptive capacity of indigenous groups by formalizing areas much smaller than the land's indigenous claim under customary use. In Peru, there are no clear criteria for guiding decisions about the extent of land that should be titled to indigenous communities. This compartmentalization of rights differs from the layers of what are broadly classifiable as access and use rights in customary regimes (Wily, 2011). It also contrasts with the socio-cultural aspects of territoriality, including the capacity to adapt to changing conditions of common pool resource availability, a characteristic of pastoralist groups in parts of Africa (Wily, 2011). Formalization should not endanger the adaptive nature of indigenous commons (Berkes, 2010).

Finally, as discussed by Larson et al. (2015), the recognition of indigenous commons is linked to the constitution of collective territories, an ongoing process of negotiation in which formalization of indigenous rights to land and forest commons can also be perceived as state territorialization practices.[8] In this context, the formalization process redefines the collective (as the right holder) but also requires further discussion of representation, legitimacy and authority (Stocks, 2005). Thus, the recognition and formalization of indigenous rights to commons is the beginning of a new phase of struggle in which the construction of territory is closely linked to the maintenance of the identity of indigenous commons.

Management and governance of indigenous commons

Another important challenge stems from the assumption that indigenous people will manage and govern common resources though collective action (in common property theory collective action is inherent to the analysis of commons). Norms and regulations are often designed under the assumption that the distribution of shared resources, rights and benefits is always equal, based on custom and kinship, independent of external institutions. As argued elsewhere, indigenous commons are varied, shaped by different political conditions at the national level but also by the characteristics of indigenous peoples' identities, beliefs and livelihoods (Wily, 2011). It is not always the case that all land within a community is considered to belong to all members of the community (ibid.). For instance, in their analysis of indigenous territories, Larson and Lewis-Mendoza (2012) showed that in the process of constituting indigenous territories in Nicaragua, varied arrangements were in place to distribute land and forest rights among individuals, families and even communities within territories. Often, while internally there is distribution of land rights for agriculture to individuals and families, this is not uniform and there are still important forest, wetlands or pasture areas that are managed as commons (Barry and Meinzen-Dick, 2009).

Nonetheless, the assumed prevalence of collective action has shaped regulations to manage resources in indigenous territories, in particular forests resources (Wily, 2011; Colfer and Capistrano, 2005). For instance, in Bolivia, indigenous rights to Brazil nut extraction are allocated formally on a collective basis, although in practice, management is usually done at the family level (Cronkleton et al., 2010). In Peru, indigenous communities need to process forest management plans, even though only a handful of families will participate in activities. This assumption, derived from the academic bias assuming that governance of common pool resources is based on collective institutions, has normative implications, as recognition of indigenous rights also has been tied to the maintenance of conservation outcomes (Larson and Dahal 2012; Blackman et al., 2017; Holland et al., 2017). Governance of indigenous commons are expected to deliver management outcomes and sustain cultural and ecological services. Nonetheless, these outcomes are based on the development of complex institutional arrangements, oftentimes intertwined between customary arrangements and efforts to formalize.

Conclusion

This chapter has analyzed the evolution of indigenous claims to common pool resources and the recognition of indigenous rights to resource commons. Indigenous commons refer to resources either claimed or recognized, as well as the set of institutional arrangements that define resource access, use and control. The process of rights devolution in Latin America has resulted in the formalization of indigenous rights over land and forests, the result of historical struggles driven by indigenous and social movements. This process changed national regulations formalizing rights that also were protected and recognized in international law leading to processes of formal recognition in many countries across the region.

Nonetheless, while Latin American countries have made significant advances in the recognition of indigenous tenure rights over land and forests, challenges remain to secure and support effective governance of indigenous commons. A first challenge will be to reconcile the rights granted to indigenous commons with the existing claims over resources. A second challenge will be to address legal overlaps in rights granted to different groups for the same area or resources. A third challenge, once rights are granted, will be to create enabling conditions so that recipients can actually exercise those rights and realize benefits from them. A fourth challenge is to build equitable governance institutions that will facilitate the administration and management and use of commons by groups receiving the devolved rights. The fifth challenge will be to overcome the assumption that management and governance institutions in indigenous commons are based solely on collective action. These challenges have normative implications shaping regulation and management of forests. To overcome these will require greater attention to dialogue, negotiation and conflict transformation and greater understanding of and support for local governance and management institutions, including technical assistance for livelihoods activities.

Notes

1 This study compares data across 64 countries constituting 82% of global land area. A full list of countries and results is found in RRI, 2015: 16–18.
2 This study refers to 21 countries in total; the list of countries is found in Wily (2011: 52).
3 Following the concept proposed by Bailey and Mattei where "commons" is understood as "political" and not only as the institutional design analysis as proposed in institutional economics (2012: 1).
4 The term institution in this text refers to socially defined relations that determine the way individuals will interact with respect to benefits (Vatn, 2005).
5 Common pool resources can also be considered open access without any limitations, which may also result in depletion and overuse (Ostrom and Hess, 2008).
6 See for instance Scott on the discussion on the illegibility of communal tenure (1998: 37–44).
7 It is important to take into account that the Convention 107 on "indigenous and tribal people in independent countries" predates this convention, 1957.
8 Territoriality strategies aim at affecting, influencing and controlling people and their relationships to space as a way to control access to things and relationships among people. Embedded in the concept of territoriality are the ways in which people use the land, organize themselves in space and give meaning to a place. Territoriality refers not only to the outcome of power, but also strategies by which power is exercised and legitimized. Territoriality can be defined in a number of ways, including the way property rights to land and resources are allocated (Sack, 2010).

References

Agrawal, A. and Ostrom, E. (2001) "Collective action, property rights, and decentralization in resource use in India and Nepal." *Politics and Society* 29(4): 485–514.
Andersson, K. (2012) *CIFOR'S Research On Forest Tenure And Rights*. Bogor, Indonesia: CIFOR.

Ankersen, T. and Barnes, G. (2004) "Inside the polygon: emerging community tenure systems and forest resource extraction." In Zarin, D. J., Alavalapati, J. R. R., Putz, F. E. and Schmink, M. (eds.), *Working Forests in the Neotropics: Conservation through Sustainable Management?* New York: Columbia University Press, 156–177.

Antinori, C. and Bray, D. B. (2004) "Community forestry enterprises as entrepreneurial firms: economic and institutional perspectives from Mexico." *World Development* 33(9): 1529–1543.

Antinori, C. M. (2000) Vertical Integration in Mexican Common Property Forests. Unpublished PhD Dissertation. Berkeley, CA: Department of Agricultural Economics, University of California.

Arifin (1995) *Indigenous Knowledge and Sustainable Commons. The Contested Commons: Conversations between Economists and Anthropologists.* https://dlc.dlib.indiana.edu/dlc/bitstream/handle/10535/316/Indigenous_Knowledge_and_Sustainable_Commons_The_Case_of_an_Indonesian_Subak.pdf?sequence=1&isAllowed=y (last accessed March 2018).

Asociación Interétnica de Desarrollo de la Selva Peruana (AIDESEP) (2013) *Territorialidad y Titularidad en la Amazonía Norte del Perú, Alto Amazonas y Datem del Marañón – Pueblos.* Lima: Achuar, Kukama, Shapra, Kandozi, Shiwilo, Aidesep/corpi.

Bailey, S. and Mattei, U. (2012) "Social movements as constituent power: the Italian struggle for the commons." Available at www.uninomade.org/wp/wp-content/uploads/2012/10/Social-Movements-as-Constituent-Power.pdf (Last accessed September 23, 2017).

Baldovino, S. (2016) *Una Primera Mirada: Situación Legal de la Tenencia de Tierras Rurales en Perú.* Lima: Sociedad Peruana de Derecho Ambiental.

Barry, D. and Meinzen-Dick, R. (2009) "The invisible map. Community tenure rights." In Hecht, S., Morrison, K. and Padoch, C. (eds.) *The Social Life of Forests: Past, Present, and Future of Woodland Resurgence.* Chicago, IL: University of Chicago Press.

Beck, T. and Ghosh, M. (2000) "Common property resources and the poor. Findings from West Bengal." *Economic and Political Weekly* 35(3): 147–153.

Beck, T. and Nesmith, C. (2001) "Building poor people's capacities: the case of common property resources in India and West Africa." *World Development* 29(1): 119–133.

Berkes, F. (2010) "Devolution of environment and resources governance: trends and future." *Environmental Conservation* 37(4): 489–500.

Berkes, F., Colding, J. and Folke, C. (2000) "Rediscovery of traditional ecological knowledge as adaptive management." *Ecological Applications* 10(5): 1251–1262.

Berkes, F. and Folke, C. (eds.) (1998) *Linking Social and Ecological Systems: Management Practices and Social Mechanisms for Building Resilience.* New York: Cambridge University Press.

Blackman, A., Corral, L., Santos Lima, E. and Asner, G. (2017) "Titling indigenous communities protects forests in the Peruvian Amazon." *Proceedings of the National Academy of Sciences of the United States of America* 114(16): 4123– 4128.

Bollier, D. and Helfrich, S. (eds.) (2012) *The Wealth of the Commons. A World Beyond Market and State.* Amherst, MA: Levellers Press.

Bray, D. (2013) "When the state supplies the commons: origins, changes and design of Mexico common property regime." *Journal of Latin American Geography* 12(1): 33–55.

Bray, D., Duran, E. and Molina, O. (2012) "Beyond harvest in the commons: multi-scale governance and turbulence in indigenous/community conserved areas in Oaxaca Mexico." *International Journal of the Commons* 6(2): 151–178.

Bray, D. and Merino, L. (2002) *The Rise of Community Forestry in Mexico: History, Concepts, and Lessons Learned from Twenty Years of Community Timber Production.* Project Report. México, DF: Ford Foundation, 133.

Chapin, M., Lamb, Z. and Threlkeld, B. (2005) "Mapping indigenous lands." *Annual Review of Anthropology* 34: 619–638.

Charters, C. and Stavenhagen, R. (eds.) (2009) *Making the Declaration Work: The United Nations Declaration on the Rights of Indigenous Peoples.* Copenhagen: International Work Group for Indigenous Affairs.

Chhatre, A. and Agrawal, A. (2008) "Forest commons and local enforcement." *Proceedings of the National Academy of Sciences of the United States of America* 105(36): 13286–13291.

Chhatre, A. and Agrawal, A. (2009) "Trade-offs and synergies between carbon storage and livelihood benefits from forest commons." *Proceedings of the National Academy of Sciences of the United States of America* 106(42): 17667–17670.

Clech, L. M. (2004) "Remembering the country of their birth: indigenous peoples and territoriality." *Journal of International Affairs* 57(2): 129–150.

Clinton, R. (1990) "The rights of indigenous peoples as collective group rights." *Arizona Law Review* 32(4): 739–748.
Colfer, C. J. P. and Capistrano, D. (eds.) (2005) *The Politics of Decentralization: Forests, Power and People.* London: CIFOR and Earthscan.
Cronkleton, P., Albornoz, M. A., Barnes, G., Evans, K. and de Jong, W. (2010) "Social geomatics: participatory forest mapping to mediate resource conflict in the Bolivian Amazon." *Human Ecology* 38(1): 65–76.
Cronkleton, P. and Larson, A. (2015) "Formalization and collective appropriation of space on forest frontiers: comparing communal and individual property systems in the Peruvian and Ecuadoran Amazon." *Society & Natural Resources: An International Journal* 28(5): 496–512.
Cronkleton, P., Pulhin, J. M. and Saigal, S. (2012) "Co-management in Community Forestry: How the Partial Devolution of Management Rights Creates Challenges for Forest Communities." *Conservation and Society* 10(2): 91–102.
Davis, S. and Wali, A. (1994) "Indigenous land tenure and tropical forest management in Latin America." *Ambio* 23(8): 485–490.
Dell Angelo, J., D'Odorico, P., Rullie, M. C. and Marchand, P. (2017) "The tragedy of the grabbed commons: coercion and dispossession in the global land rush." *World Development* 92: 1–12.
De Soto, H. (2006) "The challenge of connecting informal and formal property systems: some reflections based on the case of Tanzania." In De Soto, H. and Cheneval, F. (eds.) *Realizing Property Rights.* Zürich: Rüffer & Rub, 18–36.
Feiring, B. (2013) *Indigenous Peoples' Rights to Lands, Territories and Resources.* Rome: International Land Coalition.
Finley-Brook, M. (2016) "Territorial 'fix'? Tenure insecurity in titled indigenous territories." *Bulletin of Latin American Research* 35: 338–354.
Fitzpatrick, D. (2005) "Best practice: options for the legal recognition of customary tenure." *Development and Change* 36(3): 449–475.
Gnych, S., Lawry, S., Monterroso, I. and Adhikary, A. (2018) "Common benefits: is community tenure facilitating investment in the commons for inclusive and sustainable development?" Available at www.conftool.com/landandpoverty2018/index.php?page=browseSessions&print=yes&abstracts=show&doprint=yes&form_session=557&presentations=show (Last accessed February 2018).
Hale, C. (1997) "Cultural politics of identity in Latin America." *Annual Review of Anthropology* 26: 567–590.
Herrera, J. (2017) *CIFOR. La tenencia de tierras colectivas en Colombia: datos y tendencias.* CIFOR InfoBrief No. 203. Bogor, Indonesia: CIFOR.
Hodgon, B., Chapela, F. and Bray, D. (2013) *Mexican Community Forestry: Enterprises and Associations as a Response to Barriers.* Bangkok: The Center for People and Forests (RECOFTC); Washington, DC: Rights and Resources Initiative (RRI); New York: Rainforest Alliance.
Holland, M., Jones, K., Naughton, L., Freire, J., Morales, M. and Suárez, L. (2017) "Titling land to conserve the forests: the case of Cuyabeno Reserve in Ecuador." *Global Environmental Change* 44: 27–38.
Hvalkof, S. (2002) "Beyond indigenous land titling: democratizing civil society in the Peruvian Amazon." In Chase, J. (ed.) *The Spaces of Neoliberalism: Land, Place and Family in Latin America.* Bloomfield, CT: Kumarian Press.
Instituto del Bien Común (IBC) (2016) *Tierras comunales: más que preservar el pasado es asegurar el futuro. El estado de las comunidades indígenas en el Perú. Informe 2016.* Lima: Instituto del Bien Común.
Instituto Socioambiental (ISA) (2018) Situación jurídica de las Tierras Indígenas de Brasil en la actualidad. Brasil. Available at https://pib.socioambiental.org.br/es/c/0/1/2/situacao-juridica-das-tis-hoje (Last accessed March 15, 2018).
International Labour Organization (ILO), Indigenous and Tribal Peoples Convention, C169, 27 June 1989, C169, available at: www.refworld.org/docid/3ddb6d514.html (last accessed March 2018).
IWGIA (2015) *The Indigenous World 2015.* Copenhagen: The International Work Group for Indigenous Affairs.
Johnson, C. (2004) "Uncommon ground: The "poverty of history" in common property discourse." *Development and Change* 35(3): 407–433.
Kwokwo Barume, A. (2014) *Land Rights of Indigenous Peoples in Africa: With Special Focus on Central, Eastern and Southern Africa.* IWGIA Document 115. Copenhagen: IWGIA.

Larson, A. (2017) "Collective forest tenure reforms: Where do we go from here?" Available at https://forestsnews.cifor.org/49477/collective-forest-tenure-reforms-where-do-we-go-from-here?fnl=en (Last accessed February 2018).
Larson, A., Cronkleton, P. and Pulhin, J. (2015) "Formalizing indigenous commons: The role of 'authority' in the formation of territories in Nicaragua, Bolivia and the Philippines." *World Development* 70(June 2015): 228–238.
Larson, A. and Dahal, G. R. (2012) "Forest tenure reform: new resource rights for forest-based communities?" Introduction to special issue, *Conservation and Society* 10(2): 77–90.
Larson, A. and Lewis-Mendoza, J. (2012) "Decentralization and devolution in Nicaragua's North Atlantic autonomous region: natural resources and indigenous peoples rights." *International Journal of the Commons* 6(2): 179–199.
Larson, A. and Pulhin, J. (2012) "Enhancing forest tenure reforms through more responsive regulations." *Conservation and Society* 10(2): 103–113.
Larson, A. and Soto, F. (eds.) (2012) *Territorialidad y Gobernanza: tejiendo retos en los territorios indígenas de la RAAN.* Managua, Nicaragua: Nitlapan-UCA.
Larson, A. M. (2010). "Making the 'rules of the game': constituting territory and authority in Nicaragua's indigenous communities." *Land Use Policy* 27: 1143–1152.
Larson, A. M., Barry, D., Dahal, G. R. and Colfer, C. J. P. (eds.) (2010) *Forests for People: Community Rights and Forest Tenure Reform.* London: Earthscan.
Larson, A. M., Monterroso, I., Mwangi, E. and Banjade, M. R. (2016) "Community rights to forests in the tropics: Progress and retreat on tenure reforms." In Graziadei, M. and Smith, L. (eds.) *Comparative Property Law: Global Perspectives.* Cheltenham, UK: Edward Elgar Publishing, pp. 435–457.
Lauriola, V. (2013) "Indigenous lands, commons, juridical pluralism and sustainability in Brazil: lessons from the indigenous lands of Raposa Serra do Sol." *Journal of Latin American Geography* 12(1): 157–185.
Lehm, Z. (1999) *La búsqueda de la Loma Santa y la Marcha por el Territorio y la Dignidad.* Santa Cruz: APCOB-CIDDEBENI-OXFAM América.
Lemos, M. and Agrawal, A. (2006) "Environmental governance." *Annual Review of Environmental Resources* 31: 297–325.
Linebaugh, P. (2008) *The Magna Carta Manifesto: Liberties and Commons For All.* Berkeley, CA: University of California Press.
Mattei, U. (2007) *Institutionalizing the Commons: An Italian Primer. Part I. Political background. Indignados Italian Style.* Available at https://commonsblog.files.wordpress.com/2007/10/mattei-italian-commons-chapter-short.pdf (Last accessed September 23, 2017).
Mattei, U. (2011) "The state, the market and some preliminary questions about the commons." No 1–11, IUC Research Commons from International University College of Turin.
Mattei, U. (2016) "First thoughts for a phenomenology of the commons." In Bollier, D. and Helfrich, S. (eds.) *The Wealth of the Commons: A World Beyond Market & State.* Amherst, MA: Levellers Press.
Monterroso, I., Cronkleton, P., Pinedo, D. and Larson, A. M. (2017) "Reclaiming collective rights: land and forest tenure reforms in Peru (1960–2016)." Working Paper No. 224. Bogor, Indonesia: CIFOR.
Monterroso, I. and Larson, A. (2013) "The dynamic forest commons of Central America: new directions for research." *Journal of Latin American Geography* 12(1): 87–110.
Notess, L., Veit, P., Monterroso, I., Mancayo, A., Sulle, E., Larson, A., Quadvlieg, J. and Williams, A. (2018) *The Price of Security: Community Land Formalization Procedures.* Washington, DC: World Resources Institute. Forthcoming.
Ostrom, E. and Hess, C. (2007) Private and common property rights. Working Paper No. 07-25. Workshop in Political Theory and Policy Analysis, Indiana University. Indiana. Available at https://surface.syr.edu/sul/24/ (last accessed March 2018).
Ostrom, E. and Hess, C. (2008) *Private and Common Property Rights. Encyclopedia of Law and Economics.* Northampton, MA: Edward Elgar.
Peña, X., Velez, A., Cardenas, J., Perdomo, N. and Matajira, C. (2017) "Collective property leads to household investments: lessons from land titling in Afro-Colombian communities." *World Development* 97: 27–48.
Red Amazónica de Información Socioambiental Georeferenciada (RAISG) (2015) *Mapa de Presiones y Amenazas sobre las áreas protegidas y los territorios indígenas de la Amazonía 2015.* Available at www.amazoniasocioambiental.org (Last accessed February 2018).

Red Amazónica de Información Socioambiental Georeferenciada (RAISG) (2016) *Cartografía histórica de Areas Naturales Protegidas y Territorios Indígenas de la Amazonía*. Available at www.amazoniasocioambiental.org (Last accessed February 2018).

Rights and Resources Initiative (RRI) (2012) *What Rights? A Comparative Analysis of Developing Countries' National Legislation on Community and Indigenous Peoples' Forest Tenure Rights*. Washington, DC: Rights and Resources Initiative.

Rights and Resources Initiative (RRI) (2014) *What Future for Reform? Progress and slowdown in forest tenure reform since 2002*. Washington, DC: Rights and Resources Initiative.

Rights and Resources Initiative (RRI) (2015) *Who Owns the World's Land? A Global Baseline of Formally Recognized Indigenous Community Land Rights*. Washington, DC: Rights and Resources Initiative.

Rights and Resources Initiative (RRI) (2016) *Closing the Gap: Strategies and Scale Needed to Secure Rights and Save Forests*. Washington, DC: Rights and Resources Initiative.

Rights and Resources Initiative (RRI) and International Timber and Trade Organization (ITTO) (2009) *Tropical Forest Tenure Assessment: Trends, Challenges and Opportunities*. Prepared for the International Conference on Forest Tenure, Governance and Enterprise: New Opportunities for Central & West Africa May 25–29 2009. Yaoundé, Cameroon.

Rights and Resources Initiative (RRI) and ISA (2015) *Advances and Setbacks in Territorial Rights in Brazil*. Washington, DC: Rights and Resources Initiative, Instituto Socioambiental.

Robinson, B., Holland, M. and Naughton-Treves, L. (2014) "Does secure land tenure save forests? A meta-analysis of the relationships between land tenure and tropical deforestation." *Global Environmental Change* 29: 281–293.

Robson, J. and Lichtenstein, G. (2013) "Current trends in Latin American commons research." *Journal of Latin American Geography* 12(1): 5–31.

Roldan, R. (2004) Models for recognizing indigenous land rights in Latin America. The World Bank. Biodiversity Series Paper No. 99. Washington, DC.

Sack, R. (2010) "Human territoriality: a theory." *Annals of the Association of American Geographers* 73(1): 55–74.

Schlager, E. and Ostrom, E. (1992) "Property rights regimes and natural resources: a conceptual analysis." *Land Economics* 68(3): 249–262.

Schmink, M. and Wood, C. (1992) *Contested Frontiers in Amazonia*. New York: Columbia.

Scott, J. C. (1998) *Seeing Like a State: How Certain Schemes to Improve the Human Condition Have Failed*. The Yale ISPS Series. New Haven, CT: Yale University.

Seymour, F., La Vina, T. and Hite, K. (2014) *Evidence Linking Community: Level Tenure and Forest Condition – An Annotated Bibliography*. Climate and Land Use Alliance. Available at https://www.climateandlandusealliance.org/wp-content/uploads/2015/08/Community_level_tenure_and_forest_condition_bibliography.pdf (last accessed September 2017).

Sikor, T., He, J. and Lestrelin, G. (2017) "Property rights regimes and natural resources: a conceptual analysis revisited." *World Development* 93: 337–349.

Stavenhagen, R. (1992) "Los derechos de los indígenas: Algunos problemas conceptuales. Nueva Antropología." *Revista de Ciencias Sociales* 43: 83–99.

Stocks, A. (2005) "Too much for too few: problems of indigenous land rights in Latin America." *Annual Review of Anthropology* 34: 85–104.

Sunderlin, W., Hatcher, J. and Liddle, M. (2008) *From Exclusion to Ownership: Challenges and Opportunities in Advancing Forest Tenure Reform*. Washington, DC: Rights and Resources Initiative.

Toledo, V., Ortiz-Espejel, B., Cortés, L., Moguel, P. and Ordóñez, M. (2003) "The multiple use of tropical forests by indigenous peoples in Mexico: a case of adaptive management." *Conservation Ecology* 7(3): article 9.

United Nations (UN) (2007) *United Nations Declaration on the Rights of Indigenous Peoples*. Available at www.un.org/esa/socdev/unpfii/documents/DRIPS_en.pdf (Last accessed February 2018).

Van de Sandt, J. (2003) "Communal resource tenure and the quest for indigenous autonomy: on state law and ethnic reorganization in two Colombian resguardos." *Journal of Legal Pluralism* 48: 125–162.

Van Laerhoven, F. and Ostrom, E. (2007) "Traditions and trends in the study of the commons." *International Journal of the Commons* 1(1): 3–28.

Vatn, A. (2005) *Institutions and the Environment*. Cheltenham, UK: Edward Elgar Publishing.

Velasquez Ruíz, M. (2017) *La tenencia colectiva de la tierra en Colombia: Antecedentes y estado actual*. CIFOR InfoBrief. No. 199. Bogor, Indonesia: CIFOR.

Velasquez Runk, J., Martinez Mauri, M., Quintero, B. and Sarsaneda del Cid, J. (2011) *Pueblos indígenas en Panamá: una bilbiografía*. Panamá: Acción Cultural Ngobe (ACUN).

Velasquez Runk, J. (2012) "Indigenous land and environmental conflicts in Panama: neoliberal multiculturalism, changing legislation, and human rights." *Journal of Latin American Geography* 11(2): 21–47.
Velez, M. (2011) "Collective titling and the process of institution building: the new common property regime in the Colombian Pacific." *Human Ecology* 39: 117–129.
White, A. and Martin, A. (2002) *Who Owns the World's Forests?* Washington, DC: Forest Trends.
Wily, L. (2011) The Tragedy of Public Lands: The Fate of the Commons under Global Commercial Pressure. Paris: CIRAD, International Land Coalition.
Yashar, D. (2005) *Contesting Citizenship in Latin America.* New York: Cambridge University Press.
Zamora, A. and Monterroso, I. (2017) Visiones regionales y locales sobre la inseguridad de tenencia en Loreto y Madre de Dios. CIFOR InfoBrief No. 198. *Center for International Forestry Research (CIFOR).* Bogor, Indonesia: CIFOR.

29

GLOBALIZATION, LOCAL COMMONS, AND THE MULTISCALE ECOSYSTEM FRAMEWORK (MEF)

Timothy O. Randhir

Local commons are vital to supporting ecosystems and livelihoods. Wetlands, rivers, riparian forests, savannas, tropical forests, coastal waters, and groundwater are examples of common pool resources that are critical to ecosystem functions of a region and livelihood of local communities. With the decline in local commons under unsustainable extraction without management, there is an increasing threat of loss in the environmental and economic sustainability of regions around the globe. While many studies have highlighted the importance of looking at management and governance at multiple scales, applications of the multiscale ecosystem framework (Randhir, 2016) remain in infancy. Without evaluating the linkages between scales, management of common pool resources will be inefficient and make them vulnerable to exploitation across scales. In addition, a multiscale framework is critical for sustainability, a complex concept related to the often-quoted definition of sustainable development by the Brundtland Commission as "development that meets the needs of current generation without compromising the ability of future generations to meet their own needs" (World Commission on Environment and Development, 1987). In this chapter, I review the role of scale in managing commons and use an MEF approach to evaluate local commons and their multiscale processes using specific case studies with the aim of evaluating management and governance opportunities.

The multiscale ecosystem framework (MEF) developed by Randhir (2016) expands on the socioecological systems (SES) approach of Ostrom (2009) to represent economic and ecological processes within and across multiple scales. Scale and interaction between scales are critical dimensions that drive the dynamics of a common pool resource. Scale characteristics and inter-scale dynamics are less studied in common pool literature, which is a focus of this chapter. The MEF approach allows the evaluation of both inter- and intra-scale feedbacks under a comprehensive systems framework. The scale interaction becomes a critical dimension for comprehensive design of policy and governance to manage a common pool resource. The general objective of this chapter is to describe the role of scale in managing commons and to review MEF as a framework to incorporate multiscale issues and apply it to specific case studies of local commons. I will discuss the role of scale, MEF framework for assessing scale dynamics, and applications to two case studies in Tanzania and Swaziland.

Scale dynamics and commons

Scale is recognized as an important dimension in the management of commons. Ecologists and geographers recognize this concept, and there is a need for studying processes and interactions within and among socioeconomic and ecological processes (Sayre, 2005) at each scale. Temporal and spatial scales are used to analyze ecosystem state and dynamics. In human geography, socio-economic interactions produce scales that can range from local to global. Thus, scale is an important aspect of SES and is useful to assess inter and intra-scale interactions. Grain (smallest spatial unit) and extent (area of investigation) are used to define components of scale. In ecological systems, smaller and larger scale can be subsystems that are nested hierarchically within the higher scale and connect with each other as a single system.

In local commons like savannas, the scale can range from a patch of a few square meters in a household backyard or at the edge of a stream bank, to the larger region of a village or multiple villages and can be connected to savanna ecosystems throughout a landscape. In this case, smaller scales are critical to the stability of the structure and function of larger savanna regions throughout the landscape. From larger landscape scales, fragmentation and other impacts of regional policies can change the regional savanna ecosystem, thereby affecting local scales. Thus, a multiscale approach becomes critical in managing savannas.

Local commons like water bodies, forest ecosystems, coastal resources, fisheries, soil resources, wetlands, watersheds, and mineral resources have scale effects that cross multiple scales. These inter-scale effects can include ecological flows, impacts of stressors, policy impacts that change the state of the ecosystem, and impacts of institutional changes that govern the commons.

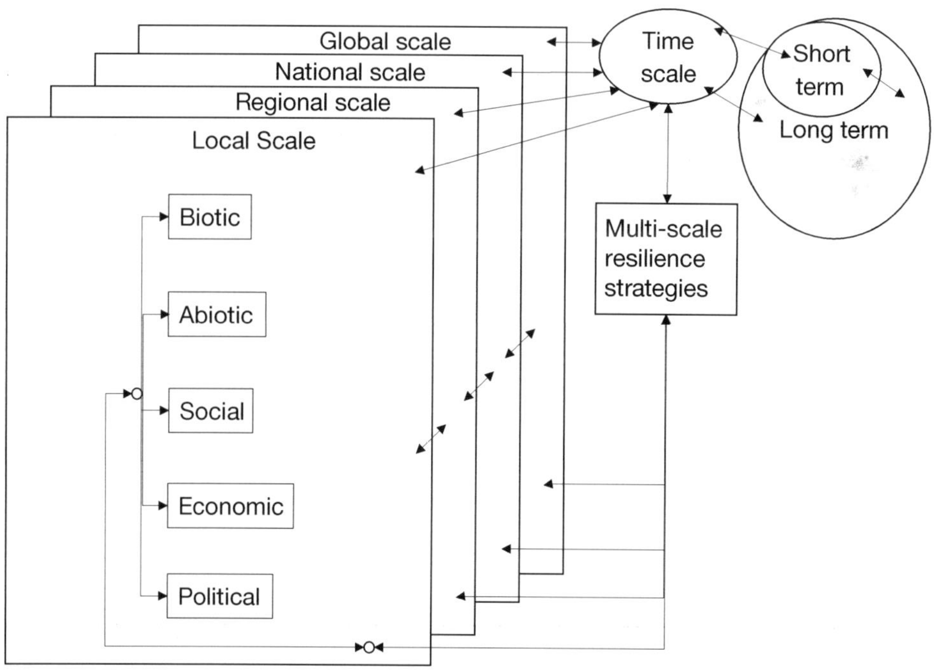

Figure 29.1 Multiscale ecosystem framework (MEF) model

Source: Adapted from Randhir, 2016.

Multiscale ecosystem framework (MEF)

The MEF (Randhir, 2016) is a conceptual model (Figure 29.1) for evaluating the multiscale nature of local commons and their influence from larger scales, especially processes like globalization and regional impacts. The framework expands on the work on SES in developing an explicit representation of geographic and temporal scales. Biotic (plants, microbes, wildlife), abiotic (land, air, water), social (institutions, people), economic (markets, capital), and political (governance) at each scale are represented in the MEF framework. These are connected to higher and lower scales of the system. Each geographic scale is also related to varying scales of time (short, medium, and long term). Multiscale strategies in the MEF connect at varying spatial and temporal scales. This framework allows understanding of the dynamics of local commons as they connect to higher scales, like the impacts of regional changes or globalization on the structure and function of these local commons.

To study the MEF in depth for commons research, I present two specific cases of local commons and discuss multiscale processes that are critical for studying system dynamics, sustainability, and governance. The first case is the case of rhinoceros populations and their habitat in the South African region. The second case is the Mount Kilimanjaro ecosystem of northern Tanzania. These two cases are chosen to review and document important problems facing these commons, the multiscale nature of impacts on them, and the governance options for managing them. The MEF can be used to assess the nature of processes, the extent of effects, and governance mechanisms surrounding these resources.

Rhinoceros populations and habitat

Black and white rhinoceroses are critically endangered in the South African region. Almost 80% of all world rhinoceroses occur in this region, with populations present in some national parks of various countries in the southern region of the African continent. It is reported that poaching has increased from 13 per year in 2007 to 455 in 2012 (CITES, 2013). Rhinoceros horns are of high demand in global markets for traditional Chinese medicine and for ornamental use (Graham-Rowe, 2011). The international ban on all trade in rhinoceros products did not bring any improvement in their populations (Leader-Williams, 2003), making illegal markets of ivory more profitable. Black rhinos are on the verge of extinction and need a new paradigm for the conservation of this species (Moodley et al., 2017). Thus, governance of rhinoceros populations and their habitat need a multiscale approach to assessment and governance.

Rhinoceros populations used to roam throughout the savannahs of the African continent. With settlements, increases in hunting, and more recently, accelerated poaching, rhinoceros populations have declined rapidly and have been reduced to a few natural parks. Major populations occur in four countries: South Africa, Namibia, Zimbabwe, and Kenya. Impacts of population numbers and reduction in their distribution can be traced to two global processes: European settlements and demand for Rhinoceros horns from Asian markets. These stressors have variable impacts on countries, regions, and specific sites of these populations depending on governance and ecosystems at each of these scales.

The ban in international trade since 1977 is a policy aimed at minimizing poaching and illegal trade but has failed in its original objective. Illegal trade and poaching have increased because of growing demand (Martin, 2012). At national and local scales, many policies including the establishment of sanctuaries (Brett, 1990), law enforcement, wildlife farming (Damania and Bulte, 2006), and auctions (Dalerum et al., 2017) are proposed. At a local scale, responses often vary with private farming of rhinoceros, poaching for private gains, and a lack of participation in conservation.

Use of the MEF approach to this common pool resource requires a comprehensive assessment of global demands, regional policies, community participation, ecological processes, and socioeconomic characteristics at each scale. Assessing the ecosystem characteristics that support rhinoceros populations and their habitat, economic conditions, social and cultural factors, existing institutions, landscape factors, and global and regional stressors is a first step in implementing an MEF approach. With this baseline information, it becomes critical to engage actors and to process information at multiple scales to develop adaptive institutions that can address current and emerging issues in managing these commons.

The need for comprehensive strategies is needed as many conservation strategies are narrow in objectives, and short-term in scope. For example, a three-day online auction of rhinoceros horns by a large private breeder in South Africa occurred in August 2017 with the approval of the High Court of Pretoria. The main argument for the sale was to offset the cost of protection and security of the rhinoceros in the farm. While the legal sale has some arguments in its favor, it also raises questions like: what are the implications for illegal global supplies? What are the effects of no demand within South Africa on global markets? Can such markets deplete rhinoceros populations with an increase in the number of wildlife farms? What are some of the ethical aspects of commodifying threatened/endangered species?

An optimal strategy is to develop a comprehensive assessment and governance at multiple scales using an MEF method as guidance. Local management needs to adapt to emerging changes in higher scales through resilience practices that prepare the system for pressures for local poaching to supply global markets. With the current populations restricted to national parks, any solution to maintain the population needs to consider governance strategies that maintain the social and ecological systems at local scales, but also include a mechanism to incorporate flows and stressors from higher scales, including global demands.

Mount Kilimanjaro

Mount Kilimanjaro in northern Tanzania is the highest mountain in Africa and highest free-standing mountain in the world. The region has distinct ecological zones: savanna/cultivates areas, rainforest, heath, moorland, alpine desert, and arctic zone (summit). Designated as a World Heritage Site in 1987, the region is home to a rich biodiversity of plants and animals. The Kilimanjaro National Park includes the mountain above the tree line and surrounding montane forest reserve and is managed by the Tanzania National Parks (TANAPA) organization. National legislation regulates park use and tourism activities through the National Parks Act, Chapter 282 (The National Parks Act, 1959) and later subsidiary legislation (Ministry of Natural Resources and Tourism, 2003). Despite these regulations, increasing and cumulative stress from adjacent land uses, pollution, invasive species, fire impacts, and climate change remain a threat to the ecosystem (UNESCO, 1987). In addition to the shrinking ice cap (Thompson et al., 2002), land use changes from settlements, cultivation (Soini, 2005), and forestry have implications for soil loss, runoff, and fragmentation in the lowland areas. An aerial review of the region (UNDP, 2002) assesses that the forest belt is disturbed by human activities, especially logging of indigenous trees, and increasing plantation forests.

Given the heterogeneous nature of ecosystem characteristics in the Kilimanjaro National Park, an MEF approach is useful in developing integrated plans for managing this threatened complex of commons. Kilimanjaro snow cover is critical to lowland plants and animals in the ecosystem, affects livelihoods of more than one million people, and is a primary source of the Pangani River. At a global scale, climate change is attributed to glacier retreat (Kaser et al., 2004). With $50 million in annual tourism revenues, Mount Kilimanjaro has become a top international tourist destination. With strict regulations in the park for park users and the tourism industry, the impacts of tourism are managed to have the least impact on the ecosystem. For example, the "Trash in Trash Out" system is strictly enforced to minimize the footprint on tourism camps and trails. With increasing tourist pressure, adaptation to future stresses and footprints is uncertain. At a regional scale, this region is part of a larger landscape ecosystem that is stressed by fragmentation of landscapes and settlements. At a local scale, deforestation for farming, soil erosion, poaching, and settlements have impacts that have cumulative and long-term effects that influence higher scales. Thus, a comprehensive approach to managing the Kilimanjaro park and regional ecosystem is to account for stressors and processes at multiple scales and develop participatory governance across scales.

Strategies for multiscale governance

Local commons are connected to multiple scales, and absence of strategies to mitigate global and regional stressors from higher scales makes them vulnerable to depletion (Randhir, 2016). Understanding the nature of interactions in common pool resources, both within and among commons and across multiple scales, is a first step in developing a baseline. Using MEF, information can be gleaned on the state of resources, users, rates of use, incentives, markets, dynamics in the resource use processes, ecosystem processes, local and global stressors, informal rules, norms, social values, and social organizations. This information is critical in developing strategies that increase the resilience of the system to stressors. The strategies can be developed with a focus on adaptive capacity, system resilience, and transformability (Walker et al., 2004) under major shocks to the system. For such strategies, there is a need for improved local participation, mutual monitoring, cooperative institutions (Vollan and Ostrom, 2010), self-organization, systems-based assessment, equity in decision-making, efficiency, and multiscale governance. In developing such institutional approaches, sustainability principles become useful in evaluating and pursuing particular trajectories. Strong sustainability and justice at each scale become critical to the common pool resource being resilient in economic, environmental, and social dimensions.

Conclusion

Local commons are under increasing pressure from regional and global processes, and there is a need for strategies to mitigate negative effects through institutions and ecosystems that are resilient to such changes. Understanding the multiscale nature of these impacts, whether from higher to lower scales or vice versa, is critical in protecting many local commons through the world. Two case studies on the African continent are used to study the interaction between local and global processes. With increasing threats from multiple scales, local commons need to be managed using a renewed, and comprehensive approach.

Expanding global markets for rhinoceros products and implications on incentives to poaching remain a critical problem. Policies and governance that focus on particular scales remain weak in addressing threats from other scales. An MEF framework could be used to develop comprehensive approaches to manage and conserve rhinoceros habitat and their populations. In the Kilimanjaro ecosystem, the global impacts through tourism and regional impacts of land use are difficult to address using governance that aims at a single scale. An MEF approach for this ecosystem can guide managers and agencies in connecting processes at local to global scales through comprehensive assessment and governance.

This chapter outlines the role of MEF in multiscale governance in developing strategies for resilience and transformation. The approach is transferable to other local commons throughout the world that face increasing pressure from global and regional processes. This approach could be useful in managing wildlife like elephants, and tigers that are influenced by poaching and global markets. It becomes critical to address environmental, economic, and social-cultural dimensions at each scale and between scales, and an MEF approach can help in developing strategies that can be useful in managing these local commons that are necessarily linked to multiple scales.

References

Brett, R.A. (1990) "The black rhinoceros sanctuaries of Kenya." *Pachyderm* 13: 31–34.

CITES (2013) *Interpretation and Implementation of the Convention. Species Trade and Conservation – Rhinoceroses.* Bern, Switzerland: Convention on International Trade in Endangered Species of Wild Fauna and Flora (CITES).

Dalerum, F., Miranda, M., Muniz, C. and Rodríguez, P. (2017) "Effects of scarcity, aesthetics and ecology on wildlife auction prices of large African mammals." *Ambio* 47(1): 78–85.

Damania, R. and Bulte, E.H. (2006) "The economics of wildlife farming and endangered species conservation." *Ecological Economics* 62(3–4): 461–472.

Graham-Rowe, D. (2011) "Endangered and in demand." *Nature* 480: 8–10.

Kaser, G., Hardy, D.R., Mölg, T., Bradley, R.S. and Hyera, T.M. (2004) "Modern glacier retreat on Kilimanjaro as evidence of climate change: observation and facts." *International Journal of Climatology* 24(3): 329–339.

Leader-Williams, N. (2003) "Regulation and protection: successes and failures in rhinoceros conservation." In S. Oldfield, editor. *The Trade in Wildlife: Regulation for Conservation.* London: Earthscan.

Martin, R. (2012) "A legal trade in rhino horn: Hobson's Choice." Johannesburg: Rhino Survival Trust.

Ministry of Natural Resources and Tourism (2003) Subsidiary Legislation: The National Parks Ordinance (CAP. 412). Government Notice 191. Supplement 28, issues 25/7/2003. http://tanzaniatourism.go.tz/uploads/National_Parks_Mountain_Regulations.pdf (accessed 20 October 2018).

Moodley, Y., Russo, I.R.M., Dalton, D.L., Kotzé, A., Muya, S., Haubensak, P., Bálint, B., Munimanda, G.K., Deimel, C., Setzer, A. and Dicks, K. (2017) "Extinctions, genetic erosion and conservation options for the black rhinoceros (Diceros bicornis)." *Scientific Reports* 7: 41417.

Ostrom, E. (2009) "A general framework for analyzing sustainability of social-ecological systems." *Science* 325(5939): 419–422.

Randhir, T.O. (2016) "Globalization impacts on local commons: multiscale strategies for socioeconomic and ecological resilience." *International Journal of the Commons* 10(1): 387–404.

Sayre, N.F. (2005) "Ecological and geographical scale: parallels and potential for integration." *Progress in Human Geography* 29(3): 276–290.

Soini, E. (2005) "Land use change patterns and livelihood dynamics on the slopes of Mt. Kilimanjaro, Tanzania." *Agricultural Systems* 85(3): 306–323.

The National Parks Act (1959) The National Parks Act – Chapter 282. http://extwprlegs1.fao.org/docs/pdf/tan155102.pdf (Accessed 28 September 2017).

Thompson, L.G., Mosley-Thompson, E., Davis, M.E., Henderson, K.A., Brecher, H.H., Zagorodnov, V.S., Mashiotta, T.A., Lin, P.N., Mikhalenko, V.N., Hardy, D.R. and Beer, J. (2002) "Kilimanjaro ice core records: evidence of Holocene climate change in tropical Africa." *Science* 298(5593): 589–593.

UNDP (2002) Aerial Survey of the threats to Mt. *Kilimanjaro forests.* Dar es Salaam, Tanzania: UNDP – The GEF Small Grants Programme.

UNESCO (1987) Kilimanjaro National Park: Outstanding Universal Value. http://whc.unesco.org/en/list/403 (Accessed 28 September 2017).

Vollan, B. and Ostrom, E. (2010) "Cooperation and the commons." *Science* 330(6006): 923–924.

Walker, B., Holling, C.S., Carpenter, S. and Kinzig, A. (2004) "Resilience, adaptability and transformability in social-ecological systems." *Ecology and Society* 9(2): article 5.

World Commission on Environment and Development (1987) *Our Common Future* Oxford, UK: Oxford University Press, p. 27.

PART III

A global context

30
BIGGER ISSUES IN A SMALLER WORLD
The future of the commons

Cheryl Chan, Fatima Noor Khan and Sajida Awan

Introduction

Commons literature is continuously expanding and shifting to reflect emerging global issues. First, drawing from our understanding of traditional commons (i.e., single, finite resources), there has been a shift towards recognizing more intangible resource types as commons (e.g., information, culture) (Bowers, 2009; Hess, 2008). Second, there is increasing attention regarding the distinction between the "tragedy of the commons" and the tragedy or dilemma of "the commoners" (De Moor, 2015; Goldman, 1993)—emphasizing not only resources but also the institutions through which they are governed. The third shift, occurring in tandem with the above shifts, is the application of an interdisciplinary and systems-based lens to examine the commons and issues of commons governance (e.g., Armitage et al., 2017; Berkes, 2006).

In this chapter, we introduce examples and problems from the global commons (e.g., outer space, deep sea, atmosphere) and the new commons (e.g., protected areas, cultural commons, knowledge commons). We then examine "commonalities" amongst these emerging commons, including shared challenges in global governance. Global governance, here, refers to "the collective effort by sovereign states, international organizations, and other non-state actors to address common challenges and seize opportunities that transcend national frontiers" (Patrick, 2014, p. 59). We argue that as technology and innovations continue to bring the world closer together, these advances have also: 1) brought formerly inaccessible commons within reach (e.g., high seas, outer space); 2) created new forms of commons (e.g., the Internet); and 3) thus produced a new suite of issues in global governance.

The global commons

We begin this chapter with an overview of the global commons, defined as "natural resources over which no single nation has a generally recognized exclusive jurisdiction" (Wijkman, 1982, p. 511). No single decision-making unit holds exclusive title to the global commons (Wijkman, 1982). International law recognizes four global commons:

1. The high seas and deep seabed—the term "high seas" is clarified in the United Nations Convention on the Law of the Sea (1982) as "all parts of the sea that are not included in the exclusive economic zone, in the territorial sea, or in the internal waters of a state, or in the archipelagic waters of an archipelagic state" (p. 57).
2. The atmosphere.
3. Antarctica.
4. Outer space (UN System Task Team, 2013).

These large-scale commons exist in contrast to the small-scale, finite resource types that informed early commons literature (e.g., Ostrom, 1990). In this section, we introduce a timeline for the global commons, including its historical roots in the Grotian interpretation of *res communis*, and current perceptions of the global commons as a common heritage of mankind. Then, we look towards the future by considering the relevance of the global commons in achieving sustainable development. We conclude with a comparison of the global commons and traditional local commons (e.g., forestry, fisheries).

Seeds for the concept of global commons have existed since ancient Rome, with applicable perspectives from philosophers like Cicero, Seneca, Thucydides, Avienus, Horace, Vergil, and Ovid (Schrijver & Prislan, 2009). In 1609, Hugo Grotius—noted for being a founder of modern international law (Parry, 2014)—synthesized these Roman ideas alongside other similar thoughts in *Mare liberum* (i.e., "The Free Sea", or "The Freedom of the Seas"), a legal brief defending international freedom of access to the sea. In *Mare liberum*, Grotius (1609) proposed that the "outer sea" was *res communis*—a term from Roman law describing common possession belonging to all (Nicholson, 2002)—"because [the sea] is so limitless that it cannot become a possession of any one, and because it is adapted for the use of all" (p. 28).

Legally, Grotius (1609) argued that *res communis* was distinct from both *res nullius* (i.e., property of no one) and *res publica* (i.e., public property) in that, first, the property could not be occupied (including on the grounds of public utility) and, second, even in serving one person the property could still be of common use for all people. As such, he suggested that no single nation or individual could establish ownership over these resources. Via Grotius, oceans are considered the original global commons (Schrijver & Prislan, 2009). Grotius' ideas maintain relevance in contemporary ideas of the commons (e.g., in relation to property rights, subtractability, excludability). However, technology has continually pushed the forefront of resource exploitation, making aspects of Grotius' original positions irrelevant. For example, in the context of resource trends (e.g., declining fish stocks), we now acknowledge that the sea is not limitless. Also, discussed later in this chapter, some new commons (e.g., protected areas) are no longer established to protect direct use values and may instead have other goals (e.g., to increase biodiversity, to protect an endangered species, bequest values) (Berge, 2006).

While oceans were the "original global commons", new forms of commons like outer space and Antarctica have emerged via improved technological capabilities (e.g., mining, transport, travel). Likewise, the atmosphere (another global commons) has become a pressing global concern for mitigating climate change (Stern, 2011)—a wicked problem stemming from many of our industrial advances. Based on international law, all four global commons share the common heritage of mankind (CHM) principle, also known as the common heritage principle (UN System Task Team, 2013). Unlike the ancient concept of *res communis*, CHM has more recent origins. This principle was introduced and brought to international attention by Maltese representative Arvid Pardo in the 1967 UN General Assembly and subsequently defined in the 1982 Law of the Sea Convention (Vogler, 2012). While there is no agreed-upon definition for CHM, most definitions contain five elements:

1. Non-appropriation, or prohibition of sovereignty: Global commons must not be appropriated either publicly or privately.
2. Common management: Global commons must be managed by representatives from all nations.
3. Benefits sharing: All nations must actively share the benefits that are obtained from exploiting the global commons.
4. Peaceful purposes: Global commons must be used for peaceful purposes (e.g., countries cannot station military personnel in these spaces).
5. Bequest values: Global commons must be preserved for the benefit of future generations, and there is a "human responsibility" for the collective good (Frakes, 2003).

In practice, resources from the global commons can be surveyed for exploratory purposes (e.g., bioprospecting), although this process is subject to international agreements and special treaties like the United Nations Convention on the Law of the Sea (UNCLOS) (Nicholson, 2002; Schrijver, 2016).

The role of the global commons in achieving sustainable development is both socially and ecologically important. The CHM principle strongly reflects a social focus through an emphasis on diverse representation, peaceful purposes, benefits distribution, and bequest values. In fact, in Pardo's (1967) original address, he emphasized recognizing the "needs of poor countries" in governance of the high seas. Contemporary scholars like Okereke (2008) highlight that CHM aims "to ensure that natural resources . . . are exploited under international governance systems informed by the idea of international distributional equity" (p. 30), although the international community's commitment to a "hegemonic neoliberal economic philosophy" (p. 46) hinders social progress. From an ecological perspective, Okereke (2008) also asserts that CHM was established more for global institutions governing resource exploitation than for the purposes of environmental protection. These and other critiques (e.g., Noyes, 2012) suggest that CHM may not suffice as the primary ideology for governing the global commons.

Despite conventions and treaties to protect the global commons (e.g., UNCLOS, the Antarctic Treaty System, Montreal Protocol), agencies of the UN have admitted numerous shortcomings of international efforts in global environmental governance (UN System Task Team, 2013). Further, much like Grotius' now moot idea of a limitless sea in *Mare liberum*, many existing formal agreements do not consider the full force of anthropogenic impacts (UN System Task Team, 2013). Many human impacts are, themselves, poorly understood. Finally, there is a lack of historical context from which to draw best practices for governing the global commons. For example, lessons learned from traditional local commons may not retain salience when applied to the global commons. Stern (2011, p. 216) suggests that while both global and local commons suffer degradation from collective human activity, they also hold key differences:

1. Geographic scale.
2. Number of resource users.
3. Actors' awareness of degradation (e.g., negative impacts on the global commons may seem intangible to resource users).
4. Distribution and divergence of interests and power (viz. resource appropriators are not the same as those most at risk from resource degradation).
5. Cultural and institutional homogeneity (viz. global commons are more heterogeneous).
6. Feasibility of learning (e.g., learning for global commons could be multi-generational, with no second chances).

While there are valuable lessons to be learned from traditional commons (Berkes, 2005), Stern (2011) argues that we cannot draw purely from knowledge of traditional commons to govern the global commons (more on governance of the global commons later in the chapter).

The first part of this chapter provided an overview of the global commons, defined as large-scale resources over which no single nation has jurisdiction. Four types of global commons exist in international law: the high seas and deep seabed, the atmosphere, Antarctica, and outer space. The global commons, as a concept, was first articulated in Hugo Grotius' (1609) *Mare liberum*, in which he applied *res communis* to defend freedom of access to the sea. Contemporary interest emerged when Maltese representative Arvid Pardo called attention to the sea as a common heritage of mankind in the 1967 UN General Assembly. Governance of the global commons has since become a pressing concern against a backdrop of large-scale environmental issues (e.g., climate change, ozone depletion, pollution) and an international focus on sustainable development. Being distinct from traditional local commons in several ways, the global commons still holds many unknowns. The next part of this chapter will introduce the new commons, another emerging commons of interest in the scientific and popular sphere.

The new commons

In this section, we introduce examples of the new commons, defined as a broadening of the commons with "various types of shared resources that have recently evolved or have been recognized as commons" (Hess, 2008, p. 1). We then examine challenges associated with the application of traditional commons thinking (e.g., excludability, subtractability) in the new commons. The emergence of the new commons can be traced to the 1995 International Association for the Study of the Commons (then known as the International Association for the Study of Common Property) conference, themed "Reinventing the Commons" (Hess, 2008). The new commons differ from traditional and global commons in that they include many intangible resource types (e.g., social, cultural). Although there are multiple types of new commons, we draw attention to three types that are particularly relevant to our analysis:

1. Protected areas, distinguished from traditional commons (i.e., the sustainable use of resources) by its non-use characteristics for the "conservation of nature and the protection of biodiversity" (Pieraccini, 2015, p. 553).
2. Cultural commons, defined as the "intergenerational knowledge, skills, and activities" that define a group of people (e.g., food, art, ceremonies, political traditions) (Bowers, 2009, p. 196).
3. Knowledge commons, defined as a body of commons where knowledge (i.e., useful ideas, information, and data in whatever form obtained or expressed) is the primary pooled resource (Hess & Ostrom, 2006).

Next, we sequentially define these new commons, assess the fit of these new commons with existing understandings of the commons, and examine potential challenges in governing the new commons.

Protected areas are considered a new commons by challenging traditional resource use values (Pieraccini, 2015). Protected areas have catalyzed the creation of new institutions for the governance of traditional commons (Hess, 2008). For example, according to designations by the International Union for the Conservation of Nature, a Category IV protected area aims "to maintain, conserve and restore [flora and fauna] species and habitats", and is therefore more ecologically than socially oriented (Dudley, 2008, p. 8). To achieve sustainability of protected areas, Berge (2006) emphasized the importance of acknowledging the differences between traditional commons values (e.g., traditional resource use) and protected area values (e.g., biodiversity, tourism, bequest). When these differences are not addressed, conflicts can arise from the contrasting objectives of "commoners" and protected areas. Without shared objectives and commons goals, mobilizing collective action in protected area commons is a challenge, and, as such, protected areas may not always succeed.

Beyond the tangible, natural world, commons thinking has been applied to analyze intangible "human-constructed" new commons, like cultural commons (Bowers, 2009) and knowledge commons (Hess & Ostrom, 2006). Cultural commons are considered by some authors to be the broader umbrella for knowledge commons (e.g., Madison et al., 2010), while others consider them as a separate and distinct type (e.g., Hess, 2008). In its broadest definition, cultural commons are the "environments for developing and distributing cultural and scientific knowledge through institutions that support pooling and sharing that knowledge in a managed way" (Madison et al., 2010, p. 659). However, here we discuss cultural commons within the scope of socially binding values and practices (e.g., Bowers, 2009; Samakov & Berkes, 2017).

An example of the cultural commons, in this context, is sacred sites. Sacred sites are "areas of land or water having special spiritual significance to people and communities" (Rutte, 2011, p. 2387). Sacred sites differ from traditional commons in that they do not always offer "intrinsic or instrumental value" (Samakov & Berkes, 2017, p. 424), although they may contain both natural (e.g., trees) and constructed elements (e.g., monuments) (Samakov & Berkes, 2017). Rutte (2011) highlighted the inherent tension between the intangible spiritual values and tangible untapped economic values of sacred sites, since spiritual values can be diminished when sacred commons are exploited for economic gain. Samakov and Berkes (2017) posed that these natural areas may have better conservation outcomes if their "sacredness" is recognized and incorporated into conservation strategies, referring to these strategies as biocultural conservation. Intangible values of cultural commons, when paired with tangible values of traditional commons, may also be a helpful lens for analyzing protected areas as a new commons (e.g., UNESCO world heritage sites). Noted earlier, a major challenge for protected area commons is reconciling old values (e.g., resource value, cultural value) with new values (e.g., biodiversity). Cultural values may offer a necessary bridge for actors to converge on common conservation goals, although capturing cultural values such as "sacredness" in traditional commons frameworks remains a challenge.

Knowledge commons, like cultural commons, are intangible human-constructed commons. Expansion of knowledge commons has illuminated prominent issues (e.g., neoliberal enclosure) in both commons literature (Hess & Ostrom, 2006) and broader society. Knowledge commons exemplify the growing interdisciplinarity of commons by showcasing a diversity of influences and inputs—like in the digital commons (e.g., the Internet, Wikis) and genetic commons (e.g., genetic seed banks) (Hess, 2008). Like the global commons, these new commons challenge traditional notions of excludability and subtractability, since knowledge lacks clear boundaries (Hess & Ostrom, 2006; Madison et al., 2010). For example, there are notable challenges to defining subtractability and exercising excludability in the digital commons, where the Internet is a platform for the proliferation of knowledge. Regulation of digital information (i.e., exercising excludability) not only requires a high degree of institutional capacity but also raises social and political concern over civil rights. Furthermore, how can knowledge—an intangible resource and "uncapturable public good" (Hess, 2008, p. 4)—be traditionally "subtracted" in a tangible way? Questions arise over how excludability and subtractability, as characteristics of traditional commons, can be distinguished from that of open access in the knowledge commons.

Outcomes of this excludability and subtractability debate will determine whether knowledge, as a global resource, can be distributed and mobilized equitably and sustainably (Hess & Ostrom, 2006). Yet, the practice of defining Intellectual Property Rights (IPR) inherently supports the privatization and commercialization of important ideas and resources. IPR turns these intangible resources into private property by creating social institutions that limit access. The privatization of the new commons is contentious, referred to by some as the "second enclosure movement" (Boyle, 2008)—with the first enclosure being the privatization of traditional commons (i.e., natural resources). These enclosures support Okereke's (2008) critique of the international community's "hegemonic neoliberal economic philosophy" (p. 46), discussed earlier. Examples of IPR disputes also appear in genetic commons discourse (Scharper & Cunningham, 2006). Despite the rapid diversification of genetic knowledge, access to this knowledge has become increasingly exclusive (e.g., limited to corporations, academia) (Scharper & Cunningham, 2006). The genetic commons movement (e.g., communal seed libraries) is an emerging response to the neoliberal second enclosure movement (Scharper & Cunningham, 2006). However, like other knowledge and cultural commons, application of excludability and subtractability in the genetic commons remains uncertain.

Neither knowledge nor culture are subtractive, but rather are cumulative and synergistic in nature (Hess & Ostrom, 2006). As such, open access in the new commons is distinct from open access of traditional and global commons (e.g., water, land, fish), where an open access regime would lead to overconsumption and resource degradation (Hess & Ostrom, 2006). Numerous authors have attempted to create frameworks for analysis of the new commons, although there is a lack of historical context from which to glean insights (see Hess & Ostrom, 2006; Madison et al., 2010; Rutte, 2011). The new commons share many of the same differences that can be distinguished between the global commons and traditional commons, outlined in the previous section. For example, the new commons can differ from traditional commons in geographic scale (e.g., digital world has no geographic boundaries), number of resource users, distribution and divergence of interests and power (e.g., corporations may lack sympathy for place values in the cultural commons), cultural and institutional homogeneity, and feasibility of learning (e.g., many aspects of the new commons are constantly changing). Therefore, conceptualizing new frameworks and testing applications of these frameworks are opportunities for future research in the new commons.

Despite many unique traits of the new commons, these commons also share two characteristics with other commons discussed in this chapter. First, new commons are still examples of "jointly used [resources], managed by groups of varying sizes and interests" (Hess & Ostrom, 2006, p. 5)—like both traditional and global commons. However, the values of resources in the new commons may be more intangible. For example, in the case of a no-take protected area with recreational (e.g., aesthetic), wilderness (e.g., biodiversity) and/or cultural value, the resource provided may not be exportable or extractable (Berge, 2006). Second, like other traditional (e.g., fisheries) and global commons (e.g., oceans, atmosphere), the new commons are complex systems. For example, while individual protected areas may be spatially small in scale, many protected areas are regionally or internationally networked (Dudley, 2008). Likewise, knowledge commons (e.g., digital commons) are a shared global resource. Governance of these new commons can be jurisdictionally multi-level and transboundary in practice, thereby sharing many of the same challenges associated with governing traditional and global commons.

This section featured an introduction to three types of new commons: 1) protected area commons; 2) cultural commons (e.g., sacred commons); and 3) knowledge commons (e.g., digital, genetic commons). Despite sharing some characteristics with traditional and global commons, we argued that the new commons challenge certain notions of conventional commons thinking, such as the application of excludability, subtractability, and collective action. Further, while conceptualization of the new commons has been largely interdisciplinary (e.g., influenced by shared discourse in biology, technology, law), there are still more opportunities for interdisciplinary and transdisciplinary contribution to understanding the challenges of new commons governance (e.g., knowledge co-production in the knowledge commons, cultural dimensions of natural resources in the cultural commons). In the next section, we examine linkages between the global commons and the new commons, while discussing challenges pertinent to the governance of both.

Governing the "commonalities"

In this section, we begin with a synthesis of insights by outlining "commonalities" (i.e., shared attributes) between the global commons and the new commons. Since Hardin's *Tragedy of the Commons*, the narrative of the commons has shifted in many ways (van Laerhoven & Ostrom, 2007). For example, we now think of the commons not only as a resource but also as a negotiation of social and ecological relationships, a process, a way of thinking, and a complex system (van Laerhoven & Ostrom, 2007). Despite the continued application of commons thinking to new disciplines—and potentially in new ways like in the new commons—all of these applications share a cocentricity and "common" origin. All commons, old or new, are shared resources governed by a group of resource users with common interests and objectives. However, our findings propose that the commons and our thinking of the commons are also evolving, particularly as resources and our ability to access these resources have changed (e.g., space travel, connecting to the Internet). Reflected in global commons and the new commons, the commons has become increasingly: 1) intangible; 2) transboundary; and 3) complex. Next, we discuss these commonalities and their implications for global governance in greater detail.

The global commons and the new commons both exhibit elements of intangibility. Many of these resource types lack physical presence (e.g., knowledge, culture, information, atmosphere), exist at a large conceptual scale (e.g., outer space), or are "out of sight, out of mind" (e.g., Antarctica, the high seas)—these traits present several challenges to governance. For example, some resources are depleted or degraded cumulatively at a global scale (e.g., migratory marine resources, atmosphere) and yet are governed at multiple levels (e.g., local, national, international) (Berkes, 2006). The intangibility of these commons makes individual and cumulative impacts difficult to monitor and evaluate, thus leading to issues in defining subtractability (i.e., outlining resource units) and exercising excludability (i.e., controlling access). In addition, some new commons (e.g., cultural commons, knowledge commons) are rich with intangible values that may not translate across sociopolitical and geographic boundaries (Berge, 2006; Samakov & Berkes, 2017). Such values may be overlooked in decision-making, leading to unintended user group conflicts. Finally, commons that are immaterial and "limitlessly" distributed (e.g., knowledge) can complicate the number and range of actors involved in governance. The global commons and the new commons highlight opportunities for reframing existing commons frameworks to capture the intangible values of resources (e.g., social and cultural values, immaterial nature of knowledge) and to recognize that resource types in the commons are no longer immutable.

While commons literature originated in small-scale traditional commons with finite user groups, the transboundary nature of the global commons and the new commons yields new and complex problems (e.g., marine pollution, climate change, abuse of open data and technologies). To overcome such challenges, governance of these commons requires collective action across multiple sovereign jurisdictions. Yet, collective action depends on numerous factors, including the size of user groups, dependence on resources, and capacity of groups to manage resources (Agrawal, 2007). This context can be difficult to establish in the global commons and new commons. Discussed earlier, many of the resources in these commons are intangible and large scale—obscuring user groups (e.g., non-point source pollution, anonymity on the Internet) and resource dependence (e.g., sacred sites), while also demanding immense managerial capacity. Thus, while understanding institutions for collective action was an early focus of commons literature, this topic remains ever relevant today (Ostrom, 1990, 2007). De Moor (2011) suggested that by analyzing the evolution of the commons (e.g., global commons, new commons) and different types of institutions for collective action, we can enrich our knowledge with institutional "tools" for managing the commons more effectively.

Ostrom (1990) highlighted the importance of cross-scale institutional linkages in strengthening collective action, but the transboundary nature of the global commons and new commons can make collective action challenging. For example, climate change strategies are often prepared at the global level and largely exclude the knowledge and lived experiences of local and indigenous people (Byg & Salick, 2009). Yet, climate change can have profound cultural and spiritual impacts on communities, which are difficult to capture via conventional metrics. Failure of scientists and policy makers to acknowledge these intangible impacts could lead to uninformed decision-making, which could subsequently produce actions deemed inappropriate or unacceptable at the local level (Byg & Salick, 2009). Efforts to strengthen institutional linkages in the commons have led to cross-pollination between the commons and systems thinking (Berkes, 2006; van Laerhoven & Ostrom, 2007). Nonetheless, there is currently a lack of empirical evidence to support how the commons applies to large-scale systems, since commons literature is populated mostly with small-scale case studies (Fleischman et al., 2014). This gap highlights an opportunity for future research to clarify the institutions around transboundary and large-scale commons.

Dynamic large-scale systems, fast-changing technologies, and an ever-growing population are increasing the complexity of the commons (e.g., number of actors and interactions, rapidly changing and unknowable resource conditions). This evolution makes it difficult to govern the global commons and new commons with old approaches. Synthesized by Armitage (2008, p. 16), numerous authors have contributed to identifying a suite of governance attributes required to govern the commons in a complex, multi-level world. These include:

- *Participation, collaboration, and deliberation* could serve as opportunities for the inclusion of diverse interests, perceptions, and interpretations.
- *Multi-layered organizational structures*—outlining these complex structures will facilitate a better understanding of cross-scale interactions, institutions, and incentives.
- *Accountability* in governance processes can reduce threats to vulnerable groups and build adaptive capacity.
- *Interactivity* (i.e., mutually influencing relationships) between two or more actors within the governing system can impact governance outcomes.
- *Leadership* plays a key role in "vision" development; leaders can be important facilitators and catalysts of change.
- *Knowledge pluralism* (i.e., drawing knowledge from multiple sources) can build a more holistic, integrated, and systems-based understanding of the commons.
- *Learning*—in this case, referring to a social process and outcome—can occur through collaboration and knowledge sharing among actors and can enhance adaptive capacity.
- *Trust* is often underestimated in top-down management, but can play a key role in successful collaboration.
- *Networking* of actors across scales (e.g., among local, regional, and international organizations) can confer institutional resilience in multi-level commons.

Effective governance of the global and new commons must therefore involve dialogue among multiple actors as part of a continuous and adaptive learning process (Armitage, 2008;

Table 30.1 Summary of shared attributes between the global commons and the new commons, alongside challenges and opportunities for global governance of these commons.

Commonalities	*Challenges and opportunities*
Intangible: Resources are difficult to conceptualize, some are immaterial (e.g., knowledge, culture), others exist on a large scale (e.g., outer space, digital world), social and cultural value (e.g., sacred site)	• Improper "fit" of existing definitions of excludability and subtractability • Difficult for resource users and appropriators to understand their individual impacts • Opportunity to re-evaluate conventional commons frameworks and to reimagine these frameworks to better capture intangible values
Transboundary: Resources cross conventional geophysical and political boundaries (e.g., global atmosphere, digital information, knowledge)	• Can hinder collective action if multi-level and transboundary collaboration is weak • Opportunity to apply a systems lens to enhance our understanding of the institutions for collective action in the global commons and new commons
Complexity: Resource conditions and contexts are unknowable, constantly changing (e.g., climate) and/or are proliferating rapidly (e.g., technology, knowledge)	• Can be difficult to identify resource users, resource dependence, resource condition and can require a huge amount of managerial capacity • Opportunity for interdisciplinary and transdisciplinary collaboration to enhance knowledge and learning

Armitage et al., 2017). Scholars have recognized the importance of incorporating different knowledge types (e.g., local, traditional, expert) and coordinating action at all institutional levels for solving emergent global issues (Armitage et al., 2017). Accordingly, there is significant scope and opportunity for interdisciplinary and transdisciplinary contribution to enhance global governance of the commons.

In this section, we examined "commonalities" between the global commons and the new commons and unpacked how these shared attributes pose challenges to the global governance of these commons. We drew particular attention to the intangibility and transboundary fluidity of these resources and the complexity of the contexts and institutions through which they are governed. However, we also highlighted several bright spots and opportunities for enhancing global governance of the commons, including: 1) the application of a systems lens to better understand the institutions around collective action; 2) the role of diverse knowledge and learning in enhancing the resilience of the commons; and 3) the potential for interdisciplinary and transdisciplinary collaboration to strengthen global governance. This broad analysis is summarized in Table 30.1.

Conclusion

In this chapter, we examined: 1) the global commons (i.e., Antarctica, the high seas and deep sea bed, outer space, atmosphere); 2) the new commons (e.g., protected area commons, knowledge commons, cultural commons); and 3) shared attributes between these commons which impact the governance of these resources. Both the global commons and the new commons have been brought within reach by decades of technological advances. For example, we now have the capacity to explore and drill in the deep sea, to travel to outer space, and to generate knowledge and cultural content in the digital world. In effect, we have made the world "smaller" while simultaneously enlarging and creating new global problems.

In our analysis, we highlighted three challenges and opportunities for global governance. First, commons frameworks must expand and evolve to account for intangible resources in the global commons and the new commons. The intangibility of certain resource types—for example, knowledge, the atmosphere, and culture—illuminate the improper "fit" of subtractability and excludability from conventional frameworks and make the global commons and the new commons difficult to monitor and manage. Second, transboundary challenges in the global commons and new commons (e.g., climate change, marine pollution, digital privacy issues) suggest that a systems lens may help to better understand and strengthen the institutions for collective action in these large-scale commons. Finally, we emphasized the importance of taking an interdisciplinary and transdisciplinary approach to addressing challenges in the growing complexity of the commons (e.g., increasing number of actors, interactions, rapidly changing contexts). Moving forward, practice and study of the commons must continue to be an adaptive and collaborative process so that our understanding of the commons can evolve alongside the emergence of new institutions and resource types.

Acknowledgements

This chapter was catalyzed in part by a series of conversations we had with a wide range of commons scholars, including: Arun Agrawal, Derek Armitage, Fikret Berkes, Michael Cox, Craig Johnson, Derek Johnson, Ruth Meinzen-Dick, Bonnie McCay, Prateep Nayak, Michael Schoon, and Ganesh Shivakoti. We thank the scholars and practitioners who not only inspired and informed this chapter but also generously provided their time, knowledge, and experience.

References

Agrawal, A. (2007). Forests, governance, and sustainability: Common property theory and its contributions. *International Journal of the Commons*, 1(1), 111–136.

Armitage, D. R. (2008). Governance and the commons in a multi-level world. *International Journal of the Commons*, 2(1), 7–32.

Armitage, D. R., Charles, A., & Berkes, F. (eds.). (2017). *Governing the Coastal Commons: Communities, Resilience and Transformation*. New York: Earthscan Routledge.

Berge, E. (2006). Protected areas and traditional commons: Values and institutions. *Norsk Geografisk Tidsskrift [Norwegian Journal of Geography]*, 60(1), 65–76.

Berkes, F. (2005). Why keep a community-based focus in times of global interactions? *Topics in Arctic Social Sciences*, 5, 33–44.

Berkes, F. (2006). From community-based resource management to complex systems: The scale issue and marine commons. *Ecology and Society*, 11(1). http://doi.org/45.

Bowers, C. (2009). Educating for a revitalization of the cultural commons. *Canadian Journal of Environmental Education*, 14, 196–200. Retrieved from http://cjee.lakeheadu.ca/index.php/cjee/article/view/896.

Boyle, J. (2008). *Public Domain: Enclosing the Commons of the Mind*. New Haven, CT: Yale University Press.

Byg, A., & Salick, J. (2009). Local perspectives on a global phenomenon-climate change in Eastern Tibetan villages. *Global Environmental Change*, 19, 156–166.

De Moor, T. (2011). From common pastures to global commons: A historical perspective on interdisciplinary approaches to commons. *Natures Sciences Sociétés*, 19, 422–431.

De Moor, T. (2015). *The Dilemma of the Commoners: Understanding the Use of Common-Pool Resources in Long-Term Perspectives*. Cambridge, UK: Cambridge University Press.

Dudley, N. (ed.). (2008). *Guidelines for Applying Protected Area Management Categories*. Gland, Switzerland: International Union for the Conservation of Nature (IUCN) Publications Services. Retrieved from https://books.google.co.uk/books?id=pq4oEg58_08C.

Fleischman, F. D., Ban, N. C., Evans, L. S., Ban, N. C., Evans, L. S., Epstein, G., . . . Villamayor-Tomas, S. (2014). Governing large-scale social-ecological systems: Lessons from five cases. *International Journal of the Commons*, 8(2), 428–456.

Frakes, J. (2003). The common heritage of mankind principle and deep seabed, outer space, and Antarctica: Will developed and developing nations reach a compromise. *Wisconsin International Law Journal*, 21, 409–434.
Goldman, M. (1993). Tragedy of the commons or the commoners' tragedy: The state and ecological crisis in India. *Capitalism, Nature, Socialism*, 4(4), 49–68.
Grotius, H. (1609). *Mare liberum. The Free Sea* (translated by Richard Hakluyt, edited by David Armitage) (2004th ed.). Dutch Republic: Liberty Fund, Inc.
Hess, C. (2008). *Mapping the New Commons*. Syracuse University, NY: SURFACE, (July), 14–18.
Hess, C., & Ostrom, E. (eds.) (2006). *Understanding Knowledge as a Commons: From Theory to Practice*. Cambridge, MA: The MIT Press.
Madison, M. J., Frischmann, B. M., & Strandburg, K. J. (2010). Constructing commons in the cultural environment. *Cornell Law Review*, 95(4), 657–709.
Nicholson, G. (2002). The common heritage of mankind and mining: An analysis of the law as to the high seas, outer space, the Antarctic and World Heritage. *New Zealand Journal of International Law*, 6, 177–198.
Noyes, J. E. (2012). The common heritage of mankind: Past, present, and future. *Denver Journal of International Law & Policy*, 40(1–3), 447–471.
Okereke, C. (2008). Equity norms in global environmental governance. *Political Science*, 8(3), 25–50.
Ostrom, E. (1990). *Governing the Commons: The evolution of institutions for collective action*. Cambridge: Cambridge University Press. http://doi.org/10.2307/1243016.
Ostrom, E. (2007). Challenges and growth: The development of the interdisciplinary field of institutional analysis. *Journal of Institutional Economics*, 3(3), 239–264.
Pardo, A. (1967). *Address to U.N. General Assembly*. New York: United Nations.
Parry, J. T. (2014). What is the Grotian tradition in international law? *University of Pennsylvania Journal of International Law*, 35(2), 299–377.
Patrick, S. (2014). The unruled world: The case for good. *Foreign Affairs*, 93, 58–73.
Pieraccini, M. (2015). Democratic legitimacy and new commons: Examples from English protected areas. *International Journal of the Commons*, 9(2), 552–572.
Rutte, C. (2011). The sacred commons: Conflicts and solutions of resource management in sacred natural sites. *Biological Conservation*, 144(10), 2387–2394. Retrieved from www.sciencedirect.com/science/article/pii/S0006320711002448.
Samakov, A., & Berkes, F. (2017). Spiritual commons: Sacred sites as core of community-conserved areas in Kyrgyzstan. *International Journal of the Commons*, 11(1), 422–444.
Scharper, S. B., & Cunningham, H. (2006). The genetic commons: Resisting the neo-liberal enclosure of life. *Social Analysis*, 50(3), 195–202.
Schrijver, N. (2016). Managing the global commons: Common good or common sink? *Third World Quarterly*, 37(7), 1252–1267.
Schrijver, N., & Prislan, V. (2009). From *Mare liberum* to the global commons: Building on the Grotian heritage. *Grotiana*, 30, 168–206.
Stern, P. C. (2011). Design principles for global commons: Natural resources and emerging technologies. *International Journal of the Commons*, 5(2), 213–232.
UNGA. (1982). United Nations Convention on the Law of the Sea, 3rd Sess United Nations Conference on the Law of the Sea 208. http://doi.org/10.1163/15718089720491594.
UN System Task Team. (2013). *Global Governance and Governance of the Global Commons in the Global Partnership for Development Beyond 2015*. Thematic Think Piece. www.un.org/en/development/desa/policy/untaskteam_undf/thinkpieces/24_thinkpiece_global_governance.pdf (accessed 20 October 2018).
Van Laerhoven, F., & Ostrom, E. (2007). Traditions and trends in the study of the commons. *The International Journal of the Commons*, 1(1), 3–28.
Vogler, J. (2012). Global commons revisited. *Global Policy*, 3(1), 61–71. Wijkman, P. M. (1982). Managing the global commons. *International Organisation*, 36(3), 511–536.

31

PROTECTING THE GLOBAL COMMONS

The politics of planetary boundaries

Oran R. Young and Falk Schmidt

As we move deeper into the Anthropocene, it has become increasingly clear that the Earth itself is a complex system subject to use and abuse as a consequence of a wide range of human activities. (Steffen et al. 2004). This system has many of the attributes we associate with the idea of the commons. For the most part, the principal elements of the system are open to entry on the part of human users; uncoordinated human uses can produce harmful impacts at the systemic level. The system lacks effective protection against the side effects or unintended consequences of human activities launched in pursuit of a variety of goals. As is the case with other commons, humans can take steps to introduce cooperative arrangements to protect various elements of the Earth system from the impacts of anthropogenic activities. By almost any measure, for example, the effort to protect the stratospheric ozone layer through the introduction and progressive strengthening of the 1987 Montreal Protocol on Substances that Deplete the Ozone Layer has been a success (Parson 2003). Yet this is an exceptional case. Most elements of the Earth system (e.g. the climate system, the planet's biological endowment, the nitrogen cycle, the open oceans beyond the limits of national jurisdiction) lack effective governance systems designed to protect them from the impacts of a wide range of human activities.

Depending on the specific issue in question and on the nature of our concerns, we can approach the challenge of protecting the global commons either as a matter of controlling overuse on the part of human users or as a matter of limiting the unintended consequences of actions taken to fulfill other goals. Consider the case of the Earth's climate system for illustrative purposes. One way to think about the problem of climate change is to treat the atmosphere as a repository for wastes in the form of anthropogenic greenhouse gases (GHGs) that humans are allowed to use freely.[1] From this perspective, climate change is a consequence of overuse of the atmospheric commons on the part of those allowed to do so without any obligation to pay for the service of waste disposal. This approach suggests that climate change is a collective-action problem and that the proper response is to introduce mechanisms (e.g. carbon taxes, cap-and-trade systems, principles of stewardship) that require human users to pay for the use of the atmosphere's waste treatment services, to find other ways to dispose of wastes, or to adopt practices that reduce the flow of wastes. Alternatively, we can think of climate change as a product of externalities in the sense of side effects or unintended byproducts of activities undertaken in pursuit of a variety of goals. In effect, agricultural practices and industrial activities generate emissions of GHGs that are side effects not included in the cost of doing business. The implication of this way of thinking is that we need to introduce regulations designed to control these externalities either through outright prohibitions or through the imposition of costs aimed at altering the behavior of relevant actors. Neither approach constitutes the correct way to frame the issue of protecting the global commons; each has merit under some circumstances. Between them, the two approaches provide the entrée to a range of analytic tools that we can make use of to organize our thinking about strategies for protecting various elements of the global or Earth system commons.

The concept of planetary boundaries

A recent contribution to thinking about protecting the global commons that has generated considerable interest centers on the concept of planetary boundaries. It emphasizes the importance of adopting policies designed to prevent anthropogenic forces from transgressing these boundaries or to encourage us to pull back to a "safe operating space" in cases where we have inadvertently crossed the boundaries. As Johan Rockström and his colleagues, the principal architects of this idea, have put it, "planetary boundaries are values for control variables that are either at a 'safe' distance from thresholds . . . or at dangerous levels" (Rockström et al. 2009; Steffen et al. 2015).[2] So long as we stay on the right side of these boundaries, humanity will be occupying "a safe operating space." Crossing one or more of the boundaries may not lead to catastrophic results right away. But it is undoubtedly dangerous to be on the wrong side of one or more of the planetary boundaries for any length of time. It is like skating on thin ice. It is always risky, and eventually we are likely to fall through the ice with costly consequences. On this account, staying within a safe operating space on a planetary scale is a necessary (though certainly not sufficient) condition for achieving all our other goals.

This idea is intuitively appealing. We all have a sense of the importance of recognizing critical boundaries and realize that transgressing such boundaries can have harmful consequences. We may get away with it in any given case, but we know that behaving irresponsibly in such situations is dangerous and almost certain to catch up with us eventually with severe consequences. But what happens when we seek to use the appealing idea of planetary boundaries as an operational guide to policymaking regarding the protection of the global commons and the pursuit of sustainability on a global scale? Will such a strategy provide a timely reminder of the need to regulate human actions in such a way as to avoid passing tipping points leading to nasty nonlinear changes regarding matters like climate change or the loss of biological integrity? Or will it divert attention from urgent concerns in the realms of human development, including eradicating poverty, achieving food security, improving health, or providing better education, and as a consequence generate pushback on the part of those whose top priority is to address a range of issues dealing with health, education, welfare, and the rights of minorities embedded in the UN's 2030 Agenda for Sustainable Development?

To answer these questions, we need to delve into matters of governance associated with the idea of planetary boundaries. Can we agree on the identity of the boundaries, the specification of the control variables, and workable procedures for measuring current values of these variables as well as tracking changes in these values over time? Even if we succeed in this endeavor, can we develop processes that will allow us to make and implement collective choices regarding actions needed to avoid transgressing one or more of these boundaries on a planetary scale? Without questioning the intuitive appeal of the idea of planetary boundaries, we direct attention in this chapter to a range of issues relating to these questions and ask about their implications for the achievement of sustainability in the Anthropocene.

Locating the boundaries

What are the planetary boundaries, and how can we define them in ways that will facilitate quantitative measurements of the sort needed to make them useful to those concerned with matters of governance? Rockström and his colleagues (2009) identified nine distinct boundaries; Steffen and others (2015) have added a tenth in their effort to update the original formulation. These boundaries range from familiar concerns like climate change and biodiversity loss (now called change in biosphere integrity) to issues like chemical pollution and changes in land use. These authors acknowledge that some of the boundaries are hard to operationalize; much of the work done since the late 2000s has focused on strategies for operationalization (e.g. by taking into account regional–global interrelationships constituting some of the boundaries). Nevertheless, we must come to terms with this challenge if the idea of planetary boundaries is to have more than rhetorical significance.

Rockström and colleagues assert that "three of the Earth-system processes—climate change, rate of biodiversity loss and interference with the nitrogen cycle—have already transgressed their boundaries" (Rockström et al. 2009: 473). Interestingly, the updated version that Steffen et al. (2015) provide suggests that while climate change has left the "safe operating space," it is not "beyond the zone of uncertainty." That is, we may still be on solid ground in the climate system, a somewhat puzzling adjustment given the fact that concentrations of GHGs in the Earth's atmosphere have continued to rise steadily in recent years.[3] Still, conclusions like this coming from highly respected members of the science community deserve to be taken seriously not only among scientists themselves but also among policymakers responsible for making collective or social choices about such matters.

This is a good start. But here is where we begin to run into difficulties. In discussing the location of the boundaries or the thresholds of safety, Rockström and his colleagues are candid enough to observe that for each boundary "determining a safe distance involves normative judgments of how societies choose to deal with risk and uncertainty" (Rockström et al. 2009: 473). This is a critical point; it amounts to an acknowledgement that policymakers have a variety of pressing concerns on their minds and that environmental threats – however serious – must compete with other matters for their attention at any given time as they go about making collective choices.[4] But there are deeper issues here as well that make it difficult to know how to extract policy implications from the analysis of planetary boundaries. These center on the selection and operationalization of the "control variables" and on the identification of the threshold separating safe and unsafe spaces with regard to each of the control variables.

Consider the case of climate change, which is arguably the easiest case for those endeavoring to operationalize a *planetary* boundary because the climate system is an archetypical global system, because scientists have studied the Earth's climate system intensively for some time, and because the location of the boundary has been a prominent focus of attention in the policy community. The critical point is this: if we cannot reach consensus within the science and policy communities regarding this case, the usefulness of the planetary boundaries concept in "guiding human development" will be called into question.

The place to start in the case of climate is with the language of Article 2 of the UN Framework Convention on Climate Change (UNFCCC), which calls for the "stabilization of greenhouse gas concentrations in the atmosphere at a level that would prevent dangerous anthropogenic interference with the climate system" (UNFCCC 1992). Determining the meaning of this goal, widely known as avoiding dangerous anthropogenic interference or DAI, has emerged as a major challenge in the context of efforts to respond to the problem of climate change.

The first question concerns the proper currency. Should we approach the issue in terms of average temperatures at the Earth's surface, calling for stabilization of the system as the 2015 Paris Agreement puts it at "well below 2°C above pre-industrial levels," or in terms of concentrations of CO_2 or CO_{2e} in the Earth's atmosphere, calling for establishing the limit at some number of parts per million (ppm) of CO_2 or CO_{2e} (Paris Agreement 2015)? It should come as no surprise that members of the policy community have a strong preference for discussing the issue in terms of increased temperatures at the Earth's surface. It is rising temperatures rather than some measure of the amount of CO_2 or CO_{2e} in the atmosphere that people can understand and respond to in thinking about how climate change will affect their well-being. Policymakers who cast the issue in terms of ppm are unlikely to be understood and even less likely to be taken seriously when it comes to proposals for launching costly measures to address the problem of climate change.

Yet this approach to the meaning of DAI has major drawbacks that are well-known to members of the science community and that are documented in the assessment reports of the Intergovernmental Panel on Climate Change (IPCC). What is the relationship between increases in concentrations of GHGs and temperature increases? Would raising the level of CO_{2e} in the atmosphere to two times pre-industrial levels (i.e. ~540–550ppm) be compatible with keeping average temperature increases to well below 2°C? We do not have clear-cut answers to these questions. If anything, the error bars in the IPCC assessments regarding these questions are growing rather than shrinking from one report to the next.

Equally troubling is the fact that increases in surface temperatures associated with rising levels of GHGs in the atmosphere are expected to vary substantially from one region to another. There is a credible case to be made that a global average increase of 2°C will translate into 3°C or more in much of sub-Saharan Africa and that temperature increases of that magnitude coupled with population growth could trigger biophysical changes causing mass starvation or mass migration in the developing countries of Africa. In the light of these concerns, it seems reasonable to ask whether the formula of well below 2°C, used in policy circles, is acceptable as a way to think about DAI.

Recognizing these problems, members of the science community exhibit a strong preference for approaching the measurement of DAI in terms of concentrations of GHGs in the atmosphere. Rockström and colleagues (2009) select concentrations of CO_2 as the best indicator of this control variable; they argue that the boundary should be set at 350ppm and that the "zone of uncertainty" regarding this variable extends from 350 to 550ppm. In their update, Steffen and colleagues (2015) introduce the idea of "radiative forcing" and propose to narrow the zone of uncertainty to 350–450ppm. This choice has obvious attractions. It allows us to measure with considerable precision where we are at any given moment and where we are headed with respect to this boundary. It is this procedure that leads to the conclusion that we are already moving beyond the safe operating zone in the case of climate change, since atmospheric concentrations of CO_2 are currently above 400ppm and rising at a rate of ~2–3ppm per year.[5]

Still, this procedure has problems of its own. The boundary of 350ppm is far below the level of 540–550ppm widely emphasized by those who think in terms of two times pre-industrial levels as an appropriate boundary. Rockström and colleagues justify the 350ppm boundary by saying that they "have taken a conservative, risk-averse approach to quantifying our planetary boundaries, taking into account the large uncertainties that surround the true position of many thresholds" (Rockström et al. 2009: 473). This is in line with the thinking of many of those engaged in the policy debate about climate change—especially those who belong to the environmental community and espouse the precautionary principle—who are working energetically to mobilize support in the policy community for 350 or less as the proper operationalization of DAI.

Note, however, that setting the boundary at 350ppm is a matter involving not only attitudes toward risk but also a variety of social tradeoffs, rather than simply a matter of scientific judgment. What would it take to move back across this boundary into a safe operating space? Are developing countries to be told that they cannot increase their emissions of CO_2 and in some cases (e.g. China) must actually cut back on their current emissions? Are countries, such as the United States, whose emissions are already far too high in per capita terms, to be told that they must take steps to reduce their emissions drastically? It does not take much political acumen to realize that policy prescriptions of this sort based on the concept of planetary boundaries are apt to be non-starters in political terms. It is hard to imagine that the Conference of the Parties (COP) of the UNFCCC or the Meeting of the Parties (MOP) of the Paris Agreement could arrive at decisions along these lines justified by the concept of planetary boundaries or that policymakers in key countries could hope to implement such decisions domestically without losing their political lives. The Paris Agreement is certainly ambitious. Implementing it requires ratcheting up "nationally determined contributions" until we end up, in the best of all worlds, adhering to the 350ppm standard. The fact that 350ppm involves a judgment call only reinforces these conclusions. In some generic sense, being risk averse seems like an appealing posture when it comes to dealing with potentially disruptive processes like climate change. But policymakers faced with extreme pressures relating to other issues (e.g. fulfilling many of the UN's Sustainable Development Goals (SDGs) or very different goals such as defining a "safe operating space" for the world's financial system) will not have the luxury of being able to adopt and defend successfully an extremely risk averse strategy with respect to determining a safe operating space regarding climate change.

Climate change is a comparatively easy case when it comes to setting planetary boundaries. We have a more-or-less straightforward way to quantify the control variable, and it is possible to envision the likely consequences of climate change conceptualized in terms of sensitive public issues like human health and welfare. In other cases, one or both of these advantages is missing. Take the case of the erosion of the Earth biological endowment, which Rockström and colleagues (2009) identify as the control variable that has transgressed its boundary farthest. They propose a boundary set at an extinction rate of ten species per million per year, whereas ecologists believe that we are actually losing species at a rate of 100–1000 per million per year. No one doubts that the loss of biodiversity is a severe problem; many now assert that we are experiencing the sixth great extinction event (Kolbert 2014). But we do not know how many species are becoming extinct per year and what the trend is in this regard. In fact, we can only guess at the total number of distinct species there are on the planet. What is more, it is hard to convince people preoccupied with the pressures of daily life to take this boundary seriously. Over time, the loss of biodiversity may prove hugely disruptive in ways that affect human well-being. But with the exception of those who think in ethical or spiritual terms or feel drawn to charismatic megafauna (e.g. elephants, panda bears, polar bears, whales), it is difficult to mobilize effective support at the policy level for the actions that would be needed to stem the rising tide of biodiversity loss. The intense battles over the implementation of the Endangered Species Act in the United States, for example, loom large in these terms; the fight may well be even tougher in countries that are struggling to lift their human residents out of poverty. The prospects for effective coordination at the international level in addressing this issue are not bright.

In an effort to address these concerns, Steffen and colleagues (2015) propose a second control variable to "capture the role of biosphere in Earth System functioning." This yields a proposal to use a Biosphere Intactness Index assessing "change in population abundance as a result of human impacts . . . using pre-industrial era abundance as a reference point" (Steffen et al. 2015). But they acknowledge "a lack of evidence on the relationship between BII and Earth System responses," a situation that leads to the conclusion that there is a wide uncertainty range regarding this variable.

For the very risk averse, this suggests locating the boundary for the BII at 90% or, in other words, minimizing changes in population abundance. Those preoccupied with other issues (e.g. extreme poverty, food security, human health), on the other hand, may have no trouble in justifying the placement of the boundary somewhere else in the uncertainty range.

In short, setting the boundaries is an *intensely political process* that will engage the efforts of a wide range of actors who have both well-defined interests and the resources needed to pursue them vigorously. This does not mean that input from the science community is unimportant in this realm. It also should not mean that the proper course for scientists is to pull back to an exclusive focus on biophysical considerations, a tendency we see in the effort by Steffen and colleagues (2015) to refocus the planetary boundaries concept. Where can we obtain useful guidance about matters like the meaning of DAI or the course of biodiversity loss? We must rely on science in seeking to address these concerns. But we must acknowledge as well that policymakers respond to a variety of other interests and values in making social choices about such matters. This may be bad news for the well-being of many humans in the future as we move deeper into the Anthropocene. But a posture of denial with regard to the nature of the processes involved in arriving at social choices will get us nowhere when it comes to locating planetary boundaries at thresholds we are likely to be able to respect.

Living within the boundaries

Assuming we can reach some measure of agreement on setting the boundaries and finding ways to come to grips with measuring changes in the control variables, what are the prospects that we can make and implement the collective or social choices needed to stay within "safe operating spaces" or, in cases where we have already transgressed the boundary, alter our behavior to cross back to safety? Here, too, progress will depend critically on understanding the nature of political processes and finding ways to frame and present the issues that will encourage multiple actors with overlapping but far from identical interests to enter into effective coalitions dedicated to taking the steps needed to achieve these goals.

Assuming that we adopt the perspective of global sustainability and seek to set values for the control variables associated with the planetary boundaries in a manner that requires some actions and prohibits others, two considerations stand out.[6] First, acting in such a way as to live within these boundaries will amount to taking steps to supply collective or public goods (see, in general, Barrett 2007). Whether we think in terms of maintaining a climate system that is benign from a human perspective, minimizing the loss of species, preventing anthropogenic disruptions of the nitrogen cycle, or limiting ocean acidification, the desired outcomes will be non-rival as well as non-excludable for the most part. The enjoyment of a benign climate on the part of one actor or human group does not detract from the benefits others derive from such a climate; the survival of a species is not a finite good to be divided among a number of beneficiaries on the basis of some allocation rule. This is encouraging. But non-excludability, the other defining feature of public goods, raises the classic concern known as the free-rider problem.[7] Although climate variability is an important concern, all the members of international society will benefit from a benign planetary climate if it is available to any of them. Much the same is true with respect to the survival of species. In such cases, individual actors can hope to free ride on the efforts of others.

It is an intriguing thought that individual societies might be able to take steps to protect themselves from the ravages of climate change or the disruption of the nitrogen cycle, while acting to exclude others from enjoying the same benefits. Some such thinking does surface from time to time in the debates about climate. Some participants even purport to believe that climate change will prove beneficial for their societies, regardless of its consequences for others. But once we move beyond the early stages of climate change, the consequences will constitute a public "bad" in the sense that they will affect everyone adversely. Conversely, maintaining a benign climate will work to the advantage of everyone. Note also that the public goods in question are of the sort that will require ongoing efforts to ensure that they are supplied at an appropriate level. This is not a situation in which a "single best effort" will suffice to supply the relevant public goods on a once-and-for-all basis (Barrett 2007). Clearly, the actions of some actors are more important than the actions of others in this connection. But we will need everyone to take steps to keep the Earth within the safe operating space when it comes to maintaining a benign climate or addressing the issues embedded in the other planetary boundaries.

Second, and implicit in what we have said in the preceding paragraph, is the fact that we are dealing with public goods that are planetary in scope. It is true, of course, that the impacts of climate change will vary substantially from one region to another, and the loss of species will be particularly pronounced in those countries that are regarded as hot spots from the perspective of biological diversity. Nonetheless, GHGs emitted anywhere add to the concentrations of these gases in the Earth's atmosphere as a whole. Although the situation is more complex in the case of biological diversity, the forces leading to the loss of species are closely tied to the forces of globalization: global supply chains and global vectors relating to the movement of invasive species underlie the loss of species located in mega-diverse countries, like Brazil or Madagascar. The effect of this is a need to think about supplying safe operating spaces for humanity as public goods on a global scale. Given the nature of international society, many of the specific steps needed to live within the boundaries will have to occur at the domestic level and require initiatives on the part of public agencies, private enterprises, and civil society actors operating within individual societies. But decisions about what steps to take (e.g. the adoption of targets and timetables regarding reductions in current emissions of GHGs) will have to be made in international fora. That is why we pay attention to the actions of the COPs of the UNFCCC and the Convention on Biological Diversity and argue that the UN's SDGs, unlike the earlier Millennium Development Goals (MDGs), are universal in nature rather than focusing only on the concerns of the developing world (Kanie and Biermann 2017). It is also why we engage in debates about the deficiencies of these efforts and ask ourselves whether there are other approaches to governance that may prove more effective in producing results, especially in those cases (e.g. climate change) where we have already transgressed the planetary boundary and are therefore living dangerously (Young 2017).

What sorts of processes can lead to success in efforts to supply public goods on a global scale? In seeking an answer to this question, we can turn to the rich tradition of institutional dimensions research and look to perspectives that emphasize the roles of power, interests, and ideas as driving forces as we think about how the concept of planetary boundaries may make a successful transition from paper to practice.

Starting with power, we want to know whether there is a dominant actor or a hegemon possessing both the capacity and the incentive to take steps to ensure that we remain within safe operating spaces for humanity. Such an actor would need to be able to impose order with respect to human-environment relations in much the same way that we refer to the Pax Britannica in thinking about the security order of the second half of the 19th century or the Washington consensus in characterizing the economic order of the second half of the 20th century. As these examples suggest, there is no reason to expect that everyone's interests will be treated fairly or equitably in systems of this sort. But the order they produce is a public good which turns out to be an important consideration in systemic terms, despite the fact that some will find it repugnant and do what they can to overturn it.

The problem with this line of thinking when it comes to planetary boundaries is that there is no hegemon in sight (Rachman 2016). The United States lacks the will and increasingly the capacity to impose an (environmental) order for the 21st century. Despite its rapidly growing role in international society, China is not yet in a position to take on the mantle of hegemon, especially when it comes to environmental issues that would require significant changes in existing policies within China. Per capita emissions of CO_2 in China, for example, are already well above those that will be acceptable on a worldwide basis if we are to bring concentrations of CO_2 in the atmosphere back to 350ppm or even to set a course that will allow us to stabilize at ~450ppm.

An alternative approach is to direct attention to interests, asking about the efficacy of a "first mover advantage" strategy (as in the field of renewable energy where sustainability experiments are developing into good business models) or about the prospects of building a coalition of the willing to address problems like climate change and the loss of biological integrity. The core strategy here is to find actors who stand to benefit in taking the lead in addressing environmental issues and to work toward identifying a relatively small number of key players and forging alliances among them to produce the commitment and the resources needed to supply the relevant public goods.

Sometimes known as what Schelling (1977) calls "k groups," the members of such alliances can take steps among themselves to deal with major problems and initiate processes that generate the momentum needed to bring others into line over time.[8] In their work on "tipping sets," Heal and Kunreuther (2011) provide an analysis of how a process of this sort might work. With regard to climate change, for example, the United States, the European Union, Japan, and the BRICS (Brazil, Russia, India, China, and South Africa) account for over 70% of emissions of CO_2 resulting from the burning of fossil fuels. If these eight players were to form an effective alliance, reaching agreement on a concrete action program to tackle the problem of climate change, the effect would be dramatic.

But what are the prospects of something of this sort coming to pass? Only the European Union has exhibited consistent leadership with regard to climate change. The United States and China were laggards regarding this issue for a long time. Despite their recent bilateral agreement on climate, it remains to be seen whether these largest emitters of GHGs will exert effective leadership in this realm going forward. The United States under President Trump, in particular, seems to be abdicating this role. Russian policymakers are fond of asserting that their country may actually benefit from the onset of climate change. Brazil and India regard themselves as developing countries that have a right to adopt whatever strategies are needed to fuel rapid economic growth and that cannot be expected to take major steps toward addressing the problem of climate change until the advanced industrial countries and especially the United States demonstrate a clear-cut willingness to take the lead in this realm.[9] Taken together, this does not add up to an optimistic scenario with regard to the supply of a safe operating space for humanity in the case of climate treated as a global public good.

Another and possibly more realistic way to come to terms with climate change may feature large-scale transformations of energy systems that individual countries or powerful corporations undertake for their own reasons. A process of this sort could trigger a snowball effect or a race to the top in the sense that at some stage additional actors would find it in their interests to join such a movement rather than resisting it. Whether this is a realistic scenario is certainly open to question, though it is interesting to see leaders of individual cities, states, and industries in the United States maintaining their commitment to reducing emissions, even as the Trump Administration acts to disown the commitments made by its predecessor in Paris. Still, it is worth noting that it is doubtful whether global concepts like the idea of planetary boundaries will play an important role in the calculations of individual countries and corporations about such matters (Underdal 2008).

What about the role that ideas play in efforts to supply global public goods? Students of politics, who like to adopt tough-minded views emphasizing the roles of power and interests, tend to dismiss the role of ideas too easily. In fact, ideas in the form of prevailing worldviews, discourses, paradigms, or narratives can exercise an enormous influence on the thinking and behavior of actors ranging from individuals on up to multinational corporations and nation states. With regard to the issue of creating a safe operating space for humanity, however, the power of ideas may emerge as part of the problem rather than part of the solution. We live in a world in which decisionmakers at all levels see natural capital as a source of commodities to be extracted and used to accommodate materialistic lifestyles that reflect little concern for issues of sustainability. What is needed to make progress toward securing a safe operating space for humanity are principles of stewardship in which humans assume an obligation to secure the well-being of the planet rather than seeing the planet largely as a source of natural wealth to be plundered to achieve or maintain a lifestyle of the sort prevalent in advanced industrial societies today.[10]

Even if a new discourse foregrounding stewardship for the Earth is not at hand, however, there is another mechanism through which ideas may contribute to the achievement of global sustainability. Comprehensive thinking of the sort reflected in the concept of planetary boundaries may help to identify alternative solutions or at least trigger innovative thinking in the search for solutions. Similarly, the current debate about "nexus governance" (Müller et al. 2015), stressing linkages among issues like energy, water, and food, suggests that striving proactively to promote cross-sectoral thinking could help to unveil possible win-win situations untapped so far (Oberthür and Stokke 2011).

Accordingly, it would be a mistake simply to reject the possibility of a transformation in prevailing discourses or narratives of the sort needed to bring about a turn toward sustainability. Cognitive change has a dynamic of its own that even the most self-interested and imposing power brokers may be unable to control once it gets underway. Still, it is hard to detect the early stages of a wave of change in ideas about human-environment relations that will transform our ways of thinking sufficiently to put us on a new path regarding the treatment of planetary boundaries anytime soon. There is every reason to provide encouragement to those seeking to mobilize social movements (e.g. 350.org) based on new ways of thinking about the impact of human actions in the Anthropocene. But the challenges are too severe and too urgent to adopt the complacent view that this process will evolve and become effective in guiding our actions in due course.

Moving forward

What we have said so far suggests that efforts to establish and live within planetary boundaries will give rise to intensely political processes that could easily fail to produce socially beneficial results. Still, the concept of planetary boundaries has great appeal; we are convinced that it is important to make a concerted effort to find ways to use it effectively in seeking to come to grips with the challenges of the Anthropocene. In a human-dominated world, it seems sensible to make a concerted effort to avoid sharp non-linear changes that are anthropogenic in origin. But how can we make progress toward addressing the various planetary boundaries, even while recognizing the severity of the pitfalls lurking in this realm? Can the idea of planetary boundaries be used to good advantage in energizing efforts to fulfill the SDGs? We have no simple answers to these questions. But we do have some suggestions about next steps that may prove helpful to those looking for practical ways to bring the concept of planetary boundaries to bear in efforts to protect the global commons. In this final substantive section, we identify five steps that seem worthy of consideration, either individually or in combination.

The first step is to recognize that living within the boundaries will require substantial alterations in the lifestyles of people living in the advanced industrial societies as well as in the aspirations of those living in the developing world. There is no way to level up the lifestyles of those living in developing countries today—much less the 10+ billion people expected to be alive by the end of the century—to the current lifestyles enjoyed by Europeans; US lifestyles are even more out of the question. Yet it would be preposterous to suppose that 7–9 billion people will be prepared to settle for diets, housing, and transportation systems that are distinctly inferior to those enjoyed by the top billion. Something has to give in these terms if we are to have any hope of living within the boundaries. The universal nature of the SDGs provides a good starting point for movement in this direction, a development that Rockström and colleagues could not foresee when they published their initial articles on planetary boundaries in 2009. Juxtaposing the idea of planetary boundaries with the framing and implementation of the SDGs provides an opportunity to move from a general discussion of the boundaries to a focused consideration of their application to the pursuit of specific goals.

Given what we have said about the politics of providing a safe operating space for humanity, our second suggestion is to recognize that there can be no substitute for finding ways to get the incentives right. The problems of providing (global) public goods can only be overcome if those ready to act do not have to take steps that critics will oppose as running counter to their best interests for too long. An approach featuring global regulations or domestic regulations developed to implement global agreements within individual societies will not work. The antagonism toward local regulations dealing with matters like land and water use is severe enough. Outright hostility toward national and international regulations justified by the need to live within planetary boundaries would be unavoidable. This suggests the importance of promoting a transition toward concepts like the "green economy." The basic idea here is to restructure economic and political institutions in such a way that individuals and businesses will find it in their interests to act in a manner that promotes sustainability. This will require some recasting of the rules governing economic activities (e.g. introducing rules designed to require full-cost accounting and to eliminate implicit subsidies), and it may require the strategic use of subsidies or tax incentives to influence calculations of costs and benefits in critical industries. But the important thing is to make full use of arrangements that give actors an incentive to change their behavior now as well as an ongoing incentive to innovate in the interests of reducing costs. Some progress has been made since 2009 in fostering what some have described as a "green race."

Our third proposition is that there is also a need to acknowledge considerations of fairness or equity, especially when it comes to persuading those in the Global South to join and participate loyally in new arrangements justified by the need to live within planetary boundaries. Any plan that leaves developing countries with the impression that they are being asked to make sacrifices while those in the advanced industrial societies continue to enjoy their current lifestyles simply will not work. The obvious implication of this observation is that those living in the industrialized North must be prepared to make serious changes in their current lifestyles in return for agreement on the part of those living in the Global South to make adjustments that are essential to the success of any efforts to live within the boundaries. Justice demands that we take this issue seriously and confront it directly in responding to challenges like making drastic reductions in emissions of GHGs over the next two to four decades. There is no way to enforce standards regarding matters like emissions of GHGs in the face of dissent or even open hostility on the part of 5–6 billion people living in the developing world. Our best hope is to craft arrangements supporting necessary sustainability transformation processes (e.g. energy transitions) that people all over the world will buy into and comply with on the grounds that they are acknowledged to be fair or just. Translating this observation into concrete practices relevant to the SDGs will require a good deal of imagination. If done poorly, introducing the concept of planetary boundaries will backfire, further complicating already extremely difficult negotiations regarding issues like climate change and the sustainable use of natural resources.

A fourth step is more concrete; it focuses on the value of spelling out goals in operational terms and developing quantitative indicators to measure progress in the effort to fulfill the goals (Kanie and Biermann 2017). A number of observers have commented on the usefulness of such indicators in conjunction with the campaign to fulfill the MDGs (Sachs 2015). This logic applies to the SDGs as well. Measuring progress on a continuous basis does not ensure success. But as those who run various charities know, it is helpful to have a way to measure progress that is useful in exhorting relevant actors to join in the effort to achieve common goals. Several analysts proposed that it would be desirable to develop a set of Global Environmental Goals even before the effort to formulate the SDGs began (Perrez and Ziegerer 2008). This idea has been taken up in conjunction with the SDGs. The problem here is that it is difficult to decide what to include in this set of goals and how to devise suitable indicators to track progress toward fulfilling them. As Rockström and colleagues (2009) acknowledge, several of the boundaries do not have quantifiable control variables; there may well be debate about the specification of these variables in other cases. But seven of the nine boundaries do have measurable control variables, and it is possible to measure some of these variables (e.g. levels of CO_2 in the atmosphere, nitrogen removed from the atmosphere) precisely. This opens up the prospect that the science community can make a major contribution by offering robust and credible measurements pertinent to a number of the planetary boundaries. In this regard, the initiative of the International Institute for Applied Systems Analysis (IIASA) in launching a project on *The World in 2050: Pathways Toward a Sustainable Future* provides reason to be optimistic about the feasibility of this effort (IIASA ongoing). This suggests that the idea of planetary boundaries can provide a way forward for improving our understanding of the Earth as a complex socioecological system. We are becoming aware that pushing too hard on one boundary (e.g. climate) may interfere with efforts to address others or with efforts to pursue other critical goals (e.g. food security). The concept of planetary boundaries may provide a starting point for efforts to avoid putting individual goals into silos and neglecting important interactions between them (Schmidt 2013: 228–231).

Finally, there is a compelling case to be made for the proposition that we should invest time and energy in the practice of institutional diagnostics in considering the relevance of planetary boundaries (Ostrom 2007; Young 2008). There are common features among the boundaries when it comes to achieving safe operating spaces. They all involve the supply of public goods on a global scale. But there are also important differences between them that need to be taken into account in designing institutional arrangements relating to the boundaries on a case-by-case basis. Protecting biodiversity hotspots is a challenge that is very different from the challenge of reducing concentrations of GHGs in the Earth's atmosphere. Designing institutional arrangements that are well matched to the individual boundaries does not mean automatically that it will be possible to implement these designs during the near future. The obstacles to institutional change in most cases are too great for that. But the phenomenon of tipping points followed by the onset of non-linear changes is just as prominent with regard to governance systems as it is with regard to biophysical and socioeconomic systems. What this means is that relatively long periods of stasis are punctuated from time to time by the opening of windows of opportunity in which changes that would have seemed totally unrealistic in earlier times come into play as realistic options. Such windows seldom stay open for long. They close quickly, whether or not key actors have been able to take advantage of them to bring about constructive changes in prevailing governance systems. It follows that we should think carefully about matters of institutional design far enough in advance to be prepared when opportunity knocks. In the case of planetary boundaries, what is needed is a collaborative effort on the part of natural scientists whose expertise is focused on the nature of the boundaries themselves and social scientists whose expertise is focused on the nature and operation of governance systems. Among other things, the results could play a critical role as we strive during the coming years to fulfill the SDGs.

Concluding thoughts

The idea of maintaining a "safe operating space for humanity" is not referenced explicitly in the SDGs. Nevertheless, the concept of planetary boundaries can play a constructive role in efforts to conceptualize and implement the idea of sustainability on a planetary scale. The essential point is we must recognize that the onset of the Anthropocene makes it imperative to pursue worthy goals like eradicating poverty, achieving food security, or improving human health in ways that do not exceed the capacity of the planet with regard to overarching considerations like the stability of the climate system, the integrity of the nitrogen cycle, and the maintenance of biological diversity. In a world of 8–10 billion people, this will require adjustments in the aspirations and lifestyles of people everywhere, rather than simply improving living conditions in the developing world. In qualitative terms, the concept of planetary boundaries can play an influential role in making this realization credible and convincing. Beyond this, if we are able to make progress in operationalizing individual boundaries and measuring the control variables quantitatively, the idea of planetary boundaries may prove useful in ensuring that strategies designed to fulfill specific SDGs do not leave us living dangerously beyond the limits of the "safe operating space for humanity."

Notes

1 Greenhouse gases occur naturally in the Earth's atmosphere. The problem arises when human actions raise levels of these gases in ways that lead to changes harmful to human welfare.

2 The so-called "doughnut model" proposed by Raworth has coupled natural and social systems toward "A safe and just space for humanity. Can we live within the doughnut?" (Raworth 2012). Schmidt 2013 elaborates this idea. We recognize the scientific improvements made since 2009, and we share the view of the authors of the 2015 updates that (better) science can inform the public and policymakers. However, we will often reference the contributions of Rockström et al. 2009, since this paper remains the seminal formulation of the idea of planetary boundaries.

3 This highlights our view that setting boundaries is itself a political act that would be compromised severely if substantial changes in judgments were to occur often and if changes in the specification of "control variables" were to lead to different outcomes.

4 Steffen et al. (2015) seem to pull back, leaving normative judgments entirely to policymakers. But we are skeptical about carrying this view too far. The classic idea of "speaking truth to power" may have worked in the past for some issue areas. However, whether it still works can be questioned, and whether it works for the planetary boundaries concept at all should be explored rigorously.

5 Those who prefer to include all GHGs in such calculations will note that we are already above 450ppm CO_{2e}.

6 There is another way of framing the challenge of global sustainability, which is not explored in this chapter. First, build on science informed and socially determined goals and targets as an orientation for action. Second, investigate actions (including R&D) conducive to meeting these goals and targets. For example, if much of the climate challenge is due to fossil-fuel-based systems of energy production and consumption, the focus may not only be on strict legally binding rules and regulation for achieving a given goal but also on actions that improve the competitiveness of other types of energy with the long-term goal that they become the dominant energy regime. The current success story of renewables, which now represent the majority of investments in new electricity capacity globally, presents a case in point of sustainability transformation management. It takes ambitious goals and massive governance intervention to destabilize the incumbent energy regime. But the point is that once the new, renewables-based energy regime is in place, it is also economically favorable and hence does not need the constant interventions determined by externally set control variables. See IIASA's *The World in 2050* project.

7 Olson (1965). The massive investments needed that led to the situation currently known as the "German Energiewende" were exactly about this. Very low prices for renewables enjoyed by many countries across the globe today—e.g. becoming increasingly competitive with *new* conventional power plants—are the results of a public consensus in Germany to make this desirable goal a reality. As an aside, much hints at the fact that Germany is about to reap the "first mover advantages" in this global race toward

the energy systems of the 21st century, indicating that arguments like "free-riding" cannot be seen in a static manner but change themselves over time.

8 On the idea of "k groups," see Schelling (1977).

9 Brazil, China, and India are not UNFCCC Annex 1 countries. This means that they did not have commitments to reduce GHG emissions under the terms of the Kyoto Protocol.

10 On principles of stewardship, see Chapin et al. (2009).

References

Barrett, Scott (2007) *Why Cooperate? The Incentive to Supply Collective Goods*. Oxford, UK: Oxford University Press.

Chapin, F. Stewart, Gary P. Kofinas, and Carl Folke eds. (2009) *Principles of Ecosystem Stewardship: Resilience-Based Natural Resource Management in a Changing World*. New York: Springer.

Heal, Geoffrey and Howard Kunreuther (2011) "Tipping Climate Negotiations," Working Paper 16954, NBER Working Paper Series, National Bureau of Economic Research.

IIASA ongoing. *The World in 2050 (TWI 2050): Pathways toward a Sustainable Future*. www.iiasa.ac.at/>Research>Research.Projects.

Kanie, Norichkia and Frank Biermann eds. (2017) *Governing through Goals: Sustainable Development Goals as Governance Innovation*. Cambridge, MA: MIT Press.

Kolbert, Elizabeth (2014) *The Sixth Extinction: An Unnatural History*. New York: Henry Holt.

Müller, Alexander, Hannah Janetschek, and Jes Weigelt (2015) "Towards a Governance Heuristic for Sustainable Development" *Current Opinion in Environmental Sustainability*, 15: 49–56.

Oberthür, Sebastian and Olav Schram Stokke eds. (2011) *Managing Institutional Complexity: Regime Interplay and Global Environmental Change*. Cambridge, MA: MIT Press.

Olson, Mancur Jr. (1965) *The Logic of Collective Action*. Cambridge, MA: Harvard University Press.

Ostrom, Elinor (2007) "A Diagnostic Approach to Going Beyond Panaceas," *Proceedings of the National Academy of Sciences USA*, 104: 15181–15187.

Paris Agreement (2015) Agreement Adopted at UNFCCC COP21. FCCC/CP/2015/L.9/Rev.1.

Parson, Edward A. (2003) *Protecting the Ozone Layer: Science and Strategy*. New York: Oxford University Press.

Perrez Franz Xaver and Daniel Ziegerer (2008) "A Non-institutional Proposal to Strengthen International Environmental Governance," *Environmental Policy and Law*, 38: 253–261.

Rachman, Gideon (2016) *Easternization: Asia's Rise and America's Decline from Obama to Trump and Beyond*. New York: Other Press.

Raworth, Kate (2012) *A Safe and Just Space for Humanity: Can We Live Within the Doughnut?* Oxfam Discussion Paper, Oxford, UK.

Rockström, Johan et al. (2009) "A Safe Operating Space for Humanity," *Nature*, 461: 472–475.

Sachs, Jeffrey (2015) *The Age of Sustainable Development*. New York: Columbia University Press.

Schelling, Thomas C. (1977) *Micromotives and Macrobehavior*. New York: W.W. Norton.

Schmidt, Falk (2013) "Governing Planetary Boundaries: Limiting or Enabling Conditions for Transitions Towards Sustainability?" in L. Meuleman ed., *Transgovernance: Advancing Sustainability Governance*. Heidelberg, Germany: Springer Verlag, 215–234.

Steffen, Will et al. (2004) *Global Change and the Earth System: A Planet under Pressure*. Heidelberg, Germany: Springer Verlag.

Steffen, Will et al. (2015) "Planetary Boundaries: Guiding Human Development on a Changing Planet," *Science*, doi:10:1126/science.1259855.

Underdal, Arild (2008) "Determining the Causal Significance of Institutions: Accomplishments and Challenges," in O.R. Young, L.A. King, and H. Schroeder eds., *Institutions and Environmental Change*. Cambridge, MA: MIT Press, 49–78.

United Nations Framework Convention on Climate Change (1992) Text available at: www.unfccc.int.

Young, Oran R. (2008) "Building Regimes for Socioecological Systems: Institutional Diagnostics," in O.R. Young, L.A. King, and H. Schroeder eds., *Institutions and Environmental Change*. Cambridge, MA: MIT Press, 115–144.

Young, Oran R. (2017) *Governing Complex Systems: Social Capital for the Anthropocene*. Cambridge, MA: MIT Press.

INDEX

FSC
www.fsc.org
MIX
Papier | Fördert gute Waldnutzung
FSC® C083411

Kolibri 360 GmbH
Ferdinand-Jühlke-Straße 7
99095 Erfurt, Deutschland
produktsicherheit@kolibri360.de